Library of
Davidson College

BIOLOGICAL MEMBRANES

Physical Fact and Function

BIOLOGICAL MEMBRANES
Physical Fact and Function

Edited by

DENNIS CHAPMAN

General Research Division
Unilever Research Laboratory
The Frythe, Welwyn, Hertfordshire
England

1968

ACADEMIC PRESS · London and New York

ACADEMIC PRESS INC. (LONDON) LTD.
24–28 Oval Road,
London NW1

U.S. Edition published by
ACADEMIC PRESS INC.
111 Fifth Avenue
New York, New York 10003

Copyright © 1968 by ACADEMIC PRESS INC. (LONDON) LTD.
Second Printing 1969
Third Printing 1973

All Rights Reserved

No part of this book may be reproduced in any form by photostat, microfilm, or any other means, without written permission from the publishers

Library of Congress Catalog Card Number 68-24696
SBN: 12 168540 3

Printed in Great Britain by Unwin Brothers Limited, Old Woking, Surrey

Contributors

D. CHAPMAN, *Molecular Biophysics Unit, Unilever Research Laboratory, The Frythe, Welwyn, Herts, England.*

R. M. C. DAWSON, *Agricultural Research Council, Institute of Animal Physiology, Babraham, Cambridge, England.*

S. FLEISCHER, *Department of Molecular Biology, Vanderbilt University, Nashville, Tennessee, U.S.A.*

R. B. LESLIE, *Molecular Biophysics Unit, Unilever Research Laboratory, The Frythe, Welwyn, Herts, England.*

J. A. LUCY, *Royal Free Hospital School of Medicine, University of London, London, England.*

V. LUZZATI, *Centre de Génétique Molecalaire, C.N.R.S., Gif-sur-Yvette, France.*

G. J. NELSON, *Bio-Medical Division, Laurence Radiation Laboratory, University of California, Livermore, California, U.S.A.*

G. ROUSER, *Department of Biochemistry, City of Hope Medical Center, Medical Research Institute, Duarte, California, U.S.A.*

G. SIMON, *Departments of Neurology and Biochemistry, University of Illinois College of Medicine, Chicago, Illinois, U.S.A.*

I. SINGER, *Laboratory of Neurobiology, National Institute of Mental Health, Bethesda, Maryland, U.S.A.*

I. TASAKI, *Laboratory of Neurobiology, National Institute of Mental Health, Bethesda, Maryland, U.S.A.*

D. F. H. WALLACH, *Biochemical Research Laboratory, Massachusetts General Hospital, Boston, Massachusetts, U.S.A.*

Foreword

SIR RUDOLPH A. PETERS

University Department of Biochemistry, Cambridge

This book, edited by Prof. Dennis Chapman, contains a series of valuable essays upon membranes and membranous structure in cells interpreted in the light of modern theory and chemical technology. Early speculations on membrane structure are now being much extended and transformed.

Many years ago, I stated my view that, as a matter of logic, there must be present in living cells a co-ordinating structure, like a nervous system, which has been termed a cytoskeleton. The view was based upon the obvious microcompartmentation in living matter and it was thought that the newly developing knowledge of surface chemistry at this time could provide the clue to the molecular basis of the integration. The presence of cytoskeletal systems, endothelial reticulum and otherwise, has been put on a proper experimental basis by the beautiful work of Palade, Porter and others, as shown by the electron microscope. This has left the way clear for the next step which should be the exploration of the molecular structure of the endoplasmic reticulum in microsomes and of membranes in general. It has become imperative to recognize that membranes have different functions in different parts of the cell and that there are likely to be differences between those at the surface and those surrounding organelles. We may expect, with confidence, exciting new developments in the near future to which this volume will be a great contribution.

Acknowledgements

We wish to acknowledge with thanks permission from the following publishers: Biochemical Society, London; Elsevier Publishing Co.; Cambridge University Press; Faraday Society; Birkhaeüser Verlag; American Oil Chemists' Society; Federation of American Societies for Experimental Biology; Cambridge Philosophical Society; Society of Experimental Biology.

Contents

List of Contributors	v
Foreword by Sir Rudolph A. Peters	vii
Acknowledgements	viii

1. Introduction
D. CHAPMAN 1

2. Lipid Composition of Animal Cell Membranes, Organelles and Organs 5
GEORGE ROUSER, GARY J. NELSON, SIDNEY FLEISCHER AND GERALD SIMON

I.	Introduction	6
II.	Lipid Classes Occurring in Animal Cells	7
III.	General Considerations in Membrane Lipid Composition Studies	9
IV.	Arrangement of Lipids in Cellular Membranes	10
V.	General Methodology	12
VI.	Mammalian Erythrocytes	21
VII.	Subcellular Organelles	36
VIII.	Whole Organ Composition and Species Variations	44
IX.	Miscellaneous Cell Types	51
X.	Glycolipids	55
XI.	Changes in Membrane Components and Pathological States	57
XII.	Summary and Conclusions	61
XIII.	Acknowledgements	63
	References	64

3. X-Ray Diffraction Studies of Lipid-Water Systems ... 71
VITTORIO LUZZATI

I.	Introduction	71
II.	Structure Analysis	73
III.	Methods	78
IV.	Simple Lipids: Soaps and Detergents	81
V.	Lipids of Biological Interest	103
VI.	Discussion and Conclusions	115
VII.	Acknowledgements	117
	Appendix 1: The disordered conformation of the paraffin chains	118
	Appendix 2: Some crystallographic verifications	119
	References	121

4. Recent Physical Studies of Phospholipids and Natural Membranes — 125
D. CHAPMAN with a contribution by D. F. H. WALLACH
- I. Introduction 125
- II. The Physical Properties of Phospholipids 127
- III. The Physical Properties of Membranes 153
- IV. Conclusions and Future Studies 196
- References 199

5. The Nature of the Interaction Between Protein and Lipid during the Formation of Lipoprotein Membranes — 203
R. M. C. DAWSON
- I. Introduction 203
- II. Physical Chemistry of the Phospholipid-Water Interface . . 205
- III. Reactions of Proteins Other than Phospholipids at the Lipid-Water Interface 209
- IV. Reaction of Phospholipases at the Phospholipid-Water Interface 213
- V. The Formation of Lipoprotein Membranes . . . 225
- References 231

6. Theoretical and Experimental Models for Biological Membranes — 233
J. A. LUCY
- I. Introduction 283
- II. Bimolecular Leaflet Models 234
- III. Micellar Models 244
- IV. Phase Changes Within Membranes 261
- V. Experimental Models for Membranes 265
- VI. Conclusions 284
- References 285

7. Membranes and Bioenergetics — 289
R. B. LESLIE
- I. Introduction 290
- II. Structural Aspects and Gross Morphology of Lamellar Systems Involved in Bioenergetics 291
- III Functional Aspects of Membranes and Bioenergetics . . 305
- IV. The Evolution of Physical Bioenergetics 312
- V. Physical Bioenergetics 314
- VI. Energy and Electron Migration in Bioenergetic Lamellar Systems 325
- VII. Some Model Systems 336
- VIII. Conclusion and Outstanding Problems 340
- IX. Acknowledgement 342
- References 342

8. Nerve Excitability and Membrane Macromolecules — 347
IRWIN SINGER AND ICHIJI TASAKI

CONTENTS

I. Introduction 347
II. Materials and Methods 349
III. Results 356
IV. Excitation as a Physico-chemical Process 396
References 406
Addendum 410

Author Index 411
Subject Index 423

Chapter 1

Introduction

D. CHAPMAN

*Molecular Biophysics Unit, Unilever Research Laboratory,
The Frythe, Welwyn, Herts, England*

Interest in the structure and function of biological membranes has increased greatly in recent years. As the details of the replication processes of the cell have been gradually revealed, so attention has increasingly turned to problems associated with cell organization. Replication and organization are two of the outstanding important features of living systems and a greater part of the organization of the cells appears to be determined by the cell membranes.

For many years the idea of cell membranes was demanded by physiological experiments and existed almost entirely at the conceptual level. These experiments suggested that a barrier existed to diffusion between the interior of the cell and its surroundings. The barrier was called a membrane because it was thought to be a thin layer, completely enclosing the cytoplasm, providing a controlled overall exchange of molecules between the cytoplasm and its surroundings. The barrier was considered to have certain definite mechanical and physical properties and, even before any direct analyses of individual or isolated membranes were admitted, suggestions were being made as to its structure and organization.

The light microscope was unable to resolve such tenuous structures as cell membranes. Nevertheless, the polarization microscope did prove to be a useful technique for the study of myelin and a model for myelin was suggested which accounts for its strong, positive, intrinsic birefringence and its weak negative form birefringence.

The introduction of the electron microscope in the 1930's provided a tool which can form images of structures previously considered sub-microscopic. The electron microscope showed clearly defined layer structures at the border of the cell boundary. In addition to this, similar structures were observed to occur in a variety of internal structures within the cell wall. Inside the cell, structures called mitochondria and lysosomes were discovered which were seen to be bounded by thin, layered structures. Paired membranes called endoplasmic reticulum were also observed. Similar structures were also observed with the chloroplast of plant cells and in the rods and cones of the eye. It is a similarity in the appearance of the electron micrographs which has led to the idea that these various structures are all associated with membranes.

For many years the basic structure of cell membranes appeared settled. The bilayer of lipid was thought to be the important structural unit and the concept of the unit membrane was introduced. It is only recently, as a result of new techniques and new ways of thinking, that the detailed structure of cell membranes has suddenly become a live issue. Now there are very many different views and theories about the structure of cell membranes and about the organization and interaction of the membrane components. Uncertainty about the structure of biological membranes naturally carries with it uncertainty about their mode of function.

This seemed, therefore, an opportune time to attempt to try to bring together, in one volume, recent studies of cell membranes made by a number of authors so as to show the progress being made in our present understanding. This has been my aim in the present work. There are so many different aspects to membranes that, even with a range of authors, not all their important features can be discussed in one volume and so some important properties of biological membranes, such as cell adhesion, are not included in the present work. Nevertheless, the book does try to encompass a fairly wide field of study, ranging from recent physical studies of the components of membranes right through to their function in nerve excitability and energy transport systems.

The chapters in the book are arranged in a particular order. We begin with a discussion of recent analyses of the constitution of biological membranes to see the way in which the constituents vary from one type of membrane to another. We follow this by recent physical studies using X-ray techniques and spectroscopic techniques of the properties of the individual constituents such as the lipids, cholesterol and protein. A short chapter is included to discuss what we know in physical terms about lipid-protein interactions. Also included in the first half of the book is a discussion of structural studies of membranes using electron microscopy, X-ray and various spectroscopic methods.

In the middle of the book there is a discussion of recent theories and new models which have been put forward to explain membrane structure and its function. This chapter is intended to show the wide variation in views which exist at the present time.

In the latter part of the book the chapters are arranged so as to show how the structure of biological membranes and their functions are interwoven. Here the authors show the way in which the structure of membranes is involved, in energy conversion processes, such as are involved in photosynthesis and also in the important processes involved in nerve excitability. As will be seen from the various chapters throughout the book, the emphasis is on the physical and biophysical aspects of membrane structure.

The field of study of biological membranes at the present time is a very exciting one and scientists involved in this area have the impression that considerable progress to our understanding will come in the very near future. I hope that the chapters written by the present contributors will provide additional stimulus to other scientists interested in membrane structure and function to produce further advances in our knowledge of these interesting and important structures.

Chapter 2

Lipid Composition of Animal Cell Membranes, Organelles and Organs

G. ROUSER

*Department of Biochemistry, City of Hope Medical Center,
Duarte, California, U.S.A.*

G. J. NELSON

*Bio-Medical Division, Lawrence Radiation Laboratory,
University of California, Livermore, California, U.S.A.*

S. FLEISCHER

*Department of Molecular Biology, Vanderbilt University,
Nashville, Tennessee, U.S.A.*

AND

G. SIMON*

I.	INTRODUCTION	6
II.	LIPID CLASSES OCCURRING IN ANIMAL CELLS	7
III.	GENERAL CONSIDERATIONS IN MEMBRANE LIPID COMPOSITION STUDIES	9
IV.	ARRANGEMENT OF LIPIDS IN CELLULAR MEMBRANES	10
V.	GENERAL METHODOLOGY	12
	A. General Comments	12
	B. Sampling Procedures	13
	C. Prevention of Postmortem Enzymatic Degradation	13
	D. Isolation and Characterization of Cellular Organelles and Membranes	15
	E. Characterization of Lipids	15
	F. Precise Determination of Lipid Class and Fatty Acid Composition	16
	G. Presentation of Polar Lipid Composition Data	20

*Permanent address: Departments of Neurology and Biochemistry, University of Illinois College of Medicine, Presbyterian-St. Luke's Hospital, Chicago, Illinois, U.S.A.

VI. MAMMALIAN ERYTHROCYTES 21
   A. Introduction 21
   B. Less Polar (Neutral) Lipids 22
   C. Phospholipids 23
   D. Fatty Acids, Fatty Aldehydes and Glycerol Ethers. . . 28
   E. General Conclusions 36
VII. SUBCELLULAR ORGANELLES 36
   A. Total Lipid Content 36
   B. Mitochondria 37
   C. Endoplasmic Reticulum 42
   D. Nuclei 43
VIII. WHOLE ORGAN COMPOSITION AND SPECIES VARIATION 44
   A. Viscera and Muscle 44
   B. Brain and Myelin 47
   C. Vascular Structures 49
   D. The Sea Anemone 50
   E. Conclusions 51
IX. MISCELLANEOUS CELL TYPES 51
   A. Platelets 51
   B. Leukocytes 53
   C. Cultured Amnion Cells 54
   D. Comparison of Animal and Yeast Cells 55
X. GLYCOLIPIDS 55
XI. CHANGES IN MEMBRANE COMPONENTS AND PATHOLOGICAL STATES . . 57
XII. SUMMARY AND CONCLUSIONS 61
XIII. ACKNOWLEDGEMENTS 63
REFERENCES 64

I. Introduction

Knowledge of the composition of cellular membranes will ultimately provide the information which can serve as the basis for formulation of membrane structure. While there is a considerable body of data on lipids of membranes and membrane-associated enzymes, it was decided to limit this chapter to considerations of quantitative lipid composition of animal cell membranes. Work on plants and microorganisms in general is not as advanced and is still largely qualitative in nature. The literature on lipids of subcellular particles was reviewed recently (Fleischer and Rouser, 1965). The literature on erythrocyte lipids is considered in detail here in the light of recent advances in methodology. These data are important to evaluate since most of the knowledge of cell surface membrane composition has come from studies of red cells. Data on the composition of whole organs now provide the most satisfactory basis for species comparisons in particular. It was decided that a detailed consideration of the deficiencies of many earlier studies of organ lipid composition would be less valuable than presentation of data obtained recently with improved analytical procedures.

Most lipids of animal cell membranes occur in five organelles: plasmalemma (cell surface), endoplasmic reticulum, Golgi complex, mitochondria and nuclei. The Golgi apparatus makes up a substantial portion of the smooth microsomal fraction in secretory cells and is isolated together with the smooth microsomal fraction. The types of membranes found in or around particular cells are highly variable. Erythrocytes, e.g. have only plasma membranes while neurones in the nervous system, at the other extreme, are very complex. Neurones contain the normal organelles and myelin, a plasma membrane derivative, is wrapped around the cell body extension (axone). Myelin is the most abundant membrane in the central nervous system of some animals. The neurone is terminated by a nerve ending that is a highly specialized lipid-containing structure.

II. Lipid Classes Occurring in Animal Cells

Animal cells in general contain three types of lipids: phospholipids, glycolipids (with one or more carbohydrate moieties) and sterol, usually cholesterol. Animal cells differ from bacterial and plant cells in several respects. Sterol is absent from most bacterial cell membranes. Membranes of plant cells differ from those of animals in that the glycolipids present are glycerol lipids, whereas animal cell membrane glycolipids are sphingolipids. The typical plant cell glycolipids (glycosyldiglycerides) are found in some bacteria and occur as minor components of some animal cells, whereas sphingolipids typical of animal cells occur as minor components of plant cells and are found in some microorganisms.

There are two types of phospholipids, glycerol phospholipids and sphingo-phospholipids. The simplest glycerol phospholipid is phosphatidic acid, a phosphorylated 1,2-diglyceride. The other glycerol phospholipids are most commonly named using "phosphatidyl" as a general term to indicate derivatives of phosphatidic acid regardless of length, degree of unsaturation, or mode of linkage of the two carbon chains. Thus, phosphatidyl choline (lecithin) is the choline ester of phosphatidic acid; phosphatidyl ethanolamine is the ethanolamine ester of phosphatidic acid, etc. The term phosphatidyl thus includes the plasmalogen (vinyl ether linked) and alkoxy (ether linked) forms of the glycerol phospholipids. The glycerol phospholipid frequently termed cardiolipin is a double unit, i.e. it has two molecules of phosphatidic acid esterified to a central glycerol and is therefore appropriately designated as diphosphatidyl glycerol. When only one hydrocarbon chain is present, one hydroxyl group of glycerol being free, the prefix "lyso" is used, e.g. lysophosphatidyl choline (lysolecithin). Sphingolipids contain sphingosine (2-amino-1,3-dihydroxyloctadec-4-ene) or a similar long chain base. Ceramide is the fundamental unit of the sphingolipids and consists of long chain base (usually sphingosine) with a fatty acid in amide linkage. Sphingomyelin and ceramide aminoethylphosphonate are both phosphosphingolipids. The former name has been retained through long usage and designates a ceramide (the fatty acid amide of a long chain base such as sphingosine) with phosphoryl choline esterified at carbon one. Ceramide aminoethylphosphonate is a ceramide with 2-aminoethylphosphonic acid esterified at carbon one (Rouser et al., 1963; Simon and Rouser, 1967).

There are four general types of lipids containing carbohydrate moieties (glycolipids):
1. cerebrosides (monoglycosylceramides),
2. sulfatides (cerebroside sulfate in brain and ceramide dihexoside sulfate in kidney),
3. ceramide polyhexosides (ceramide with two or more carbohydrate moieties, e.g. ceramide dihexoside or ceramide trihexoside, etc.),
4. gangliosides, ceramide polyhexosides that contain in addition one or more sialic acid (N-acetyl neuraminic acid) moieties.

The abbreviations for the lipid classes used in the tables and figures in this chapter are: CAEP, ceramide aminoethylphosphonate; Cer, cerebroside; CPH, ceramide polyhexosides; DPG, diphosphatidyl glycerol (cardiolipin); PA, phosphatidic acid; PC, phosphatidyl choline; PE, phosphatidyl ethanolamine; PG, phosphatidyl glycerol; PI, phosphatidyl inositol, PS, phosphatidyl serine; Sph, sphingomyelin; Su, sulfatide. Lysophosphatides are designated using L as a prefix. Uncharacterized lipids are designated X. When a lipid class was not detected (although sought) the abbreviation ND is used. Components that are detectable but below the level for accurate measurement are designated T (trace). Lipids not referred to in a report are listed as NR (not reported).

Despite many reports suggesting the presence of sterol esters and triglycerides as components of cell membranes, these lipids do not appear to be membrane components. Triglycerides occur as aggregates (fat droplets, vacuoles) of various sizes within cells, principally those of adipose tissue. Triglyceride droplets in some cells are closely associated with mitochondria that actively metabolize them and thus triglyceride may be present in organelle preparations as contaminants. Triglycerides and sterol esters are minor components of whole brain or brain subcellular particles and are clearly not components of cellular membranes from this organ. Erythrocytes that are carefully freed of leukocytes and platelets and well washed to remove plasma are also free of triglycerides and sterol esters. This has been found in the authors' laboratories for humans and various mammalian species. Sterol esters are probably involved in sterol metabolism, particularly in liver, and in sterol transport in blood rather than as components of the cellular membranes.

Steroid hormones, their conjugates, and related metabolites including bile acids occur as minor components of organ extracts and in blood, urine, bile and/or feces. These substances do not appear to be cell membrane components. Substances such as coenzyme Q and the tocopherols are minor components of some cellular membranes.

III. General Considerations in Membrane Lipid Composition Studies

It is essential to know whether the lipid composition of the different membranes of one cell type are the same or different, whether membranes of different cell types of one species differ, and the nature of species variations. Also of interest are the effects of diet, hormones, drugs and pathological states. The various areas must be studied by precise quantitative analysis of membranes isolated in pure and unaltered form.

Progress in the field has been greatly accelerated recently by improvements in methodology. There are, however, many problems associated with development of procedures for isolating pure, unaltered membranes as well as their characterization, both morphologically and chemically, and analysis of their lipid compositions. While recent new developments have been very useful, only a beginning has been made in the field. Additional procedures must be developed and very carefully applied. It appears most appropriate to view the entire field of isolated membrane lipid composition as existing in a relatively primitive state and undergoing relatively rapid development. Many previous investigations have been concerned with the study of relatively impure and incompletely characterized preparations, most commonly obtained from rat liver. Future studies will emphasize highly purified preparations from many sources. For the present, important problems can be defined and relatively neglected areas of study pointed out. Reliable data to answer many important questions are not yet available. In some cases, only the most tentative of conclusions based on fragments of data of questionable reliability can be presented.

The state of the art of membrane isolation is far removed from the ideal situation. A major deviation from the ideal is immediately obvious. Isolations almost invariably begin with whole organs rather than with the separate cell types of an organ. Also, recovery of pure organelles and membranes is not accomplished in a quantitative manner, purity being achieved by accepting loss of some of the sample, sometimes a major portion. In addition, the different membranes of organelles with more than one membrane (mitochondria, nuclei) should be separated for individual study. Although it appears probable that polar lipids occur only in the membranes, available data do not firmly establish this fundamental point.

IV. Arrangements of Lipids in Cellular Membranes

The arrangement of lipids and other components in membranes has yet to be determined and is one of the major goals of membrane studies. Our knowledge of this important area is very limited and most concepts are largely speculative and unproved. The most popular concept, i.e. that of the "unit membrane" or trilaminar arrangement, depicts membrane structure as a lipid bilayer between two layers of protein. Lipids are visualized as attaching to protein by interaction of polar functional groups and hydrocarbon chains of the lipid bilayers are visualized as interacting with each other, perhaps by interdigitation of chains (Vandenheuvel 1963; 1965). Recent evaluation of the evidence upon which the unit membrane concept is based makes it clear that there is only limited direct proof for it (Korn 1966). Furthermore, recent studies of Fleischer (1964; Fleischer et al., 1967a and Fleischer et al., unpublished observations) and Green and Fleischer (1964) are difficult to reconcile with this concept as representing the arrangement in all membranes since addition of lipid to mitochondrial preparations from which lipid has been extracted indicated binding of lipid to protein through hydrocarbon chains and the trilaminar arrangement persists after extraction of lipid from the mitochondrial (inner) membrane. Hence the simple model of the membrane described above is inadequate to explain the more recent data from the mitochondrial inner membrane. Alternate models involving lipoprotein subunits and mosaic arrangements of lipids and protein have been proposed and must be considered carefully.

It has been suggested that cholesterol does not occur in membranes adjacent to molecules of phospholipid or glycolipid but rather as a complex in which the polar lipid is wrapped around cholesterol (Finean, 1953). Molecular model studies have shown that the cholesterol molecule can form such a complex with phospholipids or glycolipids (Vandenheuvel, 1963). There is no direct evidence for the occurrence of such complexes in membranes. Another suggestion is that a cholesterol–polar lipid complex may add stability to membranes. Sterol is not essential for all membrane structures. Sterols are absent from bacterial membranes and cholesterol is a minor component of mammalian mitochondria. Isotope studies have shown that the lipids of myelin retain label for prolonged periods suggesting that membranes containing cholesterol are metabolically, and hence structurally, very stable. The erythrocyte membrane contains an amount of cholesterol about equal to that of myelin but the sterol is, however, readily exchangeable with sterol in blood plasma (Ashworth and Green, 1964; Basford et al., 1964; Eckles et al., 1955; Hagerman and Gould, 1951; Murphy, 1962), although erythrocyte phospholipids other than phosphatidyl choline (lecithin) show little or no exchange (see Section VI-C). Membrane-bound cholesterol is thus not necessarily inert. Furthermore, studies with labeled cholesterol have indicated that it is not evenly distributed over the surface of the erythrocyte (Murphy, 1965). Perhaps the differences in exchange of myelin and erythrocyte cholesterol are related to geometric and physical barriers since erythrocytes are in intimate contact with plasma, whereas myelin occurs as tightly wrapped concentric rings.

The more complex glycolipids, ceramide polyhexosides and gangliosides, have additional polar groups that require an arrangement different from the simpler molecules considered thus far in the literature. This problem does not appear to have been dealt with previously. Several arrangements can be imagined. The polar end can be visualized as a tightly packed ball formed by extremely close association of the different carbohydrate moieties or as curling back along hydrocarbon chains in position to interact with similar chains of adjacent moecules. Alternatively, the polar groups can be visualized as an extended chain lying along the inner surface of the protein of a bimolecular leaflet or protruding through the protein layer and extending into the medium external to the membrane. The present data do not provide any basis for selection among these possibilities.

The lipid class composition of normal cells and subcellular organelles appears to be highly reproducible with little or no change related to diet, although the hydrocarbon chains may vary in length and degree of unsaturation depending upon dietary and other conditions. The constancy of polar lipid class composition suggests that membranes are composed of regular repeating units. Mitochondria from bovine organs have been found to contain diphosphatidyl glycerol (cardiolipin), phosphatidyl ethanolamine and phosphatidyl choline in the approximate molar ratio 1/4/4 with phosphatidyl inositol and phosphatidyl serine being minor components (Fleischer and Rouser, 1965; Fleischer et al., 1967b). This suggests a major repeating unit composed of one molecule of diphosphatidyl glycerol between which are four molecules each of phosphatidyl ethanolamine and phosphatidyl choline with molecules of phosphatidyl inositol and phosphatidyl serine occurring either as a less abundant repeating unit or interspaced separately, possibly at regular intervals between the more abundant repeating units. Membranes generally appear to be rather constant in their lipid class composition, although this constancy may be altered in pathological states and there are species variations. The fact that the length and degree of unsaturation of hydrocarbon chains of lipid classes can be varied is indicative of a less rigid requirement for the arrangement of the nonpolar part of lipid molecules, although such variations should not be construed as being without effects on membrane properties.

Membranes may be visualized as being stabilized by several types of lipid–lipid interactions. The free hydroxyl group on the third carbon of the sphingosine moiety of sphingolipids and the 2-hydroxy group of the hydroxy fatty acids of cerebrosides and sulfatides may add stability by hydrogen bonding with neighboring groups, e.g. carbonyl groups of esters and amides. Other forms of cross-linking of lipid molecules that may lead to stabilization of structure are suggested by the widespread occurrence of acidic lipids. Thus, phosphate and carboxyl groups of phosphatidyl serine may interact with adjacent molecules through positive charges from $-NH^+_3$ or divalent ions (Ca^{2+}, Mg^{2+}). The hydroxyl groups of phosphatidyl inositol could stabilize membranes by hydrogen bonding to adjacent groups.

Stabilization of membrane structure may occur through interaction of hydrocarbon chains through van der Waals forces and dipoles (from double bonds). Although entirely speculative, the concept that the fatty acid composition of a lipid may be maintained in relation to its neighbors in the membrane is suggested as a possibility for increased interaction and hence structure stabilization. Like chains may lie adjacent to each other, i.e. saturated chains next to other saturated chains and unsaturated chains next to those with unsaturation. This placement of chains could result in the most intimate contact and stronger interaction between adjacent chains. Perhaps chains with the same number of double bonds tend to lie adjacent to each other, i.e. chains with one double bond next to each other, etc.

Most sphingolipids have another structural feature that may contribute to membrane stability. For the most part sphingolipids have longer chain fatty acids. The molecules thus have one shorter chain (from, e.g. sphingosine) and one longer chain (from fatty acid). This structure can be seen to increase tail-to-tail interaction of molecules that is easily visualized as stabilizing a lipid bilayer in a unit membrane (Vandenheuvel, 1963, 1965).

Lipid composition data will ultimately be correlated with detailed knowledge of the structures of the protein components of membranes to allow more exact structural formulations. It is essential to determine lipid class composition in full detail, including the molecular species composition of each class, for the earliest realization of the goal of full knowledge of membrane structure. Knowledge of the structure of a membrane can then be correlated with its functions.

V. General Methodology

A. GENERAL COMMENTS

Understanding the relationships of lipid composition to the structures and functions of cell membranes requires attention to proper sampling procedures, prevention of postmortem enzymatic changes, use of reliable methods for characterization of membranes and lipids and finally, analysis with precise and specific quantitative procedures. These problems are considered first since data must be judged on the basis of adequacy of methodology.

B. SAMPLING PROCEDURES

Sampling procedures must be chosen with as much care as analytical procedures. Thus, erythrocyte samples should be washed free of plasma and leukocytes and platelets removed by aspiration after repeated centrifugation. The total lipid and lipid class composition of various portions of complex organs such as brain are different. An average whole brain value is obtained by homogenization of the whole brain or one-half the brain after cutting longitudinally. Analysis of parts of brain and other heterogenous organs can provide valuable information but the values should not be thought to represent the exact composition of other parts of the organ or the organ as a whole. Whole organs are generally used in studies of small animals but with large animals (bovine, human), where only a portion of the organ is used for analysis, regional variations may be important and must be considered.

C. PREVENTION OF POSTMORTEM ENZYMATIC DEGRADATION

Autolysis may change rapidly and significantly the lipid composition of organelles and membranes. Little information on such changes was found in the literature. Recent unpublished studies by G. Rouser and coworkers disclosed postmortem enzymatic degradation to be an important factor and the general nature of the process was defined as well as measures to be taken to reduce the effects. Autolytic changes are most rapid at room temperature (20–25°C) or above. Liver organelles and liver lipids are very labile and represent one extreme. At the other extreme is adult mammalian brain with a relatively stable lipid composition (Table I). Lipids of both heart and lung are relatively stable to autolytic change but kidney and spleen show changes more rapidly although lipids of these organs are more stable than those of liver. When cooled to 4°C, membrane integrity and lipid composition of organs are more nearly preserved. Lipid composition changes are very small even for periods of several days (Table II). Rapid cooling is thus essential for isolation of organelles with maintenance of native lipid composition. Rapid freezing and storage at $-20°C$ or below is used for preserving lipid composition for periods of weeks to months prior to analysis although other classes of compounds are not necessarily preserved.

During the early phases of postmortem autolysis, sphingolipids, cholesterol and triglycerides show little or no change. The glycerol phospholipids are, however, degraded rapidly to lysophosphatides. Upon more prolonged standing, the glycerol lipids undergo further degradation with release of

TABLE I
Bovine brain autolysis*

	Fresh frozen	48 hr 24°C
PC	28.9	29.1
PE	35.1	35.7
PS	16.0	16.3
PI	2.5	2.5
PA	1.2	0.9
DPG	0.7	0.3
Sph	12.4	12.8

* Rouser and Kritchevsky, 1967.

TABLE II
Postmortem changes of phospholipids of human heart*

	Initial	7 days +4°C	24 hr 24°C
PC	38.6	37.9	35.2
PE	26.1	25.4	23.6
PS	1.8	1.5	0.9
PI	5.4	4.6	4.9
DPG	10.6	10.3	9.9
PA	ND	ND	ND
PG	0.3	0.4	0.5
X_1	0.5	0.4	0.3
LPC	2.3	2.6	5.5
LPE	0.9	1.9	5.3
LPS	ND	0.7	0.4
LPI	1.9	1.1	1.1
Sph	5.0	4.5	4.8
Origin	4.4	4.8	5.9
PC + LPC	40.9	40.5	40.7
PE + LPE	27.0	27.3	28.9
PS + LPS	1.8	2.2	1.3

* Rouser and Kritchevsky, 1967.
ND, not detected.

water soluble phosphate esters. The free fatty acid level rises in proportion to the extent of degradation of the phospholipids and the extent of autolysis can be judged by the free fatty acid content of a lipid extract. Two-dimensional thin layer chromatography is useful for this purpose since free fatty acids and polar lipids can be visualized on the same chromatogram. The same general autolytic pattern has been observed for organs of higher animals and the sea anemone (Simon and Rouser, 1967). The pattern may thus be characteristic for animals of all species.

D. ISOLATION AND CHARACTERIZATION OF CELLULAR ORGANELLES AND MEMBRANES

Highly purified and carefully characterized preparations of organelles and membranes whose morphology is characteristic of the state in the living animal are essential for studies of specificity of lipid composition. The details of general isolation procedures must be determined and adapted for each organ or cell type studied. The general considerations with reference to the specialized literature have been presented recently (Fleischer and Rouser, 1965) and need not be repeated here. Care in isolation must be followed by careful characterization. Electron microscopy is the fundamental means for characterization of organelles and membranes. It provides information on morphological integrity and the presence of contaminating structures. Morphological observations should be supplemented by chemical analysis of characteristic components and assay of enzymes characteristic for the particular organelle or membrane.

E. CHARACTERIZATION OF LIPIDS

Proper identification of lipids is of obvious importance. Ideally each lipid class should be isolated in pure form, the functional groups of the intact lipid established by specific quantitative reactions and the amounts and sequences of the different moieties established by hydrolysis and quantitative analysis. In practice, this approach is seldom used. Instead, more rapid procedures are employed. These include determination of chromatographic properties, reaction with specific spray reagents, infrared spectrophotometric examination and hydrolysis to characteristic products. In broad surveys, tentative identifications should include at least two-dimensional thin layer chromatography (TLC) with two different solvent pairs combined with specific spray reagents. Identifications are made more certain if column chromatography characteristics (see Section V-F) are determined. Still more conclusive identifications are obtained when other data are supplemented by infrared and hydrolytic product data. In the strictest sense most identifications are tentative rather than conclusive since assumptions are made. Full characterization is seldom attempted. It is common, e.g., to assume that the polar groups of sphingolipids are all attached to carbon one of the long chain base although this has only been proven for bovine brain sphingomyelin (Marinetti et al., 1953) and cerebroside (Carter and Greenwood, 1952) and ceramide aminoethylphosphonate from the sea anemone (Simon and Rouser, 1967).

Confusion arises from poor handling of lipids with production of artifacts and identification of lipid classes by one dimensional paper or thin layer chromatography only. In our studies, lipids have been extracted using a nitrogen atmosphere and low temperatures to avoid decomposition (Rouser et al., 1963) with precautions to avoid contaminants (Rouser et al., 1966b). Even in survey studies involving many samples, tentative identifications have been based on:

1 two-dimensional TLC with two different solvent pairs and specific spray reagents (for phosphorus, free amino groups, carbohydrates and sialic acid);
2 alkaline deacylation and phospholipase A degradation followed by two-dimensional TLC to determine the lipids degraded and the products formed;
3 column chromatography (usually on DEAE or TEAE celluloses, see Section V-F).

F. PRECISE DETERMINATION OF LIPID CLASS AND FATTY ACID COMPOSITION

Precise quantitative analysis of the lipid classes in a complex mixture presents several problems. Progress has been made toward the development of relatively complete schemes of analysis that provide relatively precise values but all procedures have limitations. In the broadest sense the problems of lipid analysis are:

1 There are many components in some mixtures and specific determination of each component is thereby made more difficult.
2 Mixtures may be composed of many different types of substances; characterization and analysis thus necessitate use of a variety of procedures and knowledge of chemical characteristics of very different types of organic compounds.
3 Some components occur in small amounts and are thus more difficult to determine.
4 Some components have not been characterized and thus their determination cannot always be carried out with certainty.
5 Polar lipid classes are generally complex mixtures of molecular species differing in length and degree of unsaturation of carbon chains. In addition branched chains and various substitutions along the chain may

be present. These different molecular species cannot be determined accurately at this time.

6 Some lipids are easily oxidized or otherwise altered.

7 Nonlipid contaminants are extracted with lipids and must be removed.

Since polar lipids (phospholipids and glycolipids) are of major concern in studies of cellular membrane composition, further discussion is restricted to these lipids. Two types of procedures must be differentiated clearly. Procedures involving separation of intact lipids are to be considered as primary methods of lipid analysis. Procedures involving degradation of lipids as an initial step are severely limited (Fleischer and Rouser, 1965) and cannot be recommended for general application. The most precise procedures for quantitative analysis involve minimum handling of the lipid extract to avoid decomposition and introduction of contaminants. The most rapid procedures that are relatively precise employ thin layer chromatography (TLC). Two-dimensional TLC with two different solvent systems (Rouser et al., 1966c) is used for separation. After TLC, spots are located and scraped or aspirated from the plates. The molar amounts of the lipid classes can be determined spectrophotometrically: phosphorus analysis for phospholipids (Rouser et al., 1966c), carbohydrate determination with anthrone or α-naphthol for glycolipids (Rouser et al., 1968), trinitrobenzene sulfonic acid for lipids with free amino groups (Siakotos, 1967) including sphingolipids after hydrolysis to release sphingosine (Rouser et al., 1968) and the zinc chloride-acetyl chloride procedure (Rouser et al., 1968) for cholesterol. Organ extracts can be analyzed directly by TLC without column chromatography. Subcellular organelle extracts usually contain sucrose or other substances added in the isolation procedure. The amounts are sufficiently large that spotting solutions may contain insoluble solids that interfere with sample application to the plate. These contaminants should be removed by Sephadex column chromatography (Siakotos and Rouser, 1965) prior to TLC. The high degree of reproducibility of the Sephadex-TLC procedure is shown by the values in Table III. The fatty acid composition of even small spots separated by TLC can be determined by gas chromatography of fatty acid methyl esters since quantitative gas phase chromatography is routinely carried out on nanogram amounts of materials (Feldman and Rouser, 1965).

Column chromatography is more tedious than TLC and precise values are obtained by column chromatography only after extensive experience. Careful elimination of nonvolatile residues from solvents and impurities from adsorbents is also desirable and usually essential. To avoid decomposition, column fractions must be handled in the cold in a nitrogen atmosphere as much as possible. These difficulties dictate the use of column chromatography for quantitative analysis only when absolutely essential.

Column chromatography is essential in quantitative lipid class analysis for determination of some minor and trace components and for substances not separated by TLC. Two-dimensional TLC resolution is high and overlap of spots does not always prevent accurate analysis, e.g. overlap of a phospholipid with a glycolipid, since the phospholipid can be determined by phosphorus analysis and the glycolipid by carbohydrate assay. In practice,

TABLE III
Reproducibility of phospholipid values of rat brains*

	I	II	III	IV	V	VI	Average
PC	37·3	37·4	38·2	37·0	36·7	36·8	37·2
PE	37·5	37·5	37·4	38·2	37·6	36·5	37·5
PS	13·1	12·6	12·4	12·6	12·4	12·6	12·6
PI	3·3	3·7	3·4	3·0	4·1	3·5	3·5
DPG	1·4	1·4	1·2	1·3	1·3	1·4	1·3
PA	1·0	1·1	0·9	1·4	1·5	1·4	1·2
PG	0·5	0·5	0·2	1·0	0·9	0·6	0·6
Sph	5·3	5·0	5·2	5·1	5·0	4·9	5·1

* G. Rouser, G. Simon and G. Kritchevsky, unpublished observations.
Average of 4 determinations after two dimensional TLC with two different solvent pairs.

therefore, column chromatography is generally essential only for minor components not determined by or below the acceptable limit of accuracy of TLC alone. Column chromatography is invaluable as an aid in identification and will become more important for quantitative analysis by automated procedures and when it becomes desirable to separate the molecular species of each lipid class

Column chromatography on ion exchange celluloses (DEAE and TEAE) or silicic acid have proved to be most valuable for quantitative analysis.

Both diethylaminoethyl (DEAE) and triethylaminoethyl (TEAE) celluloses are generally useful for separation of complex mixtures into simpler groups prior to separation into individual lipid classes by additional column chromatography or TLC (Rouser et al., 1961; 1963; 1964; 1965; 1967b and 1968 and Rouser and Fleischer, 1967). DEAE has normally been used in the acetate form but use of the borate form (Rouser et al., 1967b; 1968) provides a means for recovering a ceramide polyhexoside fraction free of phosphatidyl ethanolamine. With the borate form of DEAE, phosphatidyl ethanolamine is eluted with chloroform/acetic acid after chloroform/methanol 7/3 rather than the latter solvent mixture that elutes phosphatidyl ethanolamine from the acetate form of DEAE. TEAE in the hydroxyl form has the advantage of the borate form of DEAE and in addition has a higher capacity than DEAE for lipids with carboxyl groups as the only acidic groups (fatty acids, bile acids, gangliosides). A useful elution sequence for TEAE is shown in Table IV.

A useful quantitative procedure employing silicic acid column chromatography has been reported recently (Smith and Freeman, 1959; Rouser et al., 1967a; Vorbeck and Marinetti, 1965). As in earlier procedures, chloroform is

TABLE IV
Elution of the hydroxyl form of TEAE cellulose

Solvent	Eluting volumes*	Lipid classes eluted
(1) Chloroform	6	Neutral (less polar) lipids (cholesterol, sterol esters, mono, di- and triglycerides, etc.)
(2) Chloroform/methanol 9/1	8	Phosphatidyl choline, lysophosphatidyl choline, sphingomyelin, mono- and diglycosyl diglycerides, cerebrosides
(3) Chloroform/methanol 2/1	10	Ceramide polyhexosides
(4) Methanol	10	Inorganic compounds in samples or formed by ion exchange with acidic lipids
(5) Chloroform/methanol 2/1 plus 1% (v/v) glacial acetic acid	10	Free fatty acids, phosphatidyl ethanolamine, lysophosphatidyl ethanolamine, ceramide aminoethylphosphonate
(6) Glacial acetic acid	10	Phosphatidyl serine
(7) Methanol wash	3–4	Little or no lipid, used to remove acetic acid
(8) Chloroform/methanol 4/1 containing 20 ml/l of 28% aqueous ammonia and made 0·01 M in ammonium or potassium acetate	10	Remaining acidic lipids: phosphatidic acid, phosphatidyl inositol, cerebroside sulfate (sulfatide), cholesterol sulfate, phosphatidyl glycerol, diphosphatidyl glycerol and other lipids containing phosphate or sulfate groups
(9) Methanol	10	May contain some acidic lipid, used to clear column prior to reuse

* For a 2·5 × 20 cm column, in column volumes, flow rate 3 ml/min.
Columns are reconditioned before use by washing with 0·1 N methanolic KOH (3 column volumes), methanol (4 column volumes), chloroform/methanol 1/1 (4 column volumes), and chloroform (4 column volumes).

used for elution of neutral (less polar) lipids and methanol for elution of phospholipids. Elution with chloroform/acetone or acetone is inserted, however, between the two customary solvents as first described by Smith and Freeman (1959). Acetone has been found to elute cerebrosides (Smith and Freeman, 1959; Rouser et al., 1967a); sulfatides, ceramide polyhexosides and plant sulfolipid (Rouser et al., 1967a); mono and diglycosyl diglycerides (Rouser et al., 1967a, Vorbeck and Marinetti, 1965); glucose and many uncharacterized organic solvent soluble compounds present in fecal lipid extracts (Rouser, et al., 1967b). Of the common phospholipids, only diphosphatidyl glycerol (cardiolipin) is partially eluted with acetone, the remainder being eluted with methanol. Gangliosides are not eluted with acetone. Acetone elution is thus most advantageous where glycolipid levels are high and diphosphatidyl glycerol is low or absent. This situation is met with extracts of some bacteria (Vorbeck and Marinetti, 1965), spinach leaves and brain (Rouser et al., 1967a) in particular. Chromatography of extracts of heart and liver gives an acetone fraction composed almost entirely of diphosphatidyl glycerol and extracts of lung and kidney are mixtures of this phospholipid and glycolipids. With erythrocyte lipids the acetone fraction contains pigments. With animal tissues the procedure is thus most useful for brain extracts.

G. PRESENTATION OF POLAR LIPID COMPOSITION DATA

Analytical data can be expressed on a wet or dry weight basis, as percentage of the total lipid, as percentage of the polar lipids, or related to protein or nucleic acid values. Each method has value for specific applications. Expressed on a wet weight basis the data are related to the overall composition of cells, organelles, or membranes. This form of expression with polar lipids indicates the proportion of membranes per unit weight of cell substance. On a dry weight basis values are for the most part related to the ratio of lipid to protein, nucleic acid, mucopolysaccharides and inorganic components (salts). In practice we have found for vertebrate organs that expression of organ composition data on a wet weight basis gives rather constant values for different animals of the same species, whereas values on a dry weight basis are quite variable from one animal to the next, the extent of variability being different for different organs. Brain values are very consistent from animal to animal of the same species on both a wet and dry weight basis whereas other organs, particularly liver, are variable. With isolated organelles and membranes, expression of results on either a wet or dry weight basis may be ambiguous since residual water and solids (e.g. inorganic salts, sucrose, diodrast, ficoll, etc.) from the isolation process may be present and attempts to remove these may be accompanied by alteration of the lipid composition through enzymatic degradation or the structures may be ruptured and dispersed in the wash medium. Organelle total lipid values are thus best expressed as weight of lipid/weight of protein.

When results are expressed as percentage of the total lipid, several difficulties are encountered. If an aqueous wash of the extract is used (Folch et al., 1957) to remove water soluble nonlipids, lipid may be lost to the nonlipid (upper aqueous methanol) phase and water soluble nonlipid may remain in the lower (chloroform) lipid phase (Nazir and Rouser, 1966; Siakotos and Rouser, 1965). Sephadex column chromatography (Siakotos and Rouser, 1965) appears to remove all water soluble nonlipid but chloroform soluble nonlipid substances, when present, appear in the lipid fraction. This nonlipid portion may be significant. Thus, in erythrocyte extracts pigments are prominent and extracts of some organs, particularly liver, contain large amounts of uncharacterized organic solvent soluble nonlipid. In addition, organ extracts may show variability in the proportion of less polar lipids, particularly triglycerides, to polar lipids and thus polar lipid values may be more variable when expressed as "percentage of the total lipid".

Since water content, total solids and the amount of less polar lipid of the same organ from different species may vary greatly, one of the most reliable means of comparison is as a percentage (ratio) of the lipids to be compared. Thus, phospholipid composition can be compared in a relatively unambiguous manner as percentage of the total phospholipid or as the percentage (ratio) of a selected group of phospholipids by eliminating lipid classes that may be more variable. Values expressed as percentage of the total phosphorus in a lipid extract may vary because the phosphorus in extracts is not entirely from known phospholipids. With some procedures, water soluble nonlipid may appear in extracts and with others, even though water soluble phosphate compounds are removed, there is still some organic solvent soluble phosphorus-containing material that is not characterized. When comparing all types of animal cells and membranes, results should preferably be expressed as percentages (ratios) of the phospholipids determined. This is the form of expression used for the most part for tabular data in this chapter.

VI. Mammalian Erythrocytes

A. INTRODUCTION

It is generally agreed that all of the erythrocyte lipid is in the plasma membrane and thus either "ghosts" (the membrane) or whole cells can be studied. The red cell is devoid of other organelles containing lipid. Valid studies of red cell lipid composition are possible only when the cells are carefully washed free of plasma and contaminating leukocytes and platelets are removed.

The human erythrocyte is approximately 99% hemoglobin, water and electrolytes and 1% stroma. The stroma contains many enzymes (Pennell, 1964) and the enzymatic composition varies somewhat depending upon the method of preparation of the stroma (ghost).

Erythrocyte studies provide an insight into cell surface membrane lipid composition and its species variations. These cells contain cholesterol, phospholipids and glycolipids. The relative amounts of these groups in several species are shown in Table V.

TABLE V
Lipid distribution in erythrocytes from various mammalian species[a]

	Total lipid[b] mg/ml packed cells	gm/cell[c]	Cholesterol	Total gangliosides[d]	Other[e] glycolipids	Phospholipid
Cat	6·04	$3·45 \times 10^{-13}$	26·8	8·8	3·1	61·3
Cow	4·44	$2·58 \times 10^{-13}$	27·5	5·5	2·2	64·8
Dog	5·76	$4·84 \times 10^{-13}$	24·7	11·8	10·9	52·6
Goat	6·14	$1·23 \times 10^{-13}$	26·2	5·7	17·9	50·2
Guinea pig	5·72	$4·41 \times 10^{-13}$	27·0	2·2	15·2	55·6
Horse	5·37	$2·58 \times 10^{-13}$	24·5	15·5	8·0	52·0
Pig	4·33	$2·52 \times 10^{-13}$	26·8	3·3	10·1	59·8
Rabbit	4·57	$4·15 \times 10^{-13}$	28·9	4·5	0·8	65·8
Rat	5·08	$3·15 \times 10^{-13}$	24·7	6·3	2·0	67·0
Sheep	4·91	$1·62 \times 10^{-13}$	26·5	7·8	2·5	63·2

[a] Nelson, 1967b.
[b] Weight of the extract obtained as fractions 1 and 2 by Sephadex column chromatography. Siakotos and Rouser, 1965.
[c] Calculated using mean corpuscular volume. Ponder, 1948; Reed et al., 1960.
[d] Weight of fraction 2 from Sephadex. Siakotos and Rouser, 1965.
[e] Calculated from total fraction 1 (Sephadex) minus cholesterol plus phospholipid.

It is apparent from the data that there are large species differences in the amounts of glycolipids, although cholesterol is rather constant in amount. Unfortunately, precise quantitative values for the individual glycolipids are lacking and only qualitative or semiquantitative comparisons can be made. Glycolipids are considered in detail in Section X. The less polar (neutral) lipids and phospholipids are considered in this section.

B. LESS POLAR (NEUTRAL) LIPIDS

Cholesterol is a major component of the erythrocyte membrane. Sterol esters and triglycerides have been reported to be present in erythrocytes. The absence of these lipids from well washed cells of man, rat, rabbit, pig, dog, horse, sheep, cow and goat has been demonstrated recently by Nelson (1967d). The cells were found to contain cholesterol as the only neutral lipid and it accounted for about 25–29% of the total lipid (Table V). Sterol esters and triglycerides were not found.

C. PHOSPHOLIPIDS

The phospholipid composition of erythrocytes of humans and other mammalian species has been reported from many laboratories. Three analytical procedures have been used in general. The most variable values were obtained by silicic acid column chromatography. This is a reflection of incomplete separation and autoxidation to a great extent. Values obtained by one-dimensional paper chromatography using the Marinetti system have been more reproducible from one laboratory to the next, but distortion of values is commonly obtained (Fleischer and Rouser, 1965). Thin layer chromatography appears to provide the most reproducible and accurate values. Two-dimensional TLC is preferable since spot overlap is reduced and new lipids can be recognized. This is particularly important with species other than man. Values reported for normal human cells from 1960 through 1967 (Tables VI–VIII) show considerable variability. Highly variable values

TABLE VI
Phospholipids of normal human erythrocytes (1960–1963)

	1	2	3	4	5	6	7
PC	31·4	30·5	46·7	39·5	37·5	36·0	28·4
PE	25·8	24·8	28·9	12·3	30·2	22·7	28·4
PS	15·6	14·0	NR	12·8	10·4	10·8	18·9
PI	4·2	3·9	NR	5·8	NR	6·2	NR
PA	NR	NR	0·7	NR	NR	NR	1·1
Sph	23·0	26·9	23·7	29·6	21·9	24·3	23·2
Ref.	a	b	c	d	e	f	g

NR., not reported.
[a] Reed et al., 1960.
[b] Klibansky and Osimi, 1961.
[c] Kates et al., 1961.
[d] Blomstrand et al., 1962.
[e] Farquhar, 1962.
[f] Klibansky and de Vries, 1963.
[g] Dodge et al., 1963.

have been reported for phosphatidyl choline (27·9–46·7%), sphingomyelin (18·4–31·8%), phosphatidyl inositol (1·2–6·2%), phosphatidyl serine (4·0–19·7%) and phosphatidyl ethanolamine (12·3–30·2%). Most reports have not included values for phosphatidyl inositol or phosphatidic acid. It is clear that methodology has been variable and generally not precise. The values obtained by Turner and Rouser (column 3, Table VIII) are in rather good agreement with the values reported by Bradlow et al. (1964) (columns

TABLE VII
Phospholipids of normal human erythrocytes (1964–1965)

	1	2	3	4	5	6	7	8	9
PC	30·7	31·8	32·2	30·2	35·9	30·8	28·8	34·0	34·6
PE	26·7	28·7	29·6	24·5	19·7	30·3	23·8	25·0	23·9
PS	15·6	13·9	11·7	15·1	9·3	13·2	14·2	16·0	10·8
PI	2·3	NR	NR	3·0	3·3	NR	1·4	NR	3·4
PA	NR	NR	NR	NR	NR	NR	NR	NR	NR
Sph	24·7	25·7	26·5	27·2	31·8	25·8	31·8	25·0	27·4
Ref.	a	b	c	d	e	f	g	h	i

NR, not reported.
[a] Ways and Hanahan, 1964.
[b] Bradlow and Rubenstein, 1964.
[c] Bradlow et al., 1964.
[d] Condrea et al., 1964.
[e] Szeinberg et al., 1965b.
[f] Bradlow et al., 1965.
[g] Crowley et al., 1965.
[h] Hill et al., 1965.
[i] Szeinberg et al., 1965a.

TABLE VIII
Phospholipids of normal human erythrocytes (1966–1967)

	1	2	3*
PC	29·3	36·2	28·9 ±0·90
PE	25·3	28·4	27·2 ±0·41
PS	19·7	4·0	13·0 ±1·10
PI	7·4	2·5	1·3 ±0·24
PA	NR	NR	2·2 ±0·40
Sph	18·4	28·9	26·9 ±1·50
Ref.	a	b	c

* Average values from 8 blood samples followed by standard deviation; the percentage of lysophosphatidyl choline was 1·1 ± 0·13.
NR, not reported.
[a] Karaca and Stefanini, 1966.
[b] Williams et al., 1966.
[c] J. D. Turner and G. Rouser, unpublished observations.

2, 3, 6 of Table VII). Both groups used quantitative TLC. J. D. Turner and G. Rouser (unpublished) used Sephadex column chromatography (Siakotos and Rouser, 1965) for removal of water soluble nonlipids, two-dimensional TLC for separation and phosphorus analysis of spots for quantitative analysis (Rouser et al., 1966c). It is believed that these values are the most accurate thus far obtained. Phospholipid distribution did not appear to be influenced by age, sex or blood type. The ratio phosphatidyl choline/phosphatidyl ethanolamine/sphingomyelin/phosphatidyl serine is very close to 1/1/1/0·5. It is possible that the displacement of sphingomyelin by lysophosphatidyl choline followed by acylation to phosphatidyl choline (discussed below) is responsible for the presence of slightly more phosphatidyl choline and less sphingomyelin.

Sources of error encountered by Turner and Rouser were: 1. failure to wash cells free of plasma (giving rise to a high phosphatidyl choline value); 2. loss of lipid onto denatured hemoglobin; 3. oxidation of lipids if an antioxidant such as BHT is not added and 4. loss, particularly of phosphatidyl serine, when extracts are evaporated completely to dryness. Decomposition of lipids can perhaps be decreased by extraction of stroma (rather than whole cells) to eliminate the effects of hemoglobin and pigments.

The phospholipids listed in column 3 of Table VIII and lysophosphatidyl choline were characterized by TLC, specific spray reagents and TEAE column chromatographic properties by J. D. Turner and G. Rouser (unpublished observations). TEAE column chromatography followed by TLC also disclosed numerous uncharacterized minor components, some of which contained phosphorus. The nature of these components remains to be established and it is clear that knowledge of normal human erythrocyte polar lipid composition is incomplete.

Erythrocyte phospholipid values determined by quantitative two-dimensional TLC by Nelson (Table IX) clearly show a wide species variability.

TABLE IX
Phospholipid distribution in erythrocytes from various mammalian species*

	Rat	Rabbit	Pig	Dog	Horse	Sheep	Cow	Goat	Cat	Guinea pig
PC	47·5	33·9	23·3	46·9	42·4	ND	ND	ND	30·5	41·1
PE	21·5	31·9	29·7	22·4	24·3	26·2	29·1	27·9	22·2	24·6
PS	10·8	12·2	17·8	15·4	18·0	14·1	19·3	20·8	13·2	16·8
PI	3·5	1·6	1·8	2·2	<0·3	2·9	3·7	4·6	7·4	2·4
PA	<0·3	1·6	<0·3	0·5	<0·3	<0·3	<0·3	<0·3	0·8	4·2
Sph	12·8	19·0	26·5	10·8	13·5	51·0	46·2	45·9	26·1	11·1
LPC	3·8	<0·3	0·9	1·8	1·7	ND	ND	ND	<0·3	<0·3
X						4·8	1·7	0·8		

* Data of Nelson, 1967b as average phosphous values obtained using duplicate analysis with two pairs of TLC solvents (Rouser et al., 1966c); values as percentage of the phospholipid shown.

Phosphatidyl choline (lecithin) was not found in well washed erythrocytes of sheep, cow, and goat but represented almost 50% of the phospholipid of cells from rats and dogs. Rat and dog cells are quite similar as are those from sheep, cow and goats. All species differ from humans. Values for phospholipids in the same species studied by Nelson as well as other species have been reported (Condrea et al., 1964; de Gier et al., 1961, 1966; de Gier and van Deenen 1961, 1964; Farnsworth et al., 1965; Hanahan et al., 1960; Heemskerk and van Deenen, 1964; Kates and James, 1961; Leveille et al., 1962; Matsumoto, 1961; Roelofsen et al., 1964; Schrader and Dimopoullos, 1963; Sloviter and Tanaka, 1964; Soule et al., 1959; van Deenen and de Gier, 1964; van Deenen et al., 1961; Watson, 1963). The values from the general literature are for the most part less precise than those presented by Nelson who used more modern procedures for analysis.

Plasma phospholipid values are shown in Table X. The values are quite different from those of erythrocytes and it is clear that phosphatidyl choline occurs in plasma of species in which this phospholipid is absent from erythrocytes. Phosphatidyl ethanolamine is a minor component of plasma phospholipids except in the guinea pig. It is now well established that lysophosphatidyl choline occurs in plasma (Gjone et al., 1959; Phillips, 1957) and that it is bound to albumin (Phillips, 1959). It has been estimated that reticulocytes contain 4 to 5 times as much phospholipid as mature erythrocytes (Raderecht and Scholzel, 1962; Raderecht et al., 1960). Newly formed erythrocytes can be separated from older cells by centrifugation and appear to contain the same total amount of phospholipid (expressed on a packed cell volume basis) and to have the same lipid class composition as older cells (van Gastel et al., 1965; Westerman et al., 1963). Cells from newborn infants appear to have a somewhat different phospholipid distribution (Bentley, 1962).

Of particular interest are the reports of altered phospholipid compositions in several disorders. These studies, some of which are considered in more detail in Section XI, indicate that changes in phospholipid distribution of erythrocytes may occur with or without changes in red cell properties.

Although erythrocytes do not oxidize fatty acids (Pittman and Martin, 1966; Rosenzweig and Ways, 1966) and there is no synthesis or metabolic turnover of phospholipids except for phosphatidic acid (Hokin and Hokin, 1964; Paysant et al., 1963; Westerman and Jensen, 1965), an erythrocyte-plasma phospholipid exchange reaction involving phosphatidyl choline and lysophosphatidyl choline and fatty acid incorporation into phosphatidyl choline occur without formation of new cells. In vitro incubation of human or rat erythrocytes with free fatty acids, ATP and coenzyme A results in incorporation of added fatty acid into phosphatidyl choline of the cells (Mulder and van Deenen, 1965a; Mulder et al., 1963; Oliveira and Vaughan,

1964). Sheep cells lacking phosphatidyl choline (Nelson 1967a; Turner *et al.*, 1958; Turner and Parsons, 1957) did not show the incorporation thus showing that only this one lipid class is involved in all species examined.

TABLE X
Plasma phospholipid composition (as % total phospholipid)

	Various mammals*								
	Cat	Cow	Dog	Goat	Guinea pig	Horse	Pig	Rabbit	Rat	Sheep
PC	69·3	74·5	82·6	77·7	55·7	74·9	69·9	60·8	63·8	71·7
PE	1·3	1·2	1·3	0·8	21·7	3·3	1·9	6·7	1·3	1·0
PI	0·3	1·1	1·9	0·3	0·8	0·3	2·8	5·5	4·3	0·5
Sph	14·6	14·9	6·3	12·5	7·8	10·6	13·5	6·7	7·8	13·4
LPC	14·4	8·4	7·8	8·7	13·9	10·8	11·9	20·3	22·7	13·4

	Human adult							Fetal
PC	59	66·7	54·3	69·5		65·5	51·2	59·6
PE	}7		}8·0	2·8	4·1	}4·9	}5·7	
PS					2·6			
Sph	21	21·4	23·6	22·2		19·2	21·8†	26·4
LPC	12		14·1	5·5		6·4	17·7†	
Other		11·9					3·5	14·0
Ref.	a	b	c	d	e	f	g	h

* Nelson, 1967c.
† Includes uncharacterized lipids.
a Billimoria *et al.*, 1965.
b Crowley *et al.*, 1965.
c Cumings *et al.*, 1965.
d Vikrot, 1965.
e Nothman and Proger, 1965.
f Phillips, 1962.
g Rowe, 1960.
h Crowley *et al.*, 1965.

The incorporation appears to depend upon uptake of lysophosphatidyl choline by erythrocytes followed by acylation to give phosphatidyl choline (Mulder and van Deenen, 1965b; Mulder *et al.*, 1965). In plasma, enzymatic acyl transfer of a fatty acid from phosphatidyl choline to cholesterol occurs with the formation of cholesterol ester and lysophosphatidyl choline (Glomset, 1962; Glomset *et al.*, 1962, 1966; Stein and Stein, 1966). *In vivo*, the fatty acids of phosphatidyl choline show evidence of exchange with plasma lipids (Farquhar, 1965; Farquhar and Ahrens, 1963; Hill *et al.*, 1964, 1965;

Horwitt *et al.*, 1959). Fatty acid changes on different diets can be observed in humans within two weeks but these are probably restricted to phosphatidyl choline. In more prolonged studies where sufficiently large numbers of new cells are formed, the fatty acids of other phospholipids of red cells are also changed. Studies with other species indicate a similar phospholipid exchange phenomenon (de Gier and van Deenen, 1964; Farnsworth *et al.*, 1965; Fitch *et al.*, 1961; Greenberg and Moon, 1961; Grimmer *et al.*, 1965; Reid *et al.*, 1964; Voigt *et al.*, 1965; Walker and Kummerow, 1963, 1964a; Watson, 1963). The exchange of mature erythrocyte and plasma phospholipid and the more extensive fatty acid composition changes during formation of new erythrocytes provide means for alteration of some properties of the cells as a result of dietary changes and perhaps in disease.

Several studies have been performed in attempts to relate differences in lipid composition to permeability differences. Erythrocytes of sheep fall into two groups, high and low potassium cells and Nelson (1967a) has found that phospholipids of the two groups do not differ. Differences in permeability and osmotic fragility were studied (de Gier *et al.*, 1966; Jacobs *et al.*, 1950; van Deenen *et al.*, 1963). Erythrocytes of man, rat, rabbit and guinea pig hemolyze rapidly in the presence of glycerol, while cells from pig, dog, cat, sheep and cow do not. The species differences showed only a partial correlation with the phosphatidyl choline content (de Gier *et al.*, 1966; van Deenen *et al.*, 1963) since cells of some species (e.g. dog) that are resistant to hemolysis have a high level of phosphatidyl choline. Fatty acid composition differences between the two groups failed to show a correlation with differences in permeability and osmotic fragility. At present no correlations appear to have been established between phospholipid composition differences and permeability or fragility differences of different species. It appears that the phospholipid composition changes are not involved in instances where permeability differences are observed (high and low potassium cells, permeability to glycerol) and that changes in fatty acid composition do not necessarily affect permeability properties greatly. Walker and Kummerow (1964c) found, however, that the variations in rat erythrocyte total fatty acids on different diets correlated well with the rate of hemolysis in isotonic solutions of nonelectrolytes (and hence, presumably, permeability). Cells with the highest levels of polyunsaturated fatty acids were most resistant to hemolysis.

D. FATTY ACIDS, FATTY ALDEHYDES AND GLYCEROL ETHERS

The nature of the fatty acids of each lipid class and species variations are of particular interest since the physical and chemical properties of a lipid class depend upon the lengths of the hydrocarbon chains and the number of double bonds present. It can thus be expected that metabolic properties and physiological characteristics, e.g. permeability, may eventually be correlated with variations in fatty acid composition.

Tables XI and XII show some values reported for total fatty acids of

TABLE XI
Fatty acid composition of human erythrocytes (as % of total fatty acid) (1959–1963)

	1[a]	2[b]	3	4	5	6	7[c]	8[d]	9
12:0	0·5	0·4		0·3	0·2	0·1			
14:0	0·8	0·7		0·8	0·5	0·5			
16:0	27·2	30·2	37·5	41·0	23·6	28·2	29	31	24·9
16:1	0·5	0·6	1·5	1·1	1·7	0·7			
18:0	19·2	20·4	15·5	7·9	16·5	15·1	16	15	19·0
18:1	17·1	20·4	26·5	18·9	19·8	18·3	17	21	16·6
18:2	15·3	8·3	17·0	15·3	12·0	10·6	12	11	11·6
18:3			2·0		0·6				
20:0					2·1	0·1			
20:2						0·1			
20:3	2·0	1·9		1·5	7·1	1·6			
20:4	10·0	10·4		7·9	5·8	10·8	28[e]	31[f]	14·1
22:5	1·6	2·1		4·5		4·0			
22:6	1·1	1·2		4·5	5·1	2·1			
Others	4·7	3·4		0·9	4·0	6·3	5	5	
Ref.	g	h	i	j	k	l	m	n	o

[a] Adult on corn oil diet; [b] institution diet; [c] male; [d] female + male; [e] incl. 9% 20:4;
[f] incl. 10% 20:4;
[g] Horwitt et al., 1959
[h] Horwitt et al., 1959
[i] Kögl et al., 1960
[j] Kates et al., 1961
[k] Corsini et al., 1963; Manfredi et al., 1962
[l] Farquhar, 1962
[m] Farquhar and Ahrens, 1963
[n] Farquhar and Ahrens, 1963
[o] Ways et al., 1963

human erythrocytes. It is apparent that 16:0, 18:0, 18:1 and 20:4 are major fatty acids and that different laboratories have reported rather different values. Each lipid class appears to have a fairly specific fatty acid composition (Tables XIII and XIV). Phosphatidyl ethanolamine and phosphatidyl serine contain much more 20:4 and total polyunsaturated fatty acid than phosphatidyl choline and sphingomyelin. Phosphatidyl serine is highest in 18:0 and sphingomyelin contains much 16:0, 24:0 and 24:1. The findings in general are in keeping with those of the same lipid classes in most organs,

TABLE XII
Fatty acid composition of human erythrocytes (as % of total fatty acid) (1964–1965)

	1	2	3	4	5	6	7[a]	8[b]	9[c]
12:0	0·3			0·5	0·2			0·3	0·2
14:0	1·0	2·1	0·7	1·6	1·1	0·7	0·6	2·3	1·1
16:0	27·1	32·7	24·5	25·9	29·5	24·5	24·9	21·4	29·5
16:1	3·4	6·1	0·2	3·7	2·4	0·2	2·5	4·4	2·4
18:0	9·4	16·1	19·0	8·1	17·1	19·0	16·1	10·7	17·1
18:1	19·5	22·6	16 4	19 2	13·5	16·4	13·4	14·0	13·5
18:2	16·5	6·1	11·2	16·7	3·9	11·2	3·4	5·4	3·9
18:3	0·5			0·5	1·5			2·6	1·5
20:0	0·2			0·4	1·4		0·5	2·0	1·4
20:2					0·2			2·3	
20:3	1·4		1·5	2·1	1·1	1·5	2·6	1·7	1·9
20:4	19·5	8·1	15·1	18·2	2·3	15·1	16·4	2·1	2·3
22:5			1·6		0·4	1·6	0·5	0·5	0·4
22:6			3·5		0·4	3·5	5·4	0·5	0·4
24:0			2·4		2·1	2·4	5·8	1·7	2·3
24:1			3·3		2·1	3·3	3·2	1·7	2·1
Others	1·2	7·5	0·2	2·7	21·0		3·9		
Ref.	d	e	f	g	h	i	i	j	j

[a] Fetal; [b] children, 1-10 yr, mean; [c] adult, mean.
[d] de Gier et al., 1964
[e] Cardi et al., 1964
[f] Ways and Hanahan, 1964
[g] van Gastel et al., 1965
[h] Introzzi et al., 1965
[i] Crowley et al., 1965
[j] Notario et al., 1965

TABLE XIII
Fatty acid composition of individual phospholipids of human erythrocytes (as % of total fatty acid)

	Phosphatidyl ethanolamine						Phosphatidyl serine			
	1[a]	2	3	4[b]	5	6[c]	7	8[d]	9	10[c]
12:0			1·5							
14:0	3·7	0·2	6·0	0·6		0·3				0·3
16:0	37·6	18·9	23·0	15·5	29	24·4	7·1	4·4	14	4·9
16:1		0·6	4·0	0·1	3	2·1	0·4		2	1·0
18:0	10·5	8·0	6·5	14·1	9	12·0	41·6	39·7	36	46·4
18:1	20·4	25·2[e]	11·5	17·2	22	18·3	13·0[f]	9·8	15	5·4
18:2	5·6	7·0	9·3	5·6	6	2·5	2·8	2·6	7	1·4
20:2		0·1				1·2				0·8
20:3		1·0	3·3	1·0		2·4	2·1	2·6		
20:4	21·6	21·9	21·0	21·8	18	20·0	19·7	23·5	21	21·7
22:0		4·7								2·6
22:4				7·8		1·9				3·5
22:5		3·1		4·6			2·9	2·9		
22:6		3·9		8·9		8·6	4·2	7·0		7·6
24:0						6·3		4·1		4·3
24:1								3·7		
Others		2·1	14	2·8	13		2 8		5	
Ref.	g	h	i	j	k	l	m	n	o	p

[a] Includes PS; [b] 86% PE, 11% PS; [c] fetal; [d] 78% PS, 14% PI; [e] includes 3·6% trans; [f] includes 5·1% trans.
[g] Balint *et al.*, 1961
[h] Farquhar, 1962
[i] Roelofsen *et al.*, 1964
[j] Ways and Hanahan, 1964
[k] Hill *et al.*, 1965
[l] Crowley *et al.*, 1965
[m] Farquhar, 1962
[n] Farquhar and Ahrens, 1963
[o] Hill *et al.*, 1965
[p] Crowley *et al.*, 1965

TABLE XIV
Fatty acid composition of individual phospholipids of human erythrocytes

	Phosphatidyl choline					Sphingomyelin			
	1[a]	2[b]	3	4	5[c]	6	7[d]	8	9
12:0	0·1						0·2		
14:0	0·5				0·6		0·9	1·2	
16:0	33·0	34·7	39	34	39·0	24	41·3	44·9	28
16:1	1·0			2	1·8		0·1	0·7	3
18:0	11·7	13·8	11	13	10·4	7	9·1	12·9	7
18:1	20·6[e]	21·1	21	22	17·5	5	5·2	2·4	6
18:2	18·2	21·9	20	18	7·0	2	3·7	0·7	2
20:0	0·2						1·2	2·8	2
20:2	0·2								
20:3	1·6	1·0			3·7				
20:4	5·0	6·7		6	13·8		0·1		
22:0			8[f]			34[f]	8·0	5·5	8
22:4					1·1			0·4	
22:5	5·4							1·3	
22:6	1·1				4·0				
24:0					1·4		15·0	14·4	20
24:1							15·5	11·9	14
Others	1·5			5		23		0·5	
Ref.	g	h	i	j	k	l	m	n	o

[a] Includes Sph; [b] 97% PC + some traces Sph; [c] fetal; [d] 91% Sph, 7% choline glycerophosphatides; [e] includes 2·7% trans; [f] contains C_{20}-C_{22} unsat.
[g] Farquhar, 1962
[h] Ways and Hanahan, 1964
[i] Farquhar, 1965
[j] Hill et al., 1965
[k] Crowley et al., 1965
[l] Farquhar and Ahrens, 1963
[m] Ways and Hanahan, 1964
[n] Crowley et al., 1965
[o] Hill et al., 1965

although the report of a large amount of 22:4 in sphingomyelin (Hanahan et al. 1960) is different and probably reflects analytical difficulties. In general the values for fatty acid composition of individual phospholipids can be expected to be in error since isolation of pure and unaltered lipids requires good procedures, experience and very careful work. Use of DEAE or TEAE column chromatography followed by TLC for isolation and fatty acid analysis (Rouser et al., 1961a, 1964, 1965a, 1967b, 1968) will probably provide more precise quantitative values.

Tables XV and XVI show the values reported for total fatty acids of erythrocytes of various species. Variations of values from different laboratories are expected since both dietary (see Table XV) and analytical method differences can influence composition. The very high 20:4 content of rat cells and the almost complete absence of polyunsaturated fatty acids from sheep

TABLE XV
Fatty acid composition of total erythrocyte lipids from various species (as % of total fatty acid)

	Rat								Sheep		Guinea pig		Dog	
	1	2	3	4*a	5*b	6*c	7*d	8	9	10	11	12	13	14
12:0	0·3			1·5	0·7	0·4	0·4		0·6					
14:0	2·3		0·3	1·8	1·4	0·9	0·6		1·6					
16:0	16·7		31·1	22·4	20·6	25·2	24·2	28·7	16·0	15·7	13·0	11·1	16·1	16·9
16:1	6·4		2·2	2·7	2·3	1·7	0·4	1·8	2·3	1·6	3·0		2·6	1·7
18:0	11·9		13·7	14·2	9·8	13·5	13·5	12·1	9·7	9·6	34·6	27·4	26·6	19·0
18:1	12·1	20·3	18·5	15·7	16·9	13·5	8·6	14·5	49·6	52·3	23·8	7·0	12·3	14·2
18:2	5·7	0·9	7·2	2·2	4·3	5·3	11·5	6·4	11·5	14·6	2·8	19·5	15·7	12·9
18:3									0·7					
20:0	0·7			0·4	0·3	0·2	0·2		0·9					
20:2				0·2	0·1	0·1	0·3							
20:3	1·0	12·7	2·3	15·0	5·1	1·3	0·1		1·9		6·9	2·8		
20:4	11·4	5·6	24·0	15·4	24·5	26·3	31·1	33·8	1·4	2·9	7·1	23·2	16·1	30·8
22:5	2·4	0·3		1·0	0·3	0·7	1·4							
22:6	1·7			1·1	5·0	2·7	0·8							
24:0	3·1													
24:1				1·3	0·8	3·0	1·3							
Other	8·1	60·2	0·7	14·0	22·5	16·2	17·0				2·7	10·2		
Ref.	e	f	g	h	i	j	k	l	m	n	o	p	q	r

* High-fat diets—a coconut oil, b butter fat, c castor oil, d corn oil.
e Witting et al., 1961
f Mohrhauer and Holman, 1963
g de Gier and van Deenen, 1964
h Walker and Kummerow, 1964b
i Walker and Kummerow, 1964b
j Walker and Kummerow, 1964b
k Walker and Kummerow, 1964b
l de Gier et al., 1966
m de Gier and van Deenen, 1964
n de Gier et al., 1966
o Reid et al., 1964
p Ostwald and Shannon, 1964
q de Gier et al., 1964
r de Gier et al., 1964

cells (G. J. Nelson, unpublished observations) are of special interest. The sheep red cells indicate that polyunsaturated fatty acids are not required for erythrocyte structure and function. Sheep cell fatty acids can be changed markedly when the rumen is by-passed by tube feeding (de Gier and van Deenen, 1964).

Fatty acids of individual phospholipid classes (Table XVII) show species variations but the same trend in composition noted for human cells is apparent. The extent to which differences in diet and analytical procedures

TABLE XVI
Fatty acid composition of total erythrocyte lipids from various species (as % of total fatty acid)

	Cat	Rabbit	Pig	Bovine	Monkey		Chicken
14:0					0·3		0·2
16:0	20·1	22·3	21·4	12·1	15·6	23·8	28·7
16:1	2·6	3·3	2·4	2·7	0·5	2·2	0·8
18:0	17·8	10·5	10·4	14·1	15·2	15·3	8·8
18:1	11·0	11·8	32·1	34·5	15·7	13·4	21·7
18:2	21·5	32·0	23·2	21·1	11·0	14·5	31·8
18:3							1·0
20:3					1·0	1·8	0·6
20:4	18·5	6·6	6·4	4·8	12·3	25·0	4·6
22:5					4·2		1·2
22:6					10·8		3·2
Other					7·5	3·4	
Ref.		a			b	c	d

[a] de Gier et al., 1966
[b] Fitch et al., 1961.
[c] Greenberg and Moon, 1961.
[d] Kates and James, 1961.

may have influenced the results cannot be determined accurately. Improved procedures are now available and should provide more accurate values in the future.

It is now well established that the fatty acid compositions of erythrocyte phospholipids of different species can be altered by feeding different diets (Carroll, 1965; de Gier and van Deenen, 1964; Farnsworth et al., 1965; Farquhar, 1965; Farquhar and Ahrens, 1963; Fitch et al., 1961; Greenberg and Moon, 1961; Grimmer et al., 1965; Hill et al., 1964, 1965; Horwitt et al., 1959; Monsen et al., 1962; Reid et al., 1964; van Deenen et al., 1963; Voigt et al., 1965; Walker and Kummerow, 1963, 1964a, b, c; Watson, 1963; Witting et al., 1961) although simultaneous changes in phospholipid class composition generally do not take place and have been reported only infrequently. As noted in the discussion of phospholipid exchange, two distinctly different phases can be distinguished. First, an exchange of phosphatidyl choline and lysophosphatidyl choline with plasma may take place without formation of new cells. Second, with the formation of new cells, fatty acids of other phospholipid classes are changed. On diets high in polyunsaturated fatty acids, these acids appear in increased amounts in phos-

2. COMPOSITION OF MEMBRANES, ORGANELLES AND ORGANS

pholipids with a corresponding decrease of other fatty acids, particularly the monoenes. The reverse situation is observed with diets low in polyunsaturated fats and trienes appear in essential fatty acid deficiency states.

TABLE XVII
Fatty acid composition of individual phospholipids of rat and bovine erythrocytes (as % of total fatty acids)

	Lecithin		Sphingomyelin			Lyso Lec	"Cephalin"		
	Rat	Rat	Rat	Rat	Bovine	Rat	Rat	Rat	Bovine
12:0								0·1	
14:0	0·4			1·5				0·4	
16:0	38·8	45·7	17·2	32·9	40	22·0	16·7	9·8	4
16:1	1·1	0·8	4·5	3·0	1	4·0	2·8	1·2	
18:0	17·2	30·3	21·6	12·8	4	24·4	15·3	10·9	9
18:1	10·3	5·1	6·9	7·5	1	9·4	6·8	11·9	59
18:2	13·8	6·7	4·8	4·0		6·3	5·7	4·9	21
18:3				1·1		1·2		0·7	
20:0				1·1		1·1			
20:2								0·7	
20:3						1·1		0·5	
20:4	7·7	5·7		2·1		3·1	27·3	32·8	6
22:0					11	1·5			
22:5	⎤	⎤	⎤				⎤	3·4	
22:6	⎟	⎟	36·8				⎟	3·4	
24:0	4·6	4·0	⎟	11·2	31	10·5	18·4	3·5	
24:1	3·8	⎟	⎟	4·9			⎟	3·5	
Other	2·2	⎦	⎦	17·4	12	14·9	⎦	15·6	
Ref.	a	b	b	a	c	a	b	a	c

^a Watson, 1963.
^b Farnsworth et al., 1965.
^c Hanahan et al., 1960

Two types of bonds other than the ester linkage are known for the hydrocarbon residues of erythrocyte lipids. The plasmalogens have a vinyl ether linked chain. In human red cells, 10% of the choline phospholipid was reported to be plasmalogen (Farquhar, 1962) but most of the plasmalogen is in the phosphatidyl ethanolamine fraction (Farquhar, 1962; Hanahan and Watts, 1961; Hanahan et al., 1960; Hill et al., 1965; Nelson, 1967b). Investigations in Hanahan's laboratory (Hanahan et al., 1963; Hanahan and Watts, 1961; Thompson and Hanahan, 1962, 1963) first disclosed the presence of a true ether linkage (rather than the vinyl ether linkage of plasmalogen) in phosphatidyl ethanolamine of bovine red cells and G. J. Nelson (unpublished) observed the same type of linkage in sheep cells.

E. GENERAL CONCLUSIONS

Studies of erythrocytes show that very large species variations in the proportions of phospholipids and glycolipids are common. Thus, there is no single composition that is alone compatible with normal structure and function. Marked species variations in fatty acid composition also exist, thus making it apparent that normal structures and functions are possible within wide limits of variation of length and degree of unsaturation of carbon chains. The fatty acids of phospholipids can be changed appreciably by dietary manipulation with retention in many cases of apparently normal functions. Thus far there are no clear correlations between lipid composition and permeability differences. Permeability differences are seen without differences in lipid composition (e.g. high and low potassium sheep cells) and differences in lipid composition and permeability to glycerol fail to show any obvious correlation. Major permeability differences can exist that do not appear to involve membrane lipids although manyin vestigations have disclosed that some substances with lipid solubility enter erythrocytes more readily.

VII. Subcellular Organelles

The literature on subcellular particle composition has been reviewed recently (Fleischer and Rouser, 1965) and is not repeated here. It was concluded that values for subcellular particle lipid composition were generally inadequate. The major problems were the use of impure and/or inadequately characterized organelles whose morphological integrity was not preserved properly and the use of faulty analytical procedures. The data presented here were selected as the most accurate for pure, characterized, well-preserved and carefully analyzed preparations. The relative homogeneity of several isolated subcellular fractions is shown in Fig. 1. The fine structure of preparations is shown at higher magnification in Fig. 2.

A. TOTAL LIPID CONTENT

The organelles of one cell type differ widely in their total lipid content (Table XVIII). The lipid content expressed as mg lipid/mg protein is lowest in the nuclei and greatest in the smooth endoplasmic reticulum (smooth microsomes). Approximately 90% or more of the lipid is phospholipid with the exception of the plasma membrane which contains appreciably greater amounts of other lipid (cf. the lower ratio of μgP/mg lipid), most of which is cholesterol. Since the erythrocyte plasma membrane and myelin (formed in peripheral nerve by invagination of the plasma membrane of the Schwann cells) are also rich in cholesterol, it appears that plasma membranes are generally rich in cholesterol.

TABLE XVIII
Total lipid content of beef liver organelles

Cell fraction	Lipid content	
	mg lipid / mg protein	µg P / mg lipid
Nuclei[a]	0·086	31
Microsomes (rough)	0·82	—
(smooth)[a]	1·15	33
Mitochondria[b]	0·18	33
Plasma membrane[a]	0·26	19

[a] S. Fleischer, B. Fleischer and G. Rouser, unpublished.
[b] Fleischer et al., 1967b.

Whether the phospholipids are localized exclusively in the membranes remains yet to be answered. It is significant in this regard that the lowest phospholipid content is in the nucleus which is devoid of membranes, except for the nuclear envelope surrounding the organelle (Fig. 2C). The concentration of membrane and phospholipid is higher in mitochondria (Fig. 2A), while smooth microsomes which consist mainly of membranes (Fig. 2B) have the highest total lipid content.

A related problem is concerned with the definition of a membrane. What portion of the isolated membrane belongs, *per se*, to the membrane? For example, the lipid content of smooth and rough microsomes is different. The difference is mainly due to attached ribosomes which significantly lower the total lipid content. Should the ribosomes be considered as part of the membrane? This is a matter of definition. Since ribosomes can be detached from the remainder of the morphological "unit" membrane, it would probably be best to classify them as "membrane-associated" or "extra-membranous". In a like manner, the preparation of membranes frequently, perhaps always, results in the release of proteins which were originally membrane-associated. Differences observed in the lipid to protein ratios may in part be referable to this problem which remains to be clarified and better defined.

B. MITOCHONDRIA

The total amount of extractable lipid from bovine heart, kidney and liver mitochondria is different (Table XIX) and appears to correlate roughly with the abundance of cristae, heart mitochondria having more cristae and

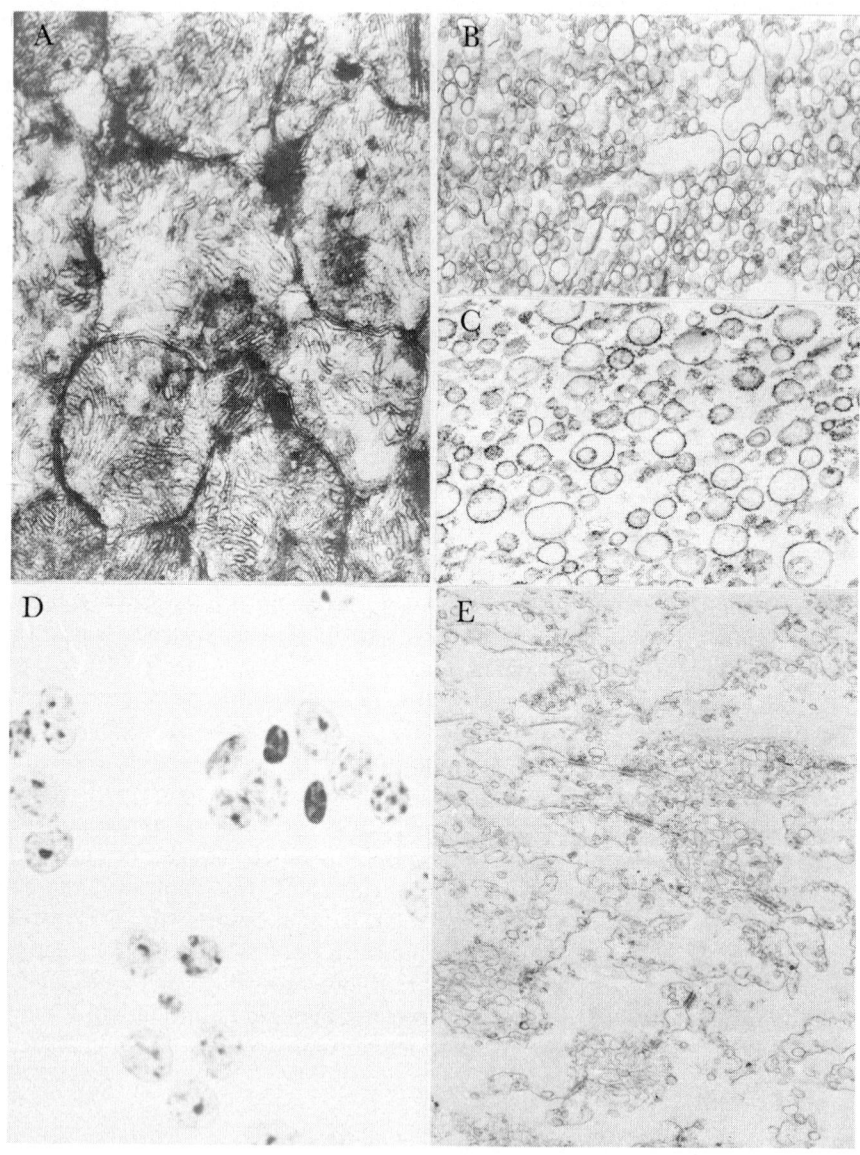

total lipid with liver representing the lower extreme. More than 90% of the lipid in mitochondria is phospholipid. Bovine mitochondria contain cholesterol but the amount is variable (1–5% of the total lipid), heart mitochondria having the smallest and kidney mitochondria the largest amount.

TABLE XIX
Lipids of mitochondria

	Heart	Beef[a] kidney	Liver	Human[b] heart
Total phosphorus (μg/mg protein)	18·5	11·5	9·7	18·1
Lipid P (μg/mg protein)	11·3	7·6	5·8	13·4
Lipid P (μg/mg lipid)	33[c]	31	33	33
Lipid (mg/mg protein)	0·32	0·24	0·18	0·4
NL (% of total lipid)[d]	5·7	7·2	9·0	—
Cholesterol (% of total lipid)	1·1	4·7	2·3	—

[a] Fleischer et al., 1967b.
[b] S. Fleischer, B. Fleischer and G. Rouser, unpublished.
[c] A value of 35 was obtained after Sephadex column chromatography.
[d] Neutral (less polar) lipid eluted from DEAE columns with chloroform.

Mitochondria thus have membranes that are characteristically low in cholesterol. Mitochondria contain coenzyme Q, a lipid oxidation-reduction component which is necessary for electron transport. Glycolipid has not been found in highly purified and carefully analyzed mitochondria.

Some of the earliest studies indicated that diphosphatidyl glycerol (cardiolipin) might be a unique component of mitochondria (Martinetti et al., 1958). The same studies indicated, however, that mitochondria contain lipid classes such as sphingomyelin also found in other subcellular particles and one unique report indicated the presence of cerebroside (Lovtrup and Svennerholm, 1963). Highly purified mitochondria from bovine heart, kidney and liver analyzed by carefully checked procedures (Fleischer et al., 1967b) were found to contain diphosphatidyl glycerol, phosphatidyl ethanolamine and phosphatidyl choline as major components approximately in the molecular

FIG. 1 (Facing) Relatively low power micrographs of isolated subcellular fractions to show homogeneity. (A) Beef heart mitochondria ($\times$ 32,000); (B) beef liver "smooth" microsomes ($\times$ 35,000); (C) beef pancreas "rough" microsomes ($\times$ 35,000); (D) beef liver nuclei ($\times$ 1650) and (E) beef liver plasma membranes ($\times$ 17,500). Part D by light microscopy; remainder are electron micrographs. Figures 1 and 2 are from joint studies of the laboratories of S. Fleischer and G. Rouser. Electron micrographs taken by A. Saito and B. Fleischer in the laboratory of S. Fleischer.

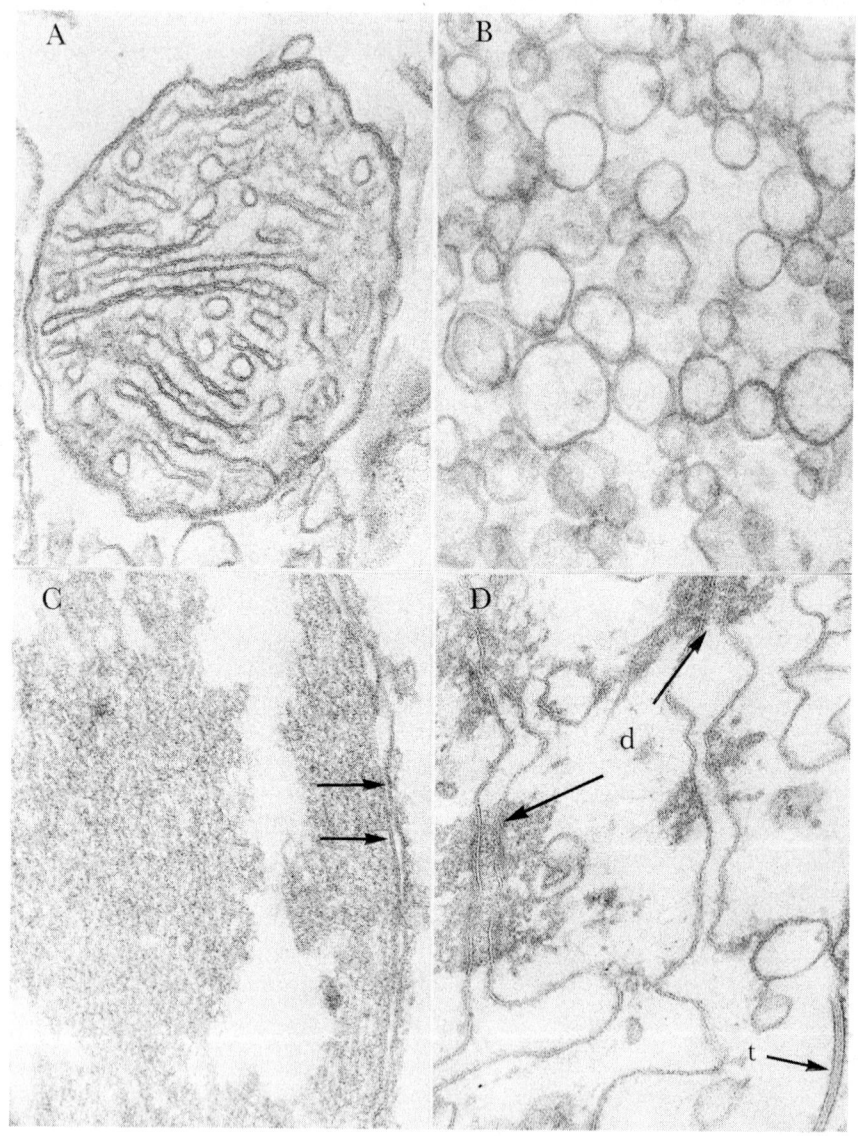

ratio 1/4/4 (Table XX). Phosphatidyl inositol was found to be a minor component representing about 3% of the total phospholipid. No trace of sphingomyelin or cerebrosides was found in mitochondria from the three organs although phosphatidyl serine was present in all as a trace (about 0·5% of the total lipid) component.

TABLE XX
Phospholipid composition of mitochondria

			Bovine[a]	Human[b]		Okasis
	Phosphorus A	Heart analysis B	Charring	Kidney Charring	Liver Charring	Heart Phosphorus analysis
PC	40·3 ±0·5	39·9 ±0·6	40·8 ±0·6	39·6 ±0·3	43·4 ±0·3	42·6
PE	35·5 ±0·4	36·8 ±0·3	37·4 ±0·4	38·1 ±0·4	34·5 ±0·4	34·2
DPG	20·6 ±0·4	20·4 ±0·8	19·1 ±0·8	19·2 ±0·4	17·2 ±0·5	18·2
Minor comp.	3·2 ±0·9	2·9 ±0·8	2·7 ±0·7	3·5 ±0·8	4·9 ±0·6	4·7

[a] Fleischer et al., 1967b.
[b] S. Fleischer, B. Fleischer and G. Rouser, unpublished.

The quantitative distribution of phospholipids of human heart mitochondria (Table XX) is very similar to that of bovine organelles. Although the data are limited, it can be concluded that mitochondrial phospholipids in general will show little or no organ or species variability. This conclusion for vertebrates in general is supported by whole organ data (see below). The higher phosphatidyl choline content of bovine liver mitochondria shown in Table XX probably arises from a difference in composition of inner and outer membranes (Table XXI) and the fact that the outer membrane is more abundant in liver mitochondria. The outer membrane may contain all the cholesterol and may not contain diphosphatidyl glycerol.

The study by Fleischer et al. (1967b) of mitochondria from bovine organs illustrated several important points. It was shown that application of a

FIG. 2. (Facing) Electron micrographs at higher magnification (all × 175,000) than Fig. 1 showing detail in isolated subcellular fractions. (A) Beef heart mitochondria; (B) beef liver "smooth" microsomes; (C) portion of a beef pancreas nucleus (arrows point to the surrounding double membrane); (D) portion of beef liver plasma membrane preparation with desmosomes (d) and a tight junction (t) clearly visible. The trilaminar arrangement of the membranes of these organelles can readily be observed.

procedure developed for one organ to another organ may not give satisfactory results. Each preparation must be carefully monitored and appropriate modifications introduced. Before a satisfactory isolation procedure

TABLE XXI

Phospholipids of guinea-pig liver fractions (as % of alkali-labile P)[a]

	Mitochondria	Inner membrane	Outer membrane	Microsomes
PC	40·0	44·5	55·2	62·8
PE	28·4	27·7	25·3	18·3
PS	ND	ND	ND	4·5
PI	7·0	4·2	13·5	13·4
DPG	22·5	21·5	3·2	0·5
X	2·3	2·2	2·5	1·1
Total PL[b]	14·4	21·4	45·1	28·0

[a] Parsons *et al.*, 1967.
[b] % phospholipid = mg phospholipid/mg phospholipid + mg protein.
ND, not detected.

for liver organelles was devised, electron microscopic examination disclosed poor morphology (loss of cristae) and contamination; the lipid analysis showed a very high phosphatidyl choline content. With very fresh liver and a modified isolation procedure, purity with well-preserved morphological integrity was achieved and lipid analysis showed close correspondence to mitochondria from heart and kidney. Distortion of mitochondrial lipid composition was also found to arise from improper procedures of lipid handling. Phosphatidyl ethanolamine content may decrease when lipid extracts are evaporated to dryness and dried to constant weight. Changes in the lipid extracts are prevented by weighing an aliquot that is discarded with the bulk of the lipid always maintained moist or in solution. Many analyses were invalidated because upon standing prior to analysis both diphosphatidyl glycerol and phosphatidyl ethanolamine had decomposed. This was traced to autoxidation that is prevented by addition of an antioxidant. Routine addition of BHT (butylated hydroxytoluene; 2,6-di-tert-butyl-p-cresol) to the extracting solvent allows prolonged storage at − 20°C without decomposition which greatly facilitates accurate lipid analysis.

C. ENDOPLASMIC RETICULUM

Highly purified preparations of total microsomal fractions and rough and smooth surfaced endoplasmic reticulum were recently prepared from bovine heart, kidney and liver and their phospholipid distributions determined

(Table XXII). Rough and smooth surfaced endoplasmic reticulum preparations from the same organ do not appear to differ significantly; however, the values for the three organs are quite different and it is clear that there

TABLE XXII

Phospholipids of bovine heart, kidney and liver endoplasmic reticulum[a]

	Total[b]	Kidney Rough	Smooth	Heart Total	Smooth	Liver Total	Rough	Smooth
PC	32.8	36.0	31.4	35.2	34.8	57.3	61.2	67.3
PE	25.2	25.4	26.3	30.2	29.0	24.5	21.6	24.3
PS	9.7	9.0	10.1	3.0	3.5	3.8	3.5	3.7
PI	8.4	7.6	9.5	4.2	4.4	10.7	10.8	10.0
DPG	ND	ND	ND	11.0	10.7	ND	ND	ND
Sph	24.0	22.1	22.8	13.6	15.0	3.79	2.83	4.53

[a] S. Fleischer, B. Fleischer and G. Rouser, unpublished.
[b] Total microsome fraction.
ND, not detected.

is organ specificity. The major variations are in the sphingomyelin content with three groups being evident: 1. kidney, high; 2. heart, intermediate and 3. liver, low. Whole organ analysis indicates that endoplasmic reticulum composition of all organs falls into one of the three groups. The preparations from heart contain in addition diphosphatidyl glycerol (cardiolipin) not found in this organelle in other organs. Organ specificity in the endoplasmic reticulum is also reflected in the analysis of lipids from whole organs in one animal. It is of special interest that there is little or no species variation for any one organ. Glycolipids were not found in endoplasmic reticulum.

D. NUCLEI

Determination of the true composition of nuclear membranes is difficult because the organelle is large and contains little lipid. Even a small contamination by other structures containing more lipid can give rise to relatively large errors. Enzymatic degradation of lipid during isolation is also a major problem. Values (Table XXIII) recently obtained by analysis of highly purified preparations from several bovine organs demonstrate that nuclei have a phospholipid distribution different from other organelles. There is little (or no) organ variability in nuclear phospholipid composition. Preparations with known contamination were distinctly different from the more highly purified preparations. Nuclei differ from both mitochondria and endoplasmic reticulum in phospholipid composition.

TABLE XXIII
Phospholipids of bovine nuclei

	Liver[a]	Thymus[a]	Pancreas[a]	Heart[a,b]	Brain[c]
PC	53·9	52·1	52·4	45·4	52·9
PE	20·6	19·6	23·5	24·6	25·7
PS	3·0	3·2	2·0	4·1	7·2
PI	9·0	9·0	8·6	8·5	7·4
DPG	1·5	1·1	0·6	3·3	—
PA	0·6	0·5	0·7	0·8	0·9
PG	0·5	0·2	0·2	0·5	0·6
Sph	2·6	2·6	2·6	5·5	5·2
LPC	4·2	7·7	5·4	4·1	ND
LPE	3·3	3·3	3·4	1·9	ND
LPS	0·7	0·6	0·9	1·4	ND

[a] S. F. Fleischer, G. Rouser, B. Fleischer and G. Kritchevsky, unpublished.
[b] Contaminated slightly with microsomes.
[c] Preparation contaminated slightly with myelin (G. Rouser, A. N. Siakotos and G. Kritchevsky, unpublished).

VIII. Whole Organ Composition and Species Variation

A. VISCERA AND MUSCLE

Species comparisons can be made rapidly by whole organ analysis. Whole organ analysis gives a composite picture of all membranes in the organ. The proportions of the membranes may vary (e.g. administration of barbiturates brings about an increase in smooth endoplasmic reticulum in liver). Despite these obvious limitations, whole organ analysis has advantages—it can be done more rapidly than organelle analysis and autolytic change can be minimized more readily. The necessity for development of modified procedures for organelle isolation and characterization from different species makes whole organ analysis the most attractive means for initial comparison of species.

Examination of liver, heart, kidney, spleen, lung, skeletal muscle and brain from several vertebrates (human, rat, mouse, bovine and frog) and skeletal muscle and brain of several invertebrates was undertaken by G. Rouser, G. Simon and G. Kritchevsky (unpublished) and G. Rouser, C. F. Baxter and G. Simon (unpublished). From these studies some important general relationships emerged. Brain is discussed separately in Section VIII-B.

The organs from any one species were found to have different phospholipid compositions. The differences are illustrated by the values obtained for bovine organs (Table XXIV). The organ variations in one species arise for the most part from variations in the lipid distribution in the endoplasmic reticulum (see Section VII-C) and the different proportions of mitochondria and endoplasmic reticulum (that contribute most of the lipid) in different organs. Thus in heart more lipid is contributed by mitochondria than in liver.

TABLE XXIV
Phospholipid composition of bovine organs[a]

	Heart	Kidney	Liver	Lung	Spleen
PC	40·7	33·8	54·0	41·9	39·3
PE	31·2	29·5	24·5	22·7	27·0
PS	3·7	7·8	3·8	10·3	12·9
PI	5·6	7·5	9·2	3·6	4·6
DPG	13·4	6·2	2·9	1·0	T
PA	T	0·2	0·4	1·5	0·6
PG	0·2	0·5	0·3	2·2	T
Sph	5·3	14·3	6·1	17·1	15·8

[a] G. Rouser, G. Simon and G. Kritchevsky, unpublished.

The values for the same organ from different vertebrates in general were found to be very similar as illustrated by the values for lung lipids (Table XXV). Values for lung were selected as an illustration because lung contains the highest phosphatidyl glycerol level of any organ thus far examined. Phosphatidyl glycerol accounts for less than 1% of the total phospholipid in other organs and is a very minor component of mitochondria and nuclei. Phosphatidyl glycerol does not appear to be present in endoplasmic

TABLE XXV
Lung phospholipids of different species[a]

	Human	Mouse	Rat	Frog
PC	50·4	49·4	48·2	46·5
PE	17·2	23·3	23·3	24·5
PS	7·4	8·7	9·7	9·1
PI	3·6	4·4	4·2	4·0
DPG	1·1	0·9	1·1	0·9
PA	0·5	0·4	0·3	0·3
PG	2·9	2·6	2·3	2·1
Sph	11·9	10·4	10·8	12·9
LPC	2·2	T	T	T
LPE	0·6	T	T	T
LPS	0·3	T	T	T
X	1·8	T	T	T

[a] G. Rouser, C. F. Baxter and G. Simon, unpublished.

reticulum or plasma membranes (erythrocytes and myelin). The close similarity of vertebrates is even more striking considering that some postmortem change is unavoidable with most human organs and that organs, particularly liver, of one species may vary appreciably from animal to animal. This latter type of variability is shown by values obtained for seven rat livers (Table XXVI). One liver was analyzed on 3 separate occasions (see 7^1, 7^2, and 7^3 Table XXVI) to demonstrate that the analytical procedures were not responsible for the variations. Good analytical reproducibility was obtained (see also values in Table III). The values for different rat livers varied as much as the values from livers of different vertebrate species indicating that there is little or no species variation among vertebrates for liver cell membrane composition.

Muscle phospholipid values for vertebrates are very similar (Table XXVII). The values in this table also demonstrate that invertebrate muscle is similar to vertebrate muscle in some respects. Thus both vertebrate

TABLE XXVI
Rat liver phospholipids[a]

	Animal number										Overall Average
	1	2	3	4	5	6	7^1	7^2	7^3		
PC	54·2	53·0	52·2	50·4	52·6	51·5	52·0	51·8	51·5	51·8	52·2
PE	24·1	26·0	24·4	26·2	25·1	25·8	25·4	25·8	25·6	25·6	25·3
PS	3·7	3·2	3·7	4·1	3·9	3·9	3·6	3·7	3·6	3·7	3·7
PI	8·5	8·5	8·9	9·5	9·0	8·9	9·1	9·5	9·5	9·4	9·0
DPG	3·8	4·5	5·7	5·0	4·8	5·1	5·2	5·1	5·0	5·1	4·9
PA	0·6	0·3	0·3	0·2	0·3	0·4	0·2	T	0·2	0·2	0·3
Sph	4·4	4·6	4·9	4·6	4·3	4·5	4·4	4·0	4·6	4·4	4·5

[a] As % of the total phospholipid G. Rouser, G. Simon and G. Kritchevsky, unpublished. 7^1, 7^3, 7^2 repeat determinations of one liver sample.

and invertebrate muscles have for the most part the same phospholipids. Lobster muscle is qualitatively the same as vertebrate muscle and differs quantitatively mainly in its sphingomyelin content. A major qualitative difference is apparent for the other invertebrate species. In some (abalone, scallop) sphingomyelin is replaced by ceramide aminoethylphosphonate. In others (sea urchin) sphingomyelin is replaced by ceramide aminoethylphosphoryl ethanolamine. In vertebrate muscle, there is more phosphatidyl inositol than phosphatidyl serine, whereas the reverse is true for most invertebrates. Since diphosphatidyl glycerol is characteristic for mitocondria and its level is low in muscles of all species, it can be judged that most of the muscle lipid is contributed by the endoplasmic reticulum. The endoplasmic reticulum of vertebrate muscle resembles that of bovine liver (Table XXII)

TABLE XXVII
Phospholipids of skeletal muscle[a]

	Vertebrates			Invertebrates		
	Human	Mouse	Lobster	Abalone	Scallop	Sea Urchin
PC	53.9	55.9	55.7	44.1	36.3	51.5
PE	26.6	26.2	23.5	29.4	26.3	25.6
PS	3.1	4.3	6.1	10.6	12.7	9.4
PI	8.4	7.1	5.1	4.7	5.0	4.5
DPG	3.4	2.7	1.2	0.9	0.9	0.3
PA	0.5	0.3	0.2	0.3	0.3	0.3
Sph	4.0	3.5	8.2	—	—	—
CAEP	—	—	—	9.9	18.5	—
X[b]	—	—	—	—	—	3.7
	|——— Sph present ———|			CAEP present		X present

[a] G. Simon and G. Rouser, unpublished.
[b] Tentatively identified as Ceramide Phosphoryle Thanolamine.

with respect to the phosphatidyl inositol/phosphatidyl serine ratio while some invertebrate muscles have the reverse ratio. Glycolipids are present in invertebrate but not vertebrate muscle.

B. BRAIN AND MYELIN

Myelin contributes most of the brain lipids in species high on the evolutionary scale and species comparisons are complicated by the fact that different species have different amounts of myelin. The fact that myelination proceeds over a period of years in humans can be used to advantage in species comparisons since human brain at different stages of myelination can be compared to brains of other species where the mature brain level of myelin is lower than in humans. The phospholipid distribution of human brain at different ages is shown in Table XXVIII. Myelination is accompanied by a marked decrease of the percentage of phosphatidyl choline and a large increase in the percentage of sphingomyelin and phosphatidyl serine. These are the changes expected since myelin (Table XXIX) contains more phosphatidyl serine and sphingomyelin and less phosphatidyl choline than most of the other cellular membranes (see Section VII).

Bovine and human brains have very nearly the same amount of myelin and thus can be compared directly for species variations. As the values for phospholipids of the two species (Table XXIX) are almost identical, it is apparent that little or no species difference exists. Cerebroside and cerebroside sulfate are characteristic components of myelin. The amount of cerebroside sulfate is less in bovine than human brain. Hence, it is thus clear

TABLE XXVIII

Phospholipid composition of normal human whole brains at different ages[a]

	Fetal	1 day	3 wk	5½ mth	8 mth	22 mth	6 yr	33 yr
PC	51·6	49·7	47·5	40·7	38·5	35·7	32·7	30·4
PE	31·1	31·5	32·3	35·3	35·0	34·9	34·6	35·1
PS	11·9	11·8	12·6	13·9	14·5	15·4	16·8	17·1
PI	2·9	3·2	3·3	2·3	2·6	2·8	3·2	2·2
Sph	1·9	3·7	4·0	7·4	9·3	11·2	12·9	14·8

[a] Rouser et al., 1966a.

TABLE XXIX

Brain phospholipids of different species[a]

	Vertebrate brains					Invertebrate brains[d]			Myelin
	Human (adult)[b]	Bovine	Rattlesnake[c]	Frog[c]	Goldfish[c]	Octopus	Fly	Lobster	Human brain[e]
PC	29·2	29·6	39·0	43·2	49·6	38·8	18·6	58·4	21·9
PE	35·0	35·7	36·7	39·0	33·6	37·7	64·5	25·7	35·9
PS	17·6	16·9	9·1	10·2	8·8	11·8	5·3	5·7	19·3
PI	2·0	2·7	3·9	2·4	3·6	2·6	2·7	1·2	3·5
DPG	0·4	0·8	0·5	1·1	0·6	2·3	1·8	0·6	ND
PA	0·5	1·0	0·6	1·5	2·0	ND	T	1·7	1·3
Sph	13·6	13·2	10·1	2·7	1·7	1·0	ND	6·7	18·0
CAEP	ND	ND	ND	ND	ND	5·9	ND	ND	ND
X	ND	ND	ND	ND	ND	ND	9·4	ND	ND

[a] Values expressed as percentage of the total phospholipid.
[b] A. Yamamoto and G. Rouser, unpublished.
[c] G. Rouser, C. F. Baxter and G. Simon, unpublished.
[d] C. F. Baxter and G. Rouser, unpublished.
[e] G. Rouser, A. N. Siakotos and G. Kritchevsky, unpublished.
ND, not detected; T, trace.

that bovine and human brain myelins are not identical in composition despite the fact that the phospholipid distributions are the same.

The cerebroside and cerebroside sulfate levels decrease in the order rattlesnake, frog and goldfish. If the levels of these myelin lipids are used as an indication of the amount of myelin and comparisons are made with human brain at various stages of development, the expected increase in phosphatidyl choline and decrease of sphingomyelin are observed. The phosphatidyl serine levels do not change in the expected manner, however. Species variation is thus apparent. The data are in keeping with the mammalian species variations in myelin composition reported in the literature (Cuzner et al., 1965; Hulcher, 1963; Korey et al., 1958; Laatsch et al., 1962; Norton and Autilio, 1965; Nussbaum et al., 1963). The apparent species differences reported cannot, however, be distinguished from variations arising from differences in methodology.

Invertebrate brains do not contain cerebroside and cerebroside sulfate detectable by thin layer chromatography, which is consistent with the absence of myelin. Most of the phospholipids found in vertebrate brains are present also in invertebrate brains. Lobster brain (like lobster muscle, Section VIII-A) is qualitatively similar to vertebrates but octopus and fly illustrate a qualitative change similar to that seen in muscle (Section VIII-A). Octopus brain contains both sphingomyelin and ceramide aminoethylphosphonate. Fly brain does not contain sphingomyelin and an incompletely characterized phospholipid is present. This lipid has been tentatively identified as ceramide phosphorylethanolamine (the ethanolamine analogue of sphingomyelin).

C. VASCULAR STRUCTURES

Highly purified preparations of bovine and human brain capillaries and human renal glomeruli (a capillary structure) have strikingly similar phospholipid distributions (Table XXX) and thus organ and species differences in capillary composition appear to be negligible. Human brain capillaries

TABLE XXX
Phospholipids of vascular structures[a]

	Glomeruli	Capillaries		Aorta[b]					
	human kidney	Bovine brain	human brain	Still-born infant	Human 24 yr	Human 54 yr	Human 72 yr	Grossly athero-sclerotic human 90 yr	Bovine
PC	32·2	33·9	36·0	36·5	32·4	22·8	23·8	31·5	28·5
PE	25·8	26·3	27·4	27·4	18·3	12·9	10·6	4·4	25·6
PS	11·1	11·5	11·4	11·1	9·4	5·5	4·7	4·6	13·3
PI	5·2	5·5	5·2	5·0	4·6	2·2	1·8	2·6	9·1
DPG	3·1	1·0	1·2	1·1	T	T	T	0·9	1·4
PA	0·2	0·2	0·3	T	T	T	T	T	T
Sph	22·5	18·5	21·6	18·5	35·4	56·5	59·0	56·0	22·1

[a] G. Rouser, A. N. Siakotos and R. Solomon, unpublished.
[b] The entire vessel from the heart to the first point of bifurcation in the abdomen.

and one-day old infant aorta are also remarkably similar in phospholipid distribution. Since diphosphatidyl glycerol is characteristic for mitochondria and the level of this lipid is low in the endothelial cells (capillaries), it can be assumed that most of the lipid of the endothelial cells is contributed by the endoplasmic reticulum. The endoplasmic reticulum of endothelial cells is thus similar to that in kidney (see Table XXII). In fact, all capillaries and kidney endoplasmic reticulum are similar in composition (compare values in Tables XXII and XXX).

Aortas as free as possible of atherosclerotic changes were obtained from individuals of different ages and the phospholipid distribution determined. The values obtained (Table XXX) indicate a maturation, developmental or aging process distinct from atherosclerosis. The sphingomyelin level rises with a decrease in the other phospholipids. The high sphingomyelin level found at 54 and 72 years of age suggest the presence of a membrane similar to that of the lens that is also characterized by an exceptionally high sphingomyelin level (Feldman et al., 1966). It is to be noted that plasma cholesterol and total phospholipid levels increase with age. Perhaps this phenomenon is related rather directly to the changes in phospholipid distribution in the aorta. The phospholipid distribution of a grossly atherosclerotic human aorta was not very different from that of a relatively normal appearing vessel from older individuals (Table XXX). Additional data are required before the significance of the changes in aorta lipids can be determined although the changes with age are probably associated with changes in the media and perhaps involve smooth muscle cells.

D. THE SEA ANEMONE

The phospholipids of the sea anemone (*Anthopleura elegantissima*), a coelenterate, provide useful comparisons to those of higher organisms. This low form of animal life contains phosphatidyl choline, phosphatidyl serine, phosphatidyl ethanolamine and phosphatidyl inositol typical of organs of higher animals (Rouser et al., 1963; Simon and Rouser, 1967). In addition, ceramide aminoethylphosphonate is a major component (Rouser et al., 1963; Simon and Rouser, 1967). This lipid is also found in brains and muscles of some invertebrates (see Sections VIII-A and B). As compared to vertebrate sphingolipids, ceramide aminoethylphosphonate possesses two unique features, a free amino group and a phosphonic acid linkage (Rouser et al., 1963) although it contains sphingosine and the polar group is attached to the primary hydroxyl group of sphingosine as in sphingolipids of vertebrates (Simon and Rouser, 1967). The sea anemone does not contain sphingomyelin although sphingomyelin is found in some invertebrates. It appears that ceramide aminoethylphosphonate is a substitute for sphingomyelin in the anemone as is also the case in certain other invertebrates (Sections VIII-A and B).

E. CONCLUSIONS

Whole organ data indicate little or no species variations among vertebrates for the phospholipid distribution of any one organ. A summary of very similar values for different species is shown in Table XXXI. These values indicate that apparent species variations among vertebrates may depend upon the variability of organ composition from animal to animal of the same species, a phenomenon that has in fact been demonstrated for rat organs (see Table XXVI). The comparison of liver and aminon cells from a tissue culture emphasizes the fact that organs and individual cell types appear to fall into definite groups each having a characteristic phospholipid distribution.

Both qualitative and quantitative differences in phospholipid distribution are apparent when vertebrates and invertebrates are compared.

IX. Miscellaneous Cell Types

A. PLATELETS

Platelets supply the phospholipid essential for blood coagulation and appear to be important in maintenance of *in vivo* hemostasis. The lipids of platelets are thus of special interest. Platelets are formed by fragmentation of the cytoplasm of megakaryocytes. In the process, mitochondria are included in the cytoplasmic particles, the membranes of which are, however, probably mainly derived from the plasma membrane of the megakaryocyte. Values for platelet phospholipid distribution reported from different laboratories are not in particularly good agreement (Table XXXII) and additional work is needed. Clearly these cell fragments (platelets) have the same phospholipids found generally in cells. Diphosphatidyl glycerol (cardiolipin) has not been reported even though its occurrence is expected from the presence of mitochondria in platelets. The failure to detect this lipid class is probably related to its occurrence in small amount.

Plasmalogen of platelets was reported (Marcus *et al.*, 1962) to be localized in the phosphatidyl ethanolamine fraction (66% plasmalogen) although other workers indicated choline plasmalogen to be more abundant (Blomstrand *et al*, 1962). The fatty aldehydes released from the phosphatidyl ethanolamine fraction were 18:0 (18% of the total fatty residues), 16:0 (9%) and 18:1 (4%) (Marcus *et al.*, 1961, 1962).

Phosphatidyl ethanolamine was found to contain 36% arachidonic, 4%

TABLE XXXI

Phospholipid values indicating lack of species variability in vertebrates

	Whole brain		Capillaries		Liver					Spleen		
	Bovine	Human	Bovine brain	Human[a] kidney	Bovine	Rat[b]	Rat[b]	Mouse	Frog	Rat	Mouse
PC	29.6	29.2	33.9	32.2	54.0	54.2	50.4	48.7	49.1	47.2	46.5
PE	35.7	35.0	26.3	25.8	25.3	24.1	26.2	28.6	29.4	27.2	27.8
PS	16.9	17.6	11.5	11.1	3.7	3.7	4.1	4.4	4.3	9.0	9.0
PI	2.7	2.0	5.5	5.2	9.0	8.5	9.5	8.8	8.2	6.3	6.6
DPG	0.8	0.4	1.0	3.1	4.9	3.8	5.0	5.1	2.2	2.4	2.2
PA	1.0	0.5	0.2	0.2	0.3	0.6	0.2	0.1	0.3	0.6	0.2
PG	T	T	T	T	T	T	T	0.1	0.3	T	0.4
Sph	13.2	13.6	21.6	22.5	4.5	4.4	4.6	4.2	6.1	7.3	7.3

	Heart		Lung			Kidney			Liver		Amnion cells
	Rat	Mouse	Rat	Mouse	Human	Frog	Mouse	Rat	Mouse	Human[c]
PC	41.2	43.8	48.2	49.4	37.9	36.7	38.5	36.6	48.7	50.3
PE	33.8	33.9	23.3	23.3	30.8	32.1	28.4	28.8	28.6	28.4
PS	3.2	3.0	9.7	8.7	7.0	6.5	7.6	8.1	4.4	4.5
PI	4.0	4.7	4.2	4.4	6.1	6.1	6.3	6.1	8.8	8.6
DPG	13.0	12.0	1.1	0.9	4.2	5.2	7.5	7.1	5.1	4.0
PA	0.2	0.1	0.3	0.4	0.6	0.3	0.3	0.3	0.1	0.5
PG	1.1	0.6	2.3	2.6	0.6	0.3	0.5	0.2	0.1	T
Sph	3.5	3.9	10.8	10.4	12.8	13.0	11.0	12.9	4.2	3.8

[a] Renal glomeruli.
[b] Selected analysis (see Table XXVI).
[c] From tissue culture.

palmitic and 20% stearic acids (Marcus et al., 1961, 1962). The values are similar to those reported for the same lipid class from beef heart mitochondria (Fleischer and Rouser, 1965). Phosphatidyl serine was found to have

TABLE XXXII
Lipid composition of human platelets

Lipid Class			
PC	35[a]	36·5[a]	27·3[b]
PE		11·9	9·2
PS	35	12·4	8·9
PI		5·6	3·9
PA		—	—
Sph	21	28·7	10·1
LPC	—	—	—
Misc PL	8	2·5	20·9
Chol	—	—	19·1
Total lipid[c]	—	—	3870
Total PL			2287
Ref.	d	e	f

[a] Expressed as % of total phospholipid; [b] as % of total lipid; 12% triglyceride reported; [c] mg/100 ml; [d] Bellimoria et al., 1965; [e] Blomstrand et al., 1962; [f] Karaca and Stefanini, 1966.

little plasmalogen, 24% arachidonic acid and 47% stearic acid. Phosphatidyl inositol was similar to phosphatidyl serine in fatty acid composition. Phosphatidyl choline contained 12% arachidonic acid and 34% palmitic acid. The general findings for phosphatidyl choline and phosphatidyl ethanolamine are similar to those reported for beef heart mitochondria (Fleischer and Rouser, 1965).

B. LEUKOCYTES

The white blood cells, like other formed elements of the blood, provide a relatively easily obtainable sample from normal humans and are thus of special interest. Values reported for the lipid composition of leukocytes are shown in Table XXXIII. The data are variable and additional study is required. Values for leukemic leukocytes were reported (Firkin and Williams, 1961) as similar to those of normal cells. The report of a high level (about 17%) of glycolipid in human leukocytes (Miras et al., 1966) is of special interest. Ceramide dihexosides were found to account for most of the glycolipid (98·8%) with ceramide monohexosides (cerebrosides) accounting for 3·6% and other ceramide polyhexosides only 1·6% of the total glycolipid. The glycolipids were reported to contain 16:0 (palmitic), 24:0 (lignoceric) and 24:1 (nervonic) acids as major components in keeping with findings from other tissues.

Leukocytes synthesize lipids (Kidson, 1961a, 1961b, 1962; Marks et al., 1960), incorporate acetate into neutral lipids and phospholipids (Sbarra and Karnovsky, 1960), ingest lipids and lipo-proteins (Casley-Smith and Day, 1966; Elsbach, 1962; Elsbach and Kayden, 1965; Suzuki and O'Neal, 1964),

TABLE XXXIII
Lipid composition of leukocytes

	Rabbit		Guinea pig	Human		
Chol		5				
Chol Esters		9				
TG		22				
PC	38.2[a]		35[a]	42.6	42[a]	
PE			35	25.6	29	
PS	26.3		4	16.2	9	
PI			4		6	
PA		62	4		4	
Sph	24.3		18	15.6	10	
LPC	11.1			<1		
Other						
Glycolipid						16
Total lipid (mg/100 ml)	471.0	8.7[b]				
Total PL						320
Ref.	c	d	e	f	g	h

[a] % of total phospholipid: [b] % dry weight of cell.
[c] Cumings et al., 1965
[d] Elsbach, 1959
[e] Karnovsky and Wallach, 1961
[f] Finke et al., 1966
[g] Firkin and Williams, 1961
[h] Miras et al., 1966

oxidize fatty acids (Evans and Mueller, 1963; Hrachovec, 1965; Rosenzweig and Ways, 1966), incorporate fatty acids into their lipids *in vitro* (Elsbach, 1963, 1964) and contain a relatively large amount of triglyceride and relatively little cholesterol. More thorough study of these interesting cells should be undertaken since they provide a valuable means for studying lipids of human cells.

C. CULTURED AMNION CELLS

In vitro culture provides a valuable means for study of lipids under various conditions but little use has been made of the technique. The proportions of phospholipids of a neoplastic cell line originally derived from human amnion is shown in Table XXXI. The cells evidently have phospholipids similar to those found in whole organs and subcellular structures. The cultured cells have a quantitative phospholipid pattern that is very similar to that of mouse liver (Table XXXI) and fall into the group with low sphingomyelin.

D. COMPARISON OF ANIMAL AND YEAST CELLS

From recent work on the yeast *Lipomyces lipofer* (Heick and Stewart, 1965a, 1965b; McElroy and Stewart, 1967), it is known that these cells contain mitochondria and nuclei with structural features typical of animal cells as well as endoplasmic reticulum, plasma membrane and cell wall. When the cell wall is removed, the cells are hardly distinguishable from those of animals (Heick and Stewart, 1965b). The presence of phosphatidyl ethanolamine, phosphatidyl serine, phosphatidyl choline, phosphatidyl inositol, phosphatidyl glycerol and diphosphatidyl glycerol has been indicated (McElroy and Stewart, 1967) and the amounts, except for the high level of phosphatidyl glycerol, are similar to levels in animal cells. Hence, the similarity noted between vertebrates and invertebrates extends to some microorganisms. It seems possible that whenever electron microscopy discloses a similarity of morphology between animal cells and microorganisms, the phospholipids of both will be similar. Microorganisms and plants with different morphology may be expected to have different lipid distributions.

X. Glycolipids

Glycolipids are not found in mitochondria (Fleischer *et al.*, 1967b) or endoplasmic reticulum (S. Fleischer, B. Fleischer and G. Rouser, unpublished) and thus far have not been detected in nuclei (S. Fleischer, B. Fleischer and G. Rouser, unpublished). The data available have clearly associated glycolipids with the plasma (cell surface) membrane or a membrane elaborated from a plasma membrane. This association is quite clear for erythrocytes in which all of the lipid is associated with the surface membrane and for myelin derived from the plasma membrane of Schwann cells in peripheral nerve and probably formed in a similar manner in brain. The data from other cell types and organs are in keeping with, but do not provide complete proof of, the concept that glycolipids occur exclusively in plasma membranes or membranes derived from them. Glycolipids have not been found in all organs and thus perhaps plasma membranes of some cells do not contain glycolipids.

Glycolipid patterns appear to fall into two major groups. One group (represented by myelin) contains cerebroside and cerebroside sulfate (sulfatide). In the second category, represented by the erythrocyte membrane (Booth, 1963; Handa and Handa, 1965; Handa and Yamakawa, 1964; Irie et al., 1961; Klenk and Lauenstein, 1952; Klenk and Padberg, 1962; Rouser et al., 1961b; Yamakawa, 1965; Yamakawa et al., 1956, 1960, 1962a, 1962b, 1963, 1965; Yamakawa and Suzuki, 1953) and the lens of the eye (Feldman et al. 1965, 1966), ceramide polyhexosides and gangliosides are the principal glycolipids. Cerebrosides and ceramide dihexosides have been isolated from milk lipid extracts. The polar lipids of milk are present in the fat globule membrane that may form still a third category with regard to the glycolipid pattern. It is believed that the fat globule membrane is formed from the plasma membrane of mammary gland cells (Dowben et al., 1967).

Aside from the erythrocyte, where all of the lipid is present in the plasma membrane, and the fully developed brain, in which myelin elaborated from a plasma membrane contributes most of the lipid, the contribution of the plasma membrane to the total lipid of the cell may be small. Erythrocyte lipid levels can be used as the basis for rough approximations of the range for contribution of plasma membranes to the total lipid of cells. Erythrocytes yield about 0·4% lipid on a fresh weight basis. Organs other than brain yield from 1-7% total lipid/g wet wt. Thus, perhaps 7 to 40% of the total lipid may be contributed by the plasma membrane. Again, judging from erythrocyte values where glycolipids represent 5-25% of the total lipid of different species (Table V), glycolipids can be calculated to fall in the range of 0·35 to 1·6% of the total lipid of organs at the lower level with 10% of the total lipid representing an approximate upper limit in specialized cells such as leukocytes where plasma membranes are relatively abundant. Although plasma membranes are seldom isolated in pure form for study, they are important structures that may contribute significantly to the total cell lipid.

Vertebrate liver, heart and skeletal muscle contain at most only minute traces of glycolipid (G. Rouser, G. Simon and G. Kritchevsky, unpublished), the level being below 0·05% of the total lipid. The most abundant organelles (mitochondria and endoplasmic reticulum) of these organs do not appear to contain glycolipids. Since traces of glycolipids from red cells or myelin of nerves may be found in organs, analysis of purified plasma membranes is required to establish the presence or absence of glycolipids in plasma membranes of a particular organ or cell type.

Lung and kidney from mammals clearly contain glycolipid. In lung it is reasonable to suppose that ceramide polyhexosides are associated with cell surface membranes that contribute a substantial amount of the total lipid. Isolated plasma membranes have not been studied and thus direct data to support this concept are lacking. The presence in metachromatic leukodystrophy of sulfatide in urine and the metachromatic staining material in tubular cells suggest that glycolipid may be present in renal tubular cell plasma membrane.

Gangliosides in brain are clearly not myelin components since they are present prior to myelination and are trace components, probably contaminants, of purified myelin preparations. From the relatively low level of gangliosides in purified preparations of brain myelin, mitochondria, nuclei, endoplasmic reticulum and nerve ending particles, it appears that gangliosides of brain occur primarily in a structure not yet isolated. This structure may well be the plasma membrane. This conclusion is supported by the morphological findings in Tay-Sachs disease where degradation of ganglioside is defective and gangliosides accumulate. Neurones are distended and contain membraneous cytoplasmic bodies rich in ganglioside. These bodies may well be masses of plasma membrane.

The high level of ceramide dihexoside reported for polymorphonuclear leukocytes (see Section IX-B) may well arise from its presence in the plasma membrane and the fact that the plasma membrane contributes a large proportion of the total lipid, mitochondria and endoplasmic reticulum being less abundant in this cell type. The total lipid of the polymorphonuclear leukocyte is relatively low and is in the same range as that of the red cell. This is in keeping with the concept that a large proportion of the lipid is present in the plasma membrane.

The plasma membranes of different types of cells from one species clearly differ in glycolipid composition. Species variations can also be expected since large species variations are found in red cell glycolipids. The concept of glycolipids as components principally (if not exclusively) of plasma membranes of some types of cells is significant in interpretation of the changes seen in hereditary metabolic diseases (Section XI).

XI. Changes in Membrane Components and Pathological States

Membranes are thought of as being composed essentially of structural protein to which lipid is attached. There are invariably membrane-associated enzymes and the membrane may have a polysaccharide coat. All of these should probably be considered as components of the functional membrane of living cells. Mutations giving rise to metabolic abnormalities in any of the components may change membrane properties and function and could conceivably be associated with gross abnormalities of the organism as a whole. It is apparent that a major problem in pathology may arise from the close association of membrane components. A metabolic defect in one component could give rise to quantitative changes in other components, i.e. primary defects may be difficult to distinguish from secondary effects.

Lipid metabolism is altered in many abnormal states. Changes in some disorders are clearly of a secondary nature, they result from a defect in some other area of metabolism, as e.g. the large blood lipid changes seen in some types of glycogen storage disease. True primary defects in lipid metabolism appear to be the basis for other disorders. There are two types of hereditary pathological states clearly associated with abnormalities in membranes. In one group the molecular basis is not known but membrane phenomena are indicated particularly by abnormalities of renal excretion. This group includes the Fanconi syndrome, cystic fibrosis of the pancreas, renal glycosuria, renal tubular acidosis, cystinuria, Hartnup disease and several others. A systematic study of these disorders may reveal some to be caused by hereditary defects in lipid metabolism (Rousen and Knitcheusky, 1967.

The sphingolipid disorders are a second group of membrane abnormalities where the basic defects are known to involve enzymes of polar lipid metabolism. These disorders and the lipid classes involved are: infantile and chronic Gaucher's diseases (cerebroside), Niemann-Pick disease (sphingomyelin), Tay-Sachs disease (ganglioside), metachromatic leukodystrophy (cerebroside sulfate) and Fabry's disease (ceramide polyhexosides).

It is notable that all of the known disorders of polar lipid metabolism involve sphingolipids, i.e. abnormalities associated with enzyme defects in the metabolism of the glycerol phospholipids have not been recognized. Since it seems probable that mutations affecting the glycerol lipids do arise, it can be assumed that these have not been recognized. Simple considerations lead to a reasonable hypothesis which explains the failure to recognize such defects. The sphingolipid disorders were first recognized by the accumulation in one or more tissues of a particular sphingolipid. Sphingolipid molecules have few double bonds. They are composed almost entirely of saturated fatty acids or those with one double bond. Such molecules are relatively stable to oxidation and their solubility in organic solvents and dispersability in aqueous media are generally much lower than molecules with several double bonds. It is thus apparent that unoxidized lipid would tend to accumulate in disorders affecting their metabolism.

The glycerol lipids, by contrast, contain more highly unsaturated fatty acids, some with six double bonds. These molecules are rapidly autoxidized, the rate increasing markedly with increase in the number of double bonds. It is thus to be expected that many, if not all, hereditary defects involving enzymes of glycerolphospholipid metabolism would lead to the storage of oxidized forms of the lipids. This expectation plus the fact that there are numerous conditions in which a substance commonly called ceroid accumulates in tissues suggests that defects of glycerol lipid metabolism will be traced to deposition of oxidized forms now recognizable as ceroid deposits but not yet associated with specific defects. Ceroid has been variously called hemofuscin, fat pigment, lipochrome, lipofuscin and "wear-and-tear" or "age" pigments. Lipofuscin is one of the more common names. It is clear that ceroid is not a single or simple substance and that lipid is a component of ceroid. The lipid has invariably been found to have the properties of oxidized lipid.

The conditions that have been proved to be related to membrane component pathology are concerned with hereditary defects in the metabolism of sphingolipids, both phospholipids and glycolipids. Before considering these problems in detail, however, the reader is encouraged to consult the recent discussion of diabetes mellitus by Renold and Cahill (1966). This discussion emphasizes the diverse nature of the clinical conditions and problems associated with a disorder that appears always to involve insulin in some manner. The uncertainties in such a much studied disease emphasize the caution required when dealing with the less well characterized disorders of lipid metabolism. Reference to the literature on the Hurler's syndrome (gargoylism) will further emphasize the difficulties involved in distinguishing lipid from nonlipid disorders. This syndrome was once thought to be a lipid disorder but now appears to be related to a defect in the metabolism of mucopolysaccharides with some secondary lipid changes.

Investigations in the senior author's laboratory indicate that the inherited disorders of polar lipid metabolism are more complex than a simple decrease in the level of one enzyme affecting the degradation of one lipid. The complex changes can perhaps be understood, at least partially, by relating them to modern concepts of the genetic mechanism. There are several types of mutations of genes. Mutation of a structural gene leads to synthesis of an abnormal protein which, if it is an enzyme, may have altered enzymatic activity. Such a mutation may also affect the quantity of a limited number of other proteins. Mutation of a control gene, on the other hand, influences the level of one or more enzymes of normal structure. Some portions of the chromosomal DNA are neither structural genes nor control genes but are instead templates for ribosomal RNA or transfer RNA. The phenotypic expression of one mutation may be reversed toward the normal phenotype by a second mutation. Mutations other than those of structural and control genes may have a variety of complex effects that can not be classified at present.

Enzyme rates depend upon many factors. Some of these are passage of substrates through membranes into cells, intracellular substrate concentrations, the amount(s) and general properties of the enzyme(s), presence of activators and inhibitors, availability of cofactors and cellular structural features giving rise to compartments.

Several disorders of sphingolipid metabolism have been studied sufficiently well to warrant detailed consideration of possible mechanisms. Consideration must be limited to the classical forms of these disorders. Supposed variants occurring at later ages, etc. have not been well defined and lipid composition studies in the senior author's laboratory have failed to support results of many earlier investigations employing less satisfactory methods.

There is no question that cerebroside metabolism is abnormal in Gaucher's disease. It has been frequently supposed that the disorder is related to deficiency of an enzyme for degrading the glycocerebroside deposited in some tissues. Recent measurements of the level of cerebrosidase have confirmed this concept. The facts that cerebroside is not deposited in all organs and other lipids may be present in abnormal amounts indicate, however, that the condition is more complicated and that the mutation probably involves other enzymes as well. In all cases of Gaucher's disease, cerebroside accumulates in liver and spleen. The brain is seriously affected in the infantile form but is not affected in the chronic forms that vary greatly in time of onset and severity. A trace of cerebroside is normally found in erythrocytes and this level is maintained in cells from Gaucher's disease, although the level of sphingomyelin is increased. The very large quantitative variations in the disease and the distribution of erythrocyte lipids can be explained as mutation(s) of control genes. The balance of factors involved in each tissue would thus determine which lipid is elevated and the extent of the increase. Much additional study is required and can be expected to yield fundamental information that may ultimately prove to be the basis for some form of therapy or control of the disorder.

Classical Niemann-Pick disease presents somewhat the same situation as Gaucher's disease. Sphingomyelin levels are very high in liver, spleen and brain and enzyme measurements indicate a deficiency of a degradative enzyme (sphingomyelinase). In addition, however, it has been shown in the senior author's laboratory that an uncharacterized phospholipid accumulates in liver, cerebroside is increased in spleen and a ceramide dihexoside accumulates in brain (Rouser et al., 1966a). Erythrocyte sphingomyelin in Niemann-Pick disease is decreased rather than increased (Balint et al., 1961; Rouser et al., 1961b). The changes are difficult to explain as arising from the utilization of an alternative biosynthetic pathway as the result of decreased sphingomyelin degradation. A more probable explanation appears to lie in control gene mutation affecting the levels of several enzymes. Much more study is required before the disorder can be understood.

Tay-Sachs disease is characterized by a large accumulation in brain of a ganglioside. The same ganglioside is increased in erythrocytes (Balint et al., 1963; Rouser et al., 1961b), and erythrocyte phospholipid distribution appears to be altered (Balint et al., 1963). Visual examination of thin layer chromatograms of the lipid from other organs indicates increases in other sphingolipids, but the chemical nature and precise amounts of these have not been established. Perhaps extension of these studies will demonstrate the situation to be similar to that in Gaucher's and Niemann-Pick diseases. A control gene mutation is perhaps indicated.

Metachromatic leukodystrophy appears to depend upon the deficiency of a sulfatase. Sulfatides accumulate in brain and kidney, the two organs known to contain sulfatides in normal individuals. The overall lipid composition has not been determined as yet and speculation about other possible effects are not warranted.

Several factors can be conceived of as giving rise to abnormal lipid class compositions as secondary changes. If the structure of a membrane is altered by change in one component, it seems possible that the membrane degradation in the normal process of turnover might be altered in such a way that some lipids (as well as other components) might become trapped, i.e. separated from the enzymes that normally degrade them, and the levels might thus increase. Since autolysis studies of normal organs (see Section V-C) have shown that sphingomyelin is more slowly degraded than the glycerol phospholipids, sphingomyelin could be involved in such a trapping mechanism and has been shown in fact to be disproportionately high in brain in infantile Gaucher's diesease, Tay-Sachs disease and metachromatic leukodystrophy (Rouser et al., 1966a).

XII. Summary and Conclusions

A. The systematic and careful study of the lipid composition of animal cell membranes has just begun. Procedures for isolation and characterization of individual pure membranes have greatly improved as have the procedures for isolation, characterization and quantitative determination of lipid composition. Rapid advances in the field can be expected in the near future.

B. All animal cell membranes contain phospholipids. All vertebrates have the same phospholipids including some uncharacterized lipid classes occurring in small amounts. The common and most abundant phospholipids are phosphatidyl choline (lecithin), phosphatidyl ethanolamine, phosphatidyl serine, phosphatidyl inositol, diphosphatidyl glycerol (cardiolipin) and sphingomyelin.

In general, the same phospholipid classes are found in vertebrates and invertebrates. In some invertebrates ceramide aminoethylphosphonate replaces sphingomyelin, whereas in others (snail, insects) ceramide phosphorylethanolamine is found and sphingomyelin is absent.

Some membranes contain glycolipids (e.g. myelin) whereas others do not (e.g. mitochondria). Relatively little information is available on glycolipids, particularly of invertebrates. Glycolipids including cerebrosides, sulfatides, ceramide polyhexosides and gangliosides have been shown to occur in all vertebrates.

Certain membranes contain sterol. This is apparently almost exclusively cholesterol in vertebrates and cholesterol is also found in the membranes in many invertebrates.

C. When membranes are obtained from cells of the same species, plasma (cell surface) membranes, the endoplasmic reticulum, nuclear membranes and mitochondria are found to have different lipid class compositions. All differ quantitatively and to some extent qualitatively (e.g. the presence in mitochondria of diphosphatidyl glycerol usually absent from other membranes, cerebroside and sulfatide in myelin, and the absence of sphingomyelin from mitochondria). Plasma membranes or elaborations of these (such as myelin) appear to contain most (if not all) of the glycolipid. Even where qualitative differences in lipid class distribution are not seen as in endoplasmic reticulum and nuclei, the quantitative distribution of lipid classes is different and characteristic for each membrane.

D. Plasma membranes, as shown by studies of mammalian erythrocytes, exhibit large species variations in composition. Phospholipid proportions vary greatly and the total amount as well as the types of both ceramide polyhexosides and gangliosides is very different in different species. Data from whole organs indicate that plasma membranes from different cell types of the same species vary in composition. Glycolipids occur for the most part (if not entirely) in plasma membranes or their elaborations (such as myelin) although plasma membranes from some cells (e.g. liver and heart) may not contain glycolipids.

E. Mitochondria from bovine heart, kidney and liver contain diphosphatidyl glycerol (cardiolipin), phosphatidyl choline and phosphatidyl ethanolamine as the major phospholipids in the approximate molar ratio 1/4/4. In addition, phosphatidyl inositol represents about 3% and phosphatidyl serine about 0·5% of the total phospholipid. Sphingolipids are not present in mitochondria and the cholesterol level is low. The inner and outer mitochondrial membranes differ in composition. Impure preparations of the outer membrane have been shown to have a lower diphosphatidyl glycerol content than inner membranes. Pure outer membrane preparations may well be found to be entirely devoid of diphosphatidyl glycerol and to contain all the cholesterol of the mitochondrion. Mitochondria from different organs of one species appear to have the same phospholipid distribution. Species variation in mitochondrial phospholipid composition among vertebrates is at most slight and some data indicate that there may be no species variation at all.

The total lipid of mitochondria from different organs is variable which appears to correlate with the proportion of cristae in the mitochondria, those containing more cristae also having more phospholipid.

F. Endoplasmic reticulum (microsomes) has a characteristic composition and rough and smooth surfaced reticulum from any one organ appear to have, in general, the same phospholipid composition. The microsome fractions of different organs from one species may differ widely in phospholipid composition although there appears to be little or no variation in phospholipid composition of the same organ from different vertebrate species.

G. Nuclei have a characteristic distribution of phospholipids different from that of other organelles. Glycolipids are probably absent from nuclei, although data supporting this conclusion are inadequate. There is little or no variation in the phospholipid distribution of nuclei from different organs of one species and available data suggest little or no species variations among vertebrates.

H. Among vertebrates, there seems to be little or no species variation in the phospholipid composition of individual membranes from the same organ or cell type except for variations in plasma membranes. These tentative conclusions are based principally upon the analysis of erythrocytes of mammals and whole organs and cells from several vertebrates. The composition of a cell or organ is a composite from all the membranes. The values obtained from different species might differ as a result of (1) variations in the proportions of the membranes without change in composition, (2) variations in the composition of one or more membranes without change in membrane proportions or (3) changes in both composition and proportions of membranes. When the composition of a cell type or organ from different species is very similar, the most reasonable assumption is that membrane composition and the proportions of the different membranes are both the same.

I. Invertebrates, while having generally the same lipid classes as vertebrates, have some lipids (e.g. ceramide aminoethylphosphonate) not found in vertebrates. The lack of species variability of membrane composition seen in vertebrates is not apparent for invertebrates. Invertebrate organs show wide species variations that could result from large differences in composition and/or in the relative proportions of the different membranes in cells and organs.

J. The fatty acid composition of each lipid class of organelles and organs from one species as well as from different species is variable even when the lipid class composition is the same. Individuality is thus expressed most clearly in differences in fatty acid composition.

K. Several hereditary disorders in man are clearly related to mutations giving rise to changes in polar lipid metabolism. Some of these appear to be mutations of control rather than structural genes. Other hereditary pathological states involve defects in renal transport and are quite possibly dependent upon altered membrane components. Increased progress in these areas can be expected from further investigations with recently devised, improved procedures for membrane studies.

XIII. Acknowledgements

This chapter is based upon research work supported in part by the following grants: NB-01847, NB-06237 and NB-06113-01 from the National Institute of Neurological Diseases and Blindness; GM-12831 from the National Institute of General Medical Sciences; Contract CA-18-135-335(A) from the U.S. Army, Edgewood Arsenal; by the Samuel Kanner Research Fund; the U.S. Atomic Energy Commission and an American Heart Association Grant-in-Aid. S.F. is an Established Investigator of the American Heart Association.

References

Ashworth, L. A. E. and Green, C. (1964). *Biochim. biophys. Acta* **84**, 182.
Balint, J. A., Nyhan, W. L., Lietman, P. and Turner, D. A. (1961). *J. Lab. clin. Med.* **58**, 548.
Balint, J. A., Spitzer, H. L. and Kyriakides, E. C. (1963). *J. clin. Invest.* **42**, 1661.
Basford, J. M., Glover, J. and Green, C. (1964). *Biochim. biophys. Acta* **84**, 764.
Bentley, H. P. Jr. (1962). *Proc. Soc. exp. Biol. Med.* **111**, 591.
Billimoria, J. D., Irani, V. J. and Maclagan, N. F. (1965). *J. Atheroscl. Res.* **5**, 90.
Blomstrand, R., Nakayama, F. and Nilsson, I. M. (1962). *J. Lab. clin. Med.* **59**, 771.
Booth, D. A. (1963). *Biochim. biophys. Acta* **70**, 486.
Bradlow, B. A. and Rubenstein, R. (1964). *S. Afr. J. Med. Sci.* **29**, 36.
Bradlow, B. A., Rubenstein, R. and Lee, J. (1964). *S. Afr. J. Med. Sci.* **29**, 41.
Bradlow, B. A., Lee, J. and Rubenstein, R. (1965). *Br. J. Haemat.* **11**, 315.
Cardi, E., Rezza, E., Rutiloni, C. and Bonomolo, A. (1964). *Ric. Sci.* **5**, 357.
Carroll, K. K. (1965). *J. Am. Oil Chemists' Soc.* **42**, 516.
Carter, H. E. and Greenwood, F. L. (1952). *J. biol Chem.* **199**, 283.
Casley-Smith, J. R. and Day, A. J. (1966). *Quart. J. Exp. Physiol.* **51**, 1.
Condrea, E., Mammon, Z., Aloof, S. and de Vries, A. (1964). *Biochim. biophys. Acta* **84**, 265.
Corsini, F., Salvioli, G. P., Paolucci, G., Manfredi, G. and Babini, B. (1963). *Lattante* **34**, 264.
Crowley, J., Ways, P. and Jones, J. W. (1965). *J. clin Invest.* **44**, 989.
Cumings, J. N., Shortman, R. C. and Skrbic, T. (1965). *J. clin. Path.* **18**, 641.
Cuzner, M. L., Davison, A. N. and Gregson, N. A. (1965) *Ann. N.Y. Acad. Sci.* **122**, 86.
de Gier, J. and van Deenen, L. L. M. (1961). *Biochim. biophys. Acta* **49**, 286.
de Gier, J. and van Deenen, L. L. M. (1964). *Biochim. biophys. Acta* **84**, 294.
de Gier, J., Mulder, I. and van Deenen, L. L. M. (1961). *Naturwissenschaff.* **48**, 54.
de Gier, J., van Deenen, L. L. M., Verloop, M. C. and van Gastel, C. (1964). *Br. J. Haemat.* **10**, 246.
de Gier, J., van Deenen, L. L. M. and van Senden, K. G. (1966). *Experientia* **22**, 20.
Dodge, J. T., Mitchell, C. and Hanahan, D. J. (1963). *Archs Biochem. Biophys.* **100**, 119.
Dowben, R. M., Brunner, J. R. and Philpott, D. E. (1967). *Biochim. biophys. Acta* **135**, 1.
Eckles, N. E., Taylor, C. B., Campbell, D. J. and Gould, R. G. (1955). *J. Lab. clin. Med.* **46**, 359.

Elsbach, P. (1959). *J. exp. Med.* **110**, 969.
Elsbach, P. (1962). *Nature, Lond.* **195**, 383.
Elsbach, P. (1963). *Biochim. biophys. Acta* **70**, 157.
Elsbach, P. (1964). *Biochim. biophys. Acta* **84**, 8.
Elsbach, P. and Kayden, H. J. (1965). *Am. J. Physiol.* **209**, 765.
Evans, W. H. and Mueller, P. S. (1963). *J. Lipid Res.* **4**, 39.
Farnsworth, P., Danon, D. and Gellhorn, A. (1965). *Br. J. Haemat.* **11**, 200.
Farquhar, J. W. (1962). *Biochim. biophys. Acta* **60**, 80.
Farquhar, J. W. (1965). *J. Am. Oil Chemists' Soc.* **42**, 615.
Farquhar, J. W. and Ahrens, E. H. Jr. (1963). *J. clin. Invest.* **42**, 675.
Feldman, G. L. and Rouser, G. (1965). *J. Am. Oil Chemists' Soc.* **42**, 290.
Feldman, G. L., Feldman, L. S. and Rouser, G. (1965). *J. Am. Oil Chemists' Soc.* **42**, 742.
Feldman, G. L., Feldman, L. S., and Rouser, G. (1966). *Lipids* **1**, 21.
Feldman, G. L., Feldman, L. S. and Rouser, G. (1966). *Lipids* **1**, 161
Finean, J. B. (1953). *Experientia* **9**, 17.
Finke, S. R., Phillips, G. B. and Middleton, E. Jr. (1966). *J. Lab. clin. Med.* **67**, 601.
Firkin, B. G. and Williams, W. J. (1961). *J. clin. Invest.* **40**, 423.
Fitch, C. D., Dinning, J. S., Witting, L. F. and Horwitt, M. K. (1961). *J. Nutr.* **75**, 409.
Fleischer, S. (1964). *In* "Sixth International Congress of Biochemistry," Vol. 8-S2, p. 605.
Fleischer, S. and Rouser, G. (1965). *J. Am. Oil Chemists' Soc.* **42**, 588.
Fleischer, S., Fleischer, B. and Stoeckenius, W. (1967a). *J. Cell Biol.* **32**, 193.
Fleischer, S., Rouser, G., Fleischer, B., Casu, A. and Kritchevsky, G. (1967b). *J. Lipid Res.* **8**, 170.
Folch, J., Lees, M. and Sloane-Stanley, G. H. (1957). *J. biol. Chem.* **226**, 497.
Gjone, E., Berry, J. F. and Turner, D. A. (1959). *J. Lipid Res.* **1**, 66.
Glomset, J. A. (1962). *Biochim biophys. Acta* **65**, 128.
Glomset, J. A., Parker, F., Tjaden, M. and Williams, R. H. (1962). *Biochim. biophys. Acta* **58**, 398.
Glomset, J. A., Janssen, E. T., Kennedy, R. and Dobbins, J. (1966). *J. Lipid Res.* **7**, 639.
Green, D. E. and Fleischer, S. (1964). *In* "Metabolism and Physiological Significance of Lipids" (R. M. C. Dawson, ed.) pp. 581–618, John Wiley and Son, New York.
Greenberg, L. D. and Moon, H. D. (1961). *Archs. Biochem. Biophys.* **94**, 405.
Grimmer, G., Voigt, K. D., Apostolakis, M. and Glaser, A. (1965). *Arzneimittel-Forsch.* **15**, 184.
Hagerman, J. S. and Gould, R. G. (1951). *Proc. Soc. exp. Biol. Med.* **78**, 329.
Hanahan, D. J. and Watts, R. M. (1961). *J. biol. Chem.* **236**, 59.
Hanahan, D. J., Watts, R. M. and Pappajohn, D. (1960). *J. Lipid Res.* **1**, 421.
Hanahan, D. J., Ekholm, J. and Jackson, C. M. (1963). *Biochemistry* **2**, 630.
Handa, S. and Yamakawa, T. (1964). *Japan J. exp. Med.* **34**, 293.
Handa, N. and Handa, S. (1965). *Japan J. exp. Med.* **35**, 331.
Heemskerk, C. H. T. and van Deenen, L. L. M. (1964). *Koninkl. Ned. Akad. Wetenschap. Ser. B.* **67**, 181.
Heick, H. M. C. and Stewart, H. B. (1965a). *Can. J. Biochem.* **43**, 549.
Heick, H. M. C. and Stewart, H. B. (1965b). *Can. J. Biochem.* **43**, 561.
Hill, J. G., Kuksis, A. and Beveridge, J. M. R. (1964). *J. Am. Oil Chemists' Soc.* **41**, 393.

Hill, J. G., Kuksis, A. and Beveridge, J. M. R. (1965). *J. Am. Oil Chemists' Soc.* **42**, 137.
Hokin, L. E. and Hokin, M. R. (1964). *Biochim. biophys. Acta* **84**, 563.
Horwitt, M. K., Harvey, C. C. and Century, B. (1959). *Science, N. Y.* **130**, 917.
Hrachovec, J. P. (1965). *Proc. Soc. exp. Biol. Med.* **118**, 328.
Hulcher, F. H. (1963). *Archs. Biochem. Biophys.* **100**, 237.
Introzzi, P., Notario, A., DiMarco, N., Doneda, G. and Meduri, D. (1965). *Haematologica* **50**, 323.
Irie, R., Iwanaga, M. and Yamakawa, T. (1961). *J. Biochem.* **50**, 122.
Jacobs, M. H., Glassman, H. N. and Parpart, A. K. (1950). *J. Exptl. Zool.* **113**, 277.
Karaca, M. and Stefanini, M. (1966). *J. Lab. Clin. Med.* **67**, 229.
Karnovsky, M. L. and Wallach, D. F. H. (1961). *J. biol. Chem.* **236**, 1895.
Kates, M. and James, A. T. (1961). *Biochim. biophys. Acta* **50**, 478.
Kates, M., Allison, A. C. and James, A. T. (1961). *Biochim. biophys. Acta* **48**, 571.
Kidson, C. (1961a). *Br. J. exp. Path.* **42**, 597.
Kidson, C. (1961b). *Australasian Ann. Med.* **10**, 282.
Kidson, C. (1962). *Australasian Ann. Med.* **11**, 50.
Klenk, E. and Lauenstein, K. (1952). *Z. physiol. Chem.* **291**, 249.
Klenk, E. and Padberg, G. (1962). *Z. physiol. Chem.* **327**, 249.
Klibansky, C. and Osimi, Z. (1961). *Bull. Research Council Israel, Sect. E*, **9**, 143.
Klibansky, C. and de Vries, A. (1963). *Biochim. biophys. Acta* **70**, 176.
Kögl, F., de Gier, J., Mulder, I. and van Deenen, L. L. M. (1960). *Biochim. biophys. Acta* **43**, 95.
Korey, S. R., Orchen, M. and Brotz, M. (1958). *J. Neuropath. Exptl. Neurol.* **17**, 430.
Korn, E. D. (1966). *Science, N. Y.* **153**, 1491.
Laatsch, R. H., Kies, M. W., Gordon, S. and Alvord, E. C. Jr. (1962). *J. exp. Med.* **115**, 777.
Leveille, G. A., Shockley, J. W. and Sauberlich, H. E. (1962). *Proc. Soc. exp. Biol. Med.* **109**, 345.
Lovtrup, S. and Svennerholm, L. (1963). *Exp. Cell Res.* **29**, 298.
Manfredi, G., Corsini, F., Paolucci, G., Salvioli, G. P. and Babini, B. (1962). *Boll. Soc. Ital. Biol. Sper.* **38**, 726.
Marcus, A. J., Ullman, H. L. and Ballard, H. S. (1961). *Proc. Soc. exp. Biol. Med.* **107**, 483.
Marcus, A. J., Ullman, H. L., Safier, L. B. and Ballard, H. S. (1962). *J. clin. Invest.* **41**, 2198.
Marinetti, G. V., Berry, J. F., Rouser, G. and Stotz, E. (1953). *J. Am. chem. Soc.* **75**, 313.
Marinetti, G. V., Erbland, J. and Stotz, E. (1958). *J. biol. Chem.* **233**, 562.
Marks, P. A., Gellhorn, A. and Kidson, C. (1960). *J. biol. Chem.* **235**, 2579.
Matsumoto, M. (1961). *J. Biochem.* **49**, 11.
McElroy, F. A. and Stewart, H. B. (1967). *Can. J. Biochem.* **45**, 171.
Miras, C. J., Mantzos, J. D. and Levis, G. M. (1966). *Biochem. J.* **98**, 782.
Mohrhauer, H. and Holman, R. T. (1963). *J. Lipid Res.* **4**, 346.
Monsen, E. R., Okey, R. and Lyman, R. L. (1962). *Metabolism* **11**, 1113.
Mulder, E. and van Deenen, L. L. M. (1965a). *Biochim. biophys. Acta* **106**, 106.
Mulder, E. and van Deenen, L. L. M. (1965b). *Biochim. biophys. Acta* **106**, 348.
Mulder, E., de Gier, J. and van Deenen, L. L. M. (1963). *Biochim. biophys. Acta* **70**, 94.

Mulder, E., van den Berg, J. W. O. and van Deenen, L. L. M. (1965). *Biochim. biophys. Acta* **106**, 118.
Murphy, J. R. (1962). *J. Lab. clin. Med.* **60**, 571.
Murphy, J. R. (1965). *J. Lab. clin. Med.* **65**, 756.
Nazir, D. and Rouser, G. (1966). *Lipids* **1**, 159.
Nelson, G. J. (1967a). *Lipids* **2**, 64.
Nelson, G. J. (1967b). *Biochim biophys. Acta*, **144**, 221.
Nelson, G. J. (1967c). *Lipids*, **2**, 323.
Nelson, G. J. (1967d). *J. Lipid Res.* **8**, 374.
Norton, W. T. and Autilio, L. A. (1965). *Ann. N.Y. Acad. Sci.* **122**, 77.
Notario, A., Zanetti, A., Ricotti, V. and Bosco, G. (1965). *Haematologica* **50**, 1.
Nothman, M. M. and Proger, S. (1965). *Med. Welt* **4**, 190.
Nussbaum, J. L., Bieth, R. and Mandel, P. (1963). *Nature, Lond.* **198**, 586.
Oliveira, M. M. and Vaughan, M. (1964). *J. Lipid Res.* **5**, 156.
Ostwald, R. and Shannon, A. (1964). *Biochem. J.* **91**, 146.
Parsons, D., Williams, G., Thompson, W., Wilson, D. and Chance, B. (1967). *In* Proc. Symp. "Mitochondrial Structure and Function", Bari, Italy, May 23-26, 1966. (J. M. Tager, S. Papa, E. Quagliarello and E. C. Slater, eds). *Biochim. biophys. Acta* Library (in press).
Paysant, M., Maupin, B. and Polonovski, J. (1963). *Bull. Soc. Chim. Biol.* **45**, 247.
Pennell, R. B. (1964). *In* "The Red Blood Cell" (C. Bishop and D. M. Surgenor, eds) pp. 29-69. Academic Press, New York.
Phillips, G. B. (1957). *Proc. natn. Acad. Sci. U.S.A.* **43**, 566.
Phillips, G. B. (1959). *J. clin. Invest.* **38**, 489.
Phillips, G. B. (1962). *J. Lab. clin. Med.* **59**, 357.
Pittman, J. G. and Martin, D. B. (1966). *J. clin. Invest.* **45**, 165.
Ponder, E. (1948). *In* "Hemolysis and Related Phenomena" Grune and Stratton, New York.
Raderecht, H. J. and Schölzel, E. (1962). *Folia Haematol.* **78**, 613.
Raderecht, H. J., Schölzel, E. and Rapoport, S. M. (1960). *Klin. Wochschr.* **38**, 824.
Reed, C. F., Swisher, S. N., Marinetti, G. V. and Eden, E. G. (1960). *J. Lab. clin. Med.* **56**, 281.
Reid, M. E., Bieri, J. G., Plack, P. A. and Andrews, E. L. (1964). *J. Nutr.* **82**, 401.
Renold, A. E. and Cahill, G. F. Jr. (1966). *In* "The Metabolic Basis of Inherited Disease" (J. B. Stanbury, J. B. Wyngaarden and D. S. Fredrickson, eds) 2nd Ed. pp. 69-108, McGraw-Hill, New York.
Roelofsen, B., de Gier, J. and van Deenen, L. L. M. (1964). *J. Cell. Comp. Physiol.* **63**, 233.
Rosenzweig, A. and Ways, P. (1966). *Blood* **27**, 57.
Rouser, G. and Kritchevsky, G. (1967). *In* "Phenylketonuria and Allied Diseases" (J. A. Anderson and K. F. Swaiman, eds) U.S. Govt. Printing Office, 214-225.
Rouser, G., Bauman, A. J., Kritchevsky, G., Heller, D. and O'Brien, J. S. (1961a). *J. Am. Oil Chemists' Soc.* **38**, 544.
Rouser, G., Bauman, A. J., Nicolaides, N. and Heller, D. (1961b). *J. Am. Oil Chemists' Soc.* **38**, 565.
Rouser, G. and Fleischer, S. (1967). *In* "Methods in Enzymology" (M. E.Pullman and R. Estabrook, eds), Academic Press, New York, **10**, 385-406.
Rouser, G., Kritchevsky, G., Heller, D. and Lieber, E. (1963). *J. Am. Oil Chemists' Soc.* **40**, 425.

Rouser, G., Galli, C., Lieber, E., Blank, M. L. and Privett, O. S. (1964). *J. Am. Oil Chemists' Soc.* **41**, 836.
Rouser, G., Feldman, G. and Galli, C. (1965a). *J. Am. Oil Chemists' Soc.* **42**, 411–412.
Rouser, G., Kritchevsky, G., Galli, C. and Heller, D. (1965b). *J. Am. Oil Chemists' Soc.* **42**, 215.
Rouser, G., Kritchevsky, G., Galli, C., Yamamoto, A. and Knudson, A. (1966a). *In* "Inborn Disorders of Sphingolipid Metabolism" (S. M. Aronson and B. W. Volk, eds), pp. 303–316. Pergamon Press, New York.
Rouser, G., Kritchevsky, G., Whatley, M. and Baxter, C. F. (1966b). *Lipids* **1**, 107.
Rouser, G., Siakotos, A. N. and Fleischer, S. (1966c). *Lipids* **1**, 85.
Rouser, G., Kritchevsky, G., Simon, G. and Nelson, G. J. (1967a). *Lipids* **2**, 37.
Rouser, G., Kritchevsky, G. and Yamomoto, A. (1967b). *In* "Chromatographic Analysis of Lipids" (G. V. Marinetti, ed.) Vol. 1, Marcel Dekker Inc., New York, in press.
Rouser, G., Kritchevsky, G., Siakotos, A. N. and Yamomoto, A. (1968). *In* "An Introduction to Neuropathology: Method and Diagnosis" (C. G. Tedeschi, ed.) Little, Brown and Co., Boston, in press.
Rowe, C. E. (1960). *Biochem. J.* **76**, 471.
Sbarra, A. J. and Karnovsky, M. L. (1960). *J. biol. Chem.* **235**, 2224.
Schrader, G. T. and Dimopoullos, G. T. (1963). *Am. J. Vet. Res.* **24**, 283.
Siakotos, A. N. (1967). *Lipids* **2**, 87.
Siakotos, A. N. and Rouser, G. (1965). *J. Am. Oil Chemists' Soc.* **42**, 913.
Simon, G. and Rouser, G. (1967). *Lipids* **2**, 55.
Sloviter, H. A. and Tanaka, S. (1964). *J. Cell. Comp. Physiol.* **63**, 261.
Smith, L. M. and Freeman, N. K. (1959). *J. Dairy Sci.* **42**, 1450.
Soule, D. W., Marinetti, G. V. and Morgan, H. R. (1959). *J. exp. Med.* **110**, 93.
Stein, Y. and Stein, O. (1966). *Biochim. biophys. Acta* **116**, 95.
Suzuki, M. and O'Neal, R. M. (1964). *J. Lipid Res.* **5**, 624.
Szeinberg, A., Zaidman, J. and Clejan, L. (1965a). *Biochim. biophys. Acta* **98**, 598.
Szeinberg, A., Zaidman, J. and Clejan, L. (1965b). *Israel J. Med. Sci.* **1**, 833.
Thompson, G. A. Jr. and Hanahan, D. J. (1962). *Archs. Biophys. Biochem.* **96**, 671.
Thompson, G. A. Jr. and Hanahan, D. J. (1963). *Biochemistry* **2**, 641.
Turner, J. C., Anderson, H. M. and Gandal, C. P. (1958). *Biochim. Biophys. Acta* **30**, 130.
Turner, J. C. and Parsons, E. A. (1957). *J. exp. Med.* **105**, 189.
van Deenen, L. L. M. and de Gier, J. (1964). *In* "The Red Blood Cell" (C. Bishop and D. M. Surgenor, eds), pp. 243–307, Academic Press, New York.
van Deenen, L. L. M., de Gier, J. and de Haas, G. H. (1961). *K. Ned. Akad. Wetenschap.* Ser. B, **64**, 528.
van Deenan, L. L. M., de Gier, J., Houtsmuller, U. M. T., Montfoort, A. and Mulder, E. (1963). *In* "Biochemical Problems of Lipids" (A. C. Frazer, ed.). pp. 404–414. Elsevier, Amsterdam.
Vandenheuvel, F. A. (1963). *J. Am. Oil Chemists' Soc.* **40**, 455.
Vandenheuvel, F. A. (1965). *J. Am. Oil Chemists' Soc.* **42**, 481.
van Gastel, C., van den Berg, D., de Gier, J. and van Deenen, L. L. M. (1965). *Br. J. Haemat.* **11**, 193.
Vikrot, O. (1965). *Acta Med. scand.* **178**, Supplement 435, 23 pp.
Voigt, K. D., Apostolakis, M. and Grimmer, G. (1965). *Klin. Wochschr.* **43**, 732.
Vorbeck, M. L. and Marinetti, G. V. (1965). *J. Lipid Res.* **6**, 3.

Walker, B. L. and Kummerow, F. A. (1963). *J. Nutr.* **81**, 75.
Walker, B. L. and Kummerow, F. A. (1964a). *J. Nutr.* **82**, 323.
Walker, B. L. and Kummerow, F. A. (1964b). *J. Nutr.* **82**, 329.
Walker, B. L. and Kummerow, F. A. (1964c). *Proc. Soc. exp. Biol. Med.* **115**, 1099.
Watson, W. C. (1963). *Br. J. Haemat.* **9**, 32.
Ways, P. and Hanahan, D. J. (1964). *J. Lipid Res.* **5**, 318.
Ways, P., Reed, C. F. and Hanahan, D. J. (1963). *J. clin. Invest.* **42**, 1248.
Westerman, M. P. and Jensen, W. N. (1965). *Proc. Soc. exp. Biol. Med.* **118**, 315.
Westerman, M. P., Pierce, L. E. and Jensen, W. N. (1963). *J. Lab. clin. Med.* **62**, 394.
Williams, J. H., Kuchmak, M. and Witter, R. F. (1966). *Lipids* **1**, 391.
Witting, L. A., Harvey, C. C., Century, B. and Horwitt, M. K. (1961). *J. Lipid Res.* **2**, 412.
Yamakawa, T. (1965). *J. Japan. Biochem. Soc.* **37**, 827.
Yamakawa, T. and Suzuki, S. (1953). *J. Biochem.* **40**, 7.
Yamakawa, T., Matsumoto, M. and Suzuki, S. (1956). *J. Biochem.* **43**, 63.
Yamakawa, T., Irie, R. and Iwanaga, M. (1960). *J. Biochem.* **48**, 490.
Yamakawa, T., Kiso, N., Handa, S., Makita, A and Yokoyama, S. (1962a). *J. Biochem.* **52**, 226.
Yamakawa, T., Yokoyama, S. and Kiso, N. (1962b). *J. Biochem.* **52**, 228.
Yamakawa, T., Yokoyama, S. and Handa, N. (1963). *J. Biochem.* **53**, 28.
Yamakawa, T., Nishimura, S. and Kamimura, M. (1965). *Japan. J. exp. Med.* **35**, 201.

Chapter 3

X-ray Diffraction Studies of Lipid-Water Systems

VITTORIO LUZZATI

Centre de Génétique Moléculaire, C.N.R.S., Gif-sur-Yvette, 91, France

I.	INTRODUCTION	71
II.	STRUCTURE ANALYSIS	73
	A. Phase Diagrams	73
	B. General Properties and Classification of Structures . .	73
	C. Determination of Lattice Dimensions and of Other Structure Parameters	76
III.	METHODS	78
	A. Notation and Abbreviations	78
	B. X-ray Scattering Techniques	81
	C. Partial Specific Volumes	81
IV.	SIMPLE LIPIDS: SOAPS AND DETERGENTS	81
	A. Liquid-Paraffin Anhydrous Phases	82
	B. Liquid-Paraffin Water-containing Phases: Long-range Periodically Ordered Phases	90
	C. Transition from the Liquid-Paraffin to the Crystalline Phases: Gel and Coagel	97
	D. Isotropic Micellar Solution	101
V.	LIPIDS OF BIOLOGICAL INTEREST	103
	A. Structure of the Liquid-Paraffin Phases	104
	B. Monoglycerides, Phospholipids, Sphingolipids . . .	104
	C. Brain Lipids	109
	D. Mitochondria Lipids	110
	E. Remarks	114
VI.	DISCUSSION AND CONCLUSIONS	115
VII.	ACKNOWLEDGEMENTS	117
	APPENDIX 1: THE DISORDERED CONFORMATION OF THE PARAFFIN CHAINS	118
	APPENDIX 2: SOME CRYSTALLOGRAPHIC VERIFICATIONS . . .	119
	REFERENCES	121

I. Introduction

Lipids are present at high concentration in a variety of cell organelles (membranes, chloroplasts, mitochondria, sense receptors, etc.) whose physiological functions are associated with particularly ordered physical structures. Yet the role of the lipids is not quite clear. For some time the opinion prevailed that lipids merely constitute passive barriers against diffusion and that all the other functions are exerted by protein molecules. More recently the analysis of the phenomena that take place in some of the organelles (for example the maintenance of the resting potential and the transmission of the action potential in nerve membranes) has led some physiologists to put forward lipids as possible candidates for some specific, and yet unspecified, function.

Besides, as a result of the X-ray study of lipid-water systems, it is now well established that lipid molecules may aggregate into a variety of structures by the interplay of parameters that often remain within the range of the physiological conditions. The possible biological significance of some of the polymorphic transitions is striking and has been a favourite subject of speculation in recent years; nevertheless the gap between the structural transitions and the physiological phenomena has not yet been bridged by convincing experimental evidence.

The purpose of this article is to review the structure analysis of the lipid-water systems with special emphasis on the numerous phases, both anhydrous and water-containing, the structures of which are disordered at the atomic level and yet display a high degree of long-range organization. The crystalline forms of the pure compounds are outside the scope of this article.

The developments in this area are quite recent and research is still in progress; only a small part of the field has been covered by review articles (Dervichian, 1964; Chapman, 1965; Reiss-Husson and Luzzati, 1967). An effort is made here to present the data as part of a coherent picture, embodying simple chemical compounds, like the saturated soaps, as well as the complex mixtures extracted from biological preparations.

The class of substances usually called lipids is poorly defined because no international system of nomenclature has been adopted. In this article a lipid is considered to be a molecule formed by a hydrophilic and a lipophilic moiety, the two linked together by bonds sufficiently flexible to yield a rather independent behaviour (these molecules are sometimes called amphiphiles).

In all the cases discussed hereunder the lipophilic moiety is formed by one or two paraffin chains, of various length and degree of unsaturation; the hydrophilic groups are more heterogeneous.

The results will be presented in three main sections, one devoted to the discussion of the general properties common to all the systems, one describing the soaps and detergents and the third describing the lipids of biological interest.

II. Structure Analysis

A. PHASE DIAGRAMS

The first step in the structure analysis of multicomponent systems is to characterize the different phases and to determine their range of existence: in other words, to construct the phase diagrams.

The phase diagrams of lipid-water systems, as a function of temperature and concentration (the usual variables in the X-ray study), obey simple rules, provided that the lipid is a pure chemical species and that the number of components is only two. Lipid mixtures as complex as those extracted from biological membranes could be expected to lead to intricate phase diagrams as the number of components is large. In fact extensive regions of the phase diagrams are as simple for these mixtures as for pure compounds. Several phase diagrams will be presented in this paper; the analogies and the differences between pure compounds and mixtures will be discussed.

B. GENERAL PROPERTIES AND CLASSIFICATION OF STRUCTURES

1. *Short Range Organization: Conformation of the Paraffin Chains*

In the different structures described in this chapter the paraffin chains take up a variety of conformations. We describe here the main types of these conformations and their characterization by X-ray diffraction methods. The spectroscopic analysis of the conformation of the paraffin chains is dealt with in another chapter of this book.

Three-dimensionally ordered conformation of the paraffin chains is observed in the crystals of pure lipids, for example saturated soaps. The lipid molecules are organized in planar double layers; each layer, covered by the polar groups, contains the stiff paraffin chains, often tilted with respect to the normal to the plane (Fig. 1). The X-ray diffraction diagrams contain a few reflections at high spacings and a large number of reflections at spacing smaller than 5 Å.

A liquid-like conformation is observed at high temperature or under the combined effect of water and temperature. This conformation is characterized, in the X-ray diffraction diagrams, by the presence of a diffuse band, around 4·5 Å, identical to that of a liquid paraffin (Fig. 2). It is clear, nevertheless, that although the short range organization is similar to that of a liquid paraffin, the disorder is not complete; for example the average orientation of the chains is perpendicular to the interface, as a consequence of the fact that one end is anchored at the interface. This problem is discussed in more detail in Appendix 1. In some cases a modulation of the intensity of the 4·5 Å band is observed, showing that the movements of the chains are hindered. All the phases in which the paraffin chains take up such disordered conformation will be called *liquid-paraffin phases*.

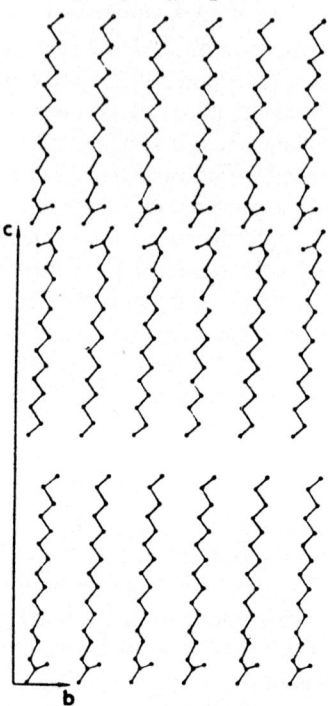

FIG. 1. The A' form of n-pentadecanoic acid. Projection along the shortest axis on the largest axis plane (from E. von Sydow, 1956).

Stiff free rotating paraffin chains are observed in several low temperature phases. The typical X-ray diffraction observation is one sharp reflection of spacing close to 4·2 Å that was interpreted by Müller (1932) as the first reflection of a two dimensional hexagonal lattice; Gulik-Krzywicki, *et al.* (1967) have confirmed the symmetry of the lattice by the observation of two weak reflections at $(4·2/\sqrt{3})$ Å and at $(4·2/2)$ Å. The paraffin chains, stiff and fully extended, are packed in a hexagonal array with random orientation around their axis. In some cases a small number of sharp reflections are observed in the 4·5 Å region; these may be interpreted as if the rotation of the chains were hindered and the two dimensional lattice became of lower symmetry.

Intermediate cases are observed as well, namely phases in which the conformation of the paraffins varies along the length of the chains (see Section V-D).

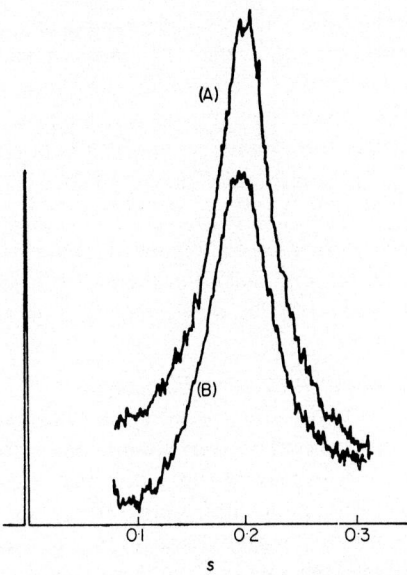

Fig. 2. Microdensitometer trace of the 4·5 Å band. (A) tetradecane, 100°C; (B) C_{14}Na-water, $c=0.68$, 100°C (from Luzzati et al., 1960).

2. Long Range Organization: Lattice Type and Symmetry

All the phases described in this article display different degrees of order in different parts of the structure. The conformation of the paraffin chains (and of the water molecules) is disordered; the long-range organization may be either of the liquid type (micellar solutions) or strictly periodic in one, two or three dimensions. The crystallographic properties of the periodically ordered phases will be described in this chapter; the structure of the micellar solutions is discussed in Section IV-D.

The periodically ordered structures may be grouped into three classes according as the periodic repeat is in *one*, *two* or *three* dimensions. Following Friedel's nomenclature (Friedel, 1922) the first two classes are *mesomorphic* (or liquid-crystalline), the third is *crystalline*. The structures may be further classified according to the symmetry of the two- and three-dimensional lattices.

Periodic in one dimension; lamellar. The lipid molecules are associated in lamellae. Each lamella is filled by the paraffin chains and is covered on both sides by the hydrophilic groups (Figs 4 and 9). The structure is of the smectic type: the lamellae are parallel and equidistant, without other correlation in position and orientation.

Periodic in two dimensions. The structure elements are long and rigid prismatic rods, all parallel to each other. The lateral position and the relative orientation of the rods are fixed; the positions are random in the direction of the rod axis. The projection on the plane perpendicular to the rod axis defines a two-dimensional lattice, the symmetry of which is bound to belong to one of a few space groups (International Tables, 1952).

Periodic in the three dimensions. Aggregates of finite size, formed either by lipid or by water molecules, are packed in three-dimensional lattices. Only a few types of such lattices will be mentioned in this article.

3. Size and Content of the Structure Elements

Besides the previous classification, based upon symmetry, other categories can be defined when the molecular structure is taken into account.

Simple and *complex* structures are those in which the structure elements, formed by aggregates of lipid molecules, either take up a simple form (lamella, cylinder, . . .) or are more complex associations of various elements. The difference is clearly illustrated in Fig. 20. These two types of structures are easily characterized when the dimensions of the structure elements are known (see below).

Type I (paraffin-in-water) and *type II* (water-in-paraffin) are the structures in which the rods (or the globular particles) are either filled by the paraffin chains and embedded in a polar matrix, or the other way around (see Figs 9A and 9C). The distinction between the two can only be made for structures in which the interior and the exterior volumes can be distinguished topologically. Furthermore the distinction cannot be based upon the X-ray data alone (as a consequence of Babinet's principle); other lines of evidence, involving chemical information, have to be used for that purpose.

C. DETERMINATION OF LATTICE DIMENSIONS AND OF OTHER STRUCTURE PARAMETERS

From the experimental standpoint the data consist of a set of reflections characteristic of one phase. As the X-ray experiments usually are carried out on unoriented samples, only the spacings and the intensities of the reflections are known. The problem is to index the reflections and to determine the symmetry of the lattice and the structure of the phase. (The cubic phases are easily characterized on the polarizing microscope, as they are optically isotropic, while all the others are spontaneously birefringent.)

The equations that define the spacings of the reflections for the different symmetry systems are given in Table I. The symmetry of the lattice is determined by finding out one equation with which all the observed spacings agree; this problem is generally easy to solve in the lipid-water phases. When the symmetry is found, the dimensions of the unit cell can be calculated.

TABLE I

Equations relevant to the different symmetry systems

s_{hkl} reciprocal spacing (Å^{-1}) of the reflection of indices h, k and l.
a^*, b^*, γ dimensions of the reciprocal unit cell
d   repeat distance of the lamellar phase
σ area of the *primitive* two-dimensional cell
V   volume of the *primitive* three-dimensional cell

PERIODIC IN ONE DIMENSION: LAMELLAR
$$s_l = l/d$$

PERIODIC IN TWO DIMENSIONS

oblique
$s^2_{hk} = h^2 a^{*2} + k^2 b^{*2} - 2hk a^* b^* \sin \gamma$
$\sigma = (a^* b^* \sin \gamma)^{-1}$

rectangular primitive
$s^2_{hk} = h^2 a^{*2} + k^2 b^{*2}$
$\sigma = (a^* b^*)^{-1}$

rectangular centred
$s^2_{hk} = h^2 a^{*2} + k^2 b^{*2}$ $(h + k = 2n)$
$\sigma = (2a^* b^*)^{-1}$

hexagonal primitive
$s^2_{hk} = a^{*2} (h^2 + k^2 - hk)$
$\sigma = (2/\sqrt{3}) a^{*-2}$

PERIODIC IN THREE DIMENSIONS (see International Tables, 1952)

If in addition the concentration, the density and the molecular weight of the lipid are known, the number of lipid molecules per unit cell can be determined (parameters N_1, N_2 and N_3, see Table II).

In order to proceed further in the structure analysis, a few hypotheses have to be made and some principles must be respected. One hypothesis, already made tacitly, is that the paraffin chains are clustered into regions from which water is excluded and that the interface between water and paraffin is covered by the hydrophilic groups of the lipid molecules. Even in the anhydrous phases the hydrophilic groups are supposed to segregate out from the paraffinic regions by analogy to what is known to occur in the crystals. The average area S available to one hydrophilic group on the interface is thus a parameter that can be defined in all the phases.

It will be shown below that in many of the phases, especially among those that contain water, the paraffin and water regions, as well as the interface, are devoid of internal rigidity. Under these conditions the shape of the structure elements is bound to be defined by the lattice interactions. For the sake of simplicity it will be often assumed that the symmetry of the structure

TABLE II
Determination of the structure parameters

$\mathcal{N}$	Avogrado number
M	molecular weight of the lipid
$\bar{v}_l$ and $\bar{v}_w$	partial specific volumes of lipid and water (cm^{3}g^{-1}), supposed independent of concentration
c	weight concentration (lipid/lipid + water)
$\phi = [1 + \bar{v}_w(1 - c)/\bar{v}_l c]^{-1}$	volume concentration
S	average area (Å^2) available to one hydrophilic group on the lipid-water interface
$N_1 = d\phi\mathcal{N}10^{-24}/M\bar{v}_l$	number of lipid molecules per unit surface of one lamella
$N_2 = \sigma\phi\mathcal{N}10^{-24}/M\bar{v}_l$	number of lipid molecules per unit length of one primitive two-dimensional cell.
$N_3 = V\phi\mathcal{N}10^{-24}/M\bar{v}_l$	number of lipid molecules per primitive three-dimensional cell.
	Lamellae
$d_l = d\phi$	thickness of the uniform lipid layer containing all the lipids of one cell.
$S = 2/N_1$	
	Cylinders
$r_\mathrm{I} = (\sigma\phi/\pi)^{1/2}$	radius of the lipid cylinder if the structure is of type I (paraffin-in-water).
$S_\mathrm{I} = 2\pi r_\mathrm{I}/N_2$	
$r_\mathrm{II} = [\sigma(1 - \phi)/\pi]^{1/2}$	radius of the water cylinders, if the structure is of type II (water-in-paraffin).
$S_\mathrm{II} = 2\pi r_\mathrm{II}/N_2$	

elements is higher than the symmetry of the lattice: for example circular cylinders in the hexagonal system.

When these assumptions are made, the thickness of the lipid layer, the diameter of the cylinders and the area S can be determined for structures of both types I and II (see Table II).

In other cases the shape of the structure elements is not fully defined by the symmetry of the lattice and has to be determined by the analysis of the intensities of the reflections.

III. Methods

A. NOTATION AND ABBREVIATIONS

The symbols and the notation are given in Tables I and II.

The chemical formulae of the lipids discussed in this chapter, and the abbreviations used, are the following.

Soaps and detergents

C_nX: saturated soaps (X is a monovalent cation): $CH_3.(CH_2)_{n-2}.COOX$
OX: oleates: $CH_3.(CH_2)_7.CH=CH.(CH_2)_7.COOX$
SLS: sodium laurylsulphate: $CH_3.(CH_2)_{11}.OSO_3.Na$
Aerosol MA: $CH_3.(CH_2)_5.OCO.CH_2$
$\quad\quad\quad\quad\quad\quad\quad |$
$\quad CH_3.(CH_2)_5.OCO.CH.SO_3Na$
C_n TAB (and C_nTACl): alkyltrimethylammonium bromide (and chloride):
$$CH_3.(CH_2)_{n-1}.\bar{N}.(CH_3)_3.Br^-$$
$TAC_{12}I$: trimethylamino dodecanoimide: $CH_3.(CH_2)_{10}.CO.\bar{N}.\overset{+}{N}.(CH_3)_3$
Arkopal n: $CH_3.(CH_2)_8.\langle\bigcirc\rangle.O.(CH_2.CH_2.O)_{n-1}.CH_2.CH_2.OH$

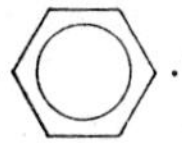

Lipids of biological interest (R_1 etc. are the paraffin chains)

Monoglycerides:
$\quad R_1.CO.OCH_2$
$\quad\quad\quad\quad\,\, |$
$\quad\quad\quad CH.OH$
$\quad\quad\quad\quad\,\, |$
$\quad\quad\quad CH_2.OH$

Lecithin: (Phosphatidyl choline)
$\quad R_1.CO.OCH_2$
$\quad\quad\quad\quad\quad\,\, |$
$\quad R_2.CO.OCH \quad\quad O$
$\quad\quad\quad\quad\quad |\quad\quad\quad \uparrow$
$\quad\quad\quad\quad CH_2\!\!-\!\!-\!\!OPOCH_2.CH_2.\overset{+}{N}.(CH_3)_3$
$\quad\quad\quad\quad\quad\quad\quad\quad\, |$
$\quad\quad\quad\quad\quad\quad\quad\quad O^-$

Lysolecithin:
$\quad R_1.CO.OCH_2$
$\quad\quad\quad\quad\quad |$
$\quad\quad\quad CH.OH \quad\quad O$
$\quad\quad\quad\quad\,\, |\quad\quad\quad\, \uparrow$
$\quad\quad\quad CH_2.\!\!-\!\!-\!\!OPOCH_2.CH_2.N^+.(CH_3)_3$
$\quad\quad\quad\quad\quad\quad\quad\quad |$
$\quad\quad\quad\quad\quad\quad\quad\quad O^-$

Phosphatidylethanolamine:

```
R₁.CO.OCH₂
     |
R₂.CO.OCH        O
     |           ↑
    CH₂.———OPOCH₂.CH₂.⁺NH₃
                 |
                 O⁻
```

Phosphatidylinositol:

```
R₁.CO.OCH₂
     |                    OH  H
R₂.CO.OCH        O                H
     |           ↑         H  HO
    CH₂.———OPO—          H  HO
                 |                OH
                OH       OH  H
```

Cardiolipids:

```
R₁.CO.OCH₂                CH₂———OPO———CH₂
     |                     |      |     |
R₂.CO.OCH        O        CH.OH   OH   CH.CO.O.R₃
     |           ↑         |            |
    CH₂.———OPO———CH₂              CH₂.CO.O.R₄
                 |
                OH
```

Sphingomyelin:

```
                                              O
                                              ↑
CH₃.(CH₂)₁₂.CH=CH.CH.CH.CH₂———OPOCH₂.CH₂.⁺N.(CH₃)₃
                    |   |         |
                   OH  NH         O⁻
                        |
                       CO.R₁
```

Cerebroside:

```
                                              H
                                              |
CH₃.(CH₂)₁₂.CH=CH.CH.———CH.CH₂.O———C———
                    |    |               |
                   OH   NH              HCOH
                        |                |
                       CO.R₁            HOCH    O
                                         |      |
                                        HOCH
                                         |
                                        HC————
                                         |
                                        CH₂.OH
```

B. X-RAY SCATTERING TECHNIQUES

The experimental techniques used for the X-ray scattering study of lipid-water systems must fulfil a few requirements: (a) the chemical composition of the sample, especially the water content, must be kept under control; (b) the experiments must be carried out at controlled temperature; (c) the small-angle region of the X-ray diagrams must be explored as the dimensions of the structure elements are fairly large.

The conventional X-ray cameras generally do not satisfy these conditions; most of the experiments described in this article were carried out with cameras of special design. In the author's laboratory Guinier-type cameras were used, operating *in vacuo*. The Cu $K\alpha_1$ line (in some cases the W $L\alpha_1$ line) is isolated and focused by a bent quartz monochromator.

The samples are held in a vacuum tight cylindrical cell provided with thin mica windows. The sample holder is kept at constant temperature either by circulation of a thermally-controlled liquid or by electric heating. The samples are prepared at the desired concentrations. The concentration is often checked by dry weight determinations, or by direct chemical analysis, at the end of the X-ray experiment.

C. PARTIAL SPECIFIC VOLUMES

The partial specific volume of the components (lipids and water) is an important parameter for the determination of the dimensions of the structure elements (see Tables I and II). In many of the systems this was determined experimentally, in others it was estimated by reference to similar compounds. The figures, reported in the original papers, will not be given in this chapter.

In all the multicomponent systems the partial specific volume of each of the components is assumed to be independent of the composition of the system.

IV. Simple Lipids: Soaps and Detergents

Soaps, the salts of the saturated fatty acids, are the simplest type of lipid easily available as pure chemical compounds. Their properties have been studied extensively and more is known about the structure of the soap-water systems than about any other lipid.

The phase diagrams of several of these systems have been established by McBain and coworkers (McBain and Lee, 1943); as an example the phase diagram of $C_{16}K$-water is given in Fig. 3. In the low temperature and high lipid region of the diagram (below the line T_c) the so called *gel* and *coagel* are found; the common feature of these phases is the ordered, or partially ordered, conformation of the paraffin chains. At high temperature and low lipid concentration (above the line T_i) the *isotropic solution* is observed, which is transparent, optically isotropic and of low viscosity; the soap molecules are dispersed in water as a molecular or micellar solution. The region between the gel-coagel and the isotropic solution contains several

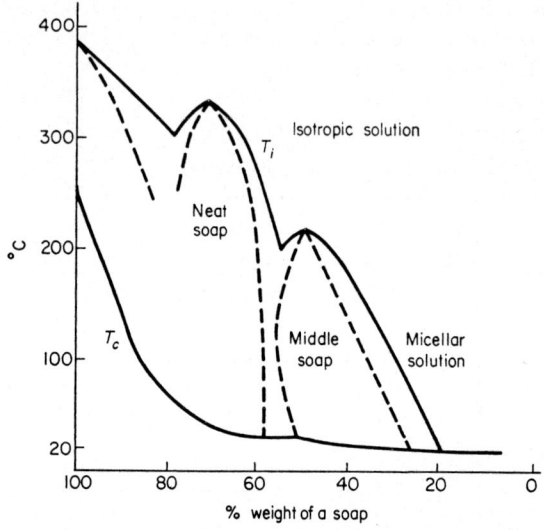

FIG. 3. The phase diagram of the system $C_{16}K$-water (from McBain and Lee, 1943).

phases, the common features of which are the spontaneous birefringence (with the exception of the cubic phases), the high viscosity, and the presence, in the X-ray diffraction diagrams, of sharp small-angle reflections and of a broad band near 4·5 Å. All these are *liquid-paraffin phases*.

A. LIQUID-PARAFFIN ANHYDROUS PHASES

Numerous phase transitions are known to occur in anhydrous soaps in the wide temperature gap between the collapse of the three-dimensional crystalline structure and the complete melting. These transitions have been investigated in the past by a variety of techniques. A thorough X-ray diffraction analysis of the different phases has been carried out over the last few years. The results are too extensive to be fully reported here; the structure of the most common phases will be described and some of the features of the phase transitions will be discussed.

1. *Structure of the Phases*

Lamellar, the structure of which is described above, is represented in Fig. 4. The lamellae are filled by the paraffin chains in a liquid conformation. The various structural parameters are determined using the equations given in Tables I and II.

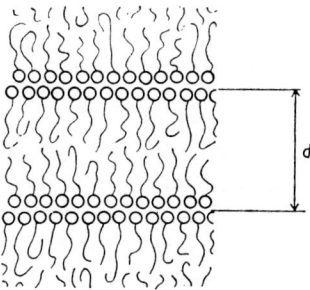

FIG. 4. Structure of the lamellar phase of anhydrous soaps: schematic representation of a section perpendicular to the lamellae. The polar group is represented by a circle, the paraffin chains by a wriggle (from Skoulios and Luzzati, 1961).

Oblique and rectangular. These phases are characterized by a large number of reflections (more than 30 in some cases) that can all be indexed in a two-dimensional lattice, either oblique or rectangular centred (see Table I). The symmetry and the dimensions of the unit cells, and the intensities of the reflections, are all consistent with structures formed by ribbon-like elements, indefinitely long in one direction, of finite width and thickness (Fig. 5). The ribbons contain the polar groups of the molecules; the disordered paraffin chains fill up the gap between the ribbons. When the cell is rectangular the ribbons are highly symmetric (two mirror planes); if the cell is oblique the symmetry is lower (one two-fold axis). For each phase the number N_2 of soap molecules per unit length of the ribbons can be determined (see Tables

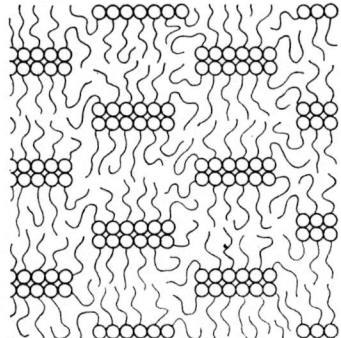

FIG. 5. Structure of the two-dimensional rectangular-centred phase of anhydrous soaps; cross-section of the prisms (see legend of Fig. 4) (from Skoulios and Luzzati, 1961).

I and II). When in addition the width of the ribbons is known, the area S can be calculated.

Hexagonal. The structure is formed by rods, containing the polar groups of the molecules, surrounded by the disordered paraffin chains and packed in a two-dimensional hexagonal array (Fig. 8A). The number N_2 can be determined in this case as well (Table I and II).

Orthorhombic. In accordance with the symmetry and the dimensions of the lattice, and with the intensities of the reflections (Skoulios, 1959), the structure of these phases appears to be formed by planar discs each containing the polar groups and embedded in a paraffin matrix (Fig. 6). The symmetry of the discs is likely to be mmm (or 222).

Body-centred tetragonal. This phase, and the following ones, are characterized by a large number of sharp reflections that can all be indexed on a three-dimensional lattice. The symmetry and the dimensions of the lattice, as well as the intensities of the reflections, are consistent with a structure formed of rods, similar to those of the hexagonal phase, but of finite length; the rods, all identical and crystallographically equivalent, are joined four by four and form square two-dimensional networks (Luzzati *et al.*, 1968 a). The networks are orderly stacked in the body-centred tetragonal lattice, space group I422 (Fig. 8B).

Rhombohedral. A crystallographic analysis (Luzzati *et al.*, 1968 a) provides strong arguments in favour of a structure analogous to that of the body-centred tetragonal phase, formed of identical rods, that join three by three to form two-dimensional hexagonal networks (Fig. 8C). The space group is R3̄m.

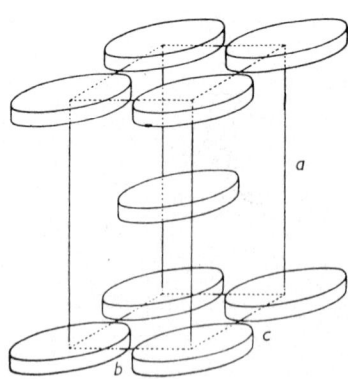

FIG. 6. Structure of the three-dimensional orthorhombic body-centred phase. The discs are the locus of the polar groups (from Skoulios and Luzzati, 1961).

Body-centred cubic, also formed by rods (Luzzati and Spegt, 1967). The rods join three by three to form two three-dimensional networks, mutually interwoven, otherwise unconnected (Fig. 8D). Space group Ia3d.

2. *Monovalent Cation Soaps*

The structure of the different high-temperature phases of several Li, Na, K, Rb and Cs soaps has been studied by X-ray techniques (Skoulios and Luzzati, 1961; Gallot and Skoulios, 1962; 1966b, c, d). Only a few examples will be discussed.

Na soaps. The high-temperature phases of the anhydrous Na soaps, from C_{12} to C_{18}, have been identified by a variety of techniques and are usually named as shown in Table III (Vold and Vold, 1939; Vold, 1941; Benton *et al.*, 1955). With the exception of the subwaxy (not reported here, see Skoulios and Luzzati, 1961), all the phases are in stable thermodynamic equilibrium and are separated from the neighbouring phases by sharp transitions. The structure and the lattice dimensions are given in Table III. All the phases are common to all the soaps with the exception of the orthorhombic which is observed in C_{12} only.

The structure of the phases waxy, superwaxy and subneat is similar, although the parameters differ. It may be noted that in each phase the number N_2 is nearly the same for all the soaps and that for one soap it decreases stepwise as the temperature rises. The width and the thickness of the ribbons can be estimated on the electron density charts and thus S can be determined, assuming that the ribbons are formed by two layers of polar groups (Fig. 5); the accuracy of these determinations is rather poor. It turns out (Table III) that S is almost the same in all the rectangular centred phases of the different soaps and that its value is similar to that of crystalline soaps.

The lamellar is the only phase whose parameters are highly dependent on temperature: d decreases continuously as the temperature rises (Fig. 7, from J to K). This observation, taken along with the fact that S is much larger than in the rectangular centred phases (Table III), indicates that the arrangement of the polar groups in the plane is disordered and provides a confirmation of the liquid structure of the paraffin chains (Appendix 1).

Other soaps. Phases with two-dimensional periodic order, similar to those of the Na soaps, are observed in some of the Li, K and Rb soaps. The lattice is rectangular centred for Li, oblique for K (Gallot and Skoulios, 1962; 1966b). As an example, the sequence of the phases of $C_{18}K$ is shown in Fig. 7. The crystalline organization of the paraffin chains breaks down at 170°C. From 170 to 272°C five oblique phases are found, the parameters of which are given in Table IV.

Table III
High-temperature phases of anhydrous Na soaps

Name of the phase	Lattice symmetry		C_{12}	C_{14}	C_{16}	C_{18}
Waxy	rectangular, centred	t (°C)	142	142	140	133
		a (Å)	75·2	80·0	86·6	80·0
		b (Å)	30·9	34·5	38·3	40·3
		N_2 (molec/Å)	2·83	2·87	3·14	2·88
		L (Å)	34	33	36	36
		S (Å²)	24	23	23	25
Superwaxy	rectangular, centred	t	183	182	176	175
		a	68·5	69·0	74·2	72·5
		b	30·3	33·6	36·0	37·9
		N_2	2·53	2·50	2·53	2·46
		L		30		
		S		24		
	orthorhombic, body-centred	t	200	210	211	210
		a	55·5			
		b	28·3			
		c	32·7			
Subneat	rectangular, centred	t	215	210	211	210
		a	49·8	53·8	56·2	62·4
		b	27·0	29·1	30·6	34·4
		N_2	1·64	1·58	1·67	1·82
		L		19	20	20
		S		24	24	22
Neat	lamellar	t	252	248	254	256
		t_1	290	271	278	285
		d	25·3	27·5	29·1	30·6
		S	36	38	40	42
		t	325	305	300	290

The temperatures of the phase-transitions are given (t). Within each phase the lattice dimensions are independent of temperature, with the exception of the lamellar phase: t_1 is the temperature for which the dimensions are given. The densities were not determined directly; thus N_2 is not very precise. L (and as a consequence S) were estimated on the electron density maps. See notation in Table I and II. (From Skoulios and Luzzati, 1961).

Table IV
High-temperature oblique phases of anhydrous C_{18}K

t (°C)	70	185	210	225	238	272
a (Å)	80·0	69·0	50·6	48·3	45·5	
b (Å)	43·9	42·2	41·2	40·8	40·0	
γ (°)	113	114	116	118	119	
N_2 (molec/Å)	5·45	4·40	3·12	2·83	2·50	

(From Gallot and Skoulios, 1962).
t is the temperature of the phase transitions.

Lamellar phases are encountered in the soaps of various cations (Fig. 7), but not of Li.

3. Divalent Cation Soaps

A large number of anhydrous soaps of the divalent cations of group II have been studied by Spegt and Skoulios (1963, 1964, 1966) and by Spegt (1964), as a function of temperature, using X-ray scattering techniques. A

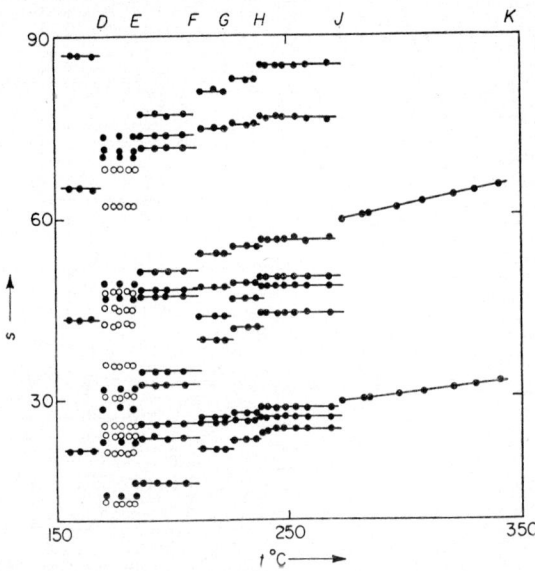

FIG. 7. High-temperature phases of anhydrous $C_{18}K$. Spacings s (unit 10^{-0} Å $^{-1}$) as a function of temperature. The various phases are easily characterized. Note that only in the lamellar phase (from J to K) s varies with temperature. (From Gallot and Skoulios, 1962).

few examples of the phase diagrams established by these authors are shown in Table V. The low temperature phases C_1 and C_2 are lamellar with ordered paraffin chains; the conformation of the chains is disordered in all the other phases.

The structure of the phases H, T, R and Q is discussed hereabove, and is shown in Fig. 8. Some of the structure parameters of the calcium and strontium soaps are given in Tables VI and VII. It may be noted that for all the soaps of the same cation, the number of cations per unit length of rod is constant for each phase; furthermore the differences between the different phases are not large (see also Spegt and Skoulios, 1963, 1964 and Spegt, 1964).

4. Discussion

An interesting result emerging from the variety of structures displayed by the anhydrous soaps is that one can construct a unified picture embodying

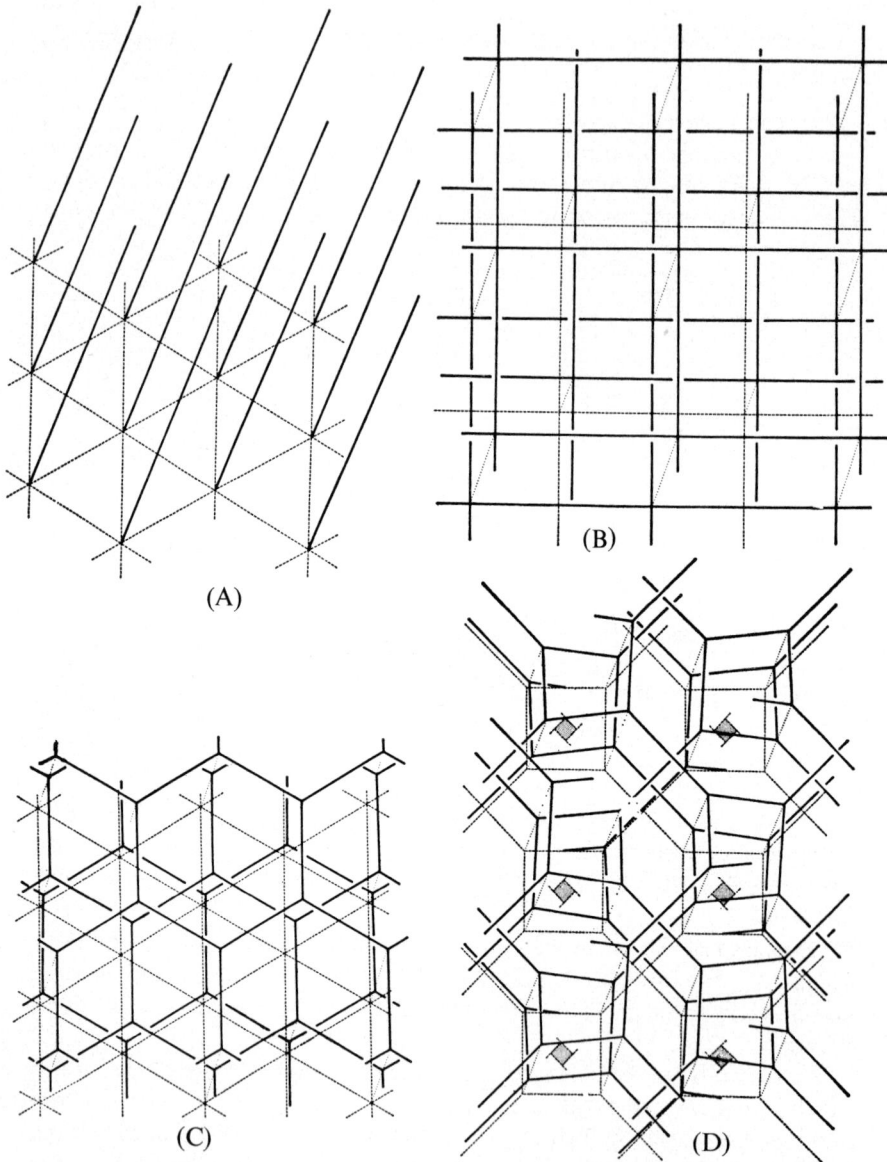

FIG. 8. Structure of the rod-like phases. The rods contain the polar groups; the heavy lines represent the rod axes. The dotted lines show the limits of the cells in projection. (A) *Phase H*: two-dimensional hexagonal array of indefinitely long parallel rods (space group p6). (B) *Phase T*: the rods, linked four by four, form planar two-dimensional square networks; the networks are orderly stacked in a three-dimensional lattice (space group I422). (C) *Phase R*: the rods, linked three by three, form planar two-dimensional hexagonal networks (space group R3̄m). (D) *Phase Q*: the rods, linked three by three, form two interwoven three-dimensional networks (space group Ia3d). (From Luzzati and Spegt, 1967, and Luzzati et al., 1968a).

TABLE V

Phase diagrams of the anhydrous stearates of the cations of group II

Mg	C_1 or C_2 (109°) H_1 (195°) H_2 (210°) melt
Ca	C_1 (100°) C_2 (123°) T (152°) ..... (179°) H (> 350°)
Zn	C_1 (130°) melt
Sr	C_1 (130°) ..... (170°) R (197°) Q (246°) H (> 400°)
Cd	C_1 (99°) H_1 (211°) H_2 (~230°) melt
Ba	C_1 (150°) ..... (220°) Q (>400°)

The transition temperatures are given in parentheses. The dotted lines show regions in which the phases have not been fully characterized. C_1 and C_2: crystalline. H, H_1 and H_2: two-dimensional hexagonal. T: body-centred tetragonal. R: rhombohedral. Q: body-centred cubic. (see Fig. 8). The phase diagrams of other soaps is similar to stearates. (From Spegt, 1964 and Spegt and Skoulios, 1963, 1964, 1966).

TABLE VI
Some data on the calcium soaps

	Phase T						Phase H	
	t °C	δ g.cm^{-3}	a Å	c Å	N_3	N_2	t °C	N_2
laurate							240	0.52_9
myristate	165	0.89_0	31.7	55.0	30.1	0.58_9	240	0.52_9
palmitate	159	0.88_5	34.1	59.1	33.2	0.59_0	240	0.53_4
stearate	152	0.88_0	35.9	62.2	35.1	0.58_7	240	0.53_4
arachidate	152	0.87_5	38.3	66.4	38.7	0.59_5	240	0.53_4

N_2 is the number of cations per unit length of rod. The data of phase H are taken from Spegt and Skoulios (1964), those of phase T from Luzzati et al. (1968a).

TABLE VII
Some data on the strontium soaps

	Phase R						Phase Q			Phase H	
	t °C	δ g.cm^{-3}	a Å	c Å	N_3	N_2	t °C	N_3	N_2	t °C	N_2
Laurate	180	0.91_0	29.0	78.0	21.4	0.52_4	235	215	0.51_1	265	
Myristate	180	0.89_3	31.0	84.8	23.3	0.52_5	235	233	0.51_0	265	
Palmitate							235	248	0.50_9	265	0.49_6
Sterate	181	0.87_2	36.0	93.3	28.0	0.52_8	235	261	0.50_7	265	
Arachidate	200	0.85_0	37.9	97.5	29.3	0.51_1	235	278	0.50_6	265	
Behenate							235	285	0.50_0	265	

N_2 as in Table VI. The data of phase H are taken from Spegt and Skoulios (1966), those of phase Q from Luzzati and Spegt (1967), those of phase R from Luzzati et al. (1968a).

them all. Indeed all the structures can be assorted into two classes that may be named "lamellar" and "rod-like". In the structures of the first of these classes the polar groups are localized in lamellar regions: the lamellae may be in the form of infinite planar sheets (Fig. 4), infinitely long ribbons (Fig. 5) or discs of finite size (Fig. 6). In the structures of the other class the polar groups are clustered in rod-like regions, and the rods, all identical and crystallographically equivalent are either infinitely long (one-dimensional network, phase H), or of finite length linked to form two- (phases T and R) and three-dimensional (phase Q) networks (Fig. 8). All the high-temperature phases of each of the anhydrous soaps studied so far belong to one or the other of the two classes, according to the nature of the cations: lamellar for the monovalent cations, rod-like for the divalent cations.

The highly developed organization of all these phases is achieved in spite of a great disorder of the paraffin chains; it is thus clear that the correlation between the chemical composition of the lipid and the class of structure must be sought in the organization of the polar groups. Although little direct evidence is available on this question, various indirect arguments suggest that the degree of order is quite high in all the phases, with the exception of the high-temperature lamellar (neat) phase of the monovalent cation soaps:

(a) The sequence of the phases is specific for the nature of the cation.

(b) In the ribbons the area S is similar to that of the crystalline soaps; in the lamellae S is much larger.

(c) The dimensions of the structure elements, and more specifically of S, are temperature dependent in the high temperature lamellae, independent in the discs, ribbons and rods.

(d) The number of polar groups per rod length, in each of the rod-like phases, is specific for the cations and is independent of temperature and of paraffin chain length.

The finite width of the ribbons can similarly be explained by the presence of both the crystalline arrangement of the polar groups and the disordered conformation of the paraffin chains. Indeed, as the paraffin chains are anchored to the polar groups that sit at fixed positions and closely packed on the surface of the ribbons, the mobility of the chains is very small near the polar groups and increases along the chains towards the CH_3 end. So the *average* orientation of the chains is fan-wise and only at some distance from the centre of the ribbon is the orientation sufficiently disordered to overtake the edge of the ribbon (Skoulios and Luzzati, 1961, see Fig. 5). As the temperature rises, the disorder increases and the width of the ribbons decreases; the change is necessarily discontinuous, one row at a time, as the number of rows of polar groups in one ribbon is quite small. Similar phenomena are likely to take place in the discs. The existence of oblique ribbons (and discs)

(Spegt, 1964) can be explained by similar arguments, assuming that the organization of the polar groups is so precisely defined that in the vicinity of the surface of the ribbons the orientation of the paraffin chains is fixed. A similar explanation can be given for the finite length of the rods (phases T, R, Q), assuming in this case that the neighbouring chains tend to keep closer than is allowed by the separation of the polar groups.

B. LIQUID-PARAFFIN WATER-CONTAINING PHASES: LONG-RANGE PERIODICALLY ORDERED PHASES

Several phases are present in the extended region of the phase diagram encompassed by the gel and the isotropic solution, delimitated in Fig. 3 by the lines T_c and T_i. Two of these phases are known under the names "neat" and "middle". A few other phases have been discovered recently which exist between these two phases.

In the past, several authors have carried out X-ray scattering study of these phases (Stauff, 1939; Doscher and Vold, 1948). Great emphasis was put on the lamellar structure; the interpretation was often biased by the *a priori* assumptions that all the structures are lamellar and that the paraffin chains take up a highly ordered conformation. Marsden and McBain (1948) reported the observation of a two-dimensional hexagonal phase, but did not appear to realize that the "middle" phase is always hexagonal.

The systematic structure study of these phases is recent (Luzzati *et al.*, 1957 and 1958; Luzzati *et al.*, 1960; Husson *et al.*, 1960; Clunie *et al.*, 1965; Gallot and Skoulios, 1966a); the main results will be reported and discussed here.

1. *Structure of the Phases*

Lamellar. This phase is similar to the lamellar phase of the anhydrous soaps, with layers of water intercalated between the lipid lamellae (Fig. 9B).

Hexagonal, type I (in short *hexagonal I*). This phase is a two-dimensional hexagonal array of rods. The analysis of the dimensions of the rods (assumed to be circular cylinders), for a variety of soaps and as a function of concentration, indicates that the structure is most likely of type I (Fig. 9A). Indeed in this case the area S turns out to be almost independent of the chain length and of concentration (Husson *et al.*, 1960). Gallot and Skoulios (1966a) have shown in fact that S is a function of the molal concentration of the polar groups in the water of the system, for all the soaps of the same cation (see Fig. 10B and discussion below). Furthermore the values of S of the hexagonal phase are found to be intermediate between those of the lamellar and the isotropic phases (Figs 10A and B and Table XI). On the contrary if the structure were of type II (Fig. 9C), S would display a complex dependence on

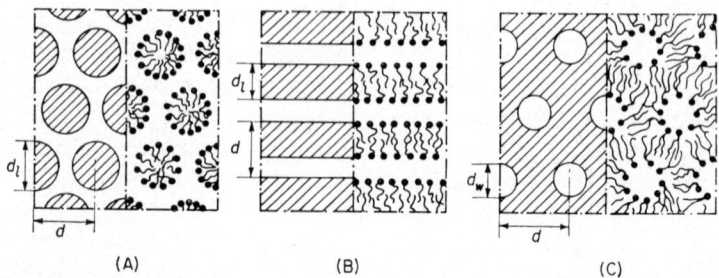

Fig. 9. Structure of some high-temperature phases of the lipid-water systems schematic representation of a cross-section of the rods and of the lamellae. The hydrophilic group is represented by a dot, the paraffin chains by a wriggle. (A) hexagonal I; (B) lamellar; (C) hexagonal II. (From Luzzati et al., 1966).

chain length and concentration (Fig. 10C). Moreover in this case the extreme values of S, for the hexagonal phase, would be lower than in the lamellar and higher than in the isotropic phases.

An additional confirmation of the structure of type I is provided by the analysis of the intensity of the reflections (Husson et al., 1960).

Complex hexagonal. The symmetry is that of the previous phase; the dimensions of the lattice are much larger and the intensities of the reflections are quite different. No simple structure can fit those data. Several models were tried (Luzzati et al., 1960; Husson et al., 1960); a cylindrical shell of

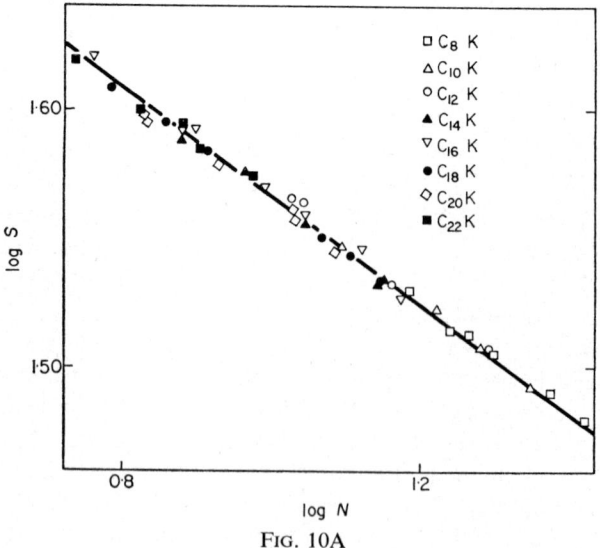

Fig. 10A

lipids with water filling the inner hole and the external gap between the cylinders was considered to be the most satisfactory.

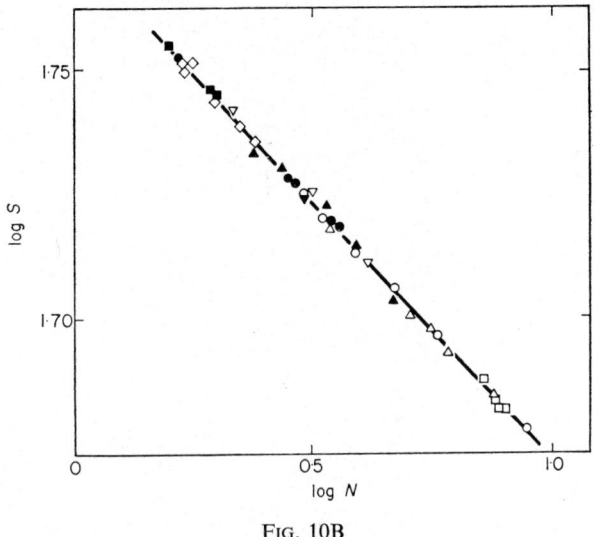

Fig. 10B

Fig. 10. Plot of the area S vs N (moles of polar groups/l of water) for several K soaps at 86°C. The straight lines represent the empirical equation $S=S_0N^{-p}$. (A) (facing page) lamellar phase; (B) hexagonal phase, type I; (C) (following page) hexagonal phase, type II (A and B from Gallot and Skoulios, 1966a; C calculated from the data of those authors).

Cubic. The number of X-ray reflections is too small to provide unequivocal support to any structure. At one time (Luzzati *et al.*, 1960), the structure was considered to be formed by lipid spheres surrounded by water; later a structure made of water spheres embedded in a paraffin matrix was considered to be more satisfactory (Luzzati and Reiss-Husson, 1966). These structures now appear quite questionable, after the structure of the cubic phase of the anhydrous soaps is known (Luzzati and Spegt, 1967; see Fig. 8D).

Other phases. These are often observed in soap-water systems. One, defined by two reflections, has been called *deformed hexagonal*; another, characterized by several reflections, all integral orders of two fundamental repeats, has been called *rectangular* (Luzzati *et al.*, 1960). The structure of these phases was not determined unambiguously.

2. *Results and Discussions*

The lipid-water phases of a variety of anionic, cationic and non-ionic soaps and detergents have been analysed by X-ray diffraction methods: some of the results are shown in Fig. 11 and in Table VIII. The six phases described above

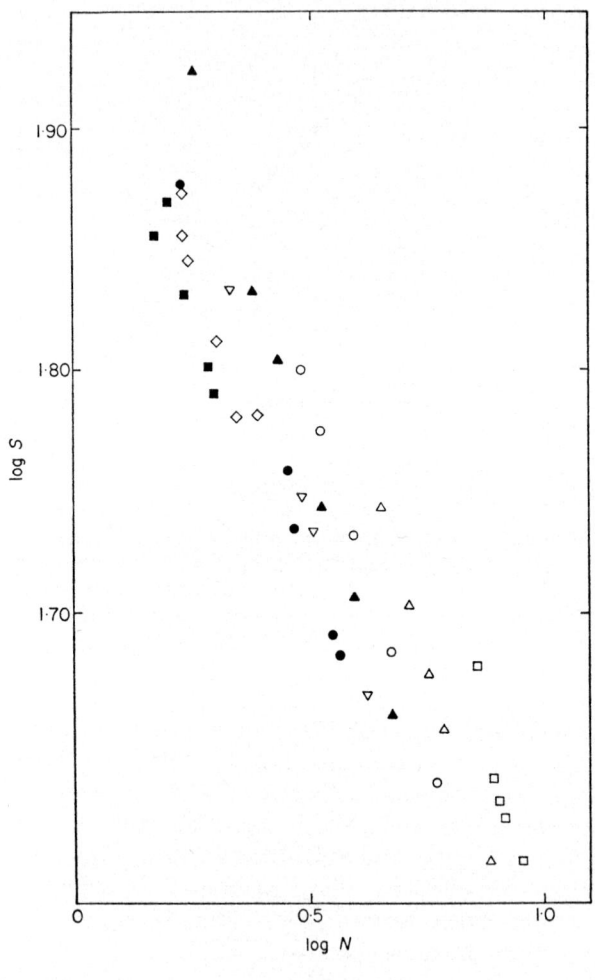

FIG. 10C

are common to all the systems. The hexagonal and the lamellar are the most frequently observed, often separated by one or more intermediate phases. The sequence of the phases, in the order of increasing concentration, is always hexagonal—deformed hexagonal—rectangular—complex hexagonal—cubic—lamellar (Table VIII); frequently gaps occur in this order but no example of inversion is observed.

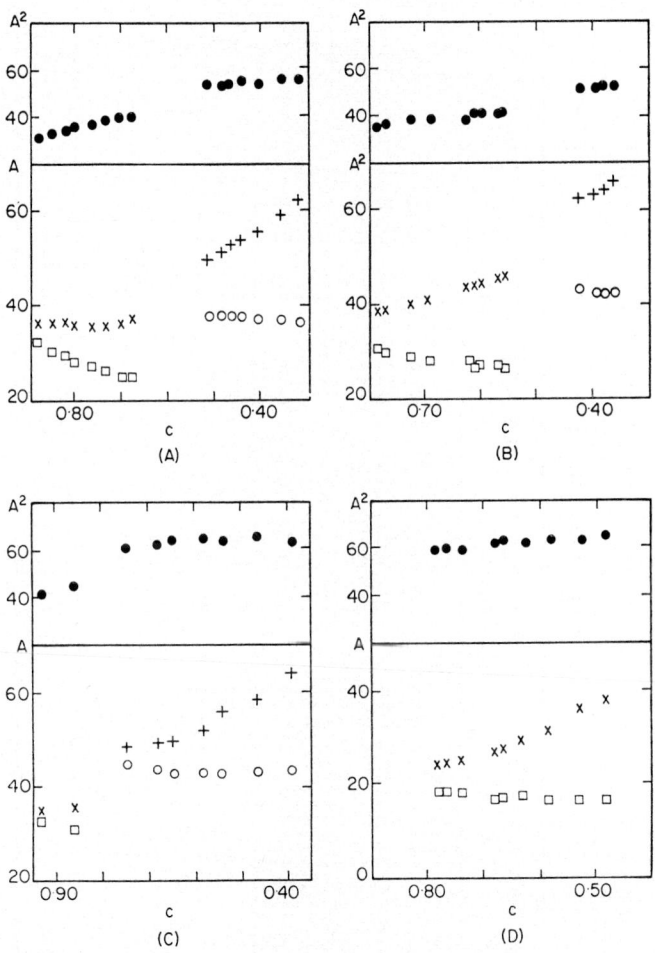

Fig. 11. Dimensions of the structural elements of the liquid-crystalline phases of some lipid-water systems. (A) $C_{16}K$, 100°C; (B) $C_{18}Na$, 100°C; (C) CTAB, 70°C; (D) Aerosol MA, 20°C.

●   average surface area per hydrophilic group
\+   distance between the cylinder axes, d ⎫ hexagonal phase, type I
○   diameter of the lipid cylinders, d_l ⎭ (see Fig. 9A)
×   repeat distance, d ⎫ lamellar phase
□   thickness of the lipid leaflet, d_l ⎭ (see Fig. 9B)

(From Luzzati and Husson, 1962).

TABLE VIII

Schematic representation of the phase diagrams of simple lipid-water systems. The concentrations of the phase boundaries are given. The phases existing in each system are shown by full lines.

Lipid	t(°C)	lamellar	cubic	complex hexagonal	rectangular	deformed hexagonal	isotropic	hexagonal
$C_{12}Na$	100	0.59	—	—	—	—	0.59	—
$C_{14}Na$	100	0.59	—	0.55	0.55	0.54	0.54	—
$C_{16}Na$	100	0.56	—	0.52	0.52	0.51	0.51	—
$C_{18}Na$	100	0.54	—	0.51	0.51	0.50	0.50	—
$C_{12}K$	100	0.69	0.61	—	—	—	0.61	—
$C_{14}K$	100	0.66	0.59	—	—	—	0.59	—
$C_{16}K$	100	0.65	0.59	0.55	0.55	0.54	0.54	—
$C_{18}K$	100	0.65	0.65	0.59	0.59	0.58	0.58	—
ONa	65	0.69	0.69	0.59	0.52	—	0.52	0.28
OK	20	0.72	0.72	0.68	0.60	—	0.60	0.21
SLS	75	0.69	0.69	0.62	—	—	0.62	0.38
MA	20	—	—	—	—	—	—	—
CTAB	70	0.84	0.78	—	—	—	0.78	0.38
$TAC_{12}I$	20	0.77	0.66	—	—	—	0.66	0.40
Arkopal 9	20	0.61	—	—	—	0.61	0.48	0.45
Arkopal 13	20	—	—	—	—	—	0.63	0.43

(From Luzzati and Husson, 1962; $TAC_{12}I$ from Clunie *et al.*, 1965).

Some features of the phase diagrams bear an obvious relation to the chemical structure of the lipid. The bulkier the hydrophilic end of the molecule, the more extended the hexagonal phase range, and reciprocally, the bulkier the hydrocarbon moiety, the more extended the lamellar phase range. The behaviour of the arkopals clearly illustrates this rule. From $n = 6$ to $n = 8$ the hydrophilic chain is relatively short and the only liquid–crystalline phase is lamellar; for $n = 10$ the hydrophilic moiety becomes large and only the hexagonal phase is present; for $n = 9$ both phases are observed. Another example is Aerosol MA in which two hydrocarbon chains join to one hydrophilic group: the only phase is lamellar. In CTAB, in which the hydrophilic moiety is fairly bulky, the hexagonal phase has a more extended concentration range than in soaps.

The factors that determine the range of existence of the intermediate phases are more obscure and do not seem to bear an obvious relation to the chemical structure of the lipids.

An inspection of the data (Figs 10 and 11) shows that for any soap and detergent, the area S decreases monotonically as the concentration increases, at least for the lamellar and the hexagonal phases in which S can be determined directly (and assuming that the hexagonal phase is of type I). We have extended the validity of this rule, at least tentatively, to the whole phase diagram.

In the hexagonal phase (Husson et al., 1960), the area S is almost independent of the concentration and of the length of the paraffin chains, whilst the concentration dependence is striking in the lamellar phase (Fig. 11). This phenomenon was explored more carefully by Gallot and Skoulios (1966a), who studied the lamellar and the hexagonal phases of a variety of Na, K, Rb and Cs soaps from C_8 to C_{22}. The results show that the area S, at constant temperature, in each phase and for one cation, is a function of the number of polar groups per volume of water irrespective of the chain length (Fig. 10A, B and Table IX). It appears, therefore, that the dimensions of the structure elements are determined by the interactions at the lipid-water interface, which in turn seem to be dependent on the molal concentration of hydrophilic groups. The conformation of the paraffin chains is sufficiently disordered to be considered to be similar to that of a liquid (see Appendix 2).

C. TRANSITION FROM THE LIQUID-PARAFFIN TO THE CRYSTALLINE PHASES: GEL AND COAGEL

If a soap-water sample, taken in one of the high-temperature phases, is cooled to a sufficiently low temperature (more precisely across the T_c line in the phase diagram, see Fig. 3), either a homogeneous and transparent preparation, the *gel*, is obtained or what appears to be a mixture of two phases, the *coagel*. The gel is often metastable and slowly transforms into a coagel.

Several authors (see review in Vincent and Skoulios, 1966a) have studied the coagel and have shown that it consists of crystalline soap (often in the hydrated form) and of water. We shall not be concerned here with the crystalline

TABLE IX
Area S in the hexagonal and in the lamellar phases of the soap-water systems

Cation	Hexagonal		Lamellar	
	S_0	p	S_0	p
Na	57·6	0·09$_5$	63·1	0·24
K	58·9	0·10$_5$	58·9	0·20
Rb	61·7	0·11	56·3	0·19
Cs	63·1	0·11$_5$	55·0	0·18

For all the soaps of the same cation, and in both the hexagonal and the lamellar phases, S is a function of the number N of moles of polar groups per litre of water, irrespective of the length of the paraffin chain (Fig. 10, A, B). The empirical equation has the form $S = S_0 N^{-p}$. The values of S_0 and p are given, for different cations, at 86°C. (From Gallot and Skoulios, 1966a).

forms of soaps (see review in Chapman, 1965); we shall devote our attention to the structure of the gel as studied by Vincent and Skoulios (1966a, b and c) for several K, Rb and Cs soaps.

1. Structure of the gel

The common feature of the X-ray diffraction diagrams of the gel and coagel is the presence of sharp reflections at spacings smaller than 5 Å and the absence of the diffuse 4·5 Å band. In the coagel the number of the high angle reflections is high; in the gel only one or two reflections are observed, at spacings near 4 Å (see below).

K soaps. The gel is a genuine phase in stable thermodynamic equilibrium over an extended region of the phase diagram: this is shown, for example, by the fact that the X-ray diffraction diagrams are independent of the previous thermal treatment of the sample. The equilibrium region is bordered, on the low temperature side, by a zone in which the structure is dependent on the history of the sample: here the gel is metastable and spontaneously gives way to a coagel.

Over an extended concentration range the X-ray diagrams of the gel are those of a pure lamellar phase, with several orders of a fundamental repeat; furthermore one sharp line is observed at 4·1 Å. Since the repeat distance varies with concentration, the phase appears to be pure and various parameters (see Tables I and II) can be determined. d_t and S turn out to be independent of concentration; the values are given in Table X. It may be noted that the thickness d_l of the lipid layer is close to the fully extended length of the soap molecule and that the thickness increment for one pair of CH_2 groups (2·5 Å) is identical to the end-to-end distance of the —CH_2—CH_2—CH_2—group of saturated paraffins. Furthermore, the surface of the hexagonal cell, defined by the 4·1 Å reflection is $\Sigma = 19·5$ Å^2 (see Section II-B1), exactly one half of the area S. A structure consistent with these observations is represented in Fig. 12; the paraffin chains are fully extended, interdigitating and hexagonally packed. More complex phase separation phenomena take place at high soap concentration: these will not be discussed here (see Vincent and Skoulios, 1966a and b).

TABLE X
The gel

	t °C	d_l Å	S Å^2
$C_{14}K$	0	20·0	39·7
$C_{16}K$	25	22·7	39·6
$C_{18}K$	25	25·2	38·8
$C_{22}K$	25	30·2	39·9
$C_{16}Rb$	25	23·6	39·6
$C_{18}Rb$	25	26·2	39·9

Note: The fully extended length of the soap molecules is approximately 18·5 Å for C_{14}, 21·0 Å for C_{16}, 23·5 Å for C_{18} and 28·5 Å for C_{22}. (From Vincent and Skoulios, 1966b).

Rb soaps. The properties of the gel and its structure are similar to those of the K soaps, with one interesting difference. At fairly low temperature, and still inside the gel region of the phase diagram, a change takes place in the organization of paraffin chains, indicated by the presence of two high angle reflections at 4·1 and 3·8 Å. Apparently the packing of the paraffin chains becomes orthorhombic (Müller, 1932). The surface of the unit cell, calculated assuming that the indices of the reflections are 110 and 200, still coincides with S (Vincent and Skoulios, 1966b).

2. *Remarks*

The gel is a mesomorphic phase, with different degrees of order in different parts of the structure. The paraffin chains are stiff and parallel, with a high rotational disorder (and probably a fairly high translational disorder parallel to the chains). Since the cross section of one paraffin chain is close to 20 Å^2, the average area available to one polar group in the interdigitating structure is 40 Å², quite larger than in the crystalline and high-temperature phases of the anhydrous soaps; as a consequence the correlations between the polar groups are loose. When the interactions become strong, for example at low

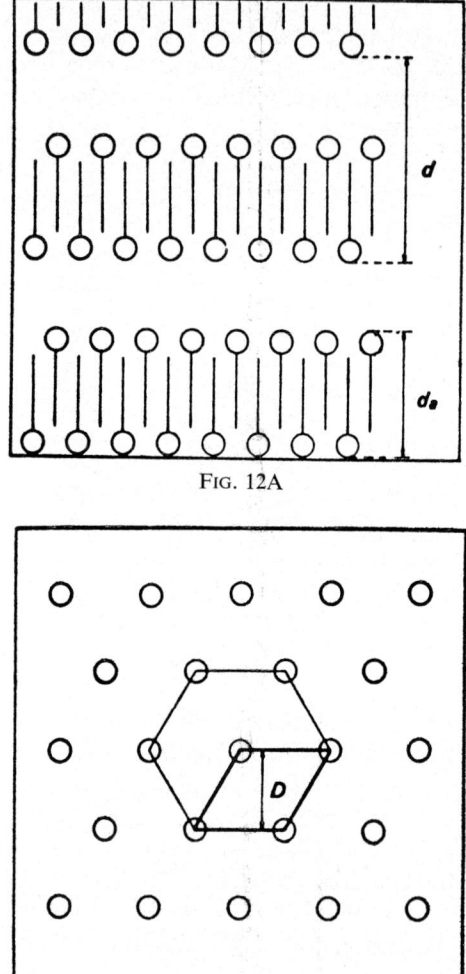

Fig. 12A

Fig. 12B

FIG. 12. Schematic representation of the structure of the gel of the K soaps. (A) section perpendicular to the lamelles, showing the stiff interdigitating paraffin chains; (B) section parallel to the lamelles, in the middle of the lipid layer, showing the hexagonal packing of the paraffin chains (from Vincent and Skoulios, 1966a).

water concentration, the stability of the gel is upset and a crystalline phase precipitates out. In spite of the disorder, the average area per polar group, and thus the surface density of the charges, is bound to remain constant.

D. ISOTROPIC MICELLAR SOLUTION

In the low concentration and high temperature region of the phase diagram, at concentrations higher than what is called the "critical micellar concentration" or c.m.c., and above the so-called Krafft temperature, the soap and detergent molecules cluster into large aggregates or *micelles*. The micellar solutions are optically isotropic and transparent and their viscosity is low; the X-ray diffraction diagrams contain one or two diffuse bands in the small angle region and the 4·5 Å band of the liquid paraffins. These solutions have been often studied by hydrodynamic and light-scattering techniques, at very low concentrations, near the c.m.c.; little information is available at higher concentrations. Structural analysis is best done in this region of concentration by small angle X-ray scattering techniques. McBain and Harkins were pioneers in this field (see McBain and Hoffman, 1949; Mattoon et al., 1947) even though their work is now only of historical interest since the theoretical basis of their analysis is unsatisfactory (Hughes, 1950). Others have attempted a more rigorous interpretation of the experimental curves, but it proved to be difficult to distinguish the effects of the shape of the micelles from that of their correlations (Brady, 1951; Andersen and Carpenter, 1953). This problem was undertaken more recently by using an improved small-angle X-ray scattering technique which is based upon intensity measurements on an absolute scale (Reiss-Husson, 1963; Reiss-Husson and Luzzati, 1964, 1966). Only a short account of the results of this work will be given here.

Several soaps and detergents were studied at different temperatures and as a function of concentration. The results are condensed in Table XI. The systems can be classified into three families:

(a) *The micelles are spherical at all concentration*. This is the case of $C_{16}TA$ Cl at 25°C. The diameter of the spheres was determined quite accurately and was shown to be independent of concentration. At the transition from the micellar solution to the hexagonal phase, the spheres are replaced by cylinders.

(b) *The micelles are rod-like at all concentration*. The only example is ONa at 27°C.

(c) *The micelles are spheres at low concentration and become rods at high concentration*. This is the most common case. The transition from spheres to rods takes place over a wide concentration range. The question whether, as the concentration rises, the spheres progressively elongate into ellipsoids

TABLE XI
Micellar solutions

	t	Spherical micelles	Onset of the sphere-rod transition	Rod-like micelles	Onset of the hexagonal phase	R_{par}	n	S micelle	S hexagonal
	°C	c	c	c	c	Å		Å²	Å²
CTACl	27	0·05	0·40		0·40	21·7	84	74	65
ONa	27				0·20				
SLS	27	0·07	0·25	0·04	0·40	17·8	67	65	59
SLS	70	0·05	0·15		0·40	17·0	57	68	
C_{12}Na	70	0·05	0·10		0·36	12·5	25	76	51
C_{14}Na	70	0·06	0·10		0·28				
C_{16}Na	80	0·05	0·18		0·23				
C_{18}Na	90	0·05	0·16		0·22				
CTAB	27		0·05	0·10	0·25				
CTAB	50	0·05	0·17		0·26				
CTAB	70	0·05	0·25		0·32				

Approximate concentration range for the different types of micelles and parameters of the spherical micelles. R_{par}: radius of the paraffinic part of the micelle; n: number of lipid molecules per micelle; S: area per polar group on the surface of the spherical micelles and of the cylinders of the hexagonal phase. (From Reiss-Husson and Luzzati, 1964).

and rods or only small spheres and very long rods are present, cannot be answered on the basis of the X-ray experiments alone.

The diameter of the spheres was determined with good accuracy in some cases and was found to agree quite well with the value determined by other techniques. It would be difficult to draw any general conclusion about the influence of the chemical structure of the detergent, of temperature and of concentration upon the shape of the micelles. It should be noted that the area per polar group is larger than in the long-range ordered phases (Table XI) in agreement with the rule discussed previously.

V. Lipids of Biological Interest

The family of lipids assembled under this headline is quite heterogeneous. The hydrophilic moieties are of the various types common among biological lipids; the paraffin chains vary in length and in degree of unsaturation. Only in a few cases pure chemical species, available in sufficiently large amounts, were used for a complete X-ray study. The majority of the samples, solvent extracts from biological materials or chromatographic fractions of these extracts, contain a variety of chemical species.

The chemical parameters, and more particularly the lipid composition and the integrity of the components, are of crucial importance for the phase diagram and yet it is by no means easy to keep them under proper control. For most of the lipids described here the extraction, the purification and the chemical analysis were carried out at the same time as the X-ray study. With mitochondria lipids the effects of some modifications of the extraction procedures and of the chemical alterations (mainly oxidation and hydrolysis) have been investigated (Gulik-Krzywicki et al., 1967).

Although the description will be limited here to a few recent results, the pioneering study of Bear. et al. (1941) and Palmer and Schmitt (1941) should be mentioned. These authors were the first to analyse by X-ray diffraction techniques the mesomorphic phases obtained by mixing water and natural lipids. They emphasized the presence of only one lamellar phase in mixed lipids and noted that different phases might be found if each component of the lipid mixture were studied separately; they interpreted the presence of a broad band near 4·5 Å as an indication of the disordered conformation of the paraffin chains. Finean (1953) has carried out more recently an X-ray study on several synthetic and natural lipids as a function of temperature. The X-ray data were interpreted within the framework of the lamellar structures with stiff paraffin chains; these should perhaps be reconsidered in the light of the more recent results described here. Chapman et al. (1966) have examined by spectroscopic and X-ray data a whole series of phosphatidylethanolamines and made different interpretations.

A. STRUCTURE OF THE LIQUID-PARAFFIN PHASES

On the high-lipid and low-temperature region of the phase diagrams the paraffin chains are at least partially ordered as shown by the presence of sharp reflections around 4·2 Å.

The liquid-paraffin phases of the anhydrous lipids and of the lipid-water systems were studied in all the cases. The phases are those described previously.

The hexagonal phase is observed over an extended concentration range in four of the systems: lysolecithin, phosphatidylethanolamine, brain and mitochondria lipids (Figs 14, 15 and 19).

By analogy with the soaps, it may be shown that in lysolecithin the structure of the hexagonal phase is of type I. Indeed for this structure S is almost independent of concentration (Fig. 14B) and its value is close to that found in the lamellar phase of lecithin over the same concentration range (Fig. 13B). On the contrary if the structure were of type II, S would be strongly concentration dependent, much more than in any other lipid at similar concentration (Fig. 16). Furthermore Reiss-Husson (1967) has shown that the intensities of the reflections agree quite well with the model type I.

In the other systems the structure of the hexagonal phase is most likely of type II. The reasons are as follows:

(a) The hexagonal phase is often observed at such high lipid concentration (Figs 14A, 15A and B) that no water is available to fill the gap between the lipid rods if the structure were of type I.

(b) The hexagonal phase is found at high lipid concentration, with respect to the lamellar phase (Fig. 15). The opposite occurs in all the cases in which the structure is of type I (Table VIII).

(c) Only if the structure is of type II does S increase as the concentration decreases in accordance with the rule previously discussed (Figs 16, 14A and 15A).

(d) An independent confirmation was provided by Stoeckenius (1962), who succeeded in fixing the hexagonal phases of the two types and observing them under the electron microscope (Fig. 17). In type I a honeycomb-like distribution of black lines is observed, in the other a hexagonal array of black spots; the two pictures look like the negatives of each other.

B. MONOGLYCERIDES, PHOSPHOLIPIDS, SPHINGOLIPIDS

The liquid-paraffin phases of the anhydrous lipids and of the lipid-water systems will be described here. The crystalline structure of these compounds, and the polymorphic transitions that take place in the crystalline state, are outside the scope of this article (see review in Chapman, 1965 and Chapman et al., 1966).

The most complete analysis of the liquid-paraffin phases is that of Reiss-Husson (1967) who has studied a few lipids of the three families either in the form of pure chemical compounds, or as chromatographic fractions of

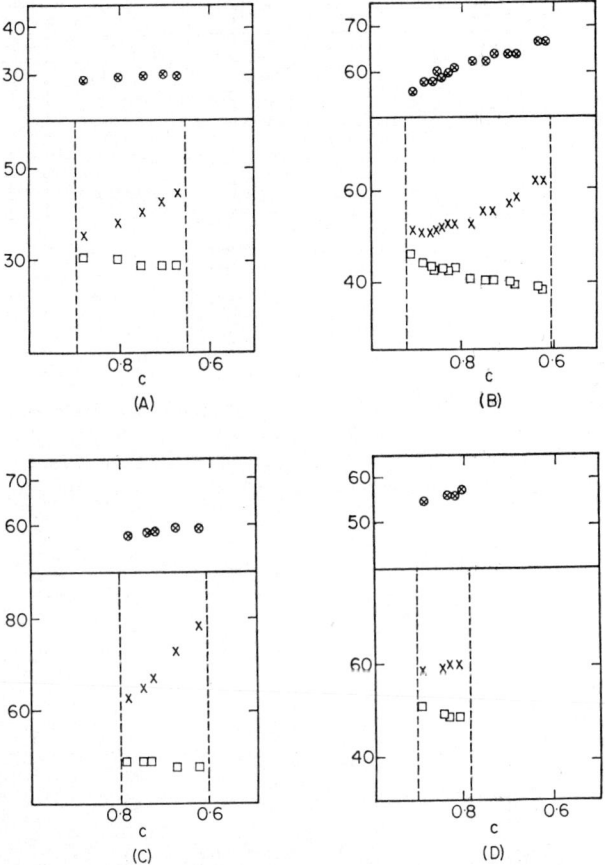

FIG. 13. Dimensions of the lamellar phase of lipid-water systems. The dotted lines show the limits of the pure phase. (A) glyceryl-1-monododecanoate, 45°C; (B) egg lecithin, 25°C; (C) beef brain sphingomyelin, 40°C; (D) beef brain cerebroside, 72°C; ⊗: area S; ×: repeat distance d; □: thickness d_l of the lipid leaflet (replotted from Reiss-Husson, 1967).

biological preparations. (Note. The chemically pure lipids (monoglycerides and dipalmitoyl-lecithin) are synthetic compounds; the chromatographic fractions contain several molecular species, all with the same hydrophilic group but with paraffin chains of different length and degree of unsaturation.) The purity of all the lipids was controlled by thin-layer chromatography. The phase diagram analysis was carried out over the temperature-concentration

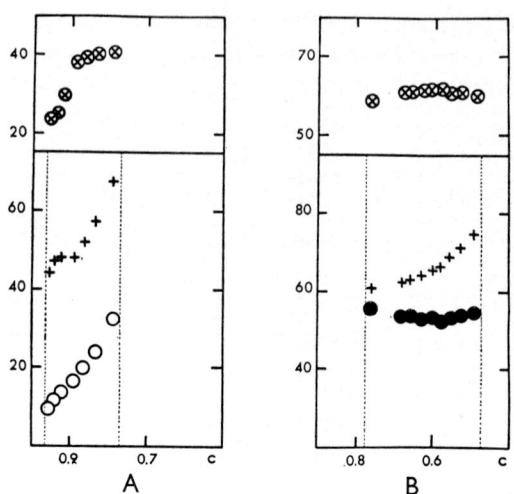

FIG. 14. Dimensions of the hexagonal phases of lipid-water systems. (A) egg phosphatidylethanolamine, 55°C; (B) egg lysolecithin, 37°C. ⊗: area S; +: distance d between the cylinder axes; ● type I, diameter d_l of the lipid cylinder; ⊙: type II, diameter d_w of the water cylinder (replotted from Reiss-Husson, 1967).

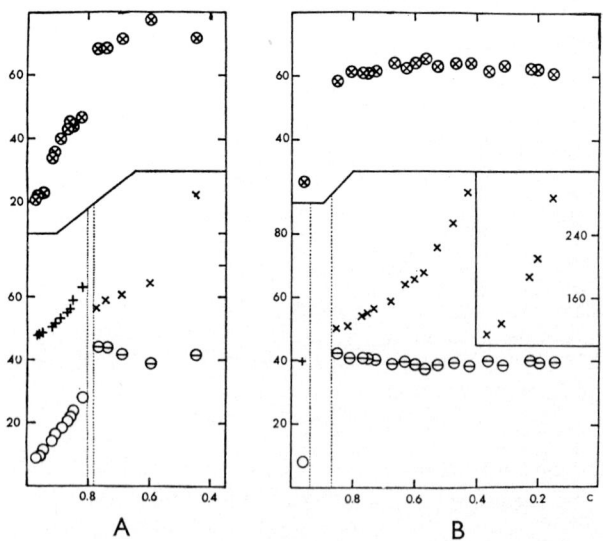

FIG. 15. Dimensions of the liquid-paraffin phases of lipid-water systems. (A) human brain, 37°C (Luzzati and Husson, 1962); (B) beef heart mitochondria, 25°C (Gulik-Krzywicki et al., 1967). Symbols as in Figs 13 and 14.

range in which the chemical degradation is not too fast (see Reiss-Husson, 1967). The results are summarized in Figs 13 and 14.

The high-lipid ($c > 0.95$) and high-temperature region of the lipid-water phase diagram of both natural (hen eggs) and synthetic lecithins has been studied recently (Luzzati *et al.*, 1968b). Several phases have been observed; three of these, namely H, R and Q (see Fig. 8), belong to the "rod-like" class discussed above (see Section IV-B).

The most common-water-containing phase is the lamellar. The structure parameters of some of the systems are plotted in Fig. 13. The hexagonal phase is observed in lysolecithin at 37°C and in phosphatidylethanolamine at 55°C (Fig. 14); it has been shown previously that the structure is of type I in the former and of type II in the latter.

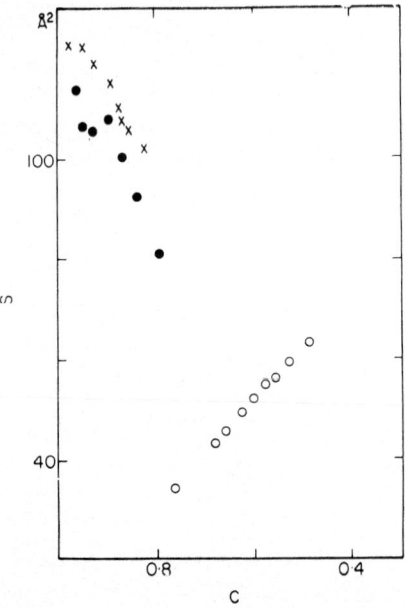

FIG. 16. Plot S *vs* c for the hexagonal phase, assuming that the structure is of the type now presumed incorrect. ×: brain lipids, type I (see type II in Fig. 15); ●: phosphatidylethanolamine, type I (see type II in Fig. 14A); ○: lysolecithin, type II (see type I in Fig. 14B).

The phosphatidylethanolamine-water system was studied at lower temperature as well: 35 and 25°C. Under these conditions two phases, one lamellar, the other hexagonal, are simultaneously observed over an extended concentration range; the relative proportion of the hexagonal to the lamellar decreases as the temperature is lowered.

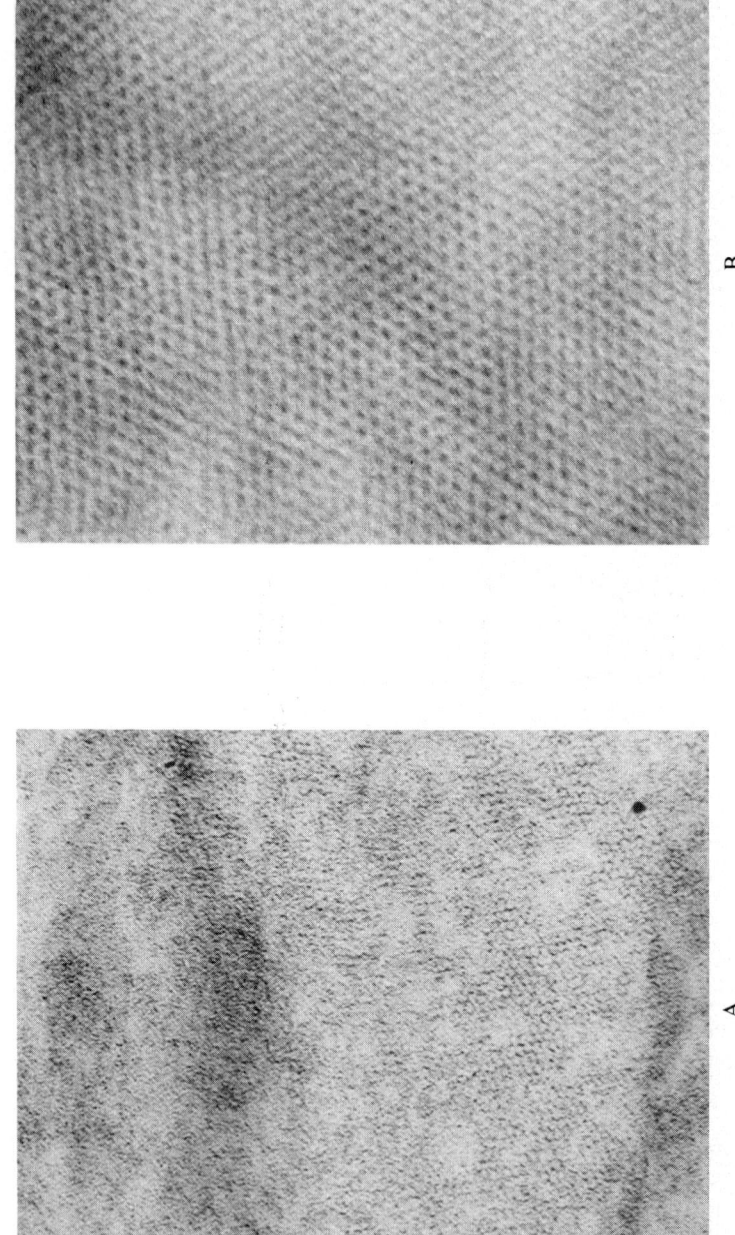

Fig. 17. Electron microscope observations of the hexagonal phases.
(A) hexagonal I: Na linolenate-water; $c = 0.46$; Os O$_4$ fixation at 22°C. Centre to centre distance approx. 40 Å (X-ray study of this system by Husson, 1961).
(B) hexagonal II: human brain lipid-water, $c = 0.97$; Os O$_4$ fixation at 37°C. Centre to centre distance approx. 44 Å (see Fig. 15A). (From Stoeckenius, 1962).

In all the systems described here the organized phase, either lamellar or hexagonal, can take up only a fairly small amount of water. At very low concentration, with the exception of lysolecithin, the preparations are heterogeneous and contain the most highly hydrated ordered phase, dispersed in water. This is shown by the aspect of the preparations and by the presence, in the X-ray diffraction diagrams, of the reflections of the ordered phase; as the amount of water increases, the intensity of the reflections decreases without any change of the spacings. With lysolecithin an isotropic micellar solution is observed, similar to that of soaps and detergents.*

C. BRAIN LIPIDS

An ether extract from human brain, provided by Dr. Stoeckenius and containing approximately 52% phosphatidylethanolamine, 35% lecithin and 13% phosphatidylinositol, was studied by X-ray scattering methods (Luzzati and Husson, 1962).

The phase diagram was explored as a function of temperature and concentration; the position of the experimental points is shown in Fig. 18. No systematic study was carried out at concentrations lower than 0·40.

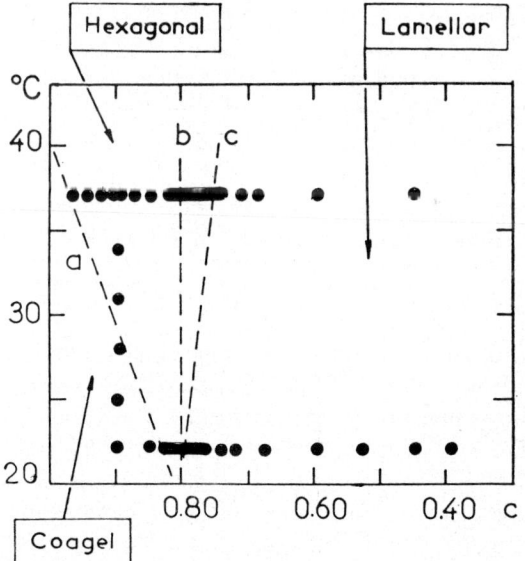

FIG. 18. Phase diagram of the brain lipid-water system, with the position of the experimental points (From Luzzati and Husson, 1962).

*Larsson (1967) has reported recently a detailed X-ray study of several glyceride-water systems as a function of temperature and concentration. The lamellar phase is common to all the systems; in addition a cubic phase was observed with 1-monopalmitin and a hexagonal II phase with 1-monobehenin.

On the left and lower side of the line a (Fig. 18) the presence of sharp reflections around 4·2 Å indicates that the paraffin chains are somewhat ordered; this region is named "coagel" in the figure. The phase encompassed by the lines a and b is hexagonal, type II (see above). The lamellar phase is present on the right side of the line c.

The dimensions of the structure elements were determined for the two phases as a function of concentration. The results at 37°C are plotted in Fig. 15.

D. MITOCHONDRIA LIPIDS

The results reported here were obtained by Gulik-Krzywicki et al. (1967). The lipids were extracted from intact beef heart mitochondria by a procedure derived from that of Folch Pi et al. (1957). The composition, as determined by quantitative thin layer chromatography, is approximately 34% lecithin, 29% phosphatidylethanolamine, 10% phosphatidylinositol, 20% cardiolipids, 2% cholesterol and 5% neutral lipids. The lipids of more than ten preparations were used for the X-ray study; the results were perfectly reproducible.

The phase digram is shown in Fig. 19. Four phases can be distinguished, each clearly characterized by X-ray diffraction; the approximate range of existence of the pure phases is shown in the figure. Only the phases H_{II}, $L\alpha$ and $L\gamma$ were obtained pure; $L\beta$ was always observed in the presence of another phase.

The structure of the phases will be described. The classification into ordered and liquid-paraffin phases, adopted here, is not as clear cut in this system as in soaps.

Hexagonal, observed at very high lipid concentration (Figs 15 and 19). The structure is most likely of type II, as previously shown. The diffuse band near 4·5 Å shows that the paraffin chains are disordered.

Lamellar $L\alpha$, *high-temperature form*, above the a-a line in Fig. 19. The structure is that of the liquid-paraffin lamellar phase of other systems. At constant temperature the thickness of the lipid leaflet, and as a consequence the area S, are almost constant over an extended concentration range; for example at 25°C d_l remains very close to 40 Å from $c = 0.85$ to $c < 0.14$, whilst the thickness of the water layer varies from 7 to 250 Å (Fig. 15). This situation is particularly favourable for a crystallographic verification of the structure model which is described in Appendix 2. The thickness of the lipid lamellae decreases with rising temperature as required by the disordered conformation of the paraffin chains (Appendix 1).

Lamellar Lα, *low temperature form.* If, at constant concentration, the temperature is lowered, above the *a-a* line, the X-ray diffraction diagrams remain unchanged (neglecting the small thermal contraction, see Appendix 1). As the *a-a* line is crossed a sharp reflection appears at 4·2 Å whose intensity increases, with respect to that of the 4·5 Å band, as the temperature decreases.

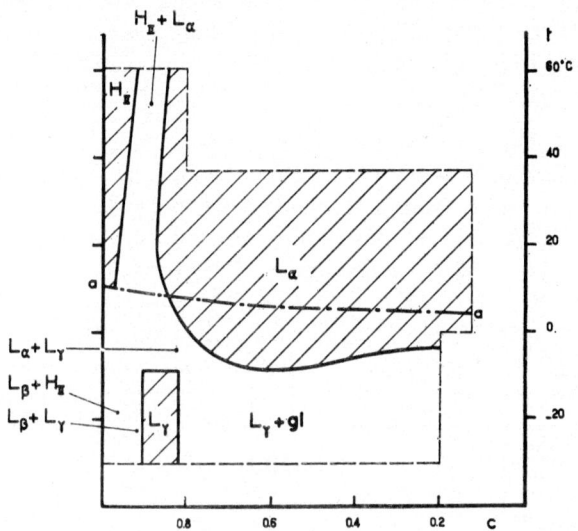

FIG. 19. Phase diagram of the mitochondria lipid-water system. The region explored is defined by the dotted line. Five phases are observed: hexagonal type II (H_{II}), three lamellar (Lα, Lβ and Lγ) and ice (gl.). Below the *a-a* line the conformation of the paraffin chains begins to become ordered. The pairs of phases observed in some of the transition regions are indicated. Other regions have more than two phases in equilibrium and were not studied in detail (from Gulik-Krzywicki *et al.*, 1967).

At the same time the number, spacing, sharpness and intensity of the small-angle reflections remain the same as above the *a-a* line. It is most unlikely, under these circumstances, that more than one phase is present; furthermore it appears that the phase contains regions with "liquid" chains, responsible for the 4·5 Å band, and regions with stiff, free-rotating and hexagonally packed chains, responsible for the 4·2 Å reflection (see Section II-B-1). The existence of lamellae (or of large patches) with all the chains stiff and hexagonally packed is incompatible with some of the experimental observations. Indeed the cross-section of the ordered paraffin chains is $\Sigma = 20\cdot4$ Å^2 and the average area per chain on the plane of the lamellae is $S/2 = 29$ Å^2. As a consequence if the chains were stiff across the whole paraffin layer, they should be tilted on the plane of the lamellae and the CH$_3$ end groups would be localized at the centre of the lipid leaflet. Such concentration of CH_3 groups of low electron density would perturbate the intensities of the reflections that, in fact, are not found to vary (see Gulik-Krzywicki *et al.*, 1967 and Appendix 2). The structure may be visualized as follows (Fig. 20C). The ordered regions are limited to a layer, in the centre of the paraffin leaflet, in which the chains issued from the polar groups sitting on the opposite faces of the lamella interdigitate over part of their length. The ordered layers are surrounded by two "liquid" layers that contain a segment of each of the chains involved in the ordered regions and some whole chains.

Lamellar $L\beta$. This phase is obtained at very low water concentration (Fig. 19). Several sharp reflections, all integral orders of one repeat, show that the structure is lamellar; a sharp and intense 4·2 Å reflection replaces the 4·5 Å band completely. At $c = 0.94$, d is 60·6 Å. Assuming $\bar{v}_l = 0.98$ and $\bar{v}_w = 1.00$, $d = 56.6$ Å and $S = 40.7$ Å². This value of S is very close to twice the cross section of one stiff chain ($\Sigma = 20.4$ Å) and is thus consistent with a model formed by a double layer of hexagonally packed stiff chains as represented in Fig. 20B (see also Vincent and Skoulios, 1966c). Since the paraffins are of unequal length, the presence, in the middle of the paraffin layer, of a thin disordered region was postulated. The agreement of the observed and calculated intensities confirms the model (Appendix 2).

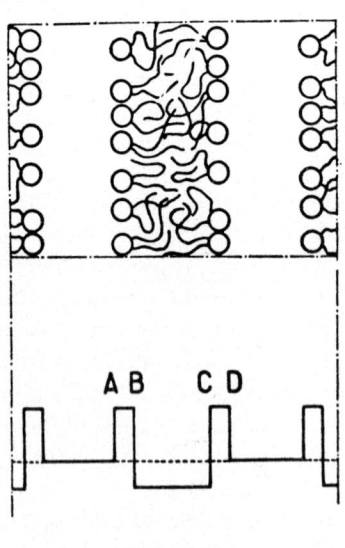

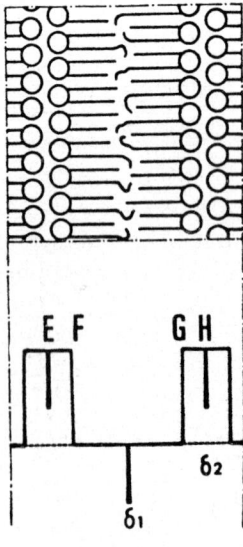

FIG. 20A   FIG. 20B

Lamellar Lγ complex. This phase is observed alone, at low temperature, over a narrow concentration range (0·9 > c > 0·8, Fig. 19). It is characterized by a strong and sharp 4·2 Å reflection and by a large number of small-angle reflections, all integral orders of one repeat, the value of which

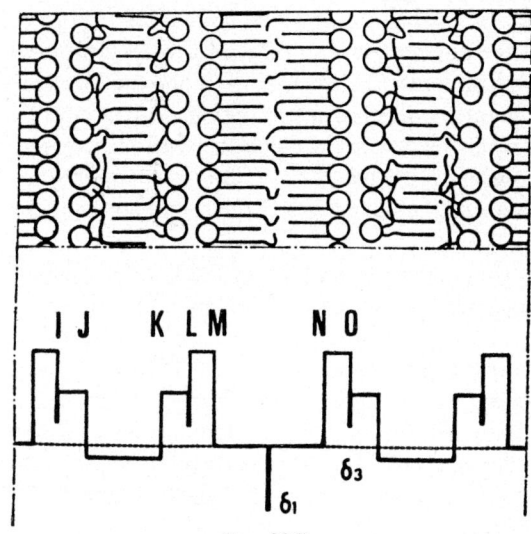

Fig. 20C

FIG. 20. Structure model and distribution of the electron density for the lamellar phases of the mitochondria lipids-water system (Figs 15B and 19). The hydrophilic groups are represented by a circle, the paraffin chains by a line.
(A) Lα, high-temperature form (above the line a-a in Fig. 19). The paraffin chains are completely disordered.
(B) Lβ, the paraffin chains are stiff and are packed in a two-dimensional hexagonal lattice. A thin disordered layer is present in the centre of the paraffin layer in order to take into account the difference in length of the chains. The thick bar represents delta functions due to the localization of the ends of the lipid molecules.
(C) Lγ, formed by the alternate sequence of Lα and Lβ layers. Lα is in the low-temperature form with some of the chains stiff and hexagonally packed over part of their length.
The dimensions adopted for the calculation of the intensities are shown in Table XII (see Gulik-Krzywicki *et al.*, 1967).

varies from 114 to 122 Å, according to concentration. It is impossible to take into account such a large repeat with one lipid leaflet, the maximum thickness of which is 57 Å (Fig. 20B), and 20% water; the structure must be complex. Various models were tried; the most satisfactory, shown in Fig. 20C, is formed by an alternate sequence of lamellae of types Lα and Lβ, as the repeat of Lγ is almost exactly the sum of those of Lα and of Lβ. A confirmation of the model is provided by the comparison of the observed and calculated intensities (Appendix 2).

E. REMARKS

Some of the salient observations can be summarized:

(a) Even when the lipid is a mixture of many chemical species, these are perfectly miscible over the extended one phase regions of the phase diagrams. On the contrary, in the regions in which more than one phase is present, some segregation of the different species takes place as the properties of the system formally require the presence of more than one lipid component.

(b) The temperature dependence of the phase diagram of the phosphatidyl ethanolamine-water system, and similar phenomena observed in other systems (T. Gulik-Krzywicki, E. Rivas and V. Luzzati, unpublished), suggests that the predominance of the hexagonal II phase at high temperature is a general rule.

(c) Unsaturation and heterogeneity of the paraffin chains lower the temperature at which the chains become ordered. As an example, dipalmitoyl lecithin yields a liquid-paraffin lamellar phase, in the presence of water, only above 60°C, whilst egg lecithin does so at 0°C (Reiss-Husson, 1967).

(d) The number of paraffin chains per hydrophilic group appears to be determinant upon the type of the hexagonal phase; type I for one chain (lysolecithin, soaps), type II for two chains (phosphatidylethanolamine, etc.). (One exception is 1-monobehenin, in which Larsson (1967) observed a hexagonal II phase.)

TABLE XII

	Δx Å	$\Delta\rho = \rho - \rho_{\text{water}}$ Electrons per Å^3	$\int \delta(x)dx$ Electrons per Å^2
AB=CD=IJ=KL	7·5	+0·095	
BC=JK	29·0	−0·044	
EF=GH=LM=NO	9·6	+0·176	
FG=MN	41·4	0	
IJ=KL	10·45	+0·098	
JK	32·2	−0·018	
$\delta_1 = \delta_2$			−0·724
δ_3			−0·362

(e) The rule discussed previously that S increases as the concentration decreases seems to apply to the lipids of this chapter as well as to soaps and detergents.

(f) One of the parameters in each phase is little dependent of concentration, at least compared to the others; the thickness of the lipid leaflet in the lamellar, the diameter of the lipid rods in the hexagonal I, the distance

(d-d_w) between the surface of the water rods in the hexagonal II (see Figs 13, 14, 15).

(g) The concentration dependence of those slowly varying parameters is much the same for the various lipids; it decreases as the amount of water increases (Figs 13, 14, 15). This phenomenon is reminiscent of the soaps (see Figs 10 and Table IX).

(h) The chemical heterogeneity of the paraffin chains does not prevent the paraffin chains from taking up a fairly ordered conformation at low temperature; it causes the order–disorder transitions to be gradual (Lα, low temperature form, mitochondria lipids) and the lipids to segregate into two different types of lamellae (Lγ, mitochondria lipids).

VI. Discussion and Conclusions

It is clear, from the results described in the previous chapters, that the number and the variety of the structures of the lipid-water systems are indeed very large. It is perhaps surprising to note that for so many years the widespread occurrence of non-lamellar phases has passed unnoticed by numerous authors who have studied those systems by X-ray diffraction techniques. The reason for this, apart from the frequent use of inadequate techniques, is the uncritical presumption that the conformation of the paraffin chains is always crystalline; this postulate inevitably restricts the choice to the lamellar structures. Such assumption is even more difficult to justify when one is reminded that as early as 1936, Hartley, and later Bear *et al.* (1941), had put forward lucid arguments showing that the conformation of the paraffin chains is disordered.

The association of a highly developed long range order with an almost complete short range disorder is one of the most remarkable features of the lipid-water phases. In fact, the liquid structure of the paraffin chains is a necessary condition for polymorphism as only liquid chains can fill up volumes of various sizes and shapes. The extensive miscibility of the different kinds of lipid molecules is another consequence of the disorder of the paraffins. Moreover, fairly large amounts of lipo-soluble water-insoluble substances are known to dissolve into the liquid paraffin regions of the lipid-water phases; a few ternary systems (ethylbenzene—C_{14}Na—water, Spegt *et al.*, 1961; cholesterol—mitochondria lipids—water) and a quaternary system (bile salt-lecithin-cholesterol-water, Small *et al.*, 1966) have been studied by X-ray techniques.

Some of the properties of the phase diagrams, and the role of the chemical parameters, have been discussed at the end of each section. One rule has been stressed in this chapter; as the concentration increases, at constant temperature, the area S decreases or remains constant but never increases even if phase boundaries are crossed. The nature and the sequence of the phases are related to the chemical structure of the lipid at least in one obvious way; phases with high surface to volume ratio are promoted by bulky hydrophilic groups, with respect to the paraffin moiety, and vice versa. Illustrations of this rule are provided by the soaps and detergents and more strikingly by the phospholipids; the diacyl compounds (lecithin, phosphatidylethanolamine) yield lamellar and hexagonal II phases, while lysolecithin, a monoacyl compound, yields the heaxagonal I phase. The chemical disorder in the paraffin chains (unsaturation, length heterogeneity) lowers the temperature of the transitions to the liquid-paraffin phases (compare egg and dipalmitoyl lecithins). Furthermore, increasing disorder, for example by raising the temperature, shifts the equilibrium towards the hexagonal phases.

The dimensions of all the structure elements vary with concentration although in each phase one of the parameters, directly related to the length of the lipid molecules, varies much less than the others; the thickness of the lipid leaflet in the lamellar phase, the diameter of the lipid rod in the hexagonal I, the distance between the surfaces of the water rods in the hexagonal II. The concentration dependence of this parameter is generally the less pronounced the larger the amount of water and eventually fades away completely if the organized phase is present at sufficiently low concentration (see Lα phase of mitochondria lipids, Fig. 15B). The role of the interfaces in the definition of the dimensions of the structure elements has been discussed in the section on soaps and in Appendix 1. It is clear that the forces are analogous to interfacial tensions and involve the activity of water and ions in the volume surrounding the lipid elements, as well as some geometric factor, specific for each type of phase.

At low concentrations the soap and the detergent molecules remain associated in micelles of different size and shape; the micelles are dispersed in water. Certain natural lipids also form micellar solutions, for example lysolecithin. In contrast, other lipids, like the diacyl phospholipids and the sphingolipids, do not yield a homogeneous phase at low concentration. With these lipids, on increasing the amount of water, the ordered phase reaches a hydration limit and remains in equilbrium with the excess water. A different phenomenon is observed with mitochondria lipids; large amounts of water can be intercalated between the lipid lamellae without upsetting the mesomorphic organization.

As the paraffin chains take up an ordered conformation, several properties of the lipid-water systems become strongly dependent upon the nature of the paraffin chains, of the hydrophilic groups and of the counterions. Only chemically pure lipids crystallize into fully ordered three-dimensional lattices; nevertheless a fairly high degree of order is observed in the paraffins even when the chains are quite heterogeneous in length and degree of unsaturation. The order–disorder transitions are sharp for the pure compounds and become gradual as a consequence of the chemical heterogeneity.

The question may now be asked, to what extent do the phenomena which take place in lipid-water systems have a bearing on the structure and function of biological membranes. Although a direct answer is not at hand, some correlations can be sought between the two orders of phenomena.

It must be noted first that the structure of membranes is still largely unknown, at least at the molecular level. The electron microscope observations, generally presented as the most direct evidence, cannot be accepted uncritically as the structures involving lipids are so labile that they are not likely to withstand the usual treatments (fixation, dehydration, embedding). As an example, only with special precautions was Stoeckenius able to observe the hexagonal phases, in spite of their widespread occurrence. It could rightly be asked whether the structure of membranes must be envisaged as a static feature or rather as a dynamic property, governed by the physiological conditions and variable in time and space. Some of the physiological functions of membranes could even be related to polymorphic transitions, analogous to those that take place in lipid-water systems, induced by the action of parameters of more direct physiological significance. It may be noted, in this connection, that the physico–chemical conditions (concentration, temperature) at which the transitions take place in the lipid-water systems are not too different from those that prevail in the living cell.

One important parameter is the conformation of the paraffin chains. In myelinated nerves, one of the rare cases in which this problem has been investigated *in situ*, the conformation has been shown to be liquid (Schmitt *et al.*, 1941). In fact the question could be asked whether the order–disorder transitions in the paraffin chains are of any biological significance, especially since the cation specificity dramatically changes at this transition; for example the structure of the gel and coagel is quite different for the Na and the K soaps, while the liquid-paraffin phases are very little sensitive to this difference of cations.

The transitions between liquid-paraffin phase probably involves drastic alterations of the properties of the system, such as permeability and electrical resistivity, similar in nature to those that occur in membranes. The hexagonal II phase suggests a particularly interesting model for permeation as it is formed by narrow water channels lined with the polar groups of the lipid molecules (Luzzati *et al.*, 1966).

VII. Acknowledgements

The author wishes to express his gratitude to Dr. F. Reiss-Husson, without whose long lasting, friendly collaboration this article could not be written, and to his past and present coworkers, Drs H. Mustacchi, A. Skoulios, P. Spegt, E. Rivas, T. Gulik-Krzywicki, R. P. Rand and A. Tardieu who generously supplied data and ideas.

This work was supported in part by grants from the "Délégation Générale à la Recherche Scientifique et Technique, Comité de Biologie Moléculaire" and from the "Direction des Recherches et Moyens d'Essais, Ministère des Armées".

Appendix I

THE DISORDERED CONFORMATION OF THE PARAFFIN CHAINS

The liquid-paraffin phases have some interesting thermodynamic properties. These will be discussed by reference to the lamellar phase; similar conclusions would be reached if any other phase were taken into consideration.

The lamellae contain disordered paraffin chains, each anchored to a hydrophilic group at the interface. Thickening of the hydrocarbon leaflet (at constant volume) increases the area S available to one hydrophilic group; thus a surface tension acting against an increase of S is equivalent to a force stretching the paraffin molecules in the direction perpendicular to the lamella.

Each lamella is thus analogous to a rubber probe, i.e. a disordered linear polymer, stretched by an external force against the thermal motion. It is clear that as temperature rises, disorder increases and the length of the probe decreases. This phenomenon is indeed observed in all the lipid-water systems; *in the lamellar and in the hexagonal I phases, the thickness of the lipid leaflet and the diameter of the lipid rods decrease as the temperature is raised* (Luzzati et al., 1960; Luzzati and Husson, 1962). In fact two conditions must be satisfied for this phenomenon to occur: (a) the conformation of the paraffin chains must be disordered and (b) the interface must be compressible or, in other words, the organization of the hydrophilic groups must be disordered.

If a precise model is adopted for the conformation of the chains, a quantitative treatment of the phenomenon can be carried out. In the case of a random conformation, the following equation is obtained (Treloar, 1958, pp 54 and 62)

$$f = 2kTb^2r = c_1Tr/L \tag{1}$$

where f is the stretching force applied to the ends of the molecules, r is the distance between the ends, T is the absolute temperature, L is the contour length, c_1 is a constant. If f is independent of r and T, the linear thermal expansion is:

$$\alpha = (1/d)(\Delta d/\Delta T) = (1/r)(\partial r/\partial T) = -1/T \tag{2}$$

which is in excellent agreement with the experimental values; for example, $\alpha = -2\cdot 7 \times 10^{-3}$ for the lamellar phase of the K soaps, from 45 to 104°C (Gallot and Skoulios, 1966a) (see also Luzzati *et al.*, 1960; Luzzati and Husson 1962).

This model leads to another prediction. If in a series of homologous compounds (for example the saturated soaps of one cation) the conformation of the paraffin chains is random, r/L is proportional to $1/S$ and Eq. (1) becomes:

$$f = c_2 T/S \tag{3}$$

The force is thus independent of the length of the paraffin chain. f, and its equivalent the surface tension, is likely to be a function of the thermodynamic properties of the aqueous regions (i.e. of the activity of the various species) and thus of the molal concentration of the polar groups in the water of the system. As a consequence it may be expected that in each phase *S is a function of the molality of the polar groups, for all the compounds of a homologous series, and at all concentration.* This phenomenon was indeed observed by Gallot and Skoulios (1966a) in the Na, K, Rb and Cs saturated soaps, from C_8 to C_{22} (see Fig. 10) (although these authors appear to misinterpret their observation).

Appendix 2
SOME CRYSTALLOGRAPHIC VERIFICATIONS

The traditional verification of the results of the X-ray structure analysis is to compare the observed and the calculated intensities of the reflections. This test can be applied to lipid-water systems (see below) whenever the number of the reflections recorded in one X-ray diffraction experiment is sufficiently high. In many cases, in fact, this number is small; nevertheless the amount of data is often quite large when the experimental study is carried out as a function of concentration. A particularly simple case, often met in lipid-water systems, is one phase formed by elements of constant dimensions, separated by variable amounts of water. In this case the amplitude of the reflections is proportional to the Fourier transform of the structure element sampled at the lattice points; the ratios of the observed amplitudes must match the Fourier transform (Husson *et al.*, 1960; Vincent and Skoulios, 1966a).

The lamellar Lα phase of mitochondria lipids is an excellent case for such verification as the thickness of the lipid leaflet is almost constant over an extended concentration range (Fig. 15B). The distribution of the electron density, shown in Fig. 20A, was determined by adopting a step function and taking into account the molecular weight and the density of the lipid molecules and of the paraffin and polar moities; the only adjustable parameter is the thickness of the polar layer (Gulik-Krzywicki *et al.*, 1967). The Fourier transform, and the visually estimated amplitude of the reflections, are plotted in Fig. 21; the agreement is excellent as evidenced, for example, by the position of the zeros.

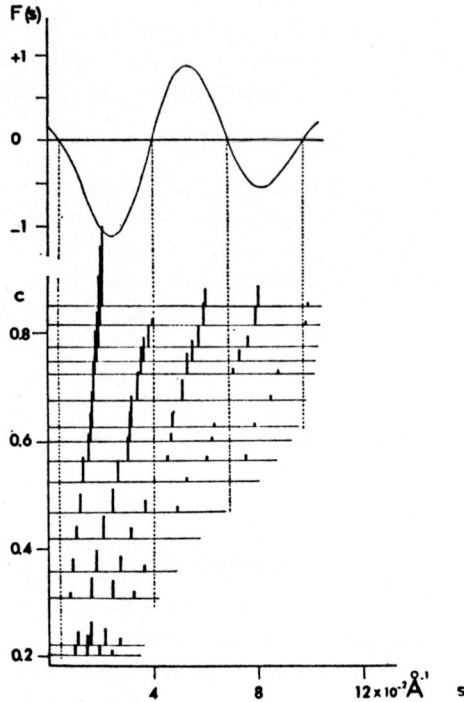

FIG. 21. Crystallographic verification for the Lα phase of mitochondria lipids. $F(s)$ is the Fourier transform of the electron density distribution shown in Fig. 20A. Each horizontal line represents one X-ray picture; the concentration is given by the ordinate. The spacings of the reflections are shown in the abscissae. The thick vertical bars are proportional to the amplitudes which were estimated optically. The fading out of the reflections, as a function of s, increases for decreasing concentration, as a consequence of disordering of the one-dimensional lattice (from Gulik-Krzywicki et al., 1967).

The phases Lβ and Lγ of mitochondria lipids are observed only over a very narrow concentration range; in both cases, nevertheless, the number of the observed reflections is sufficiently high to warrant a calculation of the structure factors. The electron density distributions are shown in Fig. 20B and C. This was determined, in the case of Lβ, with the same assumptions made for Lα; the only difference is the presence of two delta functions, one (δ_1) in the middle of the paraffin layer, the other (δ_2) between the polar groups, in order to take into account the precise localisation of the ends of the lipid molecules

(Gulik-Krzywicki et al., 1967). The electron density of Lγ (Fig. 20C) is a combination of Lα and Lβ. The calculated intensities, modulated by the function exp ($-90s^2$), are given in Table XIII; the agreement with the observed intensities is excellent.

TABLE XIII

h	Lβ				Lγ			
	s 10^{-3} Å^{-1}	I obs	I calc	sign	s 10^{-3} Å^{-1}	I obs	I calc	sign
1	16·5	vvS	7·94	−	8·8	w	0·77	−
2	33·0	vw	0·00	+	17·6	vvS	19·37	−
3	49·6	0	0·02	−	26·4	vvw	0·51	−
4	66·1	vS	3·05	−	35·2	vvw	0·59	+
5	82·6	m	0·23	+	44·0	0	0·01	+
6	99·2	S	1·12	−	52·8	0	0·22	−
7	115·7	0	0·03	−	61·6	S	4·91	−
8	132·2	m	0·24	−	70·4	m	1·43	−
9	148·7	vvw	0·00	−	79·2	w	0·54	+
10	165·2	vw	0·23	−	88·0	0	0·02	−
11					96·8	w	1·17	−

The observed intensities were estimated visually. vvS: extremely strong; vS: very strong; S: strong; m: medium; w: weak; vw: very weak; vvw: extremely weak (from Gulik-Krzywicki et al., 1967).

References

Andersen, D. E. and Carpenter, G. B. (1953). *J. Am. chem. Soc.* **75**, 850.
Bear, R. S., Palmer, K. J. and Schmitt, F. O. (1941). *J. cell comp. Physiol.* **18**, 355.
Benton, D. P., Howe, P. G., Farnard, R. and Puddington, I. E. (1955). *Can. J. Chem.* **33**, 1798.
Brady, G. W. (1951). *J. chem. Phys.* **19**, 1547.
Chapman, D. (1965). *In* "The Structure of Lipids", Methuen and Co., London.
Chapman, D., Byrne, P. and Shipley, G. G. (1966). *Proc. R. Soc.* A, **290**, 115.
Clunie, J. S., Corkill, J. M. and Goodman, J. F. (1965). *Proc. R. Soc.* A, **285**, 520.
Dervichian, D. G. (1964). *In* "Progress in Biophysics and Molecular Biology." **14**, p. 263, Pergamon Press, London.
Doscher, T. and Vold, R. (1948). *J. phys. Colloid Chem.* **52**, 97.
Finean, J. B. (1953) *Biochim. biophys. Acta* **10**, 371.
Folch Pi, J., Lees, M. and Stanley, G. H. S. (1957). *J. biol. Chem.*, **226**, 497.
Friedel, G. (1922). *Annls Phys.* **18**, 273.
Gallot, B. (1965). Thesis, University of Strasbourg.
Gallot, B. and Skoulios, A. E. (1962). *Acta crystallogr.* **15**, 826.
Gallot, B. and Skoulios, A. E. (1966a). *Kolloidzeitschrift* **208**, 37.

Gallot, B. and Skoulios, A. E. (1966b), *Kolloidzeitschrift* **209**, 164.
Gallot, B. and Skoulios, A. E. (1966c). *Kolloidzeitschrift* **210**, 143.
Gallot, B. and Skoulios, A. E. (1966d). *Molec. Cryst.* **1**, 263.
Gulik-Krzywicki, T., Rivas, E. and Luzzati, V. (1967). *J. molec. Biol.*, **27**, 303.
Hartley, G. S. (1936). "Aqueous Solutions of Paraffin Chain Salts", Hermann, Paris.
Hughes, E. W. (1950). *Nature, Lond.* **165**, 1017.
Husson, F. (1961). *Ct. r. hebd. Séanc. Acad. Sci.*, *Paris* **253**, 2948.
Husson, F., Mustacchi, H. and Luzzati, V. (1960). *Acta crystallogr.* **13**, 668.
International Tables for X-ray Crystallography, The Kynoch Press, Birmingham (1952), Vol. 1.
Larsson, K. (1967). *Z. phys. Chem.*, **56**, 173.
Luzzati, V. and Husson, F. (1962). *J. Cell Biol.* **12**, 207.
Luzzati, V., Mustacchi, H. and Skoulios, A. E. (1957). *Nature, Lond.* **180**, 600.
Luzzati, V., Mustacchi, H. and Skoulios, A. E. (1958). *Discuss. Faraday Soc.*, **25**, 43.
Luzzati, V., Mustacchi, H., Skoulios, A. E. and Husson, F. (1960). *Acta crystallogr* **13**, 660.
Luzzati, V. and Reiss-Husson, F. (1966). *Nature, Lond.* **210**, 1351.
Luzzati, V., Reiss-Husson, F., Rivas, E. and Gulik-Krzywicki, T. (1966). *Ann. N.Y. Acad. Sci.* **137**, *Art.* 2, 409.
Luzzati, V. and Spegt, P. A. (1967). *Nature, Lond.* **215**, 701.
Luzzati, V., Tardieu, A. and Gulik-Krzywicki, T. (1968a), *Nature, Lond.* **217**, 1028.
Luzzati, V., Gulik-Krzywicki, T. and Tardieu, A. (1968b). *Nature, Lond.* (in press).
Marsden, S. S. and McBain, J. W. (1948). *J Am. chem. Soc.*, **70**, 1973.
Mattoon, R. W., Stearns, R. S. and Harkins, W. D. (1947). *J. phys. Chem.* **15**, 209.
McBain, J. W. and Hoffman, D. A. (1949). *J. phys. Colloid Chem.* **53**, 39.
McBain, J. W. and Lee, W. W. (1943). *Oil and Soap* **20**, 17.
Müller, A. (1932). *Proc. R. Soc. A.* **127**, 417.
Palmer, K. J. and Schmitt, F. O. (1941). *J. cell comp. Physiol.* **18**, 385.
Reiss-Husson, F. (1963). Thesis, University of Strasbourg.
Reiss-Husson, F. (1967). *J. molec. Biol.* **25**, 363.
Reiss-Husson, F. and Luzzati, V. (1964). *J. phys Chem.* **68**, 3504.
Reiss-Husson, F. and Luzzati, V. (1966). *J Colloid and Interface Sci.* **21**, 534.
Reiss-Husson, F. and Luzzati, V. (1967). *Adv. biol. med. Phys.* **11**, 87.
Schmitt, F. O., Bear, R. S. and Palmer, K. J. (1941). *J. cell. comp. Physiol.* **18**, 31.
Skoulios, A. E. (1959). Thesis, University of Strasbourg.
Skoulios, A. E. and Luzzati, V. (1961). *Acta crystallogr.* **14**, 278.
Small, D. M., Bourgès, M. and Dervichian, D. G. (1966). *Nature, Lond.* **211**, 816.
Spegt, P. A. (1964). Thesis, University of Strasbourg.
Spegt, P. A. and Skoulios, A. E. (1963). *Acta crystallogr.* **16**, 301.
Spegt, P. A. and Skoulios, A. E. (1964). *Acta crystallogr.* **17**, 198.
Spegt, P. A. and Skoulios, A. E. (1966). *Acta crystallogr.* **21**, 892.
Spegt, P. A., Skoulios, A. E. and Luzzati, V. (1961). *Acta crystallogr.* **14**, 866.
Stauff, J. (1939). *Kolloidzeitschrift* **89**, 224.
Stoeckenius, W. (1962). *J. cell Biol.* **12**, 221.
Treloar, L. R. G. (1958). "The Physics of Rubber Elasticity", 2nd Ed., Oxford University Press, London.

Vincent, J. M. and Skoulios, A. E. (1966a). *Acta crystallogr* **20**, 432.
Vincent, J. M. and Skoulios, A. E. (1966b). *Acta crystallogr* **20**, 441.
Vincent, J. M. and Skoulios, A. E. (1966c). *Acta crystallogr* **20**, 447.
Vold, R. D. (1941). *J. Am. chem. Soc.* **63**, 2915.
Vold, R. D. and Vold, M. J. (1939). *J. Am. chem. Soc.* **61**, 808.
von Sydow, E. (1956). *Ark. Kemi* **9**, 231.

Chapter 4

Recent Physical Studies of Phospholipids and Natural Membranes

D. CHAPMAN

*Molecular Biophysics Unit, Unilever Research Laboratory,
The Frythe, Welwyn, Herts, England.*

With a contribution on

Optical Rotatory Dispersion Studies

by

D. F. H. WALLACH

*Biochemical Research Laboratory,
Massachusetts General Hospital, Boston, Massachusetts, U.S.A.*

I.	INTRODUCTION	125
II.	THE PHYSICAL PROPERTIES OF PHOSPHOLIPIDS	127
	A. Solid State Behaviour.	128
	B. The Effect of Water	139
	C. Effects Due to Cholesterol	144
	D. Interactions with Protein	152
III.	THE PHYSICAL PROPERTIES OF MEMBRANES	153
	A. X-ray Diffraction Studies of Membranes	154
	B. Electron Microscope Studies	159
	C. Spectroscopic and Optical Rotatory Dispersion Studies . .	174
IV.	CONCLUSIONS AND FUTURE STUDIES	196
	REFERENCES	199

I. Introduction

Recent studies in molecular biology have strikingly demonstrated how the application of a single physical technique can sometimes provide considerable illumination and insight into the properties and organization of important biological molecules and biological processes. We can cite as outstanding examples of this recent X-ray studies of proteins such as lysozyme, haemoglobin, myoglobin and studies of nucleic acids such as DNA. However, despite the considerable success of these studies, it is also becoming increasingly apparent that to solve many modern biological problems will require the use of a number of different chemical, biochemical and physical techniques as well as research of an inter-disciplinary nature. Biological membranes, for instance, because of their very nature, cannot provide suitable single crystals and so the powerful X-ray diffraction technique is rather more limited in its scope in this situation. The organization and mode of interaction of the complex structure of biological membranes appears to be a good example of a biological structure which will certainly require a range of techniques and disciplines for its understanding.

There is a variety of reasons for saying that such an interdisciplinary approach will be needed. The uncertainty about what really constitutes a cell membrane (see later discussion), the difficulty of separating and analysing cell membranes, the lack, until recently, of knowledge about the various components of these membranes and, particularly, their constituent protein material, the large number of different components thought to be present in a cell membrane, i.e. protein, lipid, cholesterol, polysaccharides, metal ions, water are all factors which introduce an enormous degree of complexity and with it uncertainty into any conclusions reached about the cell membrane structure. If we add to these complications the idea that dynamic rather than static conditions must occur in the cell membrane structure *in vivo*, it can soon be appreciated that an elucidation of the modes of cell membrane interactions will certainly pose difficulties and will require very careful study.

Notwithstanding these many difficulties, recent fruitful progress in physical studies of membranes seems to be being made from two main approaches. The first of these lies in a study of the physics and physical chemistry of the individual components and their interactions in simple model systems. The second approach lies in the increasing use of physical techniques, particularly spectroscopic, optical rotatory dispersion and X-ray techniques, to study cell membrane structures themselves. The junction point of these two different physical approaches should be the most illuminating for our understanding of membrane structure.

We begin this chapter by first discussing recent studies of cell membrane components pointing to recent studies of the physical properties of phospholipids in the solid state which help us to provide insight and an understanding of their condition and interaction in the presence of water. We then discuss the interaction of phospholipids with cholesterol and protein. Finally, recent physical studies of the structure and organization of biological membranes will be discussed. We shall try to illuminate our understanding of these complex structures by referring to the basic information available about the properties and interactions of the constituent molecules. In our discussions we deliberately range over a variety of modern physical techniques and particularly those which we have been using recently in our laboratory, pointing to the present applications of these techniques and also their future potential.

II. The Physical Properties of Phospholipids

Modern analytical techniques (see Chapter 1) are showing the various types of phospholipids which are thought to be basic constituents of biological membranes. The most important of these are diacylphosphatidylethanolamines, phosphatidylcholines (lecithins), phosphatidylserines and sphingomyelins. The diacylphosphatidylcholines are the major phospholipids present in mammalian tissues but these occur only in small amounts in bacterial membranes. Many biological membranes contain mixtures of these different classes and the proportions of these classes vary within a particular tissue from one animal to another (van Deenen, 1965). The reason for the occurrence of mixtures of different phospholipid classes and the variation in composition from one type of membrane to another is not yet understood.

The fatty acid residues associated with a given type of phospholipid in biological membranes usually show a distribution in chain length and also in their degree of unsaturation. Analysis of natural phospholipids shows that, in general, there is a saturated fatty acid residue in the 1-position and an unsaturated fatty acid residue in the 2-position of the glycerol moiety of the phospholipids. Stearic, palmitic and myristic acid residues are common and oleic acid is common amongst the unsaturated fatty acid residues. In some cell membranes, such as occur in bacteria, acyl groups containing branched chains are present. The reasons for this distribution of fatty acid residues associated with a given phospholipid in biological tissues are not yet fully understood, nor is it clear why, corresponding to the occurrence of certain diseases or effects due to variation of temperature, these characteristic fatty acid patterns sometimes alter.

Cholesterol occurs in many, but not all, cell membranes, e.g. there are appreciable amounts of cholesterol in myelin and red blood cell membranes but little in mitochondrial membranes. Despite many speculations, we are still uncertain of its main purpose in cell membranes, nor are we sure of the reason why it is present in some membranes but not in others. The precise way in which phospholipid molecules interact with cholesterol molecules is also uncertain. There are many theoretical models (Finean, 1958; Vandenheuvel, 1963) and speculations on this topic (see Chapter 6). Cholesterol may also complex in some way with the protein in the membrane as well as with the phospholipids. This, however, has received less attention.

Perhaps the most outstanding area of uncertainty at present concerns the way in which phospholipids and proteins interact. This is of considerable importance for understanding membrane structure. Whether the protein interacts mainly with the polar group of the lipids or whether there is some "hydrophobic interaction" between the hydrocarbon chains of the lipid and the protein is uncertain. Until recently there was little experimental information available to enable us to decide between the two possibilities and it may be that different membrane types involve each or both of these modes of interaction. There are many membrane models which are based essentially on these two alternative methods of interaction (see Chapter 6, p. 233).

A. SOLID STATE BEHAVIOUR

1. *Single Crystal X-ray Studies*

It may at first seem surprising that we should begin by discussing the structure of phospholipids in the solid state. The solid state behaviour of these molecules may appear to be too far removed from the biological situation and to be completely irrelevant to its discussion. This is not the case since information on the solid state properties provides a foundation for extrapolation to more complicated systems. If we remark that phospholipids of the type found in biological membranes are generally insoluble in water, this should go some way to suggest that information about the solid state properties are indeed relevant. We shall see that an examination of the solid state properties provides insight into their behaviour in other physical situations and, furthermore, that more detailed information about the solid state situation is desirable.

X-ray studies of phospholipids are presently in progress in various laboratories throughout the world to determine the complete crystal structure of these molecules. As yet, however (1968), no complete X-ray structure of a phospholipid fully esterified and containing fatty acid chains has been published. Preliminary X-ray studies have, however, been reported on a number of related derivatives, e.g. the structure of L-1-glycerylphosphorylcholine cadmium chloride trihydrate has been obtained (Sundaralingam and Jensen, 1965). This has shown that this molecule exists as a dipolar ion with the positive charge on the quaternary nitrogen neutralized by the negative charge on the phosphoric acid residue. The choline residue exists in the *gauche* conformation rather than the more extended form. The crystal structure of L-1-glycerylphosphorylcholine has also been reported (Abrahamsson and Pascher, 1966).

It is too early to say just how relevant these structures may be to the lipid organization in membranes. The most significant points seem to be (a) that the choline residue exists in the *gauche* conformation rather than the extended zig-zag form and (b) that the conformation of the glycerol residue can be *gauche gauche* (see Fig. 1). The molecular configuration of the deacylated plant sulpholipid has been studied via the anhydrous rubidium salt (Okaya, 1964). This has confirmed the chemical analysis of this lipid. The crystal structure of 2-aminoethanolphosphate has also been reported (Kraut, 1961).

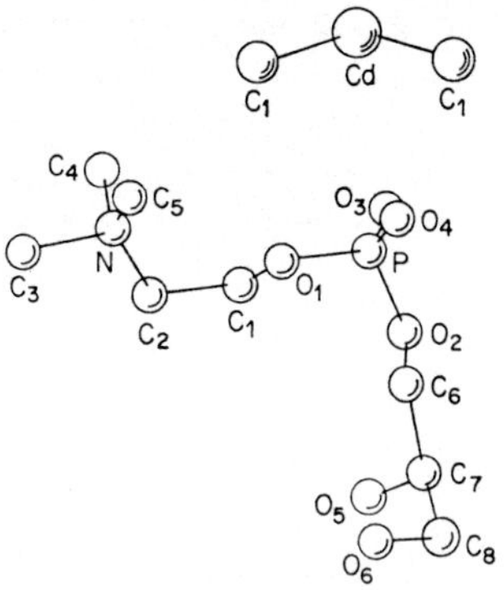

FIG. 1. The structure of L-1-glycerylphosphorylcholine cadmium chloride hydrate (from Sundaralingam and Jensen, 1965).

Single crystal X-ray studies of some diacyl-DL-phosphatidylethanolamines, e.g. 1,2-dilauroylphosphatidylethanolamine, are at present in progress in our laboratory at The Frythe. Although this study of 1,2-dilauroylphosphatidylethanolamine is, as yet, incomplete, it suggests the following important structural features: (a) the two hydrocarbon chains of the fatty acid residue present in each lipid molecule are pointing in the same direction and are parallel to each other and (b) the polar portion of the molecule containing the phosphate and amine group appears to be in line with the hydrocarbon chains.

The fact that the hydrocarbon chains in this structure are parallel to each other is not too surprising as previous X-ray diffraction studies have shown that with most long chain compounds there is a marked tendency for the hydrocarbon chains to pack in a parallel array. Long chain fatty acids, esters and glycerides all pack in this way (Chapman, 1965). This is because, beyond a certain chain length, the long hydrocarbon chains tend to dominate the crystallization processes. This in turn is related to the favourable dispersion interaction forces which arise from parallel packed chains (Salem, 1962).

In related lipid molecules, while the hydrocarbon chains within the molecule lie parallel to each other, the chains do not always point in the same direction. The single crystal X-ray studies of triglycerides, e.g. trilaurin (Larsson, 1964) and tricaprin (Jensen and Mabis, 1964), show that the two hydrocarbon chains on adjacent substituted positions, i.e. substituted at the

1- and 2-positions on the glycerol molecule, point in opposite directions. The different situation which appears to exist with the diacylphosphatidylethanolamines is important because the solid state structure can be regarded as being built up from infinite sheets of phospholipid bilayers with the polar groups organized in sheets and with all the hydrocarbon chains of adjacent lipid molecules lying parallel, i.e. it has a somewhat similar arrangement to the organization of the lipid which has been suggested to occur in many biological membrane systems.

As yet no information is available about the way in which the chains in a phospholipid pack together when the molecule contains both a saturated and and unsaturated (*cis*) fatty acid residue. A possible packing arrangement for these lipids is shown in Fig. 2.

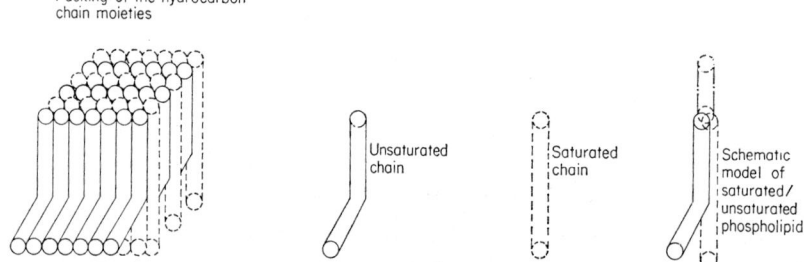

Fig. 2. A possible packing arrangement for phospholipids containing saturated and unsaturated (*cis*) chains.

Because of the difficulty of growing single crystals of phospholipids, other attempts to deduce information about the solid state properties have been made using electron diffraction methods and X-ray powder diffraction methods. Studies using electron diffraction are few and, as yet, have given only a very limited amount of information (Parsons and Nyberg, 1966). Studies using X-ray powder diffraction methods have provided more information, e.g. the way in which the hydrocarbon chains are packed in the crystal lattice, and the angle of tilt of the chains to the basal planes has been determined. The diffraction data provides information about the long and short X-ray spacings. The values of the X-ray short spacings suggest that the hydrocarbon chains are packed in similar packing arrangements to those adopted by many other long chain molecules (Chapman, *et al.*, 1966a). By examining the long spacings of an homologous series of phospholipids of various chain lengths, it is possible, by extrapolating back to zero chain length, to deduce the space taken up by the glycerol residue and the polar grouping. The long spacings of an homologous series of phospholipids are shown in Fig. 3.

Some impression of the polar group organization present in phospholipid molecules can be obtained by a consideration of their capillary melting points. The capillary melting points of a number of pure phospholipids have been determined and are shown to be quite high, e.g. the value of the capillary

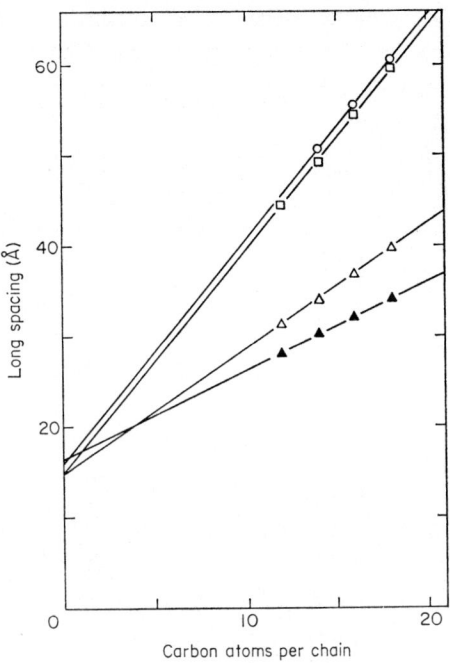

FIG. 3. The long spacings of a number of diacylphosphatidyl ethanolamines in various polymorphic forms (from Chapman et al., 1966a). ○, A form; ⊘, □ A', form; △, B form; ▲, C form.

melting points for the diacyl phosphatidylethanolamines is about 200°C, while for the phosphatidylcholines these melting points are 230°C. It is important to note that these values are independent of both the chain length and the degree of unsaturation of the fatty acid residues associated with the phospholipid. These melting points can be compared with those of other long chain molecules. The melting points of fatty acids containing the same length of hydrocarbon chain are much lower, e.g. stearic acid m.p. is 69·7°C. On the other hand, the capillary melting point of sodium stearate is ∼300°C. The high values of these capillary melting points are, therefore, consistent with the occurrence of ionic linkages existing in the crystal associated with the polar groups of the phospholipid. The higher melting points of the sodium soaps, compared with the phospholipids, suggest that there is greater ionic character associated with the polar groups of these molecules than with the diacylphosphatidylcholines and phosphatidylethanolamines.

2. Thermotropic Mesomorphism

In addition to the capillary melting point, other phase changes have been shown to occur with phospholipids at lower temperatures, for example, when a pure phospholipid, dimyristoylphosphatidylethanolamine, containing two fully saturated chains is heated from room temperature up to the capillary melting point, a number of thermotropic phase changes occur (i.e. phase

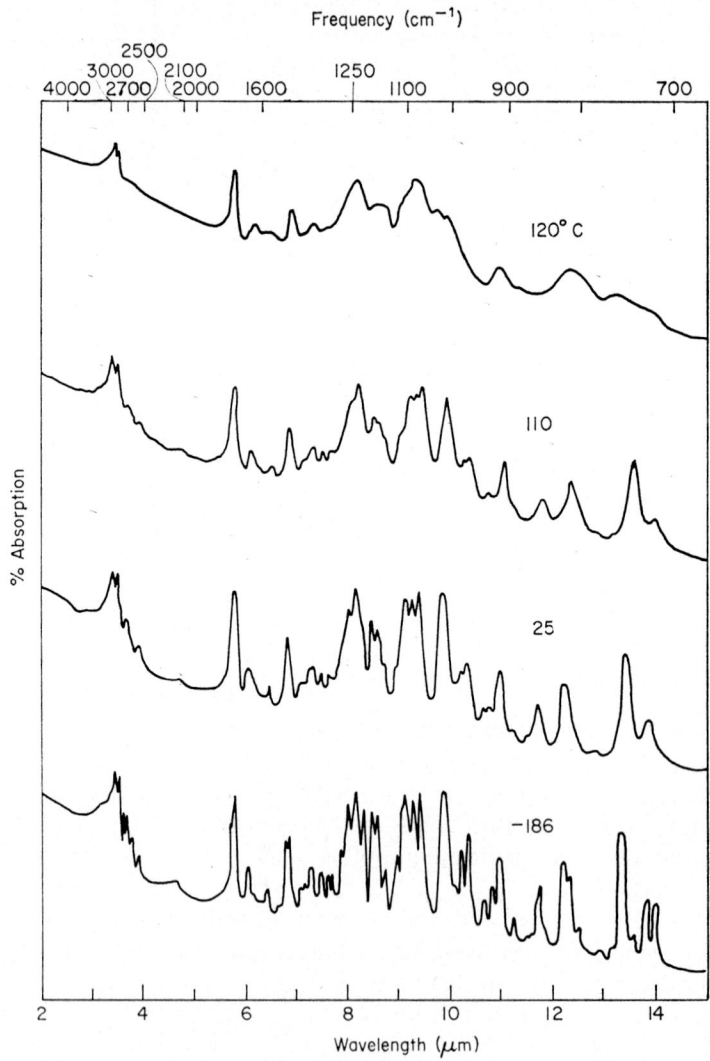

FIG. 4. The infrared spectrum of 1,2-dilauroyl-DL-phosphatidylethanolamine at different temperatures (from Chapman et al., 1966a).

changes caused by the effect of heat). This was first shown by infrared spectroscopic techniques (Byrne and Chapman, 1964), then by thermal analysis techniques (Chapman and Collin, 1965) and has now been studied by a variety of physical techniques (Chapman et al., 1966a). Optical studies show that dimyristoylphosphatidylethanolamine at room temperature is birefringent under crossed polars. On heating, three processes occur: first some loss in birefringence at the first transition temperature ~120°C; a small increase at 135°C and a pronounced overall loss of birefringence near the capillary melting point of 200°C. Above 120°C, pressure on a coverglass with a needle causes the material to flow. When the temperature of the phospholipid reaches ~120°C the i.r. absorption spectrum undergoes a remarkable change. Above this temperature the spectrum loses all the fine structure and detail which was present at lower temperatures. At the high temperature the spectrum becomes similar to that obtained with a phospholipid dissolved in a solvent such as chloroform. The i.r. spectra of this phospholipid at different temperatures is shown in Fig. 4.

Differential thermal analysis (d.t.a.) shows that a marked endothermic transition (absorption of heat) occurs at this transition temperature (Chapman and Collin, 1965). An additional heat change occurs at ~135°C and only a small heat change is involved near the capillary melting point of the lipid (see Fig. 5). This behaviour is similar to that which occurs with liquid crystals, such as p-azoxyanisole or cholesteryl acetate which form nematic and cholesteric liquid crystalline phases.

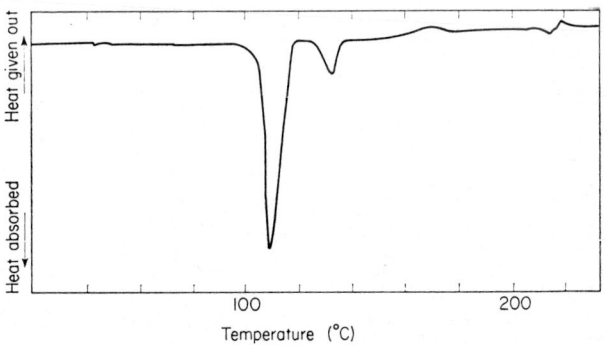

FIG. 5. Differential thermal analysis heating curve of 1,2-dimyristoylphosphatidylethanolamine (from Chapman and Collin, 1965).

The X-ray long spacings show a dramatic reduction to some two-thirds of their original value at the first transition temperature with a further small reduction at the second transition temperature. The X-ray short spacings change at the first transition point from sharp diffraction lines to a diffuse spacing at ~4·5 Å. Nuclear (proton) magnetic resonance (p.m.r.) studies show a gradual reduction in line width from about 15 gauss at liquid nitrogen temperature until, at the first transition temperature, there is a sudden reduction in the line width (to ~0·09 gauss). The p.m.r. spectrum of dimyristoylphosphatidylethanolamine is shown in Fig. 6. This shows that molecular motion increases gradually as the temperature increases until, at the transition temperature, a considerable increase in the molecular motion takes place. (Chapman and Salsbury, 1966).

The main conclusions from these various studies are that (a) even with the fully saturated phospholipid at room temperature, some molecular motion occurs in the solid. This is evident from the p.m.r. spectra and from the i.r. spectra taken at liquid nitrogen and at room temperatures. (Note the difference between the i.r. spectra at −186°C and room temperature shown in Fig. 4). (b) When the phospholipid is heated to a higher temperature, it reaches a transition point, a marked endothermic change occurs and the hydrocarbon chains in the lipid "melt" and exhibit a very high degree of molecular motion. This is evident both in the appearance of the i.r. spectrum and also in the narrow n.m.r. line width. On the one hand the broad diffuse appearance of the i.r. spectrum is consistent with the chains flexing and

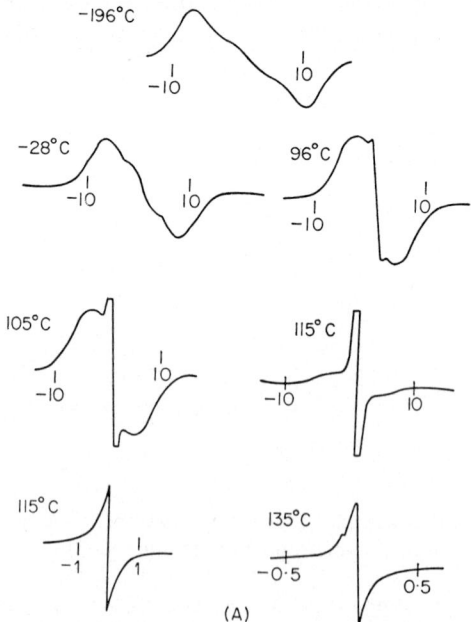

FIG. 6. (A) P.m.r. derivative absorption spectra of 1,2-dimyristoyl-DL-phosphatidylethanolamine at various temperatures. The abscissae are marked in gauss.

twisting and with a "break-up" of the all-planar *trans* configuration of the chains. (The i.r. spectrum can be regarded as giving a "rapid snapshot" showing the condition and organization of the phospholipid molecule within a period of 10^{-15} sec. The n.m.r. spectrum, on the other hand, is affected by

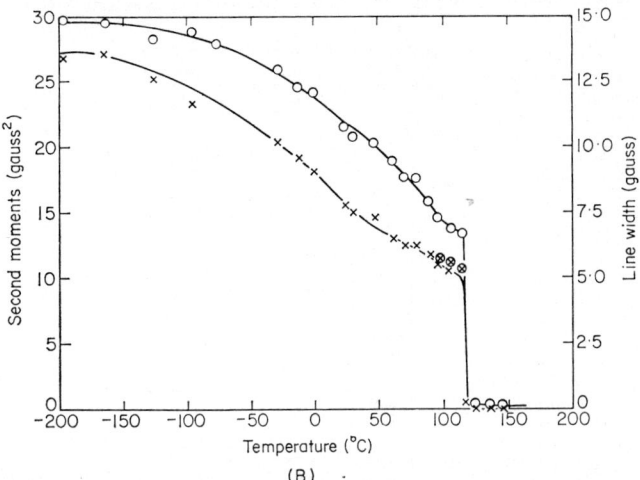

FIG. 6. (B) The line width (○) and second moment (×) of 1,2-dimyristoylphosphatidylethanolamine as a function of temperature (from Chapman and Salsbury, 1966).

slower motions. It provides an impression of any molecular motion, i.e. rotation, translation or diffusion, of the phospholipid molecule occurring within a much longer time period, some 10^{-7} sec).

The fact that the phase transition is concerned primarily with the hydrocarbon chains of the phospholipid is confirmed by the X-ray data. This shows that the space taken up by the glycerol and polar group remains essentially unchanged when this phase transition occurs. (This can be seen by extrapolation of the reduced X-ray long spacing back to zero carbon atoms in the chains in Fig. 3).

When phospholipids contain shorter chain lengths, or unsaturated bonds, those marked endothermic phase transitions occur at lower temperatures. The temperature at which these transitions occur parallels, to some extent, the behaviour of the melting point of the related fatty acids. Thus these transition temperatures are high for the fully saturated long chain phospholipids. They are lower when there is a *trans* double bond present in one of the chains and lower still when there is a *cis* double bond present. This variation of transition temperature also confirms that this phase transition is primarily associated with a "melting" of the hydrocarbon chains of the phospholipid, while this in turn is a reflection of the dispersion forces between the chains.

Only one main "melting of the chains" occurs even when there are two different types of chain present in the phospholipid. The transition temperatures for different phospholipid classes vary even though they contain exactly the same fatty acid residues. The difference between the transition temperatures for a diacylphosphatidylcholine and a diacylphosphatidylethanolamine are shown in Fig. 7. While natural phospholipids from erythro-

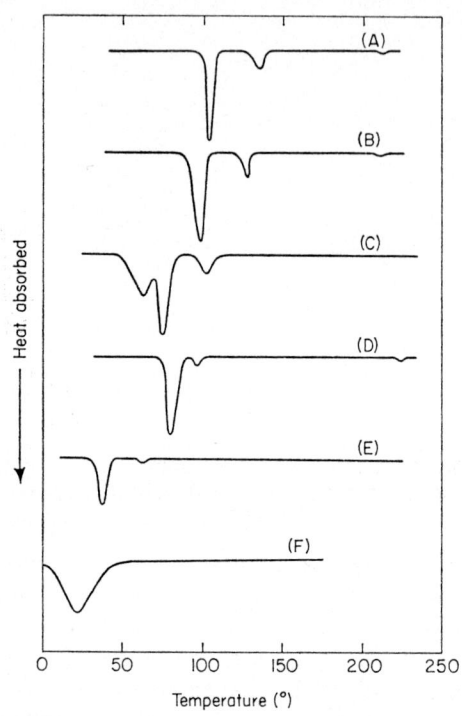

FIG. 7. Differential thermal analysis heating curves for different phospholipids (from Chapman, 1967).
(A) 1,2-dimyristoyl-DL-phosphatidylethanolamine; (B) 1,2-dielaidoyl-DL-phosphatidylethanolamine; (C) 1-stearoyl-2-oleoyl-phosphatidylethanolamine; (D) 1,2-distearoyl-DL-phosphatidylcholine; (E) -1-stearoyl-2-oleoyl-DL-phosphatidylcholine; (F) egg yolk lecithin (from Chapman, 1967).

cyte or mitochondrial membranes contain large amounts of unsaturated *cis* double bonds and, therefore, in the dry condition have endothermic transition temperatures either near or below room temperatures, the highly saturated derivatives exhibit transition temperatures much higher than room temperature.

With mixtures of phospholipids we might expect the transition temperatures to occur over a wider range of temperature than with a single phospholipid. This is illustrated in the d.t.a. heating curves of egg yolk phosphatidylcholine (see Fig. 7).

When a crystalline phospholipid approaches the first thermotropic transition temperature we can imagine that the molecular motion of the chains increases until, at the first transition point, lateral expansion of the crystal lattice is forced to take place. This expansion allows even greater chain mobility possibly involving cooperative coiling of the chains with rotation about C—C bonds occurring. Such a coiling of chains could explain the ob-

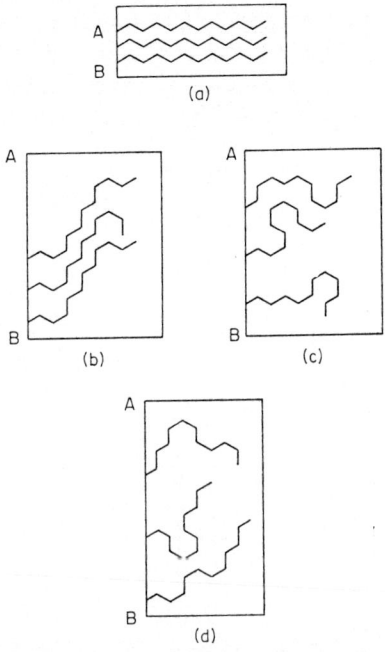

FIG. 8. Typical configurations of sets of chains on a two-dimensional hexagonal lattice as the interchain separation is increased. Separation increases from (a) to (d) (from Whittington and Chapman, 1966).

served reduction in long spacing. A similar coiling of chains probably takes place with polar long chain molecules in a monolayer at an air–water interface dependent upon the density of the chain packing.

A very simple theoretical system has been devised (Whittington and Chapman, 1966) to provide some insight for such situations. The model system is restricted to a set of simple chains in two dimensions and with the chains confined to lie on a two-dimensional hexagonal lattice. The end-to-end distance of each chain (corresponding in a real system to the distance from the polar group to the methyl group) and other properties were determined as a function of the density of packing of the chains using two simple potential functions and using the Monte Carlo computational method. Typical configurations of sets of simple chains on a two-dimensional hexagonal lattice are shown in Fig. 8. At the highest density only the fully extended configuration occurs. At lower densities other configurations are allowed. As the density of the chains decreases, the end-to-end distance of the chains suddenly falls, consistent with the occurrence of a cooperative phase transition.

An important conclusion which we obtain from these phase transition studies is that, near to the endothermic transition temperature, a given phospholipid can be in a highly mobile condition with its hydrocarbon chains flexing and twisting. The more unsaturated the chain, the lower the temperature will be at which this occurs. This is a fundamental property of the

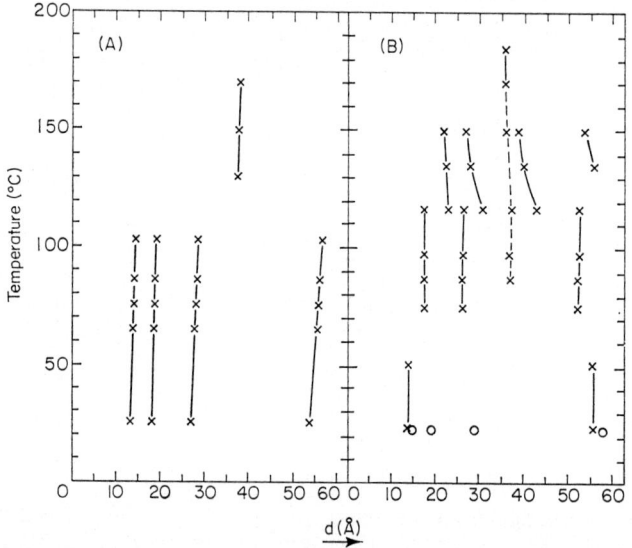

Fig. 9. Temperature dependence of the low angle X-ray diffraction patterns of (A) anhydrous 1,2-distearoyl-L-phosphatidylcholine and (B) the monohydrate of 1,2-dipalmitoyl-L-phosphatidylcholine ($\times$ form, α_1; $\bigcirc$ form, α_2) (from Chapman et al., 1967b).

phospholipid and we can expect this chain mobility to occur in whatever situation the phospholipid occurs unless, for special reasons, this motion is somehow inhibited. We can envisage that inhibition of chain motion by interaction with other molecules would provide one of these special reasons, e.g. water or protein. In other circumstances, due to less perfect packing arrangements, we might expect, at a particular temperature, even greater mobility of the chains of the lipid and indeed of the whole lipid molecules themselves.

B. THE EFFECT OF WATER

Small amounts of water can have unusual effects upon the mesomorphic behaviour. Thus the diacylphosphatidylcholines (lecithins) exhibit additional liquid crystalline forms between the first transition temperature and the capillary melting point (Chapman et al., 1967b). The intermediate liquid crystalline form is found to exhibit X-ray spacings consistent with a cubic phase organization. On the other hand, if all the water is removed from the phospholipid, the lipid will no longer exhibit this cubic phase.

The X-ray diffraction data for one of these lipids is shown in Fig. 9 where the spacings for this cubic phase are apparent. A nuclear resonance study of the lipid in this cubic phase suggests that there is considerable molecular freedom of the molecules in this phase or a reduction in the dispersion forces between the chains. The dipole–dipole broadening is averaged out sufficiently so that a high resolution spectrum is observed when the lipid is in this phase. The chemically shifted proton groupings show distinct lines similar to those observed when the phospholipid is dissolved in a solvent like chloroform. On heating to temperatures above that at which the cubic phase exists, the p.m.r. spectra suddenly broaden again and from this point up to the capillary melting point, only a broad line is observed. The p.m.r. spectra are shown in Fig. 10.

When phospholipids are examined in increasing amounts of water, the various physical techniques, such as microscopy, n.m.r. spectroscopy or differential thermal analysis, show that as the amount of water increases, the marked endothermic transition temperature for a given phospholipid falls. The transition temperature does not fall indefinitely; it reaches a limiting value independent of the water concentration. We can understand this if we regard the effect of water as leading first to a "loosening" of the ionic structure of the phospholipid crystals. This in turn affects the whole crystal structure and a reduction, up to a certain limit, of the dispersion forces between the hydrocarbon chains. Large amounts of energy are still required to counteract the dispersion forces between the chains and quite high temperatures are still required to cause the chains to melt. Some d.s.c. heating curves for a series of long chain lecithins of different chain length in water are shown in Fig. 11. These limiting transition temperatures parallel the melting point behaviour of the analogous fatty acids becoming lower with increasing unsaturation. This further reduction of the endothermic transition temperatures by water means that the natural phospholipids extracted from biological membranes usually exhibit this crystalline to liquid crystalline transition many degrees below the biological environmental temperature. At the biological environmental temperature we can expect the phospholipids which contain highly unsaturated chains to be in a highly mobile and fluid condition.

Some of the water added to the phospholipid appears to be "bound" to the lipid, e.g. 1,2-dipalmitoylphosphatidylcholine binds about 20% water. This water does not freeze at 0°C and calorimetric studies made with lipid–water

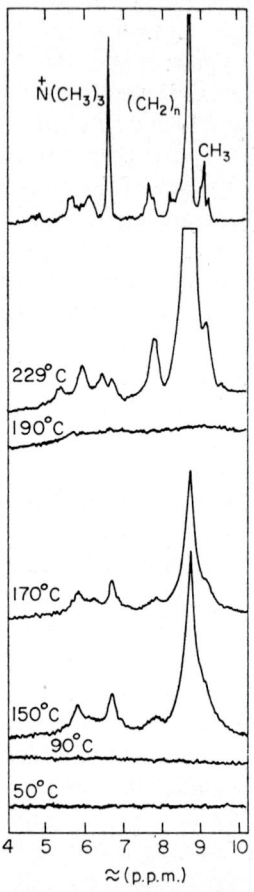

FIG. 10. High resolution p.m.r. spectra of 1,2-dipalmitoyl-L-phosphatidylcholine monohydrate at different temperatures below and above the capillary melting point. The top spectrum is of the phospholipid dissolved in chloroform (from Chapman *et al.*, 1967b).

mixtures show (Chapman *et al.* 1967b) that only after 20% water has been added to the lipid is a peak at 0°C observed. Some d.s.c. traces for lipid–water mixtures of different concentrations are shown in Fig. 12. This "bound" water may have considerable relevance to interactions of anaesthetics, drugs and ions with biological membranes. If this bound water varies either in its properties or in its total amount, dependent upon the type of ion or interacting molecule, this in turn may alter transport and diffusion properties across the membrane. The amount of bound water associated with the constituent lipids and proteins will perhaps provide a limit for the amount of water which can be removed from biological membranes before they lose their organization.

There are a number of important features associated with the transition temperature for the lipid when it is in the presence of water. The first of these is that the ability to disperse the lipid in water increases markedly above this transition temperature. Only those phospholipids which have transition temperatures, when placed in water below or near to room temperature, spontaneously form myelin figures. Fully saturated phospholipids which have

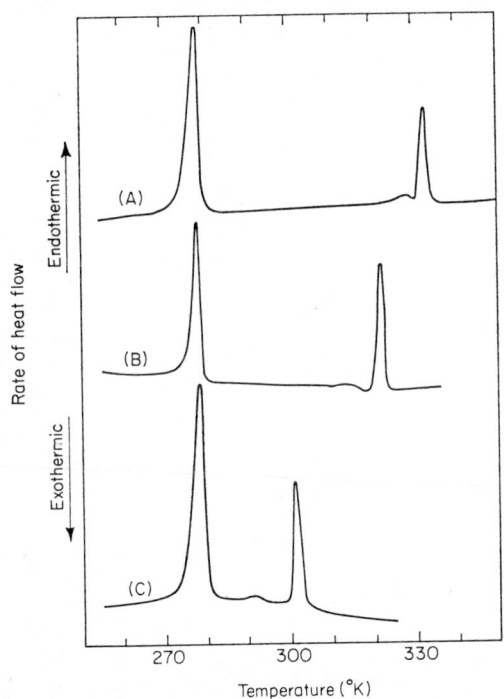

FIG. 11. Differential scanning calorimeter heating curves for some 1,2-diacylphosphatidylcholine water systems for the (A) distearoyl, (B) dipalmitoyl and (C) dimyristoyl derivative. The peak due to ice at 0°C is shown. (from Chapman et al., 1967b).

high transition temperatures do not form myelin figures at room temperature. However, if the temperature is raised to the transition point, these phospholipids form myelin figures. In the presence of an excess of water, phospholipids, such as 1,2-dipalmitoylphosphatidylcholine, spontaneously form myelin figures at 42°C.

This ability to form myelin figures with saturated phospholipids has been confirmed by microscopic and electron microscope investigations (Chapman and Fluck, 1966). It has been used to attempt to provide information about the situation of the osmium used after osmium tetroxide fixation procedures. There have been many discussions as to whether the osmium is located at a double bond of the membrane lipid rather than with the polar group. If the double bond is necessary for fixation to occur and to give the so-called

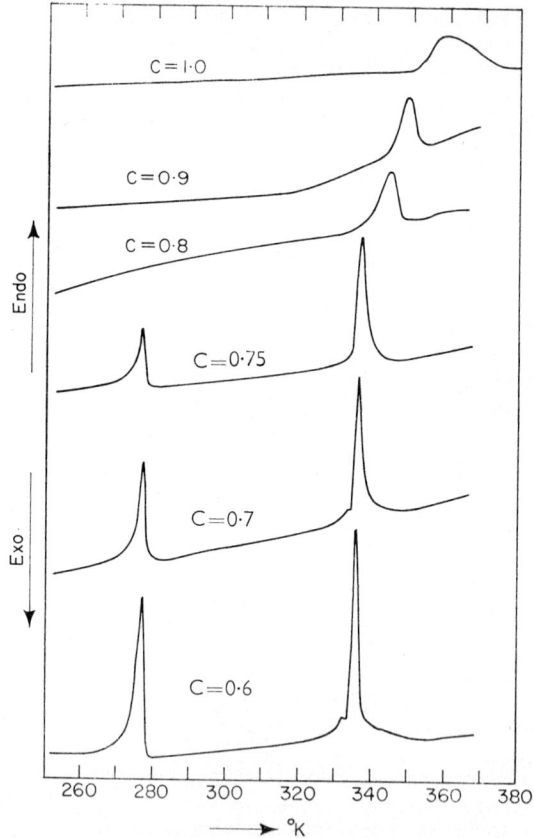

FIG. 12. D.s.c. heating curves for 1,2-distearoylphosphatidylcholine in increasing amounts of water.

"tram-line" structure (two black lines separated by a light spacing), it will not be observed with a fully saturated phospholipid. In fact, these studies show that while the fully saturated phosphatidylcholines do not show a typical membrane-like structure, even when reacted with osmium tetroxide at the high transition temperatures required for myelin figure formation, on the other hand, the fully saturated phosphatidylethanolamines do exhibit this structure in the electron micrographs.

A second feature of lipid/water systems is their monolayer behaviour. Monolayer studies of phospholipids have been carried out for a considerable number of years. Usually this work has been performed with natural phospholipid mixtures and, in the vast majority of cases, with egg yolk phosphatidylcholine. In recent years a few studies have been made with pure synthetic phospholipids. These show that the fully saturated phospholipids exhibit, at room temperature, monolayers which are more condensed than are the unsaturated phospholipids containing *cis* hydrocarbon chains, i.e. the saturated lipids occupy less area at low surface pressures than do the unsaturated derivatives.

Monolayers obtained with phosphatidylcholines are observed to be much more expanded than are the corresponding phosphatidylethanolamines containing the same acyl chains. These results can be compared with the d.t.a. results discussed earlier. A high transition temperature for liquid crystal formation is, in general, correlated with a condensed type monolayer and a low transition temperature with an expanded film (Chapman *et al.*, 1966a). The d.t.a. transition temperatures are higher for the phosphatidylethanolamines than with the corresponding phosphatidylcholines. Phospholipids containing *trans* (elaidoyl) unsaturated chains have higher transition temperatures than those containing *cis* (oleoyl) chains. Phospholipids containing one fully saturated chain and one *trans* unsaturated chain give condensed monolayers similar to those observed with completely saturated phospholipids (Chapman *et al.*, 1966b).

1. *Lyotropic Liquid Crystalline Phases*

Phospholipids in the presence of water can also form various types of lyotropic mesomorphism, i.e. they can exhibit different types of liquid crystalline organization and, in some cases, as the concentration of water varies, transitions from lamellar to hexagonal phases occur. These phases are more fully discussed in Chapter 2 by Dr. Luzzati, page 71.

In our laboratory recent studies on pure 1,2-diacylphosphatidylcholines show that there is no lamellar/hexagonal transition. The presence of impurities such as ions can have an appreciable effect upon the amount of water taken up by the lipid (Chapman *et al.*, 1967b).

It has been shown that brain phospholipid can exist in the presence of water in two different phases, the usual so-called lamellar phase and a hexagonal phase, dependent upon temperature and concentration (see page 109). It is clear that the temperature at which a particular phase of this type can exist for a given phospholipid will also ultimately depend upon the transition temperature for melting of the hydrocarbon chains. This follows from the fact that, if the phospholipid in water is in a crystalline condition at a certain temperature, it clearly cannot transform from one liquid crystalline phase to another at this temperature.

2. Model Membranes

Model phospholipid membranes of bilayer dimensions have been made with egg yolk phosphatidylcholine. These model structures are discussed in more detail in Chapter 6 by Dr. Lucy, p. 265. The formation of membranes becomes impossible as the temperature of preparation is lowered from 36° to 20°C. This is probably related to the degree of fluidity of the hydrocarbon chains of the lipid at the temperature of the experiment.

C. EFFECTS DUE TO CHOLESTEROL

An interaction which may be of considerable biological importance is that of cholesterol with phospholipids. Cholesterol occurs in many membranes, particularly in the myelin sheath and red blood cell membranes but its precise function and arrangement are not understood. The biochemical importance of the solubilization of cholesterol by phospholipid and its vehicular possibilities have often been discussed (Fleischer and Brierley, 1961).

As long ago as 1925 Leathes showed that cholesterol, when mixed with certain fatty acids in monomolecular films on water, caused a diminution of the area occupied by the fatty acids. The study of this condensing effect of cholesterol on phospholipids has continued up to the present day using monolayer techniques. Despite the many investigations carried out on this system, the interpretation of the effect is still uncertain. Monolayer studies of natural phospholipids, such as egg yolk lecithin, or with pure phospholipids containing *cis* unsaturated chains, have shown that a film of a more condensed type is produced when cholesterol is present. This is shown in Fig. 13. This has led to a variety of models and discussions of the interaction between the phospholipid and cholesterol. It has been suggested that the presence of a *cis* double bond in the 9:10 position of the hydrocarbon chain of a phospholipid is ideally suited for the combination of a cholesterol molecule (Vandenheuvel, 1963). It has also been stated that short chain fully saturated phospholipids which give expanded monolayers nevertheless do not interact with cholesterol (Demel, 1966). Recent results in our laboratory show that an apparent interaction occurs between cholesterol and phospholipids even when the double bond is in positions other than that of the 9:10 position, provided that the monomolecular film of the phospholipid is expanded. Thus phospholipids containing hydrocarbon chains with a double bond in the 5:6 position and those containing a double bond at the 11:12 position also show a condensation effect. Monolayers of the saturated phospholipids are usually already condensed and show no further condensing effect in the presence of cholesterol. Phospholipids containing a single *trans* unsaturated chain show little, if any, condensing effect but the dielaidoyl derivatives do show some condensing effect (Chapman *et al.*, 1966b). This shows that the presence of a natural "kink" in the molecule as produced by the *cis* double bond at the 9:10 position is *not* a necessary condition for condensation.

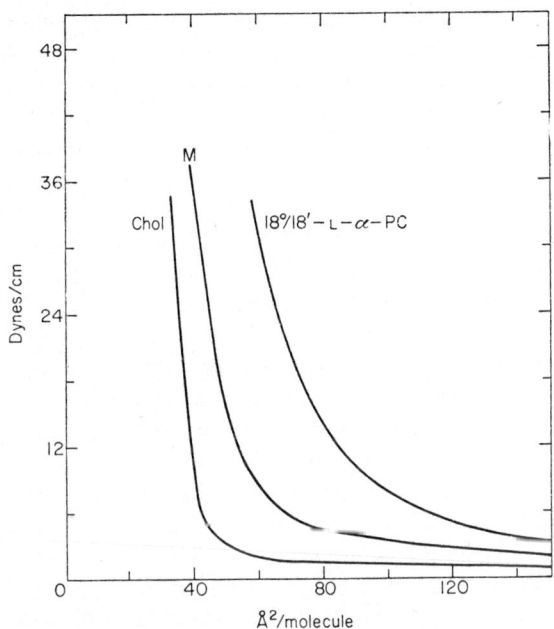

FIG. 13. Force-area characteristics of cholesterol, 1,stearoyl-2-oleoyl-L-phosphatidylcholine and a mixed film of both compounds in equimolar proportions (after van Deenen *et al.*, 1962).

Recently, further studies have been made of mixed monolayers of dicetylphosphate-cholesterol, dipalmitoylphosphatidylcholine–cholesterol and egg yolk lecithin–cholesterol, using surface pressure and surface potential measurements (Shah and Schulman, 1967). Surface potential measurements were used to provide a reliable parameter for the study of interaction in the monolayers. The additivity rule for average area was shown to be followed by monolayers of dicetylphosphate and cholesterol although lecithin–cholesterol monolayers show a deviation from this rule.

It is useful to explain the additivity rule at this point. The area available per molecule in a mixed monolayer is calculated as follows: the number of molecules of both compounds on the surface are calculated from their molecular weights and the amount of each present in the monolayer. The total area of the monolayer is divided by the total number of molecules. This gives the average area available per molecule or, more simply, the area per molecule. If the molecules of both compounds in the mixed monolayer occupy the same molecular area as in their individual monolayers, then the points for the average area per molecule of the mixed monolayer would lie on a straight line joining the two end points for the pure compounds at the same state of compression. The deviation from this additivity rule indicates condensation of the mixed monolayer and this has, in general, been regarded as an indication of interaction between the two components of the mixed monolayers. The additivity rule for average potential is found to be followed by the lecithin–cholesterol monolayers whereas dicetylphosphate–cholesterol monolayers show a deviation from this rule. This evidence has been used to suggest that there is no interaction or complex formation between lecithin and cholesterol but that there is ion-dipole interaction between dicetylphosphate and cholesterol.

Shah and Schulman argue that the apparent condensation of mixed monolayers of lecithin in the presence of cholesterol can be explained by a consideration of molecular cavities caused by the following factors: inclination, length, extent of compression and monounsaturation of the fatty acyl chains. The cholesterol molecules occupy these cavities and it is because of this that they do not cause a proportional increase in area of the mixed monolayers. The authors suggest that cholesterol rather than condensing phospholipid films actually impart fluidity to monolayers. They infer that this may also have relevance to natural membranes.

The fact that the maximum condensation at low surface pressures with lecithincholesterol occurs at a ratio of 1:1, is according to these workers, not due to the formation of a complex but represents a geometrical arrangement of these molecules for optimum packing in the mixed monolayer. The fact that a mixed monolayer of 1-palmitoyl-2-linoleoylphosphatidylcholine–cholesterol does not follow the additivity rule, although this phospholipid forms more expanded monolayers than does egg yolk lecithin, is also explained on the basis of molecular cavities. It is suggested that the linoleoyl chain, in contrast to the oleyl chain, does not cause a molecular cavity at such surface pressures and, therefore, addition of cholesterol in this case causes a proportional increase in the area of the film.

From what we have said earlier about the mobility of the chains of natural phospholipids at room temperature in the solid condition, or in the presence of water, we can appreciate that, if the hydrocarbon chains of a particular phospholipid are flexing and twisting and are in a mobile condition in the "solid" state, we can expect that these chains will be flexing and twisting *at least* as much as this in a monolayer at the same temperature. We can interpret the observed relation between the d.t.a. transition temperature for the solid phospholipid and the type of monolayer observed as being a direct reflection of the degree of flexing and twisting of the hydrocarbon chains. For a given phospholipid, the more expanded the monolayer, the greater is the molecular motion of the chains.

Now, as we have seen, cholesterol present in a monolayer with a phospholipid containing *cis* unsaturated chains causes a condensing effect on the monolayer. The explanation for the condensing effect may be, as Shah and Schulman claim, merely the result of a cavity being available with a given phospholipid because of the shape of the chain. However, another explanation is that the cholesterol, because of its flat shape when present in the monolayer, inhibits some of the freedom of motion of the lipid chains.

The fact that the fully saturated phospholipids and *trans* unsaturated phospholipids (provided that they are not in a fully condensed condition) can also give monolayers which can be condensed by cholesterol, suggests that the shape of the hydrocarbon chain is not the significant factor which these authors suggest. A more general conclusion seems to be that, providing (a) that the lipid is not in a fully condensed condition and (b) that the lipid is not in too much of a fully expanded condition, then cholesterol in a monolayer can cause condensation.

The relevance of monolayer results to biological membranes requires some cautionary remarks. These measurements are usually carried out at the air–water interface and, except for the lung membrane, this is not the situation experienced by most biological membranes. Some models of the biological membrane envisage a bimolecular layer playing a predominant role in its structure. Even if this is correct, and this is not yet certain, monolayer results, although related, may have important differences from the behaviour of phospholipids in bilayers. Even if monolayer results are relevant, there is still uncertainty which part of the force–area curve is most appropriate to biological membranes. Gorter and Grendel (1925) assumed a value of film area at the first detectable surface pressure, while Dervichian and Macheboeuf (1938) assumed that the lipid film at collapse pressures more nearly represented the membrane structure.

The interaction of egg yolk lecithin and cholesterol has also been studied in bulk systems in the presence of water using microscopic and X-ray techniques (Small and Bourgès, 1966). The three component phase-diagram of this system is shown in Fig. 14. Along the water–cholesterol side of the triangle one notes that the cholesterol is totally insoluble in water at 25°C. However, along the lecithin–water side, from about 12–45% water (A-B) lecithin forms with water a lamellar paracrystalline phase. Mixtures containing more than 45% water separate into this lamellar mesomorphic phase floating in the excess of water in the form of myelin figures and anisotropic droplets. All mixtures containing the three components appear as points within the inside of the triangular diagram.

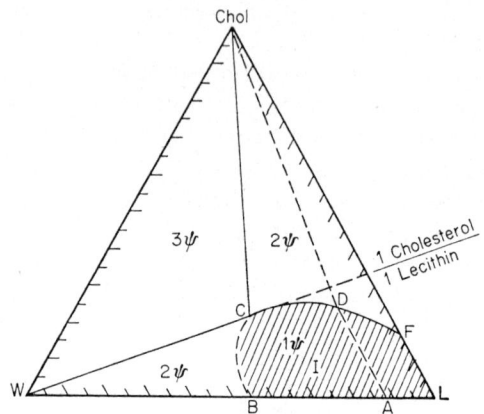

Fig. 14. Egg-yolk lecithin (L)-cholesterol (Chol)-water (W) system at 25°C. The number of phases present is indicated (from Small and Bourgès, 1966).

The microscopic appearance of lecithin or of the lecithin–cholesterol mixture is that of a neat soap and only one sharp long spacing is observed at 48 Å by X-ray examination. There is no second or third spacing to show whether it is in fact truly lamellar. Below 5% of water, several spacings are observed. From the phase diagram it appears that cholesterol can be added to lecithin up to a molecular ratio of 1:1 to give the same lamellar paracrystalline phase already given by lecithin and water. Any excess above this proportion separates as cholesterol crystals. X-ray analysis shows that, in general, the addition of cholesterol to lecithin at a constant water concentration tends to make the lipid layer slightly thicker. This slight increase of X-ray spacing, according to these authors, does not seem sufficient to account for the cholesterol being positioned between the ends of the hydrocarbon chains of the lecithin in the lamellar layer. These authors suggest that it is more probable that the cholesterol is interdigitated between the lecithin molecules and that its hydroxyl group is lying in the water layer in a similar manner to mixtures of lecithin and cholesterol spread on the surface of water.

Recent studies of lecithin-cholesterol–water interactions have also been made using differential scanning calorimetry (Ladbrooke *et al.*, 1968). This work has shown that addition of cholesterol to dipalmitoyl-L-lecithin in water lowers the transition temperature between the gel and liquid crystalline phase and decreases the heat absorbed at the transition. No transition is observed with an equimolar ratio of the lecithin with cholesterol. Unsaturated lecithins and the lipid extract of human erythrocyte ghosts exhibit similar behaviour.

The d.s.c. curves between 280 and 360°K for a series of 1,2-dipalmitoyl-L-lecithin/cholesterol mixtures each containing 50% by weight of water and varying ratios of lecithin to cholesterol are shown in Fig. 15. As the concen-

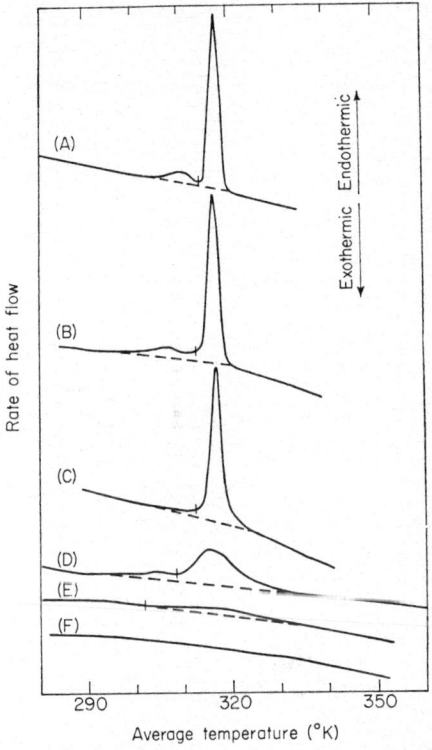

FIG. 15. D.s.c. curves of 50 wt % dispersions in water of 1,2-dipalmitoyl-L-lecithin/cholesterol mixtures containing: (A) 0·0 m %, (B) 5·0 m %, (C) 12·5 m %, (D) 20·0 m %, (E) 32·0 m % and (F) 50·0 m % cholesterol (from Ladbrooke et al., 1968).

tration of cholesterol increases the main endothermic transition remains sharp while a small peak at 35°C disappears. This is followed by a profound change in which the main transition becomes broad and decreases in area. When the concentration reaches 50 m % of cholesterol, no endothermic peak can be observed. The variations in the transition temperature and the heat associated with the transition are shown in Fig. 16A and 16B, respectively. Also shown in the diagram is the variation of the X-ray long spacing.

The effect of cholesterol is to disrupt the ordered array of the hydrocarbon chains of the lipid in the gel phase and, when cholesterol and lecithin molecules are present in equimolar proportions, all the chains are in a fluid condition. The presence of equimolar amounts of cholesterol with lipid causes the phospholipid/cholesterol mixtures to be dispersible in water over a much wider temperature range than occurs with the individual phospholipid.

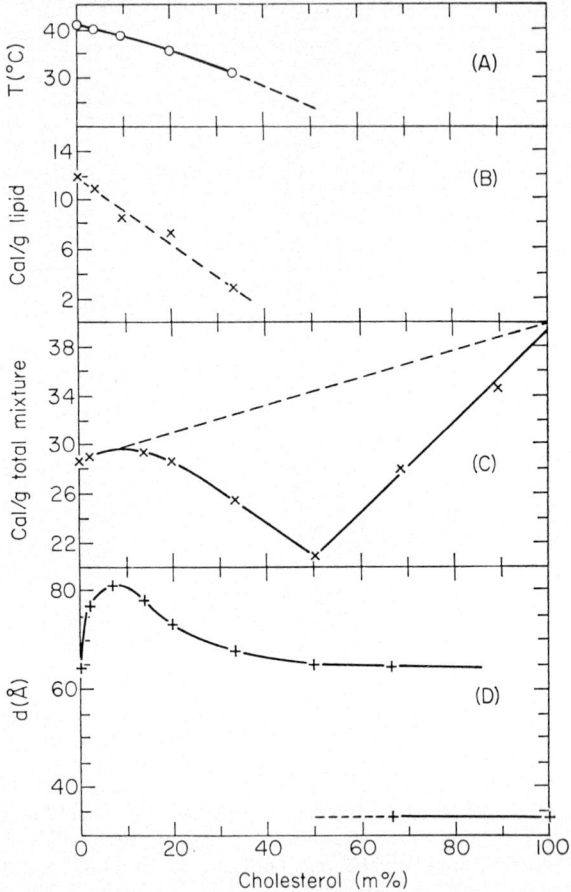

FIG. 16. The variation with 1,2-dipalmitoyl-L-lecithin/cholesterol ratio in 50 wt % aqueous dispersions of: (A) lecithin transition temperature, (B) heat absorbed in the lecithin transition, (C) heat absorbed in the ice transition at 0°C, (D) X-ray long spacing at 25°C (from Ladbrooke et al,. 1968).

The considerable mobility of the natural phospholipids at room temperature is so high that it raises the possibility that it might be possible to obtain high resolution n.m.r. spectra of phospholipids when they are dispersed in water. It should be remembered that high resolution n.m.r. spectra are usually observed when molecules are dissolved in a solvent and the molecular motion of the solute is sufficient so that all dipole interactions are averaged out. The fact that phospholipids are not soluble in water might be expected to be a limitation on the possibility of obtaining such high resolution. However, if natural phospholipids such as egg yolk lecithin or red blood cell lipids are dispersed by means of sonication in water, a high resolution n.m.r. spectrum can indeed be obtained (Chapman and Penkett, 1966; Chapman et al., 1968a). This shows the phospholipid in this phase has considerable diffusional freedom. The spectrum is similar to that which is observed when a phospholipid is dissolved in a solvent such as chloroform (Chapman and Morrison, 1966). Peaks associated with different proton groupings are observed. Peaks in the spectrum can be assigned to the $[CH_2]_n$ protons of the hydrocarbon chain of the lipid, the nine protons of the choline group $N(CH_3)_3$ and so on. This discovery led to the idea that this technique could provide a new technique for investigating the interaction of molecules such as cholesterol with phospholipids. This n.m.r. spectroscopic method is important because the organization of lipid and cholesterol should be analogous to the situation which is likely to occur *in vivo*, particularly in its vehicular situations. It therefore has advantages over the more classical monolayer type investigation.

The n.m.r. spectroscopic technique provides a method for studying the interaction of cholesterol with the various parts of the phospholipid molecule. A p.m.r. spectrum showing the effect of cholesterol on egg yolk lecithin is shown in Fig. 17. In the spectrum it can be seen that, as a result of the cholesterol interaction, the peak arising from protons in the hydrocarbon chain $[CH_2]_n$ signal is broadened out and is absent from the spectrum. The condensing effect of cholesterol with red blood cell lipids appears to be rather less than is observed with other phospholipids.

These various results seem to show that, at a particular temperature, the presence of cholesterol causes the hydrocarbon chains of differing phospholipid molecules to be in an "intermediate fluid" condition. Those lipids which would normally be above their limiting transition temperature may have a certain amount of inhibition of chain motion, while the hydrocarbon chains of those lipids which would normally be in a gel condition are given greater fluidity.

When we consider the relevance of these results to biological membranes, we know that correlations between bulk phase phenomena and membrane structure must be made with caution. We must not ignore the presence and effects due to the protein of the membrane.

They do suggest, however, that a possible role for cholesterol in membranes is to control the fluidity of the hydrocarbon chains of the phospholipids providing a coherent structure stable over a wide temperature range and permitting some latitude in the fatty acid content of the component lipids.

152   D. CHAPMAN AND D. F. H. WALLACH

D. INTERACTIONS WITH PROTEIN

It is important to note that the protein–lipid ratio of cell membranes varies (Korn, 1966). Thus the protein content of myelin appears to be sufficient to cover only 43% of the area occupied by the lipid. For erythrocyte membranes there is sufficient protein to give a monomolecular film with 2–5 times the area of the lipid. Most bacterial membranes seem to have enough protein to cover 5 times the area occupied by the lipid.

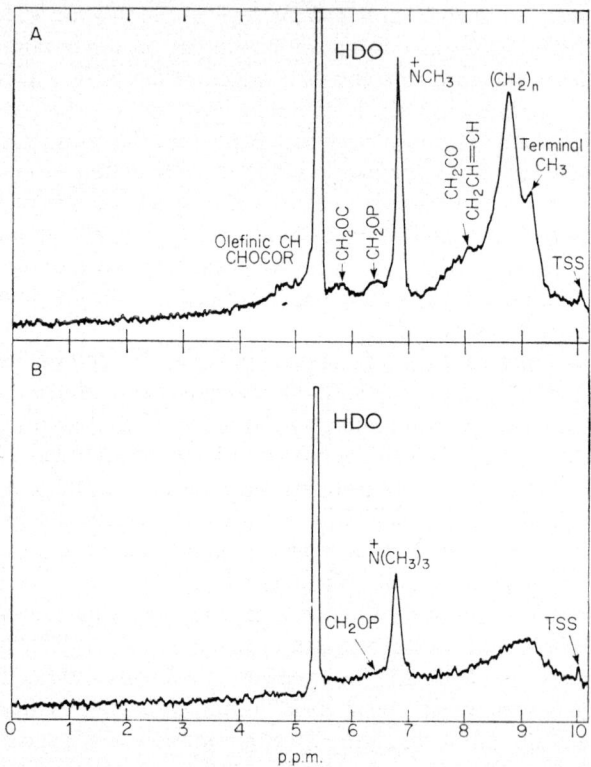

FIG. 17. High resolution p.m.r. spectrum (60 Mc/s) (A) egg yolk lecithin dispersed in D_2O, (B) an equimolar mixture of egg yolk lecithin and cholesterol dispersed in D_2O (from Chapman and Penkett, 1966).

A most important question underlying membrane organization is the way in which lipids and proteins interact. Lipid protein interactions have been studied by a variety of methods. The monolayer technique has been one of the methods used for this study. In this case, soluble protein was injected underneath a phospholipid layer spread on an air–water interface. The changes in the surface pressure of the monolayer which were produced were taken to indicate an interaction and partial penetration of the monolayer by some of the protein side chains (Matalon and Schulman, 1949). It was calculated that the protein had formed a monolayer over the polar surfaces of the phospholipid, and that the interaction involved both polar and non-polar forces. Soluble protein has also been observed to affect the properties of lipid bilayers (Mueller *et al.*, 1962). A particular sample of protein was shown to produce a marked altering of the specific response of the bilayer. Other studies have been made with protein absorbed to lipid and investigated by electron microscopy. Stoeckenius (1959) has shown that, when protein is added to an aqueous dispersion of phospholipid, a thickening of the dense line bound in large multilayered myelin figure is observed where isolated layers exceeded the dense lines at both surfaces were thickened. This thickening effect was considered to reflect the addition of protein to the exposed surfaces of the lipid structures. X-ray diffraction studies of a similar system (Schmitt and Palmer, 1940) demonstrated increases in lamellar spacings which they attributed to the insertion of monolayers of protein between lipid layers.

Many of the investigations of lipid–protein interactions have been with water soluble proteins and the interaction with phospholipid aggregates has, therefore, been predominantly electrostatic in nature. A quantitative study of the interaction between mitochondrial protein cytochrome *c* and a variety of mitochondrial phospholipids has shown that the interaction involves essentially the complete neutralization of all the positively charged groups in the protein with a corresponding number of negatively charged groups in the phospholipid micelle (Green and Fleischer, 1964). The resulting phospholipid protein complex has a molecular weight in excess of 100,000. Studies on so-called structural protein of mitochondria and its interaction with phospholipid, support the idea that, in this case, the interaction is predominantly non-polar. When the interaction is complete the resulting structural protein phospholipid complex contains about 25% phospholipid which is still capable of interacting with cytochrome *c* to an extent to be expected if the charged groups of this phospholipid were still free. It appears that both electrostatic and non-polar (sometimes called hydrophobic) interactions between lipid and protein can occur. However, it seems that there is room for many more careful experimental studies of these interactions before we can be at all certain of the precise details. Techniques such as fluorescence spectroscopy, n.m.r. spectroscopy and e.s.r. spectroscopy can be expected to produce useful information about these lipid–protein interactions (see Barratt *et al.*, 1968). Lipid protein interactions are discussed more fully in Chapter 5.

III. The Physical Properties of Membranes

It is clear that the study of the basic physics or physical chemistry of phospholipids and their interactions with water, cholesterol or protein is important information which is essential for a detailed understanding of the structure and properties of cell membranes. Once this information is available it will be possible to speculate with more certainty about the probable construction of cell membranes. Such speculations ultimately have to stand firm and to be consistent with the known facts available from chemical, biological and physical examination of biological membranes themselves. If the information about biological membranes is sparse or hazy in quality, then so will be the relevance of the information which is obtained on the individual components. We can give a simple example of this. If we discuss phospholipid monolayers and their possible relevance to biological membranes, we need to know whether the lipids in a membrane are organized in a bilayer structure and which part of the force area curve is most appropriate to the situation which exists in a membrane. Ultimately, in order to fully understand biological membranes, we must have many chemical and physical studies of the various membrane types. Only with detailed information available on both membranes and their components will a complete picture of membrane structure be possible.

Having said this, and it may appear somewhat rather platitudinous, it should be pointed out that, as recently as 1961, Ponder remarked that he was "... not convinced about the structure ore ven the necessary existence of the cell membrane as it is generally described." Ponder points out that, while there is a barrier to free diffusion of some substances between the cell and the surrounding medium, this could result when two phases meet and that no third phase or "membrane" between the two phases may be required.

According to some authors, the red cell ghost represents as nearly as possible a plasma membrane. These ghosts are obtained by haemolysis of red cells and freeing the structure by repeated washing or similar processes until they do not become thinner on further washing. These are identified with the membrane. Ponder suggests that, at best, these ghosts are representations of a surface ultrastructure. They contain lipid and protein and nearly all of the lipids of the intact red cell are found in the ghost material.

Myelin appears to be a convenient source of cell surface material being composed of Schwann cell surfaces. Membranes from other sources are also being extracted and studied by physical techniques. Accepting the spirit of criticism expressed by Ponder, let us now see what recent physical studies have revealed about these "membrane" structures.

A. X-RAY DIFFRACTION STUDIES OF MEMBRANES

Low-angle X-ray diffraction studies have been used to provide information on the thickness and spacing of adjacent membrane layers and also about membrane sub-units of size larger than 10 Å. The technique has been applied to membranes in a hydrated condition. The study of membranes using X-ray diffraction has shown that severe drying can cause a complete change in the arrangement of membrane lipids. Only a small fraction of X-rays can be diffracted by a single membrane layer and, therefore, a stack of membranes is usually used. In the myelin membranes this stack of membranes is arranged in a natural condition but the membranes of other cells need to be stacked by centrifugation. When an X-ray beam of circular cross section is used, the diffraction pattern can provide information about the spacings in two directions. In practice, in order to limit the required exposure, a rectangular cross section beam is used. The diffraction patterns give spacings perpendicular to the long dimensions of the beam cross section. Hence, if the sample is a flat stack of membrane when the beam is aligned with its length parallel to the plane of the membrane, reflections are obtained indicating the periodicity of layering. When the membranes are placed so that the beam enters normal to that plane, information can be obtained about spacings or subunits in the plane of the membrane.

The myelin sheath of nerve fibres is an example of a multilayered lipoprotein system where the application of X-ray techniques has been particularly useful. The myelin sheath is considered to be derived from a multiple folded Schwann-cell surface and may be a model system for the study of cell-membrane structure in general. However, the net enzymatic activity of myelin is essentially zero. Furthermore, the lipid:protein ratio is 4:1 whereas in other membranes it is less than 1:2.

X-ray diffraction patterns have been obtained from nerve bundles maintained in a physiologically active state in irrigation cells mounted on the X-ray diffraction camera and from nerve bundles sealed in thin-walled glass capillary tubes containing a physiological solution such as Ringer's solution. In both cases the diffraction patterns are observed to be the same.

When the nerve specimens are examined in a direction perpendicular to the fibre axis, using a symmetrically collimated X-ray beam, a meridionally accentuated ring is observed at 4·7 Å and a faint ring is observed at 9·4 Å. These have been shown to be myelin reflections. Other reflections due to myelin also occur at low angles. These low-angle reflections have been accounted for as diffraction orders from a single fundamental repeating unit varying from 150 Å to 180 Å in the different types of nerve examined. Myelin from different sources gives different low-angle reflections. Peripheral nerve from mammals gives 5 low-angle reflections showing marked alterations in intensities through the orders and indicating a fundamental repeating unit of about 180 Å. The myelin of central nerve origin, such as brain white matter, spinal cord or optic nerve, gives only two low-angle reflections corresponding to a repeat unit of about 80 Å. (The presence of the short spacing near 4·7 Å suggests that the hydrocarbon chains of the lipids present in the myelin sheath are in a liquid-like condition. This does not necessarily preclude some degree of order among the chains.) The variation in the intensities of the low-angle diffractions is interpreted to arise from variations in electron density along the axes of the myelin unit. It has been assumed that the fundamental repeating unit consists of two parts having very similar distributions of X-ray scattering power. From the intensities of the odd-order reflections, the magnitude of the difference between the two parts has been estimated. The "difference factor" is appreciable in peripheral nerve myelin but negligible in optic nerve. Diffraction patterns from the structure along the fibre axis appear as complete rings. This has been interpreted to conclude that the long axis of the rod-shaped unit cell is oriented radially in the myelin sheath.

X-ray diffraction experiments have been carried out on nerve after a variety of treatments and useful information obtained. Extraction of fresh nerve with acetone at 0°C removes about 30% of the cholesterol, leaving the other lipid components essentially intact within the still organized residual myelin sheath. The main modifications shown by both X-ray and electron microscopy are expansion of the layered structure with internal rearrangements and formation of collapsed layer systems. Low-angle diffraction patterns of the lipid extract show a strong 34·2 Å reflection, characteristic of cholesterol. A more extensive breakdown of the sheath is observed after alcohol extraction. The breakdown of myelin known to take place during *in vitro* degeneration of nerve has also been studied. The low-angle X-ray diffraction patterns show characteristic changes. There is a marked intensification of the second-order diffraction followed by the appearance of a 70 Å reflection and gradual extinction of the lower orders in later states of degeneration.

Diffraction data has been used to provide information about the molecular organization of myelin. The first ideas about this were based on a polarization optical analysis of freshly isolated nerve fibres before and after treatment with absolute alcohol. This led to the conclusion that the lipid molecules are oriented radially and the non-lipid material arranged in concentric layers. The layers of oriented lipid molecules alternate with layers of non-lipid material. X-ray data led to the further conclusion that the peripheral nerve consisted of two lipoprotein layers, each of which consists of bimolecular leaflets of 67 Å, sandwiched between protein layers of 25 Å with interposed water layers of about another 25 Å thickness. Further studies by Finean (1960), considering the contraction of the lipid layers during drying, has led to a more detailed molecular arrangement of the myelin. The changes observed with frog sciatic nerve when dried are that the myelin unit shrinks to about 145–148 Å and, at the same time, three independent diffraction lipid phases are produced. The residual myelin unit gives X-ray reflections at about 146 Å and 73 Å. X-ray diffraction studies have also been carried out on various types of nerve at different temperatures.

Additional information about the structure of myelin has been obtained from calculations of the electron density distribution in the membrane. Finean (1962) and Finean and Burge (1963) found one-dimensional electron density distribution in a direction perpendicular to the plane of the membrane to show two distinct peaks. These were assumed to represent the phosphate groups of two phospholipid molecules placed tail to tail. The separation of the peaks was about 50 Å.

Recently Finean *et al.* (1966) have also studied rat erythrocyte ghosts using low angle X-ray diffraction patterns during a controlled drying of an erythrocyte preparation. The results were compared with those obtained during a parallel study of myelin isolated from guinea-pig brain. The diffraction changes which accompany the dehydration of a sample of erythrocyte ghosts are essentially similar to those described for isolated myelin. A clearly defined point during dehydration, at which changes in low-angle X-ray diffraction patterns occur, indicates a modification of membrane structure.

Prior to this it seems probable that the membranes retain the water of hydration essential to their structural integrity. The diffraction pattern observed before this point is reached arises from a single lamellar system which has been identified as regions of close packing of erythrocyte membranes. The repeat period in this lamellar system was found to be 110–120 Å and this was taken to represent the thickness of one membrane. Values for the three spacings recorded from a fully dried sample are in the ratio of approximately $1:\sqrt{3}:\sqrt{4}$ as required for a two-dimensional hexagonal network. The observed sequences of diffraction changes can be accounted for in terms of gradual change from a lamellar to a hexagonal arrangement of structure. The observed reversibility of the changes as the system is rehydrated can be readily understood on the basis of the lamellar to hexagonal change, although there are some objections to this interpretation. The X-ray diffraction experiments do not provide any indication of the arrangement of sub-units of macromolecular dimensions within the erythrocyte membrane.

The electron density profile of the repeating unit assuming a unit membrane structure is shown in Fig. 18. The shallow trough in the electron density curve is suggested to correspond to the narrow band of low intensity which separates unit membrane features in the layered system observed in electron micrographs.

These experiments suggested that the pattern of the natural hydrated membranes can also be obtained through the rehydration of fully dried samples. Husson and Luzzati (1963) rehydrated frozen-dried preparations of human erythrocyte ghosts and detected a lamellar system with a periodicity of about 170 Å.

With regard to the wide angle reflections, the reflection at about 4·6–4·7 Å, observed with membranes and observed with myelin, has been interpreted as representing the repeat distance between parallel phospholipid hydrocarbon chains aligned perpendicular to the membrane. As we have seen, this reflection is also obtained when the phospholipid is in a liquid crystalline condition. It has been suggested that it would be useful to examine isolated myelin which

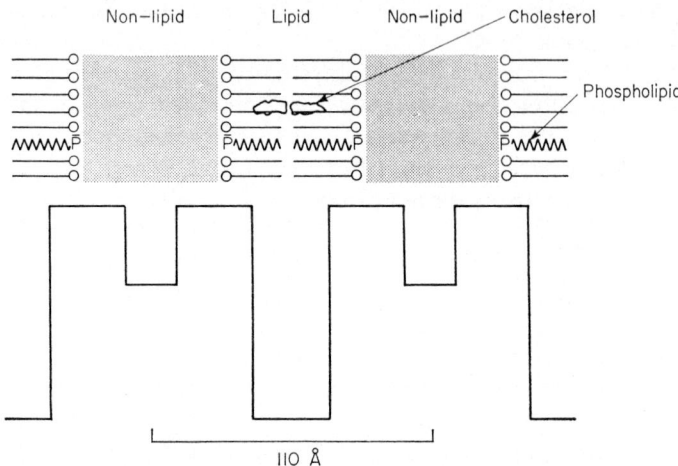

FIG. 18. The suggested form of electron density profile of the rat erythrocyte membrane with schematic drawings of phospholipid and cholesterol molecules (from Finean et al., 1966).

has been packed in stacked flat sheets rather than in the cylindrical form present in the nerve. In this case the reflection should be strong when an X-ray beam of rectangular cross section passes parallel to the membrane with the cross section length perpendicular to the membrane if the spacing is correctly associated with the hydrocarbon chain separation. On the other hand, it should be weak when the cross section length is parallel to the edges of the membrane.

In general, the X-ray diffraction studies do not support the idea that these "membranes" consist of sub-units but many other studies need to be carried out to be certain that membranes contain a bimolecular layer structure.

Recently, X-ray studies of oriented pellets of outer segments of frog retinae have been carried out (Blasie et al., 1965). The isolated membranes were packed into oriented pellets by high speed centrifugation. When the plane of the membrane was perpendicular to the X-ray beam, only diffuse reflections were obtained with fresh specimens. After mild drying, the diffraction patterns sharpened up to give distinct reflections corresponding to 40 Å sub-units packed in square array of side length 70 Å. In this case it was suggested that the membranes consisted of 40 Å sub-units of either macromicelles of phospholipid or rhodopsin molecules. However, some of these effects may be caused by the drying process.

B. ELECTRON MICROSCOPE STUDIES

The introduction of the electron microscope in the 1930s provided a tool which can form images of structures previously considered to be sub-microscopic. The observation of clearly defined layer structures at the border of cell boundaries provided strong and independent confirmation of the existence of cell membranes.

For the examination of biological specimens, the early techniques used metal shadowing and thin sectioning methods. Sectioning techniques enabled ultra thin sections to be obtained showing well preserved fine structure. This led to the observation of the regular concentric layered arrangement of the myelin sheath, to the suggestion that the myelin sheath membranes were continuous with the surrounding Schwann cell membranes (Geren, 1954) and to the conclusion that these membranes and cell membranes may be similar in molecular composition and arrangement. Thin sectioning techniques also revealed the existence of layered structures not only at the cell boundary, but also associated within the cell. Structures, such as mitochondria (Palade, 1953; Sjöstrand, 1953), lysosomes and paired membranes called endoplasmic reticulum (Porter and Thompson, 1947) were revealed by the electron microscope. Similar layered structures were also observed with the chloroplast of plant cells and in the rods and cones of the eye, suggesting that all these various structures are each associated with membrane organization. The importance of the electron microscope to the study of membrane organization is shown by the fact that electron micrographs of various biological structures are shown throughout the various chapters of this book. The earlier electron microscope studies have been extensively reviewed elsewhere, e.g. Robertson, 1959, 1964 and 1966. In this section we shall discuss briefly only some of the more important recent work using this technique. We begin with a discussion of the unit membrane concept.

1. *The Unit Membrane Concept*

The similarity in appearance of the electron micrographs of many different types of cell membranes were considered by Robertson (1957, 1958, 1959) to be evidence for a common structure at the surface of a wide variety of cells. Using a new fixative procedure involving $KMnO_4$, he was able to observe a similar three-layered unit, approximately 75 Å thick, at the surface of a number of different cell types as well as in many different cellular organelles. This unit appeared as two dense lines about 20 Å wide separated by a lighter space of 35 Å.

Robertson (1957) proposed a model of the membrane which he termed a "unit membrane". This concept appeared to clarify and unify a wide body of information including other physical data as well as other electron microscope data. By referring to the biological membrane as a unit, he emphasized not only that all three parts of the triple layered 75 Å structure seen in the electron microscope were part of one membrane, but also that all membranes had a similarity of molecular arrangement and origin.

The electron microscopic data for other plasma membranes has recently been summarized by Elbers (1964) and this data is consistent with the suggestion of Robertson.

In general with plasma membranes, triple-layered structures are also observed after fixation in $KMnO_4$ and often, but not always, after fixation in OsO_4. The endoplasmic reticulum (cytomembranes), the outer, inner and cristae membranes of mitochondria, chloroplast membranes, the two membranes of the nuclear envelope and the membranes of bacterial protoplasts and spheroplasts are also revealed as triple-layered structures in electron micrographs.

Despite these general similarities there has been some reluctance (Stoeckenius, 1966) to accept this proposal of an identical unit membrane structure for all the membranes in cells and particularly for the mitochondrial membranes. In general, membranes are thought to be about 75 Å wide. There are, however, considerable variations in dimensions of membranes.

The overall widths of triple-layered plasma membranes appear to vary from about 50 Å to perhaps 130 Å (Elbers, 1964). How much of this variation is due to differences in the methods of preparation and how much to fundamental differences in structure is not clear. Perhaps the clearest electron micrographic indication of differences among membranes was obtained by Sjöstrand (1963) who compared adjacent membranes in single cell sections of mouse kidney and pancreas fixed with OsO_4 and $KMnO_4$. The thinnest membranes (mitochondrial and α-cytomembranes) were 50 Å to 60 Å and the thickest membranes (plasma and zymogen granules) were 90 Å to 100 Å. These variations indicate the difficulty of interpreting micrographs in terms of molecular structure.

Several other observations have appeared which seem to disturb the "unit membrane" picture. Hillier and Hoffman (1953) studied the structure of the membrane of erythrocytes by using shadowing techniques which revealed a mosaic structure. They suggested that the erythrocyte envelope was composed of plaques situated on the outside of a fibrous network joined together by lipids. Recently, Glaeser et al. (1966) have studied the membrane structure of OsO_4-fixed erythrocytes viewed "face on" by electron microscopy. This supports the conclusions of Hillier and Hoffman and shows that the surface of that rat red cell membrane has a "pebbly" appearance at the level of 400 to 500 Å. These authors suggest that these bumps on the surface may be associated with a filamentous structure, the bumps representing the tops of loops of the filaments.

Also using a shadowing technique, Frey-Wyssling and Steinmann (1948) examined the structural features of the closed flattened sacs which constitute the internal membrane system of the plant chloroplast. These workers had previously noted that the chloroplast membranes failed to show any substantial intrinsic birefringence which would be expected if they contained highly oriented lipid bilayers as in myelin. These membranes show a repeating granular structure which suggest that the membranes may be composed of an array of micellar or globular sub-units.

Sjöstrand (1963) observed globular sub-units in one of the opaque layers of mitochondrial membranes and smooth endoplasmic reticulum and an asymmetry in the electron opacity of the dense lines in the plasma membrane. Sub-units, or cross-linkages bridging the gap between the two opaque bands of the triple-layered structure, have been observed by Robertson (1963) who later reinterpreted them as an electron optical artifact derived from a mosaic pattern in the plane of one or both surfaces of the triple layer. In several instances, hexagonal mosaic patterns have been seen on the surfaces of plasma membranes (Benedetti and Emmelot, 1965). Strong evidence for membrane sub-units is suggested in a recent paper by Blasie *et al.*, 1965. Outer segment membranes of frog retina were isolated and were oriented in ultracentrifugal pellets. Electron microscopic surface views of negatively stained membranes and low-angle X-ray diffraction patterns from unfixed, unstained pellets showed a square array of spherical particles. The unit cell size was about 70 Å and the particles had a non-polar core about 40 Å in diameter. Professor Lucy discusses more fully the evidence and implications of the presence of micellar units in membrane structure in Chapter 6.

When attempts are made to obtain further information about the interpretation of the electron microscope data of fixed sectioned tissues in molecular terms, we find that there are a number of difficulties. The drastic fixation techniques using reagents such as $KMnO_4$ or OsO_4, the complicated dehydration and embedding processes, the bombardment of the tissue with electrons; all these would appear *a priori* to be able to introduce artifacts of various types which could confuse interpretation. The "unit membrane" concept, however, includes such an interpretation at the molecular level. The dense lines of the triple-layered membrane structure are equated with the proteins and polar groups of the membranes, while the lighter interzone spaces are equated with the non-polar group (Fig. 19a). It is important to know what atoms are responsible for the microscope image and to be able to interpret correctly the electron micrographs to see whether they are consistent with the structure suggested by Robertson. Korn (1966) has critically discussed this question in some detail and, since it is important for all interpretations of electron micrographs of cells, we shall also discuss it here in some detail. He points out that Robertson (1959) and Fernandez-Moran and Finean (1957) have

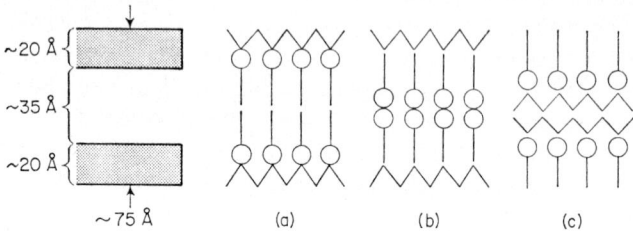

FIG. 19. Three possible molecular diagrams that could explain the observed unit membrane pattern shown on the left (after Robertson, 1966).

questioned whether the manganese atom is responsible for any of the electron opacity in micrographs of $KMnO_4$-fixed cells. He suggests that, if this is not the case, there is then no way to interpret the dense lines in micrographs of membranes fixed with $KMnO_4$ in molecular terms. This is a serious deficiency because it is in such preparations, as we have already seen, that the triple-layered structure is most reproducibly and distinctly seen. There appears to be only one study in which cells were chemically studied during fixation with $KMnO_4$. Korn and Weisman (1966) found that the lipids of amoebae were essentially unaffected by fixation with 1% $KMnO_4$ for 1 hr at 0°C. All the neutral lipids and about half of the phospholipids were extracted from the amoebae during dehydration in ethanol.

Similar triple layered structures are often observed when OsO_4 is used as a fixative and various experiments have been devised so as to ascertain the site of the fixation process. This fixation was originally thought to take place at the sites of the double bonds in the hydrocarbon chain. Wigglesworth (1957) suggested that osmium can cross-link through the ethylenic double bonds and has given evidence to show the occurrence of an insoluble polymeric complex of lipid and osmium which he considers is the basis of cytological fixation. This observation was considered to be consistent with the fact, noted by many authors, that fully saturated phospholipids, such as the phosphatidylethanolamines, do not react easily at room temperature with osmium tetroxide, although suggestions had been made that brominated or hydrogenated phospholipids can take up osmium without fixation occurring. Bahr (1954) showed that many amino acids would also react with osmium tetroxide and so some authors have considered that the dense lines observed in electron micrographs of cell membranes represent protein, whereas the light central line corresponds either to the hydrocarbon chain region of the lipid or to a gap produced by removal of lipid in the preparative stages for electron microscopy (Robertson, 1960).

Stoeckenius (1962a,b) studied natural lipids after reaction with osmium tetroxide and concluded that the dark area seen in electron micrographs corresponds to osmium located at the polar groups of the phospholipid molecules. Finean (1962), after an analysis of X-ray data of fixed and unfixed tissue, also considered that the osmium is located among the polar groups of the bimolecular sheets of lipid. Riemersma (1963), after chromatographic analysis of the intermediates formed during osmium fixation of unsaturated lecithins, considers that the initial reaction is with the double bonds, but that there is a subsequent migration of osmium derivatives to the polar groups. Despite this work, some authors are still, nevertheless, of the opinion that the osmium is located at the double bond of the lipids (Hayes *et al.*, 1963).

Korn (1966) has recently reopened this question of the location of osmium. He refers to the work of Craigee (1936) who showed that OsO_4 reacts stoichiometrically with olefins to form a stable osmic acid ester of the glycol derived from the olefin by oxidation:

$$\begin{array}{c} \diagdown\!\!C\text{---}C\!\!\diagup \\ | \quad | \\ O \quad O \\ \diagdown\!\!\diagup \\ Os \\ \diagup\!\!\diagdown \\ O \quad O \end{array}$$

This product, under reasonably strong conditions, can be hydrolysed to free glycol. In a similar reaction, starting from a glycol and tetramethyl dipotassium osmate, a dimer can be synthesized:

$$\begin{array}{c} \diagdown\!\!C\text{---}C\!\!\diagup \\ | \quad | \\ O \quad O \\ \diagdown\!\!\diagup \\ Os\!=\!\!=\!O \\ \diagup\!\!\diagdown \\ O \quad O \\ | \quad | \\ \diagdown\!\!C\text{---}C\!\!\diagup \end{array}$$

Korn points out that, despite this chemical evidence, Stoeckenius (1960) has interpreted the similiarity of electron micrographs of uranyl linolenate before and after exposure to vapours of OsO_4 and of potassium linolenate after fixation with vapours of OsO_4 as an indication of binding of the osmium to the uranyl ion or carboxylate group. Korn points out that no chemical evidence was provided for any reaction between OsO_4 and uranyl or potassium linolenate, nor was a mechanism proposed to explain the affinity of osmium for the uranyl or potassium carboxylate.

Neither Stoeckenius (1962a,b) nor Riemersma (1963) proposes a mechanism whereby osmium tetroxide, having attacked the double bond to form an osmic acid ester, can then migrate to the polar group although this is the basis of the argument. Furthermore, while Stoeckenius suggests that the osmium is bound to the anionic carboxyl group, Riemersma proposes that the osmium is bound to the cationic quaternary nitrogen. Korn (1966) also points out that none of these reactions was carried out under the conditions used for the normal fixation of biological material.

In a recent groups of papers, Korn and Weisman (1966) and Korn (1966) showed that when methyl oleate is reacted with a 2% solution of OsO_4 in water for 1 hr at 0°C, it is quantitatively converted to bis(methyl-9,10-dihydroxystearate)osmate, probably with the following structure:

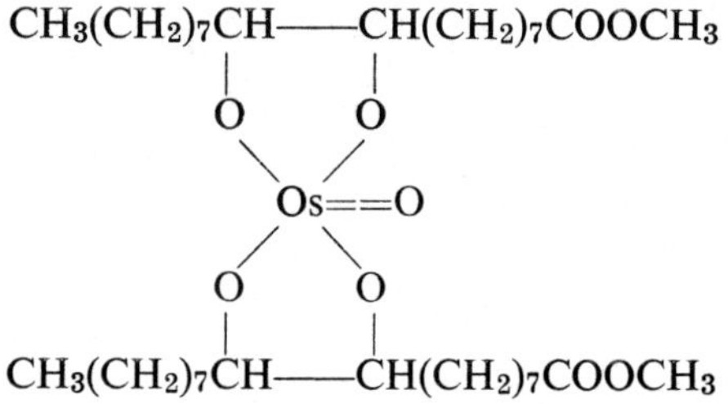

$$\begin{array}{c} CH_3(CH_2)_7CH\!-\!\!\!-\!CH(CH_2)_7COOCH_3 \\ | \quad\quad | \\ O \quad\quad O \\ \diagdown \diagup \\ Os\!=\!\!=\!O \\ \diagup \diagdown \\ O \quad\quad O \\ | \quad\quad | \\ CH_3(CH_2)_7CH\!-\!\!\!-\!CH(CH_2)_7COOCH_3 \end{array}$$

This supports the idea that OsO_4 reacts with the olefinic groups in lipids to form stable osmic acid esters of glycols and, therefore, that osmium is covalently bound to the hydrocarbon portions of lipids in membranes fixed with OsO_4. However, in the model reactions, an approximately equal amount of osmium was recovered as uncharacterized products of, presumably, lower oxides. Thus, on the basis of these experiments, it is still possible that some osmium, other than that bound at the hydrocarbon portions of fatty acids, may be deposited in tissues after fixation by OsO_4. Chapman and Fluck (1966), in an attempt to clarify the problem, attempted to fix saturated phospholipids with OsO_4, arguing in this case that, as there were no double bonds present, any successful fixation must be at the polar group. This study showed that, while fully saturated phosphatidylethanolamine reacts with OsO_4, the fully saturated lecithins do not. However, both types of lipid had to be heated above room temperature to the liquid crystalline temperature to form myelinic forms and so the chemical reaction involved need not necessarily be identical with those which occur when fixation is carried out at room temperature.

Korn (1966) concludes that the dense lines in membranes fixed with osmium tetroxide *reveal nothing* about the molecular orientation of the phospholipid in the original membrane. He also points to the fact that triple-layered membranes are seen in osmium-fixed mitochondria from which all the lipid has previously been removed by extraction with acetone (Fleischer *et al.*, 1965) (*Escherichia coli* B (van Iterson, 1965) also gives a triple-layered membrane despite the fact that this organism contains essentially no unsaturated fatty acids). If lipids are not necessary to reveal the triple-layered structure, then its explanation in terms of particular molecular configurations cannot be correct. Experiments carried out with a "structural" protein which accounts for 50% of the total protein isolated from bovine heart mitochondria (Criddle *et al.*, 1962) shows that this insoluble protein binds phospholipids, irrespective of their charge, through non-polar hydrophobic bonds (Green and Fleischer, 1963).

In addition to the question of the site of fixation there is also the important question as to whether the structures which contain lipids are unchanged after fixation with OsO_4. It seems likely that replacement of a double bond by a glycol osmate ester could have some effect on the configuration of the lipid. There have been suggestions which agree with this. Thus, Stoeckenius (1962a) has suggested that fixation with aqueous OsO_4 might change the arrangement of molecules in phospholipid structures, including membranes, while Lucy and Glauert (1964) have observed that OsO_4-converted helical arrangements of lipid micelles into stacked discs.

We see that there are still many doubts concerning a full acceptance of the "unit membrane". Many of the supporting arguments for its acceptance depend upon studies of myelin which may be chemically, metabolically and functionally, different from all other membranes.

The interpretation of electron micrographs of biological membranes has been strongly influenced by the belief that the bimolecular leaflet is the most probable form adopted by lipids in water. In Chapter 3 by Dr. Luzzati we have seen that this is not always the case and that other liquid crystalline forms occur. This raises additional possibilities of phase changes occurring during the preparative stages required for the electron microscope.

Of course, even if the bimolecular leaflet is the form always adopted by lipids this does not necessarily prove that biological membranes adopt this configuration. Both lipid and protein separately take up stable configurations in water to minimize the hydrophobic interactions and maximize the hydrophilic interactions with the water. The important question for membrane organisation is whether the combined lipid–protein complexes will adopt an entirely different arrangement.

2. The Negative Staining Technique

The negative staining technique using phosphotungstate has increased in popularity in recent years. In this method the biological membranes are immersed in a pool of electron-dense material (e.g. sodium phosphotungstate) which dries to form an electron-dense glass (Brenner and Horne, 1959). Membranes appear as regions of electron transparency against a dark background. The specimen is not sectioned.

Using this technique, Fernandez-Moran (1962) discovered a structural component of the inner mitochondrial membrane. This component is composed of a base piece embedded in the membrane attached to a stalked globular particle which protrudes from the surface. These structures have been called elementary particles and also inner membrane particles. With a few exceptions (Gompel, 1964; Ashhurst, 1965) these particles are not visible in embedded and sectioned material either because they are destroyed by the technique or because they lack sufficient contrast. It has been claimed that the particles are artifacts of the preparation technique which arise when mitochondria are broken (Sjöstrand et al., 1964). Moor et al. (1964), using the new technique of freeze-etching, have, however, demonstrated a granularity of the inner membranes of fractured mitochondria within frozen cells.

Cunningham and Crane (1964) have also used the negative staining technique to examine membrane fractions from a variety of cell types. They comment that many membrane fractions, which look uniform and homogeneous after fixation and section, can be seen, after negative staining, to be composed of a diverse population of membrane types. They claim to see three membrane types. The most common type are irregularly shaped vesicle about 200 mμ across with a rough granular surface. The edges of these vesicles often extend into blebs or tubular protrusions. A second membrane type occurs as tubular sheets with a smooth surface. Finally, a fringe of 50–60 Å particles around the smooth even edges of the tubes is observed.

Whittaker (1966a), using sub-cellular fractionation techniques, has isolated three neuronal membranes; myelin, presynaptic membranes and synaptic vesicles. With the negative staining technique, the external synaptosomes appear as non-laminated membranes 80–100 Å thick. The synaptic vesicles, under certain conditions, such as treatment with phosphate buffer, show a prominent bilayered structure in negative as well as in positive staining. Whittaker (1966a) suggests that this seems to require a hydrophilic space in the centre of the bilayer into which the polar negative stain may diffuse.

The possibility of artifacts being introduced with the negative staining technique has been considered. Bangham and Horne (1964) show a number of different phase structures with artificial lipid mixtures, some of which resemble structures seen in negatively stained natural membranes. Whittaker (1966b) has commented that the negative staining technique only reveals those structures sufficiently tightly packed to exclude the negative stains. Proteins possessing an α-helix are clearly seen, while those with an open structure are not. In some instances the fine structures observed can be generated on the electron microscope grid by interaction with negative stains during drying. There is the further possibility of spontaneous or enzyme catalysed rearrangements occurring when biological organelles are ruptured so that interpretations at the macromolecular level must be made with caution. Other electron microscope studies using the negative staining technique are discussed in Chapter 6.

3. *The Freeze-etching Technique*

Recently a new technique has been introduced to electron microscopy (Moor *et al.*, 1961) which attempts to overcome the many problems associated with chemical fixation methods and the production of artifacts associated with these methods. This is the technique of freeze-etching.

Freeze-etching involves six preparational steps: pretreatment of the object, freezing, chipping of the frozen specimen followed by etching and coating and, finally, the cleaning of the replica. One of the most important processes is to get a clean fracture plane through the frozen specimen. This can be done by chipping under high vacuum at a controlled object temperature. In this method there is no chemical treatment of the object during the whole procedure until the replica is formed on the fracture plane from the frozen specimen. By means of snap freezing and/or glycerol impregnation, cells have been frozen so as to preserve the life of the organisms.

Freeze-etching depends upon freezing a very small sample (the sample is frozen in freon cooled with liquid nitrogen) very rapidly, usually in the presence of glycerol and then, subsequently, putting the rapidly frozen specimen on to the cold state of a vacuum coating unit. After a very high vacuum has been obtained, say 2 or 3×10^{-6} torr, the surface of the specimen is cut with a cold knife. The temperature of the specimen is then raised to $-100°C$ and a thin layer of ice is sublimed off. The cut surface is therefore etched by vacuum sublimation (see Fig. 20). Immediately after this, the surface of the specimen is shadowed with a heavy metal and a carbon replica is made. This replica is then examined in the electron microscope.

Using this technique Moor and Mühlethaler (1963) reported a study, using the freeze-etching method, on yeast cells, *Saccharomyces cerevisiae*. They show that the cell wall and cytoplasmic membrane are clearly separated and suggest that the fracture plane either follows the surface of the cell wall or penetrates the whole wall perpendicularly. By combining surface views and cross fracture views, they were able to obtain a three-dimensional image of the invaginations of the yeast cytoplasmic membrane. The invaginations have an average length of 3000 Å, a width of 200–300 Å and a depth of 500 Å.

The cross fractured membranes give images the same as those shown by membranes in sections of permangante-fixed material. Thus the membrane consists of three sub-units each about 25 Å thick (see Fig. 21). They suggest that this is in agreement with the structure of the unit membrane suggested by Robertson (1957, 1958).

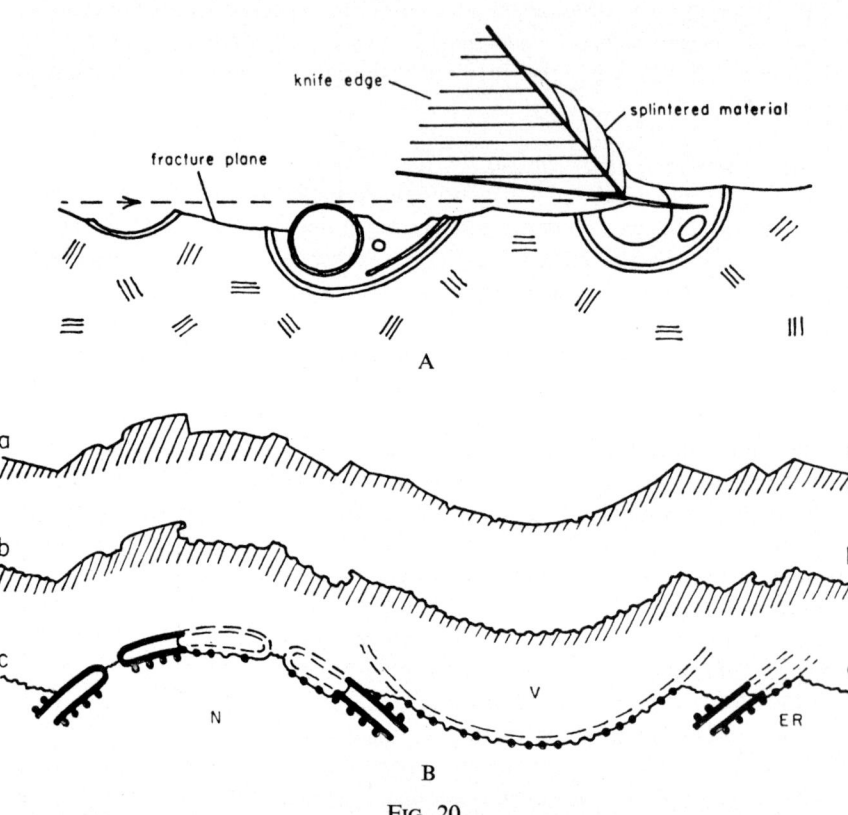

FIG. 20.
A. Diagram of the "cutting" procedure which actually consists of a fine splintering of the deep-frozen object (yeast cells).
B. Diagram of how the splintering and the etching reveal the fine structure of a frozen object. (a) Cross section through the fracture plane. (b) Etched fracture plane, showing the fine structure. (c) The reconstructed structural details of the recorded object. N, nucleus, showing a partially removed envelope. V, totally removed vacuole, rendering possible a surface view of the adjacent cytoplasmic ground substance. ER, endoplasmic reticulum, fractured at a low angle (from Moor and Mühlethaler, 1963).

Electron micrographs of the nucleus of *S. cerevisiae* show that it is surrounded by two unit membranes (Figs 21A–C). Perforations or pores are

4. STUDIES OF PHOSPHOLIPIDS AND NATURAL MEMBRANES

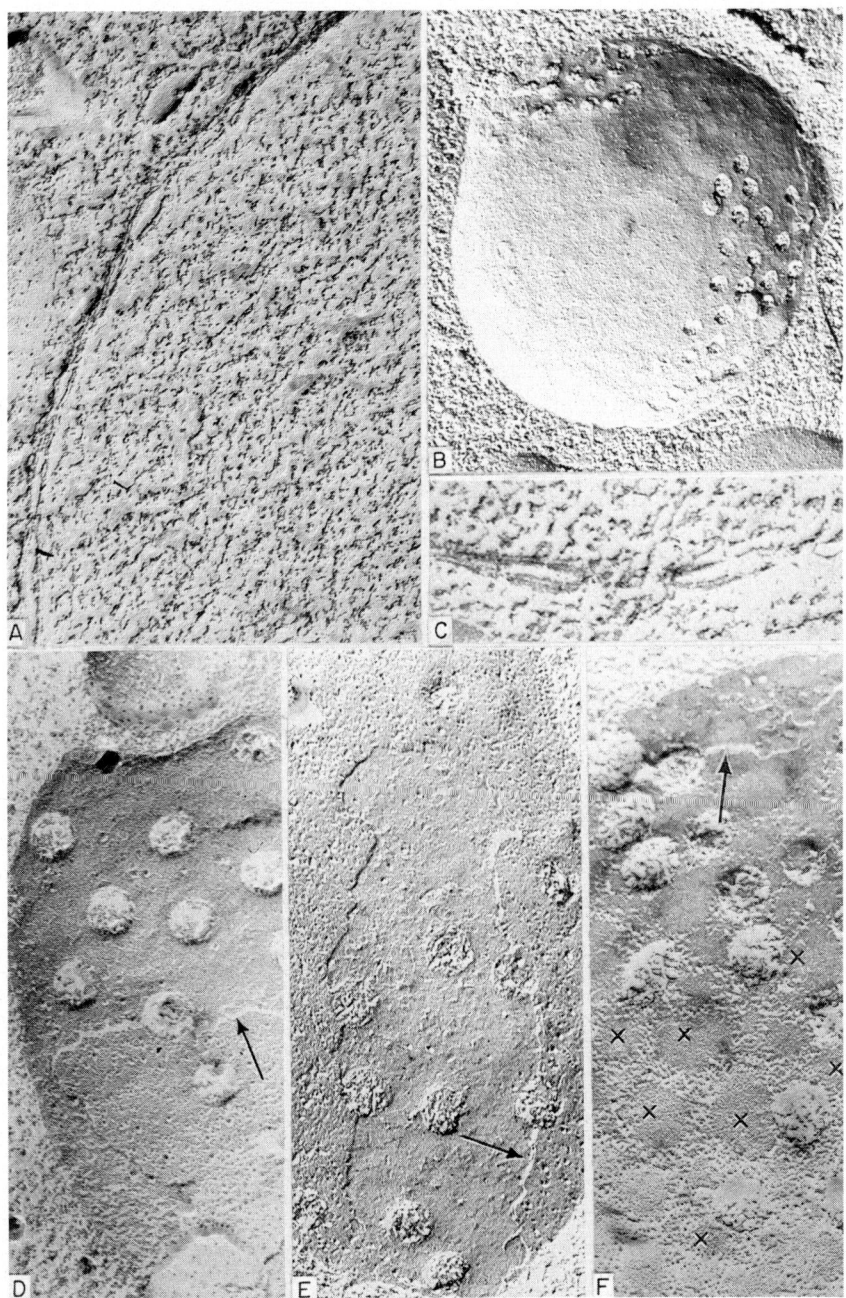

Legend for Fig. 21 at bottom of page 170

seen to be present (Figs 21B–F). Endoplasmic reticulum and vacuoles can be distinguished using the freeze-etching method. Membranes of the vesicular systems are seen to be covered by ribosomes arranged in circular patterns.

Mühlethaler (1966) has also looked at the isolated chloroplasts from *Spinacea oleracea*. The arrangement of the thylakoids as shown in spinach is shown in the electron micrograph shown in Fig. 22. The freeze-etched plastid closely resembles that of a chemically fixed specimen. In the frozen state, the thylakoid lumen is slightly larger than is observed using osmium tetroxide or potassium permanganate fixed organelles. The width of the membranes is also greater. In the frozen state they are 120 Å wide but in the ordinary sections they are 70–80 Å.

The surface views of the lamellae show them to be beset with particles which are thought to be identical with so-called "quantasomes" of Park and Biggins (1964). A different structural composition is observed between freeze-etched and chemically fixed plastids. All the plastids in the frozen state consist of a central homogenous layer of about 40 Å which is covered on both sides with particles. The globular layers are about 60 Å wide, but the total thickness of the membrane amounts to only 120 Å because the particles are partially embedded in the central region. When the usual chemical fixation method is applied, the dark strata of the membrane does not show a globular sub-structure and the thicknesses of the sub-lamellae are different.

Mühlethaler suggests that the reason why the globular substructure is absent in the chemically fixed preparations arises from the denaturation of protein molecules occurring during the fixation, leading them to uncoil. The different steps in these changes are shown in Fig. 4 in Chapter 7 by Dr. Leslie,

---

FIG. 21.

A. A cross-fractured nuclear envelope, showing the structure of the unit membranes. The nucleoplasm is visible on the lower right, the cytoplasmic ground substance on the upper left. × 90,000.

B. Surface view of a nucleus derived from an old cell. The pores are concentrated in certain areas. × 30,000.

C. Cross-fracture through a nuclear envelope which is perforated by a pore. The structure of the unit membranes is partially visible. × 135,000.

D. Outside view of a nuclear envelope. In the lower part of the figure it is splintered away, unveiling the adjacent surface of the nuclear content. × 80,000.

E. Inside view of a piece of the nuclear envelope left by the splintering on the surface of the adjacent ground plasm. × 80,000.

F. Surface view of the content of an old nucleus. The envelope is totally removed by the splintering. The position of the pores is indicated by circular depressions or elevations. The pattern on the flat part of the surface ("closed pores") (×) is created by circularly arranged particles (ribosomes). × 80,000. Figs 21D to 21F —the arrows indicate the fracture edge of the nuclear double membranes (from Moor and Mühlethaler, 1963).

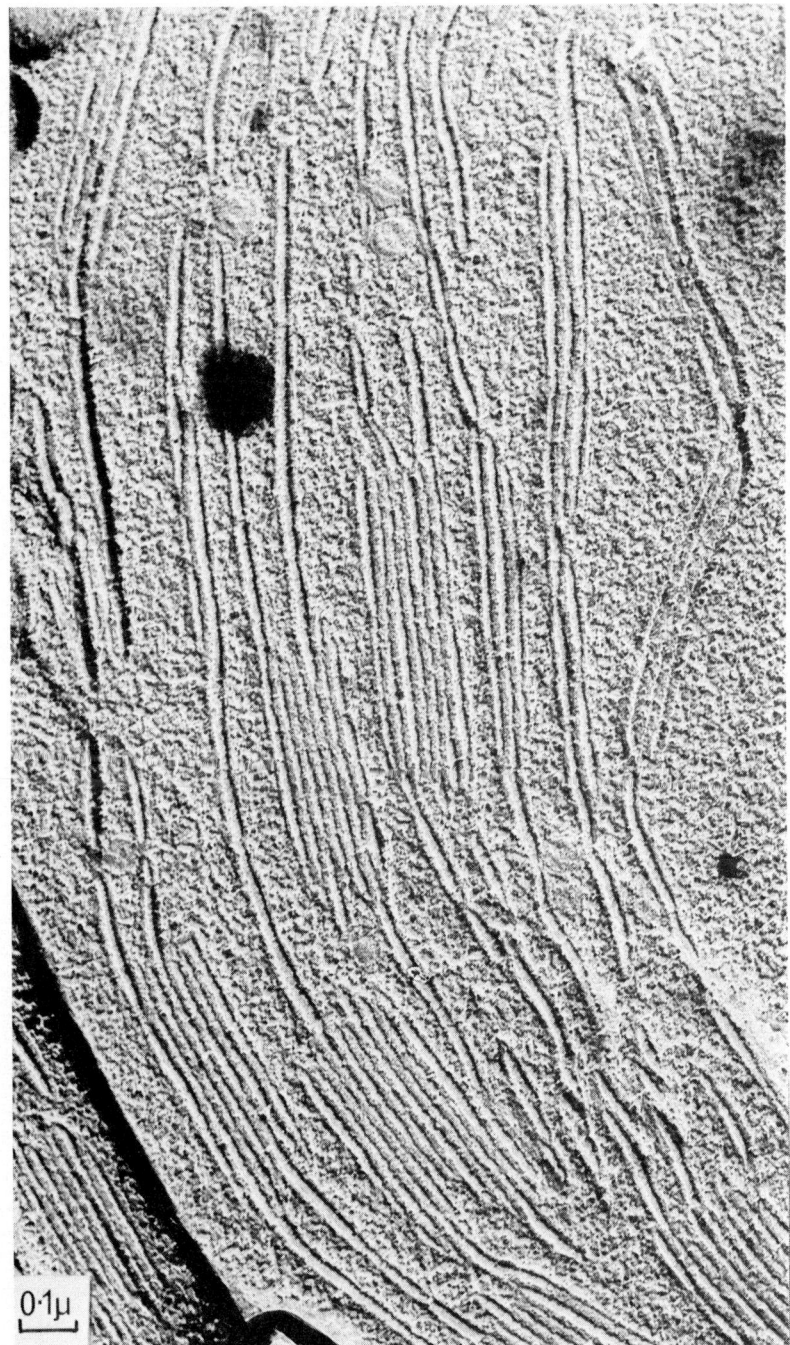

FIG. 22. Cross-sectional view of a spinach chloroplast showing the arrangement of grana and stroma thylakoids. × 74,000. (from Mühlethaler, 1966).

p. 300. The denaturation may be caused by changing the pH value, leading to breakage of the hydrogen bonds, or by altering the oxidation/reduction potential, causing destruction of the disulphide bridges. Particles seen after freeze-etching are considered to represent a mosaic structure of different enzyme complexes bound to the membranes.

Branton (1966) has examined freeze-etched root tip cells (see also Branton and Moor, 1964) and concluded that the true membrane surface is rarely seen in freeze-etched preparations. He suggests that the fracture process splits the membrane and exposes the internal membrane face. He bases this

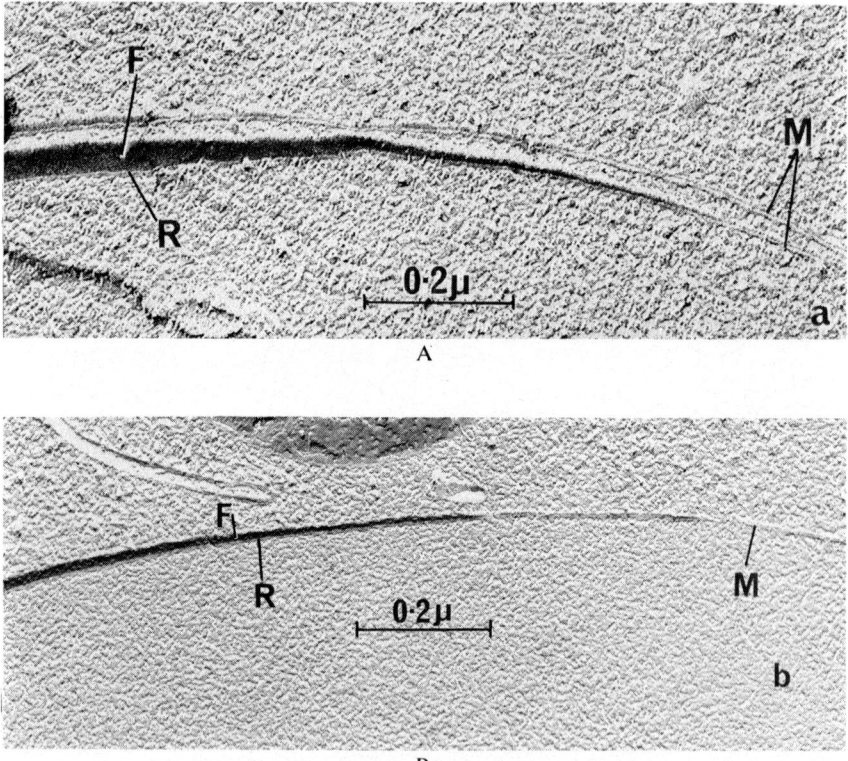

FIG. 23.
A. Endoplasmic reticulum in onion root tip.
B. Vacuolar membrane face in onion root tip; view from inside the vacuole.
In A and B the fractures are tangent to the membrane surfaces on the left and almost perpendicular to the membrane surfaces on the right. The small ridge (R) at the base of an exposed membrane face (F) on the left is continuous with one of two ridges which forms the typical freeze-etch image of a single membrane (M) on the right (from Branton, 1966).

conclusion on the presence of a small ridge in the electron micrographs of various membranes at the base of most exposed membrane faces (see Fig. 23). Branton suggests that this small ridge is continuous with and identical to one of the ridges previously assumed to represent part of a unit membrane structure (Moor et al., 1964; Branton and Moor, 1964). The same type of fracture is observed in freeze-etched preparations of the plasma, nuclear, vacuolar and dictyosomal membranes. An interpretative diagram, due to Branton, is shown in Fig. 24.

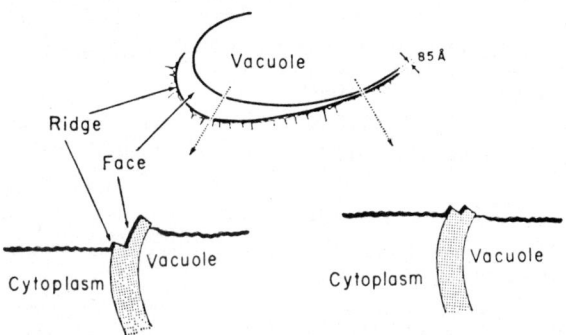

Fig. 24. Top, a representation of Fig. 23C. Bottom, diagrams of imaginary sections perpendicular to the plane of the page through the fractured tissue along the *dashed arrows*. These diagrams assume arrival of shadow-replica material from upper left, and show why fracture of an inclined *ca.* 75Å wide single membrane frequently produces the freeze-etch image seen in Figs 23 A-C (from Branton, 1966).

Branton points out that, if freeze-etching does indeed split membranes and reveals inner faces, then structural features known to exist on the other surfaces should not be visible on this membrane face exposed to freeze-etching. Consistent with this, ribosomal particles which are frequently associated with the surface of the endoplasmic reticulum in chemically fixed preparations, are not seen on the endoplasmic reticulum faces using the freeze-etching technique. Branton also points out that small particles can be seen averaging some 85 Å in diameter on many freeze-etched membrane faces. He suggests that these particles represent sub-structures within which membrane components have assumed globular or micellar configurations. The smooth region between the 85 Å particles are considered to represent ridges in which the membrane component exists as an extended bilayer, while the rest of the membrane is composed of the globular sub-units.

Moor (1966) has commented recently about this suggestion of Branton. He differs from Branton in his interpretation of the electron micrographs and states that, in many electron micrographs of freeze-etched specimens, the surface is definitely exposed. He points out that: (a) A freeze-etched myelin sheath shows two different types of surface, one finely granulated and the other covered by very flat elevations, whereas a splintering through the central lipid of the membrane should reveal only one type of smooth lipid surface. (b) Freeze-etched yeast plasmalemma show fibrils adhering to the surface. (c) Freeze-etched mitochondria show elementary particles attached to the mitochondria membrane.

Moor (1966) presents other arguments to support this view as well as that of the unit membrane concept. Despite this controversy, the freeze-etching technique appears to have important potential for the study of cell membranes.

C. SPECTROSCOPIC AND OPTICAL ROTATORY DISPERSION STUDIES

1. *Infrared Spectroscopic Studies*

Infrared spectroscopy has been applied to the study of certain membranes, including myelin (Chapman, 1965; Maddy and Malcolm, 1965, 1966) and to the erythrocyte and plasma membranes of Ehrlich ascites carcinoma (Wallach and Zahler, 1966). The infrared spectrum of a sample of myelin is shown in Fig. 25 and the infrared spectra of human erythrocyte ghost material and the total lipid extract are shown in Fig. 26.

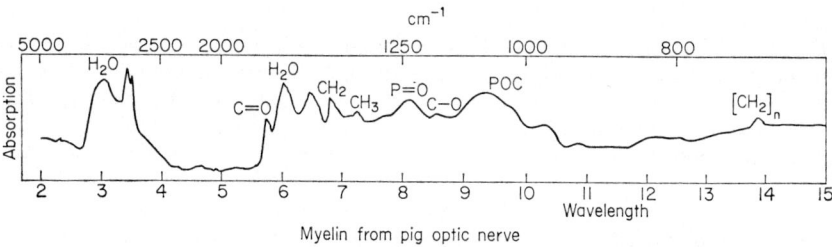

FIG. 25. The infrared spectrum of myelin (from Chapman, 1965).

A quantitative comparison of the infrared spectra of human erythrocyte membrane and its total lipid extract is shown in Fig. 26. It can be seen that the 720 cm^{-1} band is prominent in the spectrum of the lipid extract but it is extremely *weak* in the spectrum of the original membrane. This suggests that at room temperature there is little all-*trans* character of the CH$_2$ groups of the lipid hydrocarbon chains when they are organized in a membrane and that there is more all-*trans* character associated with the lipid extract at the same temperature.

This may be because, even if the arrangement of the lipids is lamellar in membrane and with the lipids, the packing of the lipids in the membrane is less compact than occurs with the bulk total lipid (even when water is present). A looser packing would allow more movement of the methylene groups to flex and rotate from the *trans* configuration. Another reason may be that, in addition to some "disorder" in the chains associated with temperature and kinetic effects, the lipid hydrocarbon chains in the membrane also tend to adopt conformations other than the all-*trans* conformation because of interaction with apolar amino acid residues of the membrane protein.

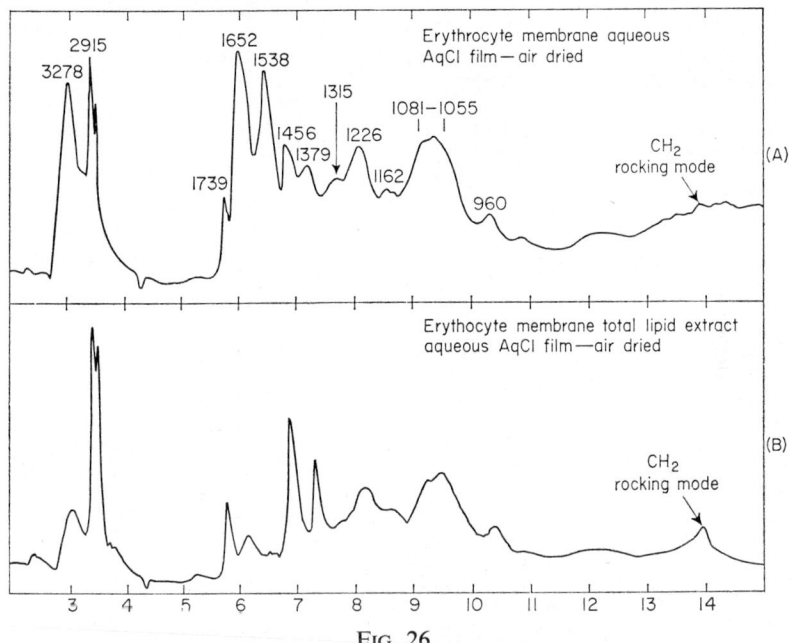

FIG. 26.

(A) The infrared spectrum of human erythrocyte ghost material. (B) The infrared spectrum of the total phospholipid from the ghost material (from Chapman *et al.*, 1968b).

The position of certain bands, referred to as amide I and amide II bands in the infrared spectra of peptide chains, differ dependent upon the conformation of these chains. Thus the amide I band is located at 1652 cm^{-1} and is associated with an α-helical and/or random coil conformation of peptide chains. The amide II band at about 1535 cm^{-1} does not allow distinction between the α-and β-conformation. A band at 1630 cm^{-1} is correlated with a β-conformation. The infrared spectra of erythrocyte and Ehrlich ascites carcinoma membranes do not show a strong band at 1630 cm^{-1}, suggesting that there is no extensive β structure in these membranes. The appearance of the spectra of a plasma membrane in the region 1800–1400 cm^{-1} is shown in Fig. 27.

Lipid extraction from the membrane removes the band at 1740 cm^{-1} due to carbonyl stretching in fatty acid esters. There is also some reduction of the two amide bands due, in part, to the extraction of sphingomyelin. However, lipid extraction does not produce a detectable transition to a β-conformation of the protein.

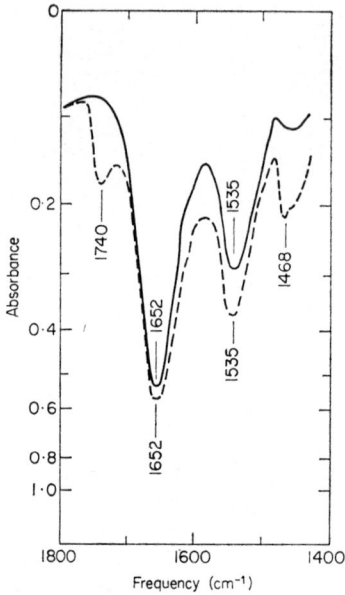

FIG. 27. The infrared spectrum of plasma membrane cast from aqueous suspension. Original film – – – –. After extraction with 2:1 chloroform:methanol ——— (from Wallach and Zahler, 1966).

It appears that the full potential of infrared spectroscopy has not yet been reached for obtaining information about membrane structure. Further studies using this technique should provide valuable information in the future.

2. *Optical Rotatory Dispersion Studies*—by D. F. H. Wallach

Substances whose constituent atoms are in an asymmetric or disymmetric array are optically active. This important physical property is manifest in two closely related phenomena, namely: (1) different absorbances for left and right circularly polarized light or *circular dichroism* (CD), (2) different refraction of left and right circularly polarized light—(circular birefringence) —causing rotation of plane polarized light; the variation of optical rotation with wavelength is called optical rotatory dispersion (ORD). While both phenomena have their origins in optically active absorption bands, CD is seen only in the wavelength interval where absorption takes place and thus allows resolution of optically active bands. Optical rotation occurs at wavelengths both near to and far from the optically active absorption band. The relationship between absorption, circular dichroism and optical rotatory dispersion is shown in Fig. 28 (Beychok, 1966).

The optical activity of peptide bonds arises from two types of electronic transitions (Shellman and Shellman, 1964), (Fig. 29): (1) promotion of electrons from the π^0 orbitals to antibonding π^- orbitals ($\pi^0 - \pi^-$ transition),

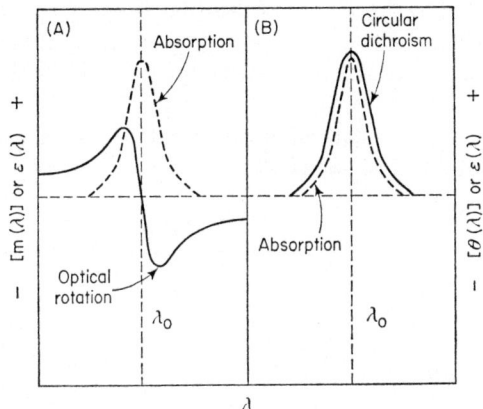

FIG. 28. Isolated absorption bands with associated negative Cotton effect in the optical rotatory dispersion spectrum (A) and positive ellipticity band in the circular dichroism spectrum (B). The symbol λ is a wavelength designation, increasing in value toward the right and decreasing toward the left in each half of the figure (from Beychok, 1966).

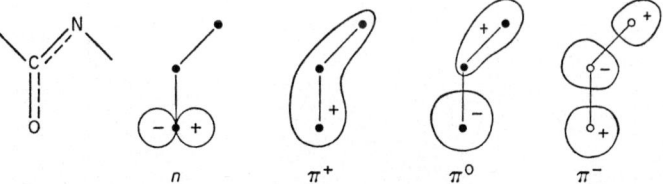

FIG. 29. Molecular orbital representation of the n- and π-orbitals of the peptide link. Only upper lobes of the π-orbitals are shown. Lower lobes are mirror images of the upper lobes in the plane of the paper but opposite in sign (from Shellman and Shellman, 1964).

responsible for the intense absorption near 190 mμ; (2) excitation of non-bonding electrons of the oxygen heteroatom to the π^- orbital (n $-$ π^- transition), responsible for the weak band near 220 mμ.

Measurement of the CD and ORD of the peptide chromophores has become extremely valuable in elucidating the conformations of soluble polypeptides and proteins because in such polymers the optical activity of the peptide transitions is not simply the summed contribution of single peptide linkages, but critically depends upon their spatial relationships. Thus, polypeptides in known α-helical, β- or unordered conformations exhibit distinctly different ORD and CD spectra (Fig. 30A, B). The relevant parameters of these spectra are listed in Table I.

When polypeptides are in an unordered conformation, their optical activity is dominated by the $\pi^0 - \pi^-$ transition at 198 mµ which is responsible for the intense, negative CD at 198 mµ and the large, negative ORD extremum at 205 mµ. The small, positive CD extremum at 223 mµ is due to the $n - \pi^-$ transition.

Polypeptides in right-handed α-helical conformation have two large, negative CD extrema at 222·5 mµ and 208 mµ and a positive extremum at 192 mµ.

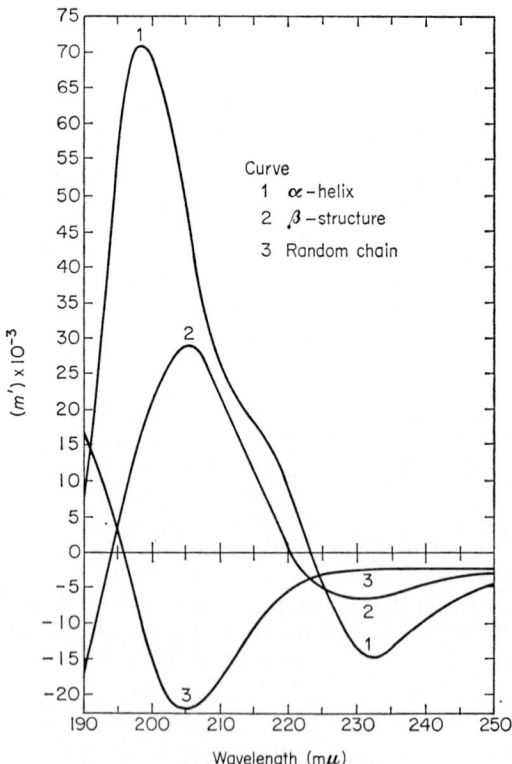

Fig. 30A. The ORD of poly-L-lysine in the α-helical, β and random conformations (from Greenfield et al., 1967).

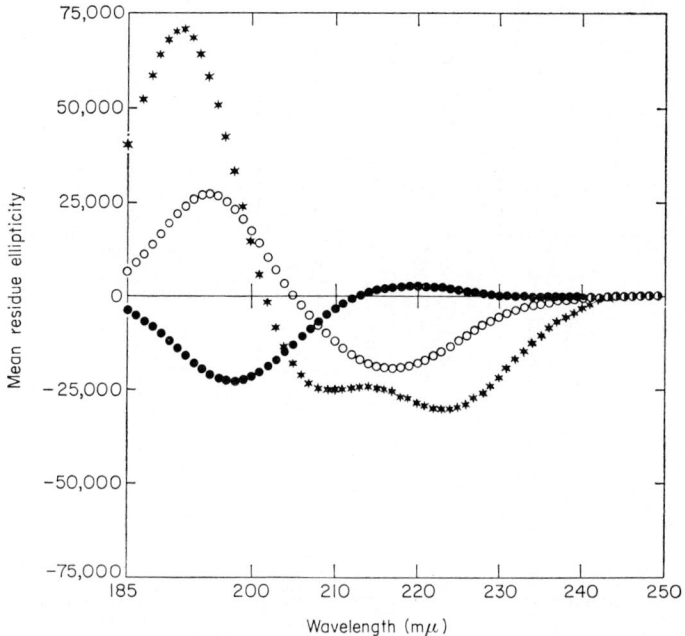

FIG. 30B. CD spectra of α-helix, β-conformation and unordered "coil". *, α-helix; ○, β-conformation; ●, unordered coil (θ) is in degrees cm^2/decimole (from Wallach and Gordon, 1967).

The two extrema at shorter wavelength arise from the $\pi^0 - \pi^-$ transition, which is split into two components (at 206 mμ and 192 mμ) in the α-helical conformation. The 222·5 mμ extremum is due primarily to the strong $n - \pi^-$ transition at 224 mμ. (The CD extremum is at slightly shorter wavelength than the transition, because of the overlapping, negative $\pi^0 - \pi^-$ component). The corresponding ORD spectra are characterized by a large negative extremum at 233 mμ, a crossover to positive rotation at 223 mμ, a shoulder near 210 mμ and a positive extremum at 198 mμ. The negative extremum is due primarily to the $n - \pi^-$ transition.

The CD spectra of polypeptides in β-conformation show a single negative band at 218 mμ, attributable to the $n - \pi^-$ transition (Pysh, 1966) and a positive extremum at 195 mμ. The corresponding ORD spectra have a negative extremum at 230 mμ, a crossover at 220 mμ and a positive extremum at 205 mμ.

Mixtures of conformations have CD and ORD spectra with intermediate band positions, shapes and amplitudes (Wallach and Gordon, 1967; Greenfield et al., 1967) as shown in Figs 31 and 32. Thus, conformational analysis of proteins should, in principle, be possible from comparisons of their ORD and CD spectra with those of synthetic polypeptides of known conformation.

TABLE I
ORD and CD parameters of synthetic polypeptides in various conformations

Conformation	CD[1]					ORD[2]		
	−Extrema (mμ)	[θ]λ[3]	Crossover (mμ)	+Extremum (mμ)	[θ]λ[3]	−Extremum (mμ)	[m']λ[4]	Crossover (mμ)	+Extremum (mμ)	[m']λ[4]
α-helix	224	29,400	201	192	73,300	233	14,723	223	198.6	70,856
	206	24,800								
β-conformation	217	19,300	205	195	27,900	230	6254	220.5	205	29,100
Unordered	198	22,800	212	217	3240	205	21,929	~195.5	—	—

[1] From Carver et al., 1966.
[2] From Greenfield et al., 1967.
[3] Mean residue ellipticity, where $[\theta]_\lambda = 3300(\varepsilon_L - \varepsilon_R)$, ε_L and ε_R being the mean residue extinction coefficients of left and right circularly polarised light, respectively. $[\theta]_\lambda$ is in degrees × cm² decimole⁻¹.
[4] Mean residue rotation in degrees × cm × decimole⁻¹.

This prediction is fulfilled reasonably well in the case of certain proteins (Greenfield et al., 1967). However, it should be noted that the overall shapes of ORD and CD spectra depend markedly on the proportion of unordered peptide, even though a very small proportion of α-helix maintains the ORD trough at 233 mμ and the long wavelength CD band at 223 mμ. Moreover, the ORD spectra of proteins can be considerably distorted by the optical activity of side-chain chromophores (Greenfield et al., 1967).

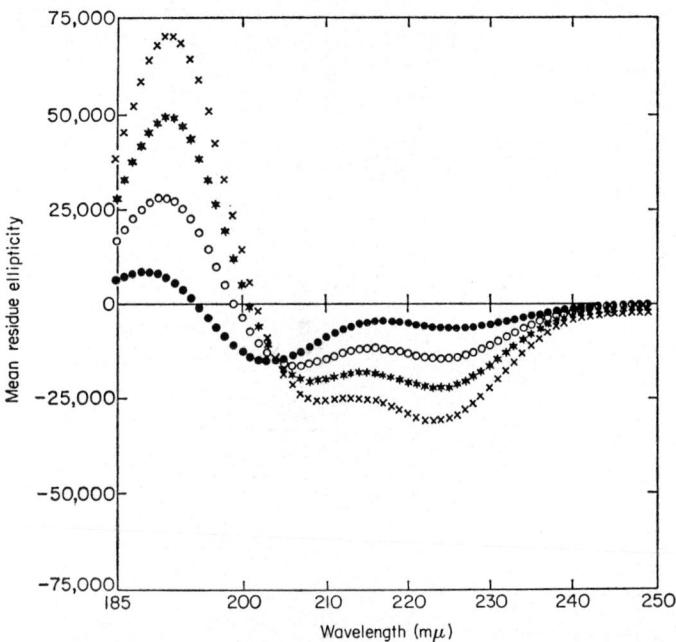

FIG. 31. CD spectra of hypothetical mixtures of α-helical and unordered conformations. ×, 100% α-helix; *, 75% α-helix, rest unordered; ○: 50% α-helix, rest unordered; ●: 25% α-helix, rest unordered. (θ) is in degrees cm²/decimole (from Wallach and Gordon, 1967).

In the light of the above experience, several investigators have examined the optical activity of cellular membranes with the aim of gaining insight into their protein architecture. Membranes from diverse sources prepared in many different ways have been examined, namely: plasma membranes and endoplasmic reticulum of Ehrlich ascites carcinoma (Wallach and Zahler, 1966; Wallach and Gordon, 1967); erythrocyte ghosts (Lenard and Singer, 1966; Steim, 1967); *B. subtilis* and mycoplasmal membranes (Lenard and Singer, 1966); membranes of halophile bacteria (Steim, 1967); mitochondria and mitochondrial fragments (Urry et al., 1967). The ORD spectra of these diverse membrane types all exhibit the following anomalous features (Table II and Fig. 33): (1) a shape closely approximating that of pure, right-handed α-helix; (2) low amplitude and (3) displacement of the entire spectrum to longer wavelengths than what is observed for α-helix.

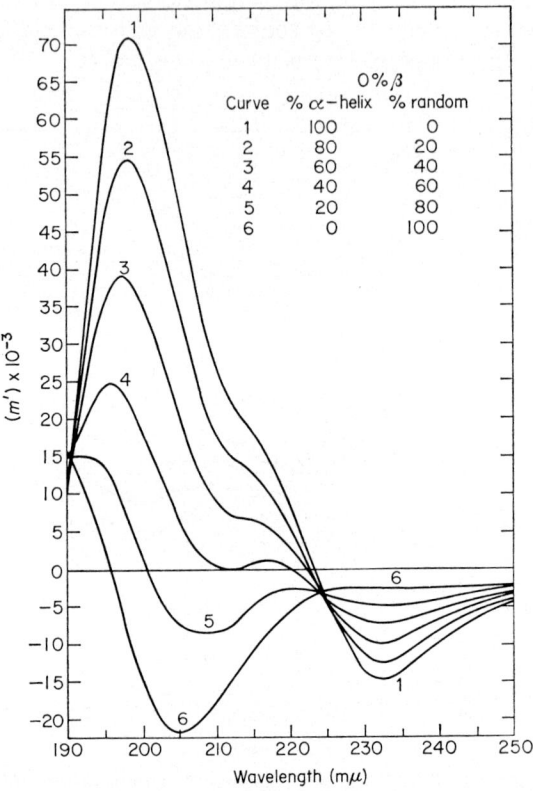

FIG. 32. The ORD of poly-L-lysine in the α-helical and random forms together with calculated combinations thereof in varying proportions as indicated (from Greenfield et al., 1967).

CD spectra have been reported by Lenard and Singer (1966), Urry et al., (1967) and by Wallach and Gordon (1967) and have the following characteristics (Fig. 34 and Table III): (1) "α-helical" shape, with a broad, negative band at 223–225 mμ; (2) low amplitude (3) a red shift of the spectrum, at least at shorter wavelengths, with a shoulder at 210–222 mμ and the crossover to positive CD at 205 mμ. (Because of the low signal to noise ratio below 200 mμ, this region has not been well studied).

The ORD and CD spectra are thus quite different from those of soluble lipoproteins (Scanu, 1965) and of ionic complexes between cytochrome c

TABLE II
ORD parameters of various membrane types

Membrane source	−Extremum (mμ)	[m']$_\lambda$[1]	Crossover (mμ)	+ Extremum (mμ)	[m'][1]	Reference
Ehrlich ascites carcinoma	235–237	−3000–4500	226	201–202	12,900	Wallach and Gordon (1967)
Erythrocytes	235	−5800	226–227	201–202	24,000	Lenard and Singer (1966)
B. subtilis	235–238	−5200	227	201–203	21,100	Lenard and Singer (1966)
M. Laidlawi	235	—	—	—	—	Lenard and Singer (1966)
Halobacterium cutirubrum or halobium	235–237	—	—	202	—	Steim (1967)
Beef heart mitochondria	237	−1000	~230	201–203	~3,000	Urry et al., (1967)

[1] Mean residue rotation.
[2] Corrected for dispersion of the refractive index of water.

TABLE III
CD parameters of various membrane types

Membrane source	Extrema (mμ)	[θ]$_\lambda$[1]	Crossover (mμ)	Reference
Ehrlich ascites carcinoma	223	−10,250	205	Wallach and Gordon (1967)
	212			
Erythrocyte	224	−13,200	205	Lenard and Singer (1966)
	210			
B. subtilis	224	−12,900	—	Lenard and Singer (1966)
	211–212			
Mitochondria	225	—	208	Urry et al., (1967)
	211	—		

[1] Mean residue ellipticity.

and acidic phosphatides (Ullmer *et al.*, 1965). The membrane ORD spectra are perplexing because, if the membrane protein is partly in right-handed α-helix and the remainder unordered, the spectra should not have the "α-helical" shape (Greenfield *et al.*, 1967). Conversely, if the proteins are

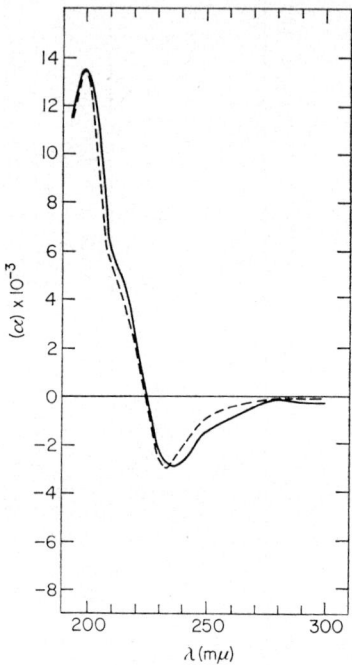

FIG. 33. Optical rotatory dispersion of plasma membrane in aqueous suspension: ———, plasma membrane; - - - - -, poly-L-glutamic acid (pH 4·25). The values of poly-L-glutamic acid have been reduced by a factor of 4·5 (from Wallach and Zahler, 1966).

primarily helical, they should show greater rotational amplitude. The same arguments apply to the CD spectra. Both the ORD and CD amplitudes and shapes are *as if* most of the optical activity of the membrane protein were masked with only right α-helix being expressed. This idea does not account for the observed red shifts, however, particularly those in the CD spectra, since, in mixtures of α-helix and unordered conformation, even small proportions of the latter displace the spectra to the blue (Greenfield *et al.*, 1967; Wallach and Gordon, 1967).

Since ORD and CD spectra have common origins and have, in principle, a precisely defined relationship (Beychok, 1966), any explanation of the anomalous optical activity of membranes must account for the peculiarities of both types of spectra. Lenard and Singer (1966) do not examine the relationships of amplitude, shape and band position of their spectra in any detail but they suggest that the red displacement of the membrane spectra might originate in optical activity arising when α-helices are packed parallel but at a slight twist (Robinson, 1961). This concept has also been invoked to explain the small red shift which accompanies aggregation of α-helical poly-L-glutamic acid (Cassim and Yang, 1967); but in this case the spectral displacement is accompanied by increased optical activity. In any event, the small displacements in the $n - \pi^-$ region of the CD spectra cannot account for the position of the trough and crossover in membrane ORD spectra.

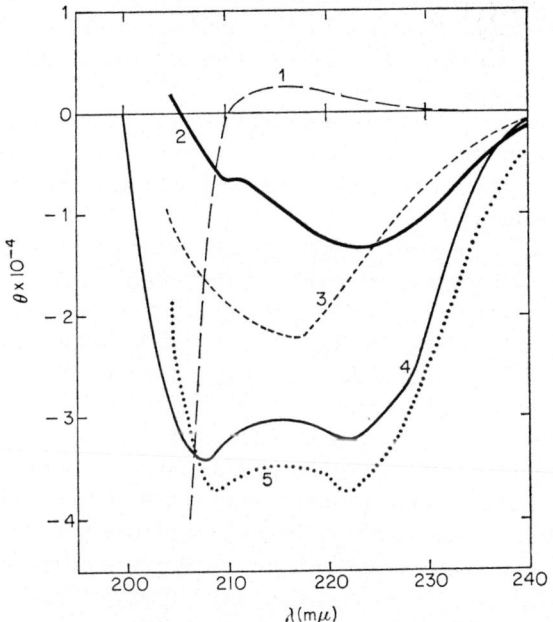

FIG. 34. CD spectra of random coil poly-L-glutamic acid (1). CD spectra of red blood cell membranes suspended in 0·008M phosphate buffer, pH 7·7 (2). The CD spectra of poly-L-lysine in the β-conformation (3) and in the α-helical form (4). CD spectra of red blood cell membranes dissolved in 2-chloroethanol (5), plotted for comparison (from Lenard and Singer, 1966).

Urry et al., (1967), much concerned with the low amplitude of their spectra, argue that the displacement of the negative ORD extremum in membranes is due to a displacement of the $n - \pi^-$ transition, giving as reason the 3mμ displacement of the long wavelength CD band. However, their reasoning would be correct only if the transition were quite isolated from the long wavelength $\pi^0 - \pi^-$ transition, which is not the case, and if their CD spectra (as well as those of Lenard and Singer (1966) and Wallach and Gordon (1967)) did not show red shifts also in the $\pi^0 - \pi^-$ region; these facts cannot be explained by changes in only the $n - \pi^-$ band. Wallach and Gordon (1967), with the aid of a rather extensive computer analysis, found a plausible solution for both the ORD and CD spectra of cellular membranes. These could arise under the following conditions (assuming identical rotatory strength for membrane proteins as for synthetic polypeptides):

(1) 50–60% α-helix;
(2) $n - \pi^-$ transition at 224 mμ with 14 mμ bandwidth;
(3) $\pi^0 - \pi^-$ transitions shifted 2 mμ to the red, with 11–12 mμ

bandwidth. The important lesson of their studies is that changes in band position and *bandwidth* can profoundly affect the position and amplitude of both ORD and CD spectra in the direction of what is observed in cellular membranes.

It should be noted here that infrared spectra, discussed elsewhere in this volume, indicate that at least the native membranes of erythrocytes and Ehrlich ascites carcinoma lack sufficient β-conformation to significantly affect their optical activity.

One explanation for these findings—which cannot yet be properly evaluated —is the presence in membranes of other, hitherto undefined protein conformations, or of an uneven mixture of right- and left-handed α-helix. The latter suggestion could account for the helical shape and low amplitude of the spectra but not for the red shifts. Moreover, L-amino acids are not known to form stable, left-handed α-helices.

A second possibility is that the ORD and CD spectra are perturbed by chromophores other than those of the peptide linkage. Indeed, the membranes of Ehrlich ascites carcinoma exhibit optical activity attributable to aromatic side chains (Wallach and Zahler, 1966) and membrane-bound RNA (Wallach and Gordon, 1967) but these effects are small and are lacking in the other membranes studied so far (Lenard and Singer, 1966; Steim, 1967; Urry *et al.*, 1967). Another possible origin of optical activity is the membrane phosphatide. Lenard and Singer (1966) looked for this but failed to find significant optical activity in 2-chloroethanol solutions of extracted membrane lipid when these were examined at concentrations corresponding to the lipid content of intact membranes. D. F. H. Wallach and A. Gordon (unpublished) examined the ORD and CD of aqueous phosphatide dispersion but did not find the optical activity significant compared with that of cellular membranes. However, Urry *et al.*, (1967) report that both lecithin and phosphatidyl ethanolamine in trifluorethanol have a weak, positive CD band at 218 mμ which they attribute to an ester $n - \pi^-$ transition. They suggest that this optical activity might be enhanced when phosphatides are in an asymmetric array, as might be the case in membranes, and that this could account for the peculiarities of membrane ORD and CD. This appealing idea is unlikely, however, since the ORD spectra of halophile membranes—which lack ester phosphatides—have the same anomalies as other membranes (Steim, 1967).

It has also been suggested that asymmetric or disymmetric packing of α-helices produces anomalous optical activity by interaction of the peptide transitions of adjacent helices. As already noted above, Cassim and Yang (1967) consider this as the explanation for the small red shifts and amplitude increases which occur in the CD and ORD spectra of poly-L-glutamate when it aggregates. However, it should be noted that in the case of α-helical poly benzyl-L-glutamate, which is known to pack disymmetrically (Robinson, 1961), aggregation is accompanied by very marked reduction in the amplitude of the ORD and CD spectra and red shifts of 7–8 mμ (Wallach and Gordon, 1967). The long wavelength CD band is in this case near 230 mμ, in contrast to its position in native membranes, but quite similar to what is observed in the case of defatted endoplasmic reticulum membranes of Ehrlich ascites carcinoma (Wallach and Gordon, 1967).

There is, however, one model system which simulates the ORD and CD spectra of cellular membranes as far as band positions, shape and amplitude are concerned, namely, the complex of poly-L-ornithine and sodium dodecyl sulfate (Grourke and Gibbs, 1967). Interestingly enough, a similar complex between sodium dodecyl sulfate and poly-L-lysine has the ORD and CD characteristics of β-conformation except for anomalously low amplitudes (Sarkar and Doty, 1966). These complexes are clearly worthy of further study.

In the preceding discussion it has been assumed that synthetic polypeptides dissolved as individual molecules in media of relatively low refractive index are adequate models for membrane proteins, which are in a rather special environment and more in a "solid state" than in solution. This question has been considered in some detail (Wallach and Zahler, 1966; Wallach and Gordon, 1967; Urry et al., 1967) and it is agreed that solvent polarity is not likely to affect the $n - \pi^-$ transition in helical polypeptides. However, Wallach and Zahler (1966) point out that this transition could be influenced by solvent polarizability and Wallach and Gordon (1967) argue that the red displacement and low amplitude of the ORD trough is due to increased bandwidth of the α-helical $n - \pi^-$ transition when this is in a medium with a refractive index as high as observed in cellular membranes (Wallach et al., 1966).

As far as the $\pi^0 - \pi^-$ transition is concerned, all of the data indicate red displacement. Wallach and Zahler (1966) and Wallach and Gordon (1967) indicate that this too could be due to high polarizability of the medium surrounding the helical portions of membrane proteins. Wallach and Gordon (1967) suggest that the bandwidths of the $\pi^0 - \pi^-$ transitions are also increased, a condition necessary to account for the observed ORD and CD amplitudes and crossover positions. Urry *et al.* (1967) do not discuss polarizability but point out that in the case of helical polypeptides, media of low polarity would also cause a red shift of the $\pi^0 - \pi^-$ transition.

Both groups come to the same tentative conclusion, namely that helical segments of membrane protein lie in a hydrophobic environment (low polarity—Urry, *et al.*, 1967; high polarizability—Wallach and Zahler, 1966; Wallach and Gordon, 1967). Some experimental support for this idea comes from the observation that membrane ORD and CD spectra shift to normal position upon treatment with phospholipase A (Wallach and Gordon, 1967) or lysolecithin (Wallach and Zahler, 1966) but not upon removal of polar moieties of the phosphatides with phospholipase C (Wallach and Gordon, 1967).

One cannot now come to firm conclusions about the conformations of membrane proteins and their helix content but Wallach and Gordon (1967) point out that the optical activity of membranes could be consistent with a proportion of α-helix near 60%.

The picture of membrane proteins emerging from the above studies, particularly the evidence for substantial α-helix content, is not easily reconciled with the Davson-Danielli-Robertson hypothesis of membrane structure (Robertson, 1964). Thus, Lenard and Singer (1966) argue that the stable, interfacial location of protein envisaged in this model would require amino acid sequences incompatible with the α-helical conformation. They therefore propose that: "(1) The ionic and polar heads of the lipid molecules, together with all of the ionic side chains of the structural protein, are on the exterior surfaces of the membrane in van der Waals contact with the bulk aqueous phase. (2) Sequences of the structural protein consisting predominantly of nonpolar side chains are in the interior of the membrane, together with the hydrocarbon tails of the phospholipids and the relatively nonpolar lipids such as cholesterol. In particular, the helical portions of the protein are interior, where they are stabilized by hydrophobic interactions. There is no definite evidence, however, to indicate whether or not single polypeptide chains traverse the entire thickness of the membrane. (3) Structural proteins are characterized by unique amino acid sequences which specifically adapt them to interact with the lipid components of the membranes and the aqueous environment; the over-all conformations of the structural proteins are determined by these interactions." Lenard and Singer also suggest that the proposed structure ". . . could aggregate in two dimensions to form an intact membrane."

This hypothesis is almost identical to that proposed independently by Wallach and Zahler (1966) except that these authors feel that it is not imperative to place all ionic side chains at the membrane surfaces and that suitable amino acid sequences could produce largely α-helical aggregates, containing aqueous channels lined with polar side chains. This scheme is analogous to what is actually found in hemoglobin (Perutz, 1965) and provides at least a conceptual basis for membrane "pores". Wallach and Zahler (1966) and Wallach and Gordon (1967) also postulate that the native conformation of membrane proteins depends upon their apolar interaction with membrane lipids, just as the conformation of hemoglobin is critically influenced by the hydrophobically bound heme groups (Perutz, 1965). The model of Wallach and Zahler (1966) implies more subtle peculiarities of amino acid sequences that that proposed by Lenard and Singer (1966) and also suggests that the primary structure of the helical sequences leads to preferential binding of phosphatides with sterically favorable fatty acid compositions.

The new models deal with general structural features which could make a protein suitable for a membrane localization. They are consistent with the known facts of membrane structure and function but are far more compatible with modern concepts of protein architecture and molecular biology than the Davson-Danielli-Robertson hypothesis. Proof of their validity will require much additional experimentation.

3. *Nuclear Magnetic Resonance Studies*

The fact that it was possible to obtain high resolution n.m.r. spectra from phospholipid dispersed in aqueous systems led to the idea that biological membranes themselves could perhaps be examined using this technique. We have seen that there is a high degree of molecular motion of the phospholipid molecules in the crystal at the environmental temperature. There is also considerable molecular motion of the phospholipid in the presence of water so that one might expect that the motion would be sufficient for an n.m.r. signal to be obtained from the lipids in the membrane itself.

The first experiments carried out (Chapman et al., 1967a) examined the proton magnetic resonance spectroscopy of membrane fragments of the erythrocyte membranes, or ghosts, when they are dispersed by ultrasonics in D_2O. These studies have shown that, despite the fact that we are dealing with insoluble materials composed of lipid, protein, water, cholesterol and sugar molecules, useful information can indeed be obtained.

The high resolution proton magnetic resonance p.m.r. spectra of membrane fragments is shown in Fig. 35. It can be seen that peaks associated with some of the functional groups present in the membrane can be clearly observed. Thus there is a peak at 6·8 p.p.m. which arises from the 9 protons present in the choline group of the phosphatidylcholine and sphingomyelins present in the membrane. Peaks are also observed which have been assigned to the protons present in sugar groupings (6·3 p.p.m.) and also to sialic acid (7·3 p.p.m.) which is known to be associated with the erythrocyte ghost material. The p.m.r. spectra of ox brain ganglioside and *N*-acetyl neuraminic

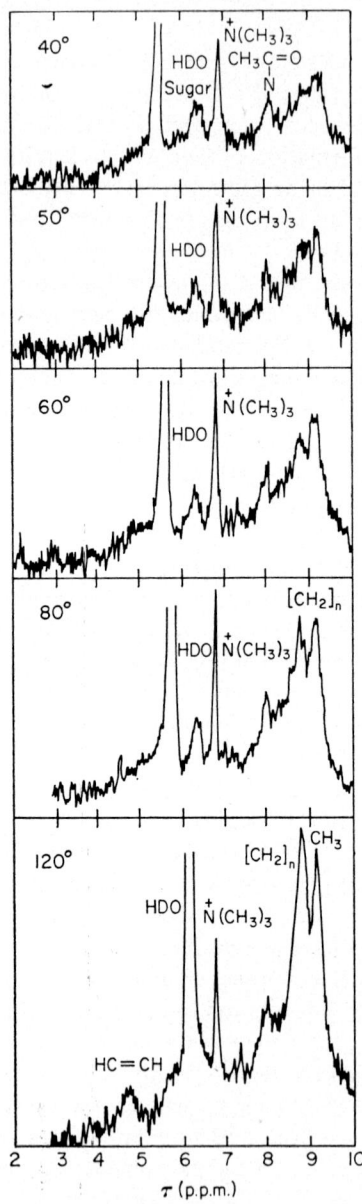

FIG. 35. The p.m.r. spectra at different temperatures of a sonicated dispersion (5% w/v) of erythrocyte membrane fragments in D_2O (64 scans) (from Chapman et al., 1968a).

acid give peaks near these positions and aid the assignment. What is interesting is that the considerable number of protons present in the hydrocarbon chains of the phospholipids do not give rise to a strong peak. A broad hump occurs in the 8·8 p.p.m. region rather than a sharp peak. It is similar to that observed when we have phospholipid-cholesterol interactions. No signal is apparent at 4·7 p.p.m. due to the HC≡CH protons which are present in the chains.

The assignment of the various peaks in the p.m.r. spectra have been supported by various studies, e.g. the fact that the signal at 6·8 p.p.m. is indeed associated with choline protons is supported by the study made of the erythrocyte membrane after treatment with phospholipase C. Phospholipase C is known to hydrolyse the whole of the polar grouping from the phospholipid molecule beyond the CH_2 group of the glycerol leaving a diglyceride molecule (van Deenen, 1965). The p.m.r. spectra of the material after this treatment is shown in Fig. 36 and it can be seen quite clearly that the peak associated with the choline group protons has been removed from the spectrum.

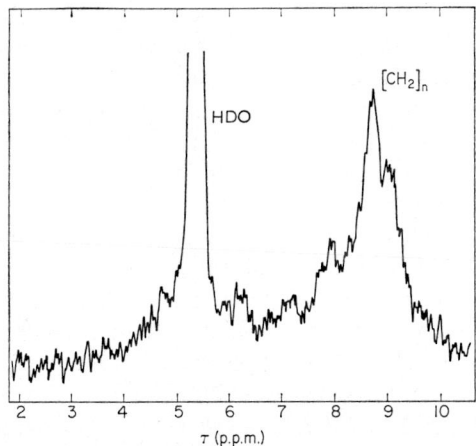

FIG. 36. The p.m.r. spectrum of erythrocyte membrane fragments pretreated with phospholipase C and sonicated in D_2O to form a 5% dispersion (512 scans) (from Chapman et. al., 1968a).

It is interesting that the choline protons show up so clearly in the n.m.r. spectra of the sonicated erythrocyte membrane material. This implies that, in the membrane, the protons in the choline group are able to move quite freely This suggests that the local viscosity about these choline protons is not high. It is also interesting that the n.m.r. peaks associated with the sugar and sialic acid groups should give rise to high resolution signals since there have been many conjectures about the presence of sugar and sialic acid on the surface of the membrane. Since the membrane contains a large quantity of protein, it is an important question to ask whether any of the broad hump in the 9 p.p.m. region arises from protons associated with the protein material. Spectroscopic studies of proteins have shown that absorption does occur in this region. This is associated with absorption arising from methyl and methylene groups in the constituent amino acids. In general, however, the n.m.r. spectrum of a protein does not show very much fine structure (Kowalsky, 1962). The first tentative conclusion obtained from this high resolution n.m.r. spectrum is that the hydrocarbon chain of the lipid appears to be hindered in its molecular motion, either as a result of being associated with cholesterol or protein, or with both of these membrane constituents. The n.m.r. spectrum is consistent with the membrane being a fairly compact structure of protein and lipid chains with sugar sialic acid groups and the choline groups situated on the outside of this compact structure.

With an increase in temperature several changes occur in the high resolution n.m.r. spectra. (a) The signal at 6·7 p.p.m. becomes sharper and the peak height increases. The area of this peak, however, remains approximately constant, (b) the signal at 8·8 p.p.m. due to $(CH_2)_n$ protons increases in intensity, while the peak at 9·0 p.p.m. becomes sharper, (c) a new peak becomes apparent at 4·7 p.p.m. which can be assigned to HC=CH protons. Above 120°C the intensities of all the signals in the spectrum fall, probably due to thermal breakdown of the membrane.

Additional high resolution p.m.r. spectra of the erythrocyte membrane and its constituents reveal further information about membrane construction. Thus, in Fig. 37 a spectrum of the phospholipid extracted from the membrane is shown at the same concentration as the phospholipid present in the membrane itself and the large signal associated with the $[CH_2]_n$ protons can be seen. The spectrum of the total lipid, including the cholesterol, is also shown and it can be seen that in this case the inhibition of the chain signal by the cholesterol is not as great as is observed when cholesterol interacts with egg yolk lecithin. This difference in effect may arise because of the much greater unsaturation present in the human red blood cell lipids and it can be compared with the monolayer results of Demel (1966) which show little condensing effect by cholesterol with phospholipids of this type.

It is suggested that the inhibition of the $[CH_2]_n$ chain signal in the membrane n.m.r. spectrum is due to a reduction in the lipid chain mobility caused by an interaction between these chains and the protein present in the membrane. Additional support for this idea comes from the n.m.r. spectrum of the erythrocyte membrane after treatment with ether. Ether treatment removes all the cholesterol as well as a small amount of the phospholipid. The effect of this treatment, although drastic, is not understood in detail and does not seem to allow chain $[CH_2]_n$ mobility because no increase in the $[CH_2]_n$ signal

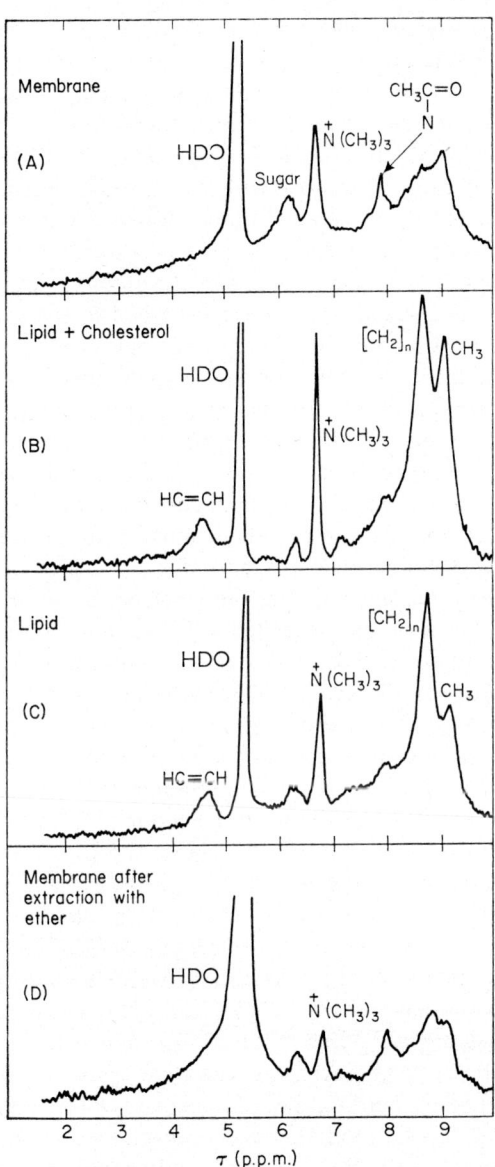

FIG. 37. P.m.r. spectrum of a 5% sonicated dispersion in D_2O. (A) erythrocyte membrane fragments, (B) total equivalent membrane lipid (phospholipid + cholesterol), (C) total equivalent phospholipid, (D) lipid deficient membrane fragments extracted with ether to remove all cholesterol and about 25% phospholipid (64 scans each) (from Chapman *et al.*, 1968a).

at 8·8 p.p.m. is observed. The n.m.r. evidence seems to favor some lipid-chain-protein interaction occurring in the human erythrocyte membrane.

Urea is known to affect the secondary and tertiary structure of proteins and high resolution n.m.r. studies of pure proteins in the presence of urea show an increase in fine structure in their spectra (Kowalsky, 1962). When the erythrocyte membrane is treated with urea, peaks appear in the n.m.r. spectra arising from the associated amino acids of the membrane protein. Some segmental motion of parts of the protein associated with the membrane appears to become possible and certain of the amino acids are free to move. Some of these n.m.r. signals can even be assigned to specific types of amino acid. However, even in this case there is no marked increase in the alkyl chain signal. Addition of trifluoroacetic acid to the membrane fragments produces an n.m.r. spectrum showing strong amino acid peaks and also a strong alkyl chain signal. This acid appears to break up the interlocked organization of amino acids of the protein and the chains of the lipid.

Addition of a molecule such as sodium desoxycholate, however, does cause a progressive increase in the alkyl chain signal to occur and it appears that the sodium desoxycholine has weakened the lipid protein interaction and increased the chain mobility. In fact the whole spectrum of the lipid can be seen, including the protons associated with the double bonds of the alkyl chains at 4·7 p.p.m., as well as other lipid groupings. What is also interesting is that the spectra of the sodium desoxycholine itself does not show up, implying that the sodium desoxycholine has entered into some complex with the membrane material and, at the same time, released the lipid. The concentration of sodium desoxycholate required for complete hemolysis of these red blood cells has been shown to be about 1%. This is just the concentration at which the alkyl chain $[CH_2]_n$ signal at 8·8 p.p.m. markedly shows an increase in intensity. These spectra are shown in Fig. 38.

Studies on the effects of other molecules, using n.m.r. spectroscopy, have also been carried out and these show interesting interaction effects. The high resolution n.m.r. spectrum of lysolecithin in D_2O shows a number of similar peaks to that of normal lecithin, i.e. it shows a choline signal and it shows a large alkyl $[CH_2]_n$ signal. When lysolecithin is added to the membrane, the p.m.r. spectra show, surprisingly, that the chain signal from lysolecithin has disappeared, although there has been an increase in the area associated with the choline peak. This shows that the interaction of lysolecithin with the erythrocyte membrane on the one hand, and sodium desoxycholate on the other, are quite different although both these materials are known to cause lysis of red blood cells.*

The application of n.m.r. spectroscopy to membranes is still in its infancy but it is already apparent that considerable penetration in our understanding of the organization of membranes and their interaction with various mole-

* These spectral effects occur with original erythrocyte ghost material as well as with sonicated membrane fragments.

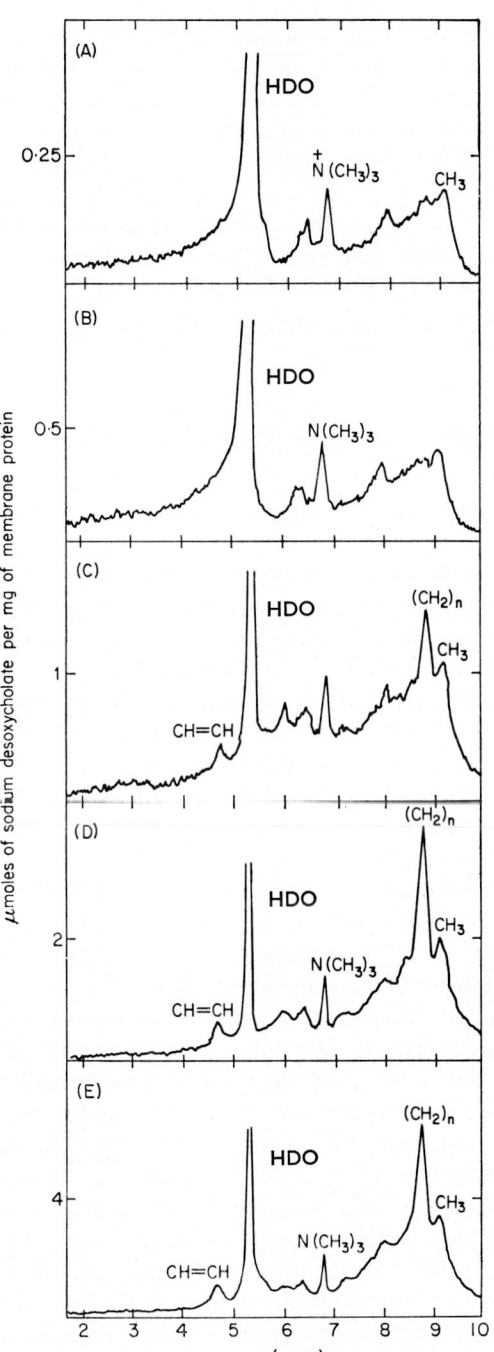

FIG. 38. P.m.r. spectra of erythrocyte membrane co-dispersed by sonication in D_2O at 5% (w/v) concentration with increasing concentrations of sodium desoxycholate (64 scans). Spectra in D and E are printed at one-half and one-quarter the sensitivity of A, B and C (from Chapman et al, 1968a).

cules, e.g. hormones, drugs, etc. may occur as a result of studies using this technique. Future nuclear resonance studies on membranes will include nuclear resonance studies involving the phosphorus nuclei and studies of the water structure concerned with membranes using deuterium resonance. The use of higher magnetic fields, such as those obtained with superconducting magnet systems and high frequencies (220 Mc/s), is expected to provide additional information. The technique of spin labelling (McConnell and Boeyers, 1967) of the constituent molecules may also be useful. Electron spin resonance spectroscopy can then be applied as a further aid to the study of membrane structure and membrane interactions.

IV. Conclusions and Future Studies

In this section we shall ask ourselves what relevance do the physical studies of the component molecules have on the possible structure of natural membranes.

We know that the pure phospholipids in water have limiting transition temperatures which occur progressively at lower temperatures as the chain length decreases. With unsaturated phospholipids these transition temperatures are very low and considerably below body temperature.

This means that, at this temperature, highly unsaturated phospholipids are in a highly mobile condition and this is particularly true of the hydrocarbon chain. This does not necessarily means that they are as mobile in a membrane. They may not be in a highly mobile condition in a cell membrane but if not then we need to have some inhibitory interaction, such as interaction with cholesterol or protein, etc. to explain this. To imagine that the unsaturated lipid chains are mainly static with only slight transient departures of relatively small amplitude without invoking inhibitory effects does not seem reasonable, yet this idea is an important point in recent theories and speculations about membrane structure (Vandenheuvel, 1966).

If we are prepared to speculate about the reason for the particular distribution of fatty acid residues observed in biological membranes, we can envisage that one of the functions of the distribution of fatty acid residues observed with these phospholipids is to provide the correct fluidity at a particular environmental temperature so as to match the required diffusion or rate of metabolic processes for the tissue. Other functions may include the possible structural requirement associated with lipid-protein interactions. Thus in membranes where metabolic and diffusion processes are required to be of a rapid nature, such as in the mitochondria, the average transition temperature of the phospholipids present will probably below compared with the biological environmental temperature, while in membranes where these processes are slow, e.g. in myelin of the central nervous system, the average transition temperature for the phospholipids will be higher and may be close to that of the biological environmental temperature.* This idea that one of the functions of the fatty acid distribution observed in membranes is to provide the correct fluidity has some support from the following observations.

Studies of the lipids of poikilothermic organisms show that they vary with the temperature of growth. Plants, insects, blow fly larvae and microorganisms all appear to contain increased proportions of unsaturated fatty acids or more highly unsaturated fatty acids if they are grown at low temperatures. House et al. (1958) have shown that the resistance of larvae of *Pseudosarcophaga affinis* (Fall) to a high temperature is affected by the diet on which they are reared. Insects fed higher saturated fatty acid diets have greater resistance to mortality to a temperature of 45°C than did those fed with unsaturated fatty acids. Homiothermic organisms also contain a greater proportion of unsaturated fatty acids in the surface lipids if the environmental temperature is low. Recently (Marr and Ingraham, 1962) it was shown that the proportion of unsaturated fatty acids of *Escherichia coli* decreases continuously as growth temperature is increased. The fatty acids in bacteria are mainly contained in phospholipids.

A study has also been made of brain lipid fatty acids of goldfish acclimatized to various temperatures (Johnston and Roots, 1964). This showed that with decreasing acclimatization temperatures, (a) the total amount of lipid increases, (b) there is an overall tendency for the degree of unsaturation of the fatty acids to increase. G.L.C. analysis showed that the major changes in the brain involved the polyunsaturated acids of twenty carbons or more.

Thus, poikilothermic organisms and bacteria appear to have a feedback mechanism linked to the environmental temperature which enables the fatty acid residue of the phospholipids to be altered so as to keep the hydrocarbon chain fluidity fairly constant.

A further example which appears to be consistent with the membrane keeping the hydrocarbon chain fluidity constant, is the effect caused by the incorporation of *trans* fatty acids into phospholipids. As the fatty acids in membranes are in a dynamic exchange system, we can envisage that *trans* fatty acid could be incorporated into this 2-position in replacement of the *cis* acid. If this does occur, our thermal studies suggest that this *trans* phospholipid will be less fluid and will give a more condensed layer than the *cis* phospholipid. Cholesterol interaction will then be considerably less.

If, however, the *trans* fatty acid is incorporated into the 1-position of the phospholipid molecule in replacement of a saturated fatty acid and a *cis* fatty acid remains in the 2-position, the resulting phospholipid will then be rather similar in behaviour to the normal *cis* type phospholipids. In view of these deductions, it is of interest to see what is at present known about the rate

* Studies in our laboratory show that the phospholipids of myelin at 37° are close to the gel/liquid crystalline condition. The cholesterol present seems to prevent the lipid from forming a crystalline phase at this temperature.

of replacement of *cis* by *trans* fatty acids in the 2-position of phospholipids and to see where *trans* fatty acids are found in a natural biological system.

Lands (1964) has suggested that with phosphatidylcholines the enzymes catalysing esterification at the 2-position do not discriminate between the *cis* and *trans* isomers. However, the enzymes which act at the 1-position discriminate markedly between isomers, but treat elaidic acid similarly to a fully saturated derivative. Studies of pigs fed with elaidenized glycerides (Raulin et al., 1965) have shown that, in the liver of the pig, the *trans* fatty acid is found in both the 1- and 2-positions of the phosphatidylethanolamine and phosphatidylcholines but that more is found in the 1-position than in the 2-position of the phospholipid. With the phosphatidylethanolamines, more than twice the amount of *trans* fatty acid is found to occur in the 1-position than in the 2-position, while with the phosphatidylcholines the ratio is about one and a half to one in favor of the 1-position, so that the *trans* fatty acids behave more like saturated fatty acids.

This enzymic discrimination between the 1- and 2-positions, observed with these *trans* fatty acids, may also be directly related through some feed-back mechanism, to the fluidity of the resultant phospholipid. At higher environmental temperatures this enzymic discrimination could be less marked for both phospholipid types.

If we now ask the question do all cell membranes have as a basic structural unit the bilayer type structure, then we can only say that this has not yet been proved, although the X-ray studies suggest that this is the case with some membranes. At the same time, the evidence that membranes consist of lipoprotein sub-units is not yet very strong.

They may be many types of lipid-protein organization varying from one membrane type to another. Many more physical studies, X-ray, i.r. and n.m.r., ORD and CD spectroscopic studies, need to be carried out on the whole range of membrane types as well as on simple lipid-protein complexes. Physical studies of simple system of this type where hydrophobic bonding is known to occur would be valuable.

Looking further into the future, the interaction of hormones, vitamins, drugs and other molecules with membranes can also be studied using these spectroscopic methods. Features of ion permeability and competition by lipid and molecules such as ATP and ADP for metal ions associated with membranes, may also be studied by some of these spectroscopic methods (Allen et al., 1966).

The way in which membranes are synthesized is another important problem for future solution. Does the lipid determine the way in which metabolically functional protein subsequently attaches, or is the bonding of the lipid dictated by the amino sequence of the proteins. Are both components synthesized simultaneously?

4. STUDIES OF PHOSPHOLIPIDS AND NATURAL MEMBRANES

Cellular organization is a function of membranes. Many of the major activities of cells occur in, or through, membranes. The laying down of memory itself may be ultimately associated with nerve membrane interactions. The future will see a growing realization of the importance of information about membrane structure and membrane interactions.

References

Abrahamsson, S. and Pascher, I. (1966). *Acta cryst allogr.* **21**, 79.
Allen, B. T., Chapman, D. and Salsbury, N. J. (1966). *Nature, Lond.* **212**, 282.
Ashhurst, D. E. (1965). *J. biophys. biochem. Cytol.* **24**, 497.
Bahr, G. F. (1954). *Expl Cell Res.* **7**, 457.
Bangham, A. D. and Horne, R. W. (1964). *J. molec biol.* **8**, 660.
Barratt, M. D., Green, D. K. and Chapman, D. (1968). *Biochim. biophys. Acta* **152**, 20.
Benedetti, E. L. and Emmelot, P. (1965). *J. biophys. biochem. Cytol.* **29**, 299.
Beychok, S. (1966). *Science, N.Y.* **154**, 1288.
Blasie, J. K., Dewey, M. M., Blaurock, A. E. and Worthington, C. R. (1965). *J. molec. Biol.* **14**, 143.
Branton, D. (1966). *Proc. natn. Acad. Sci., U.S.A.* **55**, 1048.
Branton, D. and Moor, H. (1964). *J. Ultrastruct. Res.* **11**, 401.
Brenner, S. and Horne, R. W. (1959). *Biophys. biochim. Acta* **34**, 103.
Byrne, P. and Chapman, D. (1964). *Nature, Lond.* **202**, 987.
Carver, J. P., Shechter, E. and Blout, E. R. (1966). *J. Am. chem. Soc.* **88**, 2550.
Cassim, J. Y. and Yang, T. J. (1967). *Biochem. biophys. Res. Commun.* **26**, 58.
Chapman, D. (1965). *In* "The Structure of Lipids". Methuen and Co. Ltd., London.
Chapman, D. (1966). *Ann. N.Y. Acad. Sci.* **137**, 745.
Chapman, D. (1967). *In* "Thermobiology" (A. H. Rose, ed.) Academic Press, London.
Chapman, D. and Collin, D. T. (1965). *Nature, Lond.* **206**, 189.
Chapman, D. and Fluck, D. J. (1966). *J. biophys. biochem. Cytol.* **30**, 1.
Chapman, D. and Morrison, A. (1966). *J. biol. Chem.* **241**, 5044.
Chapman, D. and Penkett, S. A. (1966). *Nature, Lond.* **211**, 1304.
Chapman, D. and Salsbury, N. J. (1966). *Trans. Faraday Soc.* **62**, 2607.
Chapman, D., Byrne, P. and Shipley, G. G. (1966a). *Proc. R. Soc A* **290**, 115.
Chapman, D., Owens, N. F. and Walker, D. A. (1966b). *Biochim. biophys. Acta* **120**, 148.
Chapman, D., Kamat, V. B., de Gier, J. and Penkett, S. A. (1967a). *Nature, Lond.* **213**, 74.
Chapman, D., Williams, R. M. and Ladbrooke, B. D. (1967b). *Chem. Phys. Lipids* **1**, 445.
Chapman, D., Kamat, V. B., de Gier, J. and Penkett, S. A. (1968a). *J. molec. Biol.* **31**, 101.
Chapman, D., Kamat, V. B. and Levene, R. J. (1968b). *Science, N.Y.*, **160**, 314.
Craigee, R. (1936). *Ann. Chem.*, **75**, 522.
Criddle, R. S., Bock, R. M., Green, D. E. and Tisdale, H. (1962). *Biochem.* **1**, 827.
Cunningham, W. P. and Crane, F. L. (1964). *Expl Cell Res.* **44**, 31.
Demel, R. A. (1966). Thesis, University of Utrecht.
Dervichian, D. G. and Macheboeuf, M. (1938). *C. r. Hebd. Séanc. Acad. Sci., Paris* **206**, 1511.

Elbers, P. F. (1964). *Recent Prog. Surf. Sci.* **2**, 443.
Fernandez-Moran, H. (1962). *Circulation* **26**, 1039.
Fernandez-Moran, H. and Finean, J. B. (1957). *J. biophys. biochem. Cytol.* **3**, 725.
Finean, J. B. (1958). *Expl Cell Res.* **5**, 18.
Finean, J. B. (1960). *J. biophys. biochem. Cytol.* **8**, 31.
Finean, J. B. (1962). *Circulation* **26**, 1151.
Finean, J. B. and Burge, R. E. (1963). *J. molec. Biol.* **7**, 672.
Finean, J. B., Coleman, R., Green, W. G. and Limbrick, A. R. (1966). *J. Cell Sci.* **1**, 287.
Fleischer, S. and Brierley, G. (1961). *Biochem. biophys. Res. Commun.* **5**, 367.
Fleischer, S., Fleischer, B. and Stoeckenius, W. (1965). *Fedn Proc. Fedn Am. Socs exp. Biol.* **24**, 296.
Frey-Wyssling, A. and Steinmann, E. (1948). *Biochim. biophys. Acta*, **2**, 254.
Geren, B. B. (1954). *Expl Cell Res.* **7**, 558.
Glaeser, R. M., Hayes, T., Mel, H. and Tobias, C. (1966). *Expl Cell Res.* **42**, 467.
Gompel, C. (1964). *J. Microscopy*, **3**, 427.
Gorter, H. F. and Grendel, F. (1925). *J. exp. Med.* **41**, 439.
Green, D. E. and Fleischer, S. (1963). *Biochim. biophys. Acta* **70**, 554.
Green, D. E. and Fleischer, S. (1964). *In* "Metabolism and Physiological Significance of Lipids" (R. M. C. Dawson and D. N. Rhodes, eds) p. 581. John Wiley and Sons Inc., New York.
Greenfield, N., Davidson, B. and Fasman, G. (1967). *Biochemistry* **6**, 1630.
Grourke, M. J. and Gibbs, J. H. (1967). *Biopolymers* (in press).
Hayes, T. L., Lindgren, F. T. and Gofman, J. W. (1963). *J. biophys. biochem. Cytol.* **19**, 251.
Hillier, J. and Hoffman, J. F. (1953). *J. cell comp. Physiol.* **43**, 203.
House, H. D., Riordan, D. F. and Barlow, J. S. (1958). *Can. J. Zool.* **36**, 629.
Husson, F. and Luzzati, V. (1963). *Nature, Lond.* **197**, 822.
Jensen, L. H. and Mabis, A. J. (1964). *Nature, Lond.* **197**, 681.
Johnston, P. V. and Roots, B. I. (1964). *Comp. Biochem. Physiol.* **11**, 303.
Korn, E. D. (1966). *Science, N.Y.* **153**, 1491.
Korn, E. D. and Weisman, R. A. (1966). *Biochim. biophys. Acta* **116**, 309.
Kowalsky, A. (1962). *J. biol. Chem.* **237**, 1807.
Kraut, J. (1961). *Acta cryst allogr.* **14**, 1146.
Ladbrooke, B. D., Williams, R. M. and Chapman, D. (1968). *Biochim. biophys. Acta* **150**, 333.
Lands, W. E. M. (1964). *J. Amer. Oil Chem. Soc. Abstracts* 38th Fall Meeting, Chicago.
Larsson, K. (1964). *Ark. Kemi* **23**, 1.
Leathes, J. B. (1925). *Lancet* **208**, 853.
Lenard, J. and Singer, S. J. (1966). *Proc. Natn. Acad. Sci., U.S.A.* **56**, 1828.
Lucy, J. A. and Glauert, A. M. (1964). *J. molec. Biol.* **8**, 727.
McConnell, H. M. and Boeyers, J. C. A. (1967). *J. phys. Chem.* **71**, 12.
Maddy, A. H. and Malcolm, B. R. (1965). *Science, N.Y.* **150**, 1616.
Maddy, A. H. and Malcolm, B. R. (1966). See Kavanau, J. L. *Science, N.Y.* **153**, 213.
Marr, A. G. and Ingraham, J. L. (1962). *J. Bact.* **74**, 1260.
Matalon, R. and Schulman, J. H. (1949). *Discuss. Faraday Soc.* **6**, 27.
Moor, H. (1966). *Balzers Report* **9**, 1.
Moor, H. and Mühlethaler, K. (1963). *J. biophys. biochem. Cytol.* **17**, 609.

Moor, H., Mühlethaler, K., Waldner, H. and Frey-Wyssling, A. (1961). *J. biophys. biochem. Cytol.* **10**, 1.
Moor, H., Ruska, C. and Ruska, H. (1964). *Z. Zellforsch. mikrosk. Anat.* **62**, 581.
Mueller, P., Rudin, D. O., Tien, H. T. and Westcott, W. C. (1962). *Circulation* **26**, 1167.
Mühlethaler, K. (1966). *In* "Biochemistry of Chloroplasts" Vol. V, (T. W. Goodwin, ed.), p. 49, Academic Press, London and New York.
Okaya, Y. (1964). *Acta crystallogr.* **17**, 1276.
Palade, G. E. (1953). *J. Histochem. Cytochem.* **1**, 188.
Park, R. B. and Biggins, J. (1964). *Science, N.Y.* **144**, 1009.
Parsons, D. F. and Nyberg, S. C. (1966). *J. appl. Phys.* **37**, 3920.
Perutz, M. F. (1965). *J. molec. Biol.* **13**, 646.
Ponder, E. (1961). *In* "The Cell" (J. Brachet and A. E. Mirsky, eds) Vol. 2, p. 1, Academic Press, New York and London.
Porter, K. R. and Thompson, H. P. (1947). *Cancer Res.* **7**, 431.
Pysh, E. S. (1966). *Proc. natn. Acad. Sci., U.S.A.* **56**, 825.
Raulin, J., Lapous, D., Dauvillier, P. and Rerat, A. (1965). *C. r. Lebd. Séanc Acad. Sci., Paris* **260**, 344.
Riemersma, J. C. (1963). *J. Histochem. Cytochem.* **11**, 436.
Robertson, J. D. (1957). *J. biophys. biochem. Cytol.* **3**, 1043.
Robertson, J. D. (1958). 4th Internat. Conf. Electron Micr., Springer, Berlin, **2**, 159.
Robertson, J. D. (1959). *Biochem. Soc. Symp.* **16**, 3.
Robertson, J. D. (1960). *Prog. Biophys.* **10**, 343.
Robertson, J. D. (1963). *J. biophys. biochem. Cytol.* **19**, 201.
Robertson, J. D. (1964). *In* "Cellular Membranes in Development" (M. Locke, ed.), p. 1, Academic Press, New York and London.
Robertson, J. D. (1966). *In* "Principles of Biomolecular Organization" p. 357 (G. E. W. Wolstenholme and M. O'Connor, eds.) J. and A. Churchill Ltd., London.
Robinson, C. (1961). *Tetrahedron Lett.* **13**, 219.
Salem, L. (1962). *Can. J. Biochem. Physiol.* **40**, 1287.
Sarkar, P. K. and Doty, P. (1966). *Proc. Natn. Acad. Sci., U.S.A.* **55**, 981.
Scanu, A. (1965). *Proc. Natn. Acad. Sci., U.S.A.* **54**, 1699.
Schmitt, F. O. and Palmer, K. J. (1940). *Cold Spring Harb. Symp. quant. Biol.* **8**, 94.
Shah, D. O. and Schulman, J. H. (1967). *J. Lipid Res.* **8**, 215.
Schellman, J. A. and Schellman, C. (1964). *In* "The Proteins". (H. Neurath, ed.). Ind Ed., Chapter 7, Academic Press, New York and London.
Sjöstrand, F. S. (1953). *Nature, Lond.* **171**, 30.
Sjöstrand, F. S. (1963). *J. Ultrastruct. Res.* **9**, 561.
Sjöstrand, F. S., Anderson-Cedergren, E. and Karlson, U. (1964). *Nature, Lond.* **202**, 1075.
Small, D. M. and Bourgès, M. C. (1966). *Molec. Cryst.* **1**, 541.
Steim, J. (1967). 153rd Meeting of the American Chemical Society, Miami Beach, Florida, April, 1967.
Stoeckenius, W. (1959). *J. biophys. biochem. Cytol.* **5**, 491.
Stoeckenius, W. (1960). *Proc. European Reg. Conf. Electron Microscopy, Delft* **2**, 716.
Stoeckenius, W. (1962a). *J. biophys. biochem. Cytol.* **12**, 221.
Stoeckenius, W. (1962b). *Circulation* **16**, 1066.

Stoeckenius, W. (1966). *In* "Principles of Biomolecular Organization". p. 418, (G.E.W. Wolstenholme and M. O'Connor, eds) J. and A. Churchill Ltd., London.
Sundaralingam, M. and Jensen, L. H. (1965). *Science, N.Y.* **150,** 1035.
Ullmer, D. D., Vallee, B. L., Gorchein, A. and Neuberger, A. (1965). *Nature, Lond.* **206,** 825.
Urry, D. W., Mednieks, M. and Bejnarowiecz, E. (1967). *Proc. Natn. Acad. Sci., U.S.A.* **57,** 1043.
van Deenen, L. L. M. (1965). *In* "Progress in the Chemistry of Fats and Other Lipids" (R. T. Holman, ed.) Vol. VIII, 1. Pergamon Press.
van Deenen, L. L. M., Houtsmuller, U. M. T., de Haas, G. H. and Mulder, E. (1962). *J. Pharmac. Pharmacol.* **14,** 429.
van Iterson, W. (1965). *Bact. Rev.* **29,** 299.
Vandenheuvel, F. A. (1963). *J. Am. Oil Chem. Soc.* **40,** 455.
Vandenheuvel, F. A. (1966). *J. Am. Oil Chem. Soc.* **43,** 258.
Wallach, D. F. H. and Gordon, A. (1967). 15*th International Symposium on the Protides of Biological Fluids* Brugge, Belgium (in press).
Wallach, D. F. H. and Zahler, P. H. (1966). *Proc. Natn. Acad. Sci., U.S.A.* **56,** 1552.
Wallach, D. F. H., Kamat, V. B. and Gail, M. H. (1966). *J. biophys. biochem. Cytol.* **30,** 601.
Whittaker, V. P. (1966a). *Ann. N.Y. Acad. Sci.* **137,** 982.
Whittaker, V. P. (1966b). *In* "Regulation of Metabolic Processes in Mitochondria". (Tager, Papa, Quagliariello and Slater, eds) Elsevier Publishing Co., Amsterdam).
Whittington, S. G. and Chapman, D. (1966). *Trans. Faraday Soc.* **62,** 3319.
Wigglesworth, V. B. (1957). *Proc. R. Soc.* B **147,** 185.

Chapter 5

The Nature of the Interaction Between Protein and Lipid during the Formation of Lipoprotein Membranes

R. M. C. DAWSON

*Department of Biochemistry,
A.R.C. Institute of Animal Physiology, Babraham, Cambridge, England*

I.	INTRODUCTION	203
II.	PHYSICAL CHEMISTRY OF THE PHOSPHOLIPID-WATER INTERFACE	205
III.	REACTIONS OF PROTEINS OTHER THAN PHOSPHOLIPIDS AT THE LIPID-WATER INTERFACE	209
IV.	REACTION OF PHOSPHOLIPASES AT THE PHOSPHOLIPID-WATER INTERFACE	213
	A. Phospholipase B (*Penicillium notatum*)	214
	B. Phospholipase C (α toxin, *Cl. perfringens*)	218
	C. Triphosphoinositide Phosphomonoesterase and Phosphodiesterase	220
	D. Phospholipase D	222
	E. Phospholipase A (*Naja Naja* Venom)	224
V.	THE FORMATION OF LIPOPROTEIN MEMBRANES	225
	REFERENCES	231

I. Introduction

In his excellent monograph on the lipids of biochemical significance published in 1954, Lovern starts his section on the structure of lipoproteins as follows, "It may be said at the outset that virtually nothing is known with certainty about the structure of lipoproteins, e.g. what types of linkage are involved between lipid and protein". In spite of the current research emphasis on the natural lipoprotein membranes of the cell which makes "membrane" the keyword most avidly searched for by biological scientists, generally this statement would be true today. Although a good deal of progress has been made in understanding how lipids form microstructures in isolation, we have little detailed knowledge of the way they are organized in natural lipoproteins and of the various forms of bonding between the individual molecular moieties of the phospholipids, neutral lipids and proteins which form lipoprotein complexes.

Because of the difficulty of examining such unstable and often ill-defined structures by known physical methods, it is necessary at the present time to resort again to the model, i.e. to examine the properties of greatly simplified systems such as, for example, the interaction of a purified protein and a single lipid and to use the knowledge gained as a help to understanding the more complex situation. Now it is true that one of the most important linkages occurring in lipoproteins is that between phospholipids and proteins. Discounting rather exceptional lipoprotein structures, such as the albumin-fatty acid complex, phospholipids are major constituents of all natural lipoproteins and/or especially those which constitute the membranes of cells. This is in contrast to triglycerides and sterols which can be virtually absent from many lipoproteins, thus brain proteolipids contain no triglycerides while lipovitellin contains practically no cholesterol or cholesterol esters. In fact many authorities (e.g. Dervichian, 1949) believe that the primary basis of lipoprotein structures is Coulombic binding between the ionic lipids, i.e. phospholipids and the ionogenic groups of the "so-called" structural proteins, while the neutral lipids are then held by a much weaker hydrophobic bonding with the hydrocarbon chains of the phospholipids and/or the lipophilic protein side chains. However, this theory may not be universal for all natural lipoprotein membranes and there is no doubt that hydrophobic bonding may play an important and even major role in the stability of the final lipoprotein structure (Salem, 1962; Green and Fleischer, 1963a,b; Chapman et al., 1967). Even in the last instance, however, the initial union between phospholipid and protein is often likely to be electrostatic in nature (Salem, 1962).

It is intended in this chapter to discuss the very limited knowledge we have concerning the associations between pure proteins and ionogenic lipids and especially the interactions between water-soluble phospholipases and their water-insoluble substrates. Although at first sight the latter association may appear to be rather a special case of phospholipid-protein association, there are two reasons why it may act as a satisfactory model for studying the factors which control the genesis of the natural lipoprotein membrane. Firstly, in such membranes the proteins are combined with lipid molecules which are themselves associated in some way as, for example, in bimolecular layers or possibly micelles. It seems possible that such an association between lipid molecules may occur before the interaction between protein and lipid, since the main lipid components of natural membranes are rarely found to exist as monomers in an aqueous environment. Secondly, the uniformity of lipoprotein composition in a particular cellular membrane or organelle suggests that the tightness and specificity of the binding of lipids to proteins may depend to a great extent on a precise steric arrangement. Indeed it seems likely that the organization of the lipoprotein microstructure will depend on the same factors of closeness of fit, multiple attachment and matching of polarity that are recognized to determine the combination of enzymes with their substrates as for example in the formation of enzyme-substrate complexes in phospholipase reactions.

II. Physical Chemistry of the Phospholipid-Water Interface

Before embarking on discussing individual cases of protein-lipid interaction, it is necessary to consider briefly some pertinent physico-chemical facts concerning the phospholipid-water interface. In a pure phospholipid-water system the lipid will almost invariably exist as an organized structure either as a micelle, a monolayer at an air-water interface or a bimolecular layer. The type of microstructure adopted will depend on a number of properties of the phospholipid molecule, for example the number, size and shape of its hydrocarbon chains and the nature and charge on its polar hydrophilic region. The electrostatic charge on the latter will vary with the degree of ionization of any ionogenic groups which can, of course, be altered by changes in the bulk pH. It is also modified by the type and concentration of oppositely-charged water-soluble cations present in the system.

In all organized lipid associations it can be assumed that when a layer of phospholipid molecules is in contact with water, the individual molecules will be orientated so that the more polar or ionic portions are directed towards the aqueous phase. The phospholipids which comprised the bulk of those present in cell membranes (lecithin, sphingomyelin, choline plasmalogen, phosphatidyl ethanolamine, ethanolamine plasmalogen) are ideally suited to forming in an aqueous environment the bimolecular leaflets or lamellae which, as we have discussed previously, are proably the basis of many, if not all, membrane microstructures. This is because at a physiological pH the choline-containing phospholipids will be electrically neutral because of the balanced charges of the zwitterion and the ethanolamine phosphoglycerides will only possess a limited negative charge (due to partial depolarization of the $NH^{3}+$ group). Thus there will be little or no electrostatic repulsion between the polar head groups and this will allow close packing of the hydrocarbon portions by hydrophobic bonding. In fact, it is likely that Coulombic interactions between the zwitterionic head groups arranged tangentially coplaner will help to stabilize the sheet structure of the phospholipid molecules (Pethica, 1965). The cohesion between single sheets back to back to form lamellae is stabilized mainly by the energetically unfavourable interaction of the hydrocarbon chains with water rather than by electrostatic forces. The stability of the lamellar structure will also be helped by the roughly cylindrical shape of the phospholipid molecules with the cross sectional area of the two hydrocarbon chains (50 Å^2 $\pm$ depending on the degree of unsaturation of the fatty acids) approximately matching that of the polar head group (Haydon and Taylor, 1963).

With phospholipids which are highly negatively charged at a physiological pH (e.g. triphosphoinositide with 5 negative charges per 2 hydrocarbon chains), the electrostatic repulsion between the head groups will not allow the molecules to form the bilayer structure so easily. Thus they tend to disperse into smaller particles which are very often in the form of spherical micelles. However, they can exist as components of a lamellar structure, in which their highly charged head groups are spaced out by zwitterionic phospholipids or depolarized by the presence of cations. Transition from leaflet to micellar form can also be brought about by changing the shape of the molecule. Thus when a fatty acid is removed from lecithin with phospholipase A, the molecule produced (lysolecithin) is more wedge-shaped and the stable configuration becomes a spherical micelle. In fact, lysolecithin will even leave a unimolecular film of lecithin at the air-water interface when this is digested with phospholipase A (Dawson, 1966). Presumably the smaller bulk of the lipophilic region of lysolecithin is more easily accommodated in the interior of a spherical micelle compared with the hydrocarbon chains of the original cylindrical lecithin molecule (Kavanau, 1965).

A consequence of the orientation of phospholipid molecules in an organized microstructure is that usually an electrostatic field is produced at the lipid-water interface. The density and sign of the surface potential Ψ_0 (Fig. 1) will, of course, depend on the relative numbers and nature of the charged groups on the interface. The electrostatic field will extend out into the aqueous bulk phase and will decrease to practically zero strength at a certain distance which depends on the ionic composition of the bulk phase. If for example, the lipid surface contains negatively-charged phosphate groups and NaCl is added to the bulk phase, sodium ions will be attracted to the surface and chloride ions repelled. The field strength will therefore decrease more rapidly. The field itself is often called the electrical double layer and it is now generally considered to consist of two parts; that nearest the surface in which the counter ions are more or less frozen in a fixed structure and an outermost diffuse portion in which all the water-soluble ions are free to move and distribute themselves according to the opposing tendencies of thermal agitation distributing the ions equally and electrostatic attraction which will tend to concentrate charges of one sign close to the surface. The effective thickness (x in Å) of the ionic double layer at 25°C will be approximately given by $x = 3 \cdot 05 \ I^{-\frac{1}{2}}$ and represents that point at which the potential Ψ has declined such that $\Psi/\Psi_0 = e^{-1} = 0 \cdot 367$. I is the ionic strength of the counter ion obtained by $I = \frac{1}{2}\Sigma C_i Z_i^2$, Ci is the ionic concentration and Zi the valence. Of course if the counter ion has, as well, a high specific affinity for the surface, a deviation from this relationship is to be expected. Thus the dimensions of the double layer increase rapidly as the ionic strength is reduced. In a uni-univalent salt solution, therefore, (e.g. NaCl) at 0·02 M the effective thickness will extend no further than about 20 Å and at 0·001 M it will extend to approximately 100 Å.

If the charged lipid-water interface is part of a phospholipid particle suspended in an aqueous electrolyte, the particle will move when an electric field is applied along with an associated cloud of counter ions. The potential at the plane of shear is called the ζ potential (Fig. 1) and as a near approxima-

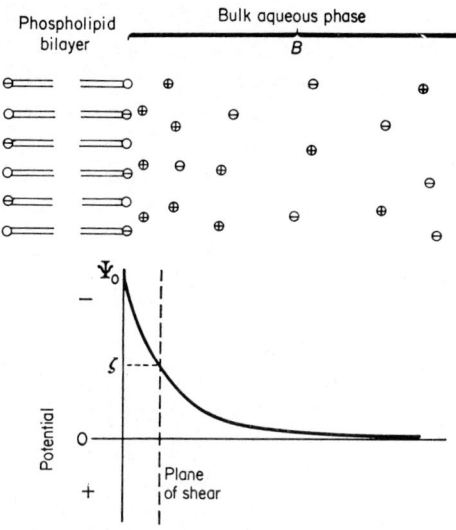

FIG. 1. The bimolecular layer of phospholipid is diagrammatically represented as containing two components, one of which has a head group which is a zwitterion and the other with a net negative charge. The negative potential at the interface (Ψ_0) falls very rapidly going out into the bulk phase due to univalent cations being preferentially attracted to the interface. This potential at the plane of shear when either the bulk phase (endosmosis) or the interface move (electrophorosis) in the plane of the interface is called the ζ potential.

tion the ζ potential is that existing at the boundary between the fixed and diffuse double layers, i.e. the potential falls as a point is moved through the diffuse double layer to a position in the bulk phase far away from the surface.

The adsorption of counter ions into the fixed part of the double layer has an enormous influence on the ζ potential and this effect increases greatly with an increase in valency of the ion. Often univalent counter ions can, if present in sufficiently high concentration, reduce the ζ potential to almost negligible proportions and thus a charged phospholipid particle will become isoelectric in an applied electric field. If sufficient divalent or polyvalent counter ions are added to the bulk phase, the fixed layer of counter ions on the phospholipid surface may have enough charges to balance those on the surface and provide a surplus. In other words, a charge reversal occurs and the sign of the ζ potential changes. Very often different ions with the same valency can vary in their affinity for a particular charged surface and thus their ability to change the ζ potential of a phospholipid-water interface also varies. Thus the divalent uranyl ion shows a remarkable affinity for phosphate groups at an interface (Kruyt, 1949), while the trivalent ferricyanide ion appears to have great affinity for choline groups on a lecithin-water interface.

As well as binding counter ions of salts present in the bulk phase, the charged phospholipid interface will also attract H^+ or OH^- depending on the sign of the charge. This asymmetry of hydrogen or hydroxyl ion distribution means that the effective pH near the surface will be different from that measured in the bulk phase. Assuming a Boltzmann distribution and no solvation effects, the pH at the boundary between fixed and diffuse double layers will be related to the bulk pH

$$\text{pH surface} = \text{pH bulk} + \frac{e\Psi}{2 \cdot 3\ KT}$$

where Ψ is the surface potential in mV, e = the electronic charge, T = absolute temperature, K = Boltzmann's constant, $e/KT = 0 \cdot 0393$ at 20°.

An approximate value of Ψ may be calculated from the Gouy equation if only univalent cations are present in the bulk phase

$$\Psi \text{ Gouy} = \frac{2KT}{e} \sinh^{-1}\left(\frac{134}{AC^{\frac{1}{2}}}\right)$$

at 20° Ψ Gouy = $50 \cdot 4 \sinh^{-1}\left(\dfrac{134}{AC^{\frac{1}{2}}}\right)$

where A = area occupied by each charge in Å^2, C = total concentration of univalent counter ions (mole/l), $\sinh^{-1}x = \log[x + \sqrt{(x^2 + 1)}]$.

It will be apparent from earlier chapters in this volume that X-ray diffraction studies and electron microscopy have shown that in an aqueous environment the bimolecular phospholipid lamellae are arranged in sheets separated by an aqueous phase. The thickness of the aqueous layer increases if a negatively charged lipid is present in the bimolecular layer. The mutual electrostatic repulsion between adjacent layers enlarges the intermediate spaces and encourages hydration of the lipid. If uni-univalent electrolytes (KCl, NaCl) are added to the system, the thickness of the aqueous layer shrinks due to the effect of the counter ions on the electric field (Palmer and Schmitt, 1941). If calcium chloride is added, the hydration is virtually eliminated and here the calcium ions may form a bridge between negative charges on the two adjacent phospholipid bilayers.

III. Reactions of Proteins Other than Phospholipids at the Lipid/Water Interface

Although a good deal of work has been done on the association of proteins with micelles and unimolecular films of lipid at the air-water interface, many of the earlier studies are difficult to interpret because of the inadequate techniques used and because the proteins and/or phospholipids employed were often grossly impure.

The studies of Matalon and Schulman (1949) on the adsorption of various proteins on unimolecular films of charged lipids suggests that the primary association which occurs is almost entirely between the charged ionic groups of the interface and the protein. Thus when albumin or haemoglobin was injected below a monolayer of cardiolipin (negatively charged), the protein was adsorbed provided the bulk pH was below the isoelectric point of the latter so that its net charge was opposite to that on the interface. This adsorption was measured as an increase in the surface pressure of the monolayer from an original pressure of 15–20 dynes/cm to a value of 27–30 dynes/cm. If after adsorption had taken place the pH of the bulk phase was adjusted to a value above that of the isoelectric point of the protein, then the film pressure was reduced to a value close to but somewhat greater than the original. As the net charge on the protein became of a similar sign to that on the phospholipid interface, desorption of the protein occurred. Thus the electrostatic binding of phospholipid and protein is a reversible phenomenom to the extent that the adsorption can be reversed by changing the pH although it does not necessarily indicate that the protein is unaltered in the process. In these experiments the unimolecular film was kept at constant surface area and the increase in surface pressure measured. Consequently, the density of molecular packing prevented anything but a limited penetration of the hydrophobic side chains of the protein into the film. If the pressure of the film was kept constant, the addition of protein caused an expansion of the monolayer. In this case there is probably extensive penetration of protein into the film and it is bound both by electrostatic and non-ionic forces. When such a penetration occurs there are usually radical changes in the molecular configuration of the protein to enable its non-polar chains to enter into the surface and associate with the non-polar portion of the film forming molecule. Consequently, the penetration is generally irreversible and often leads to partial or complete unfolding and denaturation of the protein molecule.

That structural changes in proteins occur at highly charged lipid-water interface has been indicated by studies on the interaction of proteins and emulsions. If, for example, haemoglobin is adsorbed on an oil in water emulsion stabilized by hexadecyl sulphate and then desorbed, it is recovered as parahaematin (Elkes et al., 1945). Furthermore when the enzymes catalase and trypsin are adsorbed on oil in water emulsions stabilized by anionic amphipathic substances such as anthracene-C_{21}-sulphonate, lauryl phosphate and mixed cephalin fractions, adsorption occurs and on desorption the enzymes are found to be extensively denatured (Frazer et al., 1955; Frazer and Schulman, 1956). Here again the adsorption of the enzymes on the lipid stabilized oil-water interface was a function of the charge on the interface. With a negative interface the amount of adsorption at a low bulk pH (enzyme positively charged) was proportional to the interfacial charge. Its extent however, was also a function of the type of lipid polar groups with which the protein interacted; for example, sulphonate stabilized particles adsorbed far more trypsin per charged group than those which were phosphate stabilized. Under certain conditions, for example, when the ratio between the protein concentration and the emulsion surface area was increased, the protein could be desorbed and the activity partially recovered. It can be assumed that in such a case the enzyme is adsorbed but the unfolding process of the protein molecule brought about by polar interaction between charged groups on the lipid-water interface and those on the enzyme has not proceeded to a sufficient degree to inactivate all of the enzyme. The protein is probably adsorbed on the interface in layers, the first consisting of the fully unfolded protein chain which would have no enzyme activity on desorption. This adsorption would reduce the potential at the interface so that subsequently adsorbed protein would not be subjected to the same electrostatic stresses as the initially adsorbed layer. The protein would become less and less unfolded as the multilayers built up and finally the outermost layer would probably be in almost the normal globular form. Figure 2 shows how the adsorption of trypsin onto oil droplets stabilized with anthracene-C_{21}-sulphonate or lauryl phosphate altered the electrophoretic mobility of the particles. Initially these had a fast anodic mobility but as the ratio of protein to surface was increased, then the particles became progressively less negative until charge reversal occurred. The final cathodic mobility probably approaches that of the globular enzyme molecule itself which at the bulk pH used would be slightly positively charged. The adsorption of the enzyme on surfaces stabilized with long chain phosphate (lauryl phosphate, cephalin) resulted in less uptake of enzyme than on the sulphonate-stabilized surface and there was less reduction in activity when the protein was desorbed.

The adsorption of enzymes on positively charged particles (stearyl-trimethyl-ammonium bromide) was negligible at low bulk pH values (enzyme positive) and even at high bulk pH values was very small and nearly complete recovery of activity was obtained when the enzyme was desorbed. Presumably the interaction of the quaternary N groups with the ionized carboxylic groups of the protein is less strong than that of the negative groups on the lipid with the protonated amino groups ($- NH_3 +$) of the protein. This may be because of the greater ionic radius of the quaternary N group. The lack of opportuniteis it presents for H bonding may also be a factor. Thus little adsorption occurs and the electrostatic deformation of the protein is limited, so the monolayers of adsorbed protein are diffuse and composed of native or only slightly unfolded protein. Consequently, almost full activity was found when the enzyme was desorbed.

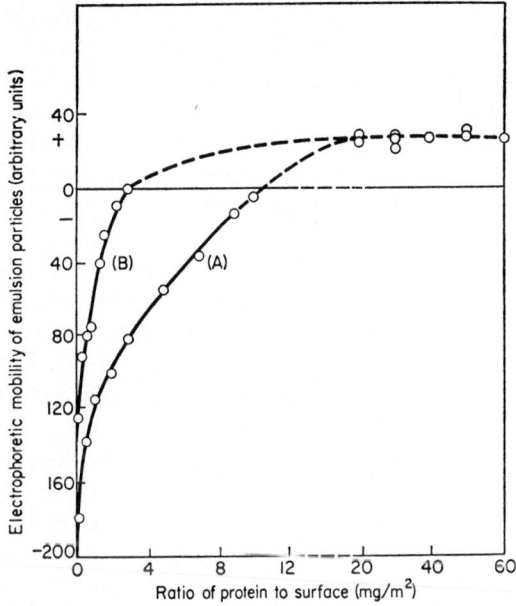

FIG. 2. Electrophoretic mobility of oil droplets stabilized with anthracene —C_{21}— sulphonate (A) and lauryl phosphate (B) in the presence of increasing amount of trypsin at pH 4.

In these experiments the enzyme adsorbed on the charged lipid particles could still act as a catalyst even though the activation energy for the enzyme-substrate reaction was greater for the lipid-enzyme complex than that of the free enzyme. Desorption of the enzyme, at least from the cephalin-stabilized surface, caused the activation energy of the reaction to return to its normal value. Of course it cannot be stated unequivocally that the activity of the adsorbed enzyme is that of the partially unfolded protein. In fact it is not known to what extent the degree of unfolding of an enzyme protein parallels that of the loss of activity. Certainly this would vary for each enzyme although it seems possible that a certain amount of structural change in the protein could take place without loss of activity especially if this alteration was located away from the active centre of the enzyme.

Eley and Hedge (1956, 1957a,b) measured the adsorption of various proteins onto monlayers of cholesterol and stearic acid (or distearin) at initial pressures of 10 dynes/cm and 2 dynes/cm, respectively. None of these lipid films would contain charged groups orientated towards the subphase since the carboxylic acid groups of stearic acid are apparently only ionized at an interface above a bulk pH of 9 (Hauser and Dawson, 1967). The increase in surface pressure on injecting protein into the subphase is interpreted as showing first the formation of a layer of completely unfolded protein with one amino acid residue under, and possibly bound by hydrogen bonds to, each polar group of the lipid followed by a sublayer of "native" protein with an area/molecule of protein which is many times less than the denatured protein layer. It is suggested that the increase in surface pressure is due to the non-polar side chains of various amino acid residues (namely valine, leucine isoleucine, proline, phenylalanine, tryptophan and tyrosine) penetrating into the film.

At such low film pressures it is difficult to exclude the possibility that whole protein molecules are "dissolving" in the film and simply denaturing at the air-water interface forming a mixed monolayer with a maximum pressure of about 16 dynes/cm the collapse pressure of the protein. However, the maximum surface pressure reached varied with the lipid film and was in fact somewhat different from the collapse pressure of the protein.

Recently a good deal of attention has been paid to the complexes formed between cytochrome c and acidic phospholipids, particularly phosphatidyl ethanolamine. When the protein and ultrasonically-dispersed phospholipids are mixed within a certain pH range, precipitation occurs and the precipitate can be extracted into iso-octane in contrast to either of the free components of the complex. When the cytochrome c is recovered from the complex by extracting an aqueous suspension with an equal volume of 2-pentanone, it is found to have the same absorption spectrum, electrophoretic mobility and enzyme activity as the original (Reich and Wainio, 1961). It seems likely that such a complex is primarily a salt formed between the highly basic cytochrome c and the acidic phospholipids. Consequently it is to be expected, and is found, that the complex formation is inhibited by metallic cations, especially those of higher valencies (Das and Crane, 1964). These would effectively reduce the negative ζ potential of the dispersed phospholipid particles and consequently electrostatic binding with the cytochrome c would be decreased. The observation (Das et al., 1964) that unsaturated phosphatidyl ethanolamine has a greater affinity for cytochrome c than saturated does not necessarily mean that other factors such as hydrophobic bonding are playing any part. This could merely be because such particles are more readily dispersed and therefore present a larger surface area for reaction.

Purified acidic phospholipids such as cardiolipin or phosphatidyl inositol can form complexes with cytochrome *c* which contain 9–10 molecules of phospholipid to one of cytochrome *c* and are insoluble in iso-octane. With the less acidic phosphatidyl ethanolamine the equivalent ratio is 24 and purified lecithin does not react at all (Das *et al.*, 1965). However, both lecithin and phosphatidyl ethanolamine can also react with the complexes formed with the more acidic phospholipids in which, presumably, all the external basic sites of cytochrome *c* are neutralized (Green and Fleischer, 1963a), and convert them from iso-octane insoluble complexes to soluble ones. Here presumably hydrophobic bonding between the very acidic phospholipid, bound by salt linkages to the protein, and the lecithin or phosphatidyl ethanolamine may occur.

Palmer *et al.* (1941b) have studied by X-ray diffraction the insoluble complexes formed when cephalin reacts with certain basic proteins such as histone. From our present knowledge of the composition of earlier cephalin fractions, it can be assumed that the product they used probably consisted of ethanolamine-containing phosphoglycerides contaminated with considerable amounts of the more acidic phospholipids such as phosphatidyl serine and phosphatidyl inositol. The X-ray diffraction data suggests that in water the cephalin forms bimolecular leaflets separated from one another by extensive aqueous layers of about 80 Å thickness. This is presumably due to the repulsion between the negatively-charged phospholipid head groups separating out the sheets of phospholipids bilayers as was discussed in the previous section. When the basic protein is added to the cephalin, a collapse of the hydrated cephalin sol occurs presumably because of the positively-charged groups on the protein reacting with the negatively-charged phospholipid head groups. This results in a dehydration equivalent to that produced in similar sols by metallic cations. The diffraction pattern of the final structure is best interpreted by assuming the complex consists of bimolecular leaflets of cephalin separated by a monolayer of slightly hydrated protein. The protein is presumably partially unfolded and electrostatically interacting with both lipid surfaces on either side of the hydrated layer.

IV. Reaction of Phospholipases at the Phospholipid-Water Interface

In considering phospholipase reactions it must be remembered that the demonstration of enzyme activity implies that the enzyme has approached the phospholipid-water interface where its active centre has become stereochemically locked with the phospholipid forming the enzyme-substrate complex. The demonstration of a phospholipase reaction also means that the enzyme protein has initially reacted or complexed with the organized lipid structure in such a way that any deformation of the protein molecule has not been sufficient to affect the enzyme activity. It is possible, although no evidence is available to support the idea, that the enzyme-substrate complex leaves the interface and enters the bulk phase before it is hydrolysed and the enzyme reformed. It is more likely that the breakdown of the complex takes place at the interface with the ejection of the water-soluble products to the bulk phase and the retention of lipoidal products in the hydrophobic matrix. However, as has been mentioned previously, with phospholipase A one of the lipoidal products, lysolecithin, can leave the interface as a micelle. Presumably, if the complete reaction occurs at the interface, the enzyme reformed by the breakdown of the enzyme-substrate complex can either diffuse into the bulk phase or remain within the electrical double layer surrounding the lipid-water interface.

The way in which the physiochemical state of the lipid-water interface can affect phospholipase reactions will now be considered. In essence two methods have been used to study such reactions. In the first, the purified enzyme has been incubated with particles of phospholipid suspended in a buffer and its activity correlated with the ζ potential of the phospholipid water-interface as determined by electrophoretic mobility measurements. The electrophoretic mobility (μ) and ζ potential of particles are related by the Helmholtz-Smoluckowski equation $\mu = (\zeta D)/(4\eta\Pi)$ where D is the dielectric constant and η the viscosity of the suspending medium. The effect on the enzymic activity of varying the ζ potential by introducing charged amphipathic substances in the interface or counter ions in the bulk phase can then be studied. In the second method, the digestion of unimolecular films of ^{32}P-labelled phospholipids has been examined (Fig. 3). The hydrolysis of the film has been measured as the loss of surface radioactivity occurring as the water-soluble ^{32}P-labelled products leave the surface and become shielded from the counter by the stopping power of the water of the bulk phase. In such experiments, the effect of changing the charge on the interface can again be studied and in addition such factors as the packing density of the phospholipid molecules and the surface potential can be varied and evaluated. The enzymic reactions investigated by these techniques will now be discussed in turn.

A. PHOSPHOLIPASE B (*Penicillium notatum*)*

The enzyme did not normally hydrolyse at a significant rate the large lecithin particles formed by shaking dry lecithin with water. Such particles consist of aggregates of many millions of lecithin molecules arranged in hydrated and tightly coiled bilayers. They were practically isoelectric at pH values between 3 and 12, the choline and phosphate groups of the balanced zwitterion being fully ionized within this range. When a small amount of anionic amphipathic substance was mixed with the lecithin, the enzymic

* Dawson, 1958a,b; Bangham and Dawson, 1959, 1960; Dawson and Hauser, 1967.

hydrolysis (lecithin→ 2 fatty acids + glycerylphosphorylcholine) of the latter was initiated. The hydrolysis only took place when sufficient long chain anion had been introduced to give the lipid water-interface a certain critical negative ζ potential (equivalent to -0.9 µ/sec/V/cm as assessed by the

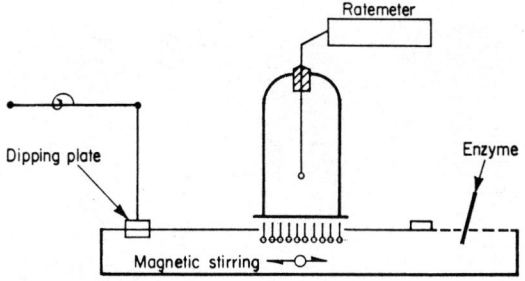

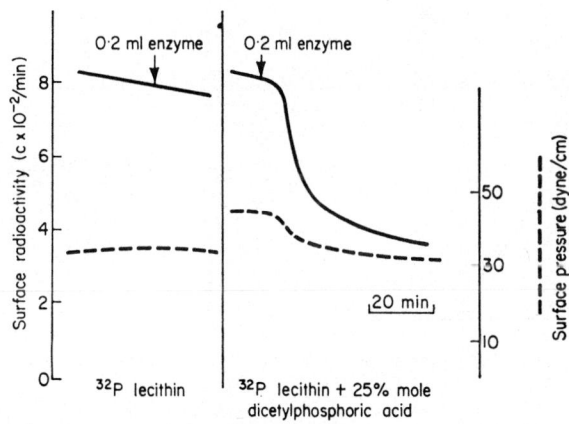

FIG. 3. The measurement of the enzymic digestion of a phospholipid monolayer using a surface radioactivity technique. The β-emission from the ^{32}P in the film is continuously integrated in the ratemeter and recorded. On the addition of enzyme to the bulk phase, any water-soluble ^{32}P-products of the hydrolysis are dispersed in the bulk phase by magnetic stirring and largely shielded from the counter by the water in the trough. The lower part of the diagram shows that a ^{32}P lecithin film at pressures above 30 dynes cm is not hydrolysed by the enzyme unless an anionic amphipathic substance is added even though the surface radioactivity indicates that in both cases the areas occupied by individual molecules of lecithin are very similar.

electrophoretic mobility of the substrate particles) (Figs 4 and 5). This threshold was independent of the chemical nature of the anionic amphipathic substance or the type of negatively charged group it possessed. As more and more amphipath was introduced, the rate of hydrolysis rose to a maximum when the particles had a mobility of -1.7 μ/sec/V/cm and subsequently declined (Fig. 4). There was no evidence that this decline was brought about by surface denaturation of the enzyme and this could mean formation of the

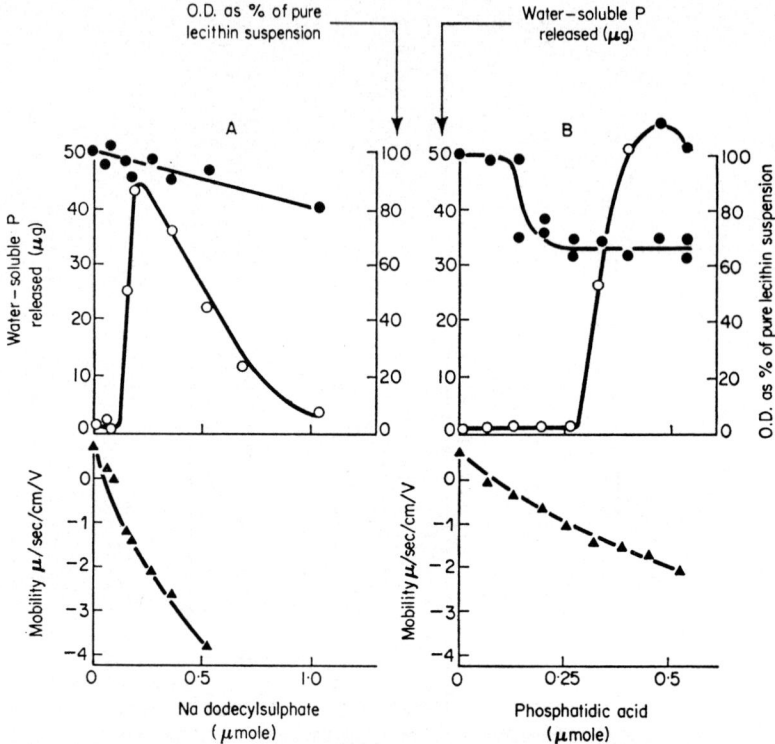

FIG. 4. Effect of adding anionic amphipathic substances to lecithin suspensions on: ○—their optical density at 660 mμ, ●—their susceptibility to phospholipase B hydrolysis, ▲—their electrophoretic mobility. The ovolecithin (3·1 μ mole) was suspended in 1·2 ml formate buffer pH 3·3.

enzyme/substrate complex requires a certain well-defined density of negative charges on the surface.

The enzymic activity was not initiated by introducing a long chain cation into the lecithin and, in fact, progressive addition of such a cation to a lecithin-anionic amphipath system resulted in total inhibition of the enzymic activity as the mobility of the particles fell below the activation threshold (-0.9 μ/sec/V/cm). A similar inhibition of the enzymic activity was obtained by adding di- or multi-valent metallic cations such as Ca^{2+}, $(UO_2)^{2+}$, Th^{4+}.

However, with these water-soluble cations inhibition of the enzyme began when the mobility of the substrate particles fell below the critical value and became complete when the lipid-water interface became isoelectric. One

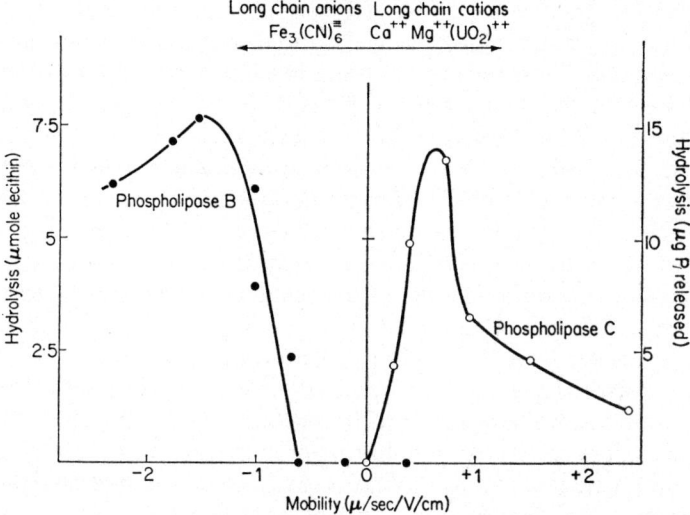

FIG. 5. A comparison of the electrophoretic mobility of lecithin particles and their susceptibility to enzymic attack by phospholipases. Phospholipase B is only active against lecithin when the particles are given a negative ζ potential by the introduction of long chain anions into the phospholipid or ferricyanide counter ions into the bulk phase. Conversely, phospholipase C is only active when the lecithin is given a positive ζ potential by mixing it with a long chain anion or on introducing divalent metallic cations into the bulk phase.

explanation of this difference between the inhibition caused by long chain cations added to the lipid and metal ions to the bulk phase could be that the binding of the latter is undoubtedly a reversible phenomenon. This could mean that at a given instant of time, individual molecular regions on the substrate will have a negative ζ potential greater than the critical one activating the enzyme even though the net potential on the interface is below it. Furthermore it has to be remembered that such water-soluble divalent cations in the bulk phase can, unlike the long chain cations, also react directly with the enzyme, binding with its carboxyl groups and making its net charge less negative.

The addition of anionic amphipathic substances to the lecithin-water interface may (but not necessarily) mean that the space occupied by the orientated lecithin molecules will increase by the physical spacing out effect of the added molecules possibly enhanced by an electrostatic repulsion between their negatively-charged head groups. The question therefore arises as to whether the stimulation is caused by such a repulsion allowing easier access of the enzyme to the susceptible acyl ester bonds by spacing out the lecithin molecules and probably causing dispersion of the particles into smaller units. The effect of such a detergent-like action of the anionic amphipaths can be observed in Fig. 4. However, it was found that there was no correlation between the dispersion produced and the initiation or rate of enzymic attack. Further evidence on this score was provided by studying the hydrolysis of unimolecular films of ^{32}P-lecithin by the enzyme. At low film pressures (i.e. < 30 dynes/cm) a slow hydrolysis of a pure lecithin film occurred. This hydrolysis was not inhibited by calcium or by the incorporation of cationic amphipathic substances into the film. Presumably, therefore, in these circumstances the enzymic hydrolysis of the substrate is not critically governed by electrostatic conditions at the lipid-water interface. It is known of course from the work of Matalon and Schulman (1949) that protein cannot readily penetrate into unimolecular films of phospholipids when these are at pressures greater than 30 dynes/cm. At pressure greater than 30 dynes/cm enzymic hydrolysis of the film occurred unless a small proportion of an anionic amphipathic substance was introduced into the film (Fig. 3). As the film was compressed there was a directly proportional increase in the amount of anionic amphipath required to initiate hydrolysis. This hydrolysis of a high pressure film was inhibited by divalent metallic cations (Ca^{2+}, $(UO_2)^{2+}$) which indicates that electrostatic conditions at the interface were now becoming of overiding importance in determining the rate at which the enzyme substrate complex was formed at the interface. These experiments again show that dispersive factors are not responsible for the activation of lecithin hydrolysis since it can be shown that the concentrations of anionic amphipathic substances required to initiate hydrolysis do not cause micellization of the unimolecular film of lecithin. The question remains as to whether the enhanced hydrolysis is due to the physical spacing out of the lecithin molecules by the activator. Careful experiments, in which the surface dilution produced by the activator was compensated for by compressing the film, suggest this is not so (Fig. 3) (see also Dawson and Hauser, 1967). In fact, if the degree of packing was the only consideration one would expect that lecithin surfaces made positively-charged by the addition of long chain cations would also be hydrolysed by the enzyme.

B. PHOSPHOLIPASE C (α *toxin*, *Cl. perfringins*)*

Highly purified phospholipase C also required definite electrostatic conditions on the lecithin-water interface before hydrolysis occurred (lecithin →

* Bangham and Dawson, 1962.

diglyceride + phosphoryl choline) but, in contrast to phospholipase A, a positive ζ potential was usually needed before the activity commenced (Fig. 5). This could be produced by introducing a long chain cation into the lecithin (stearylamine, docosanylpyridinium bromide) or by adding divalent metallic cations (Ca^{2+}, Mg^{2+}, $(UO_2)^{2+}$ to the bulk aqueous phase. However, irrespective as to how the positive ζ potential had been produced, maximum hydrolysis occurred at a point when the mobility of the lecithin particles was about + 0·5 μ/sec/V/cm and with increasing values of the ζ potential the hydrolysis sharply declined (Fig. 6). As is to be expected, the introduction of

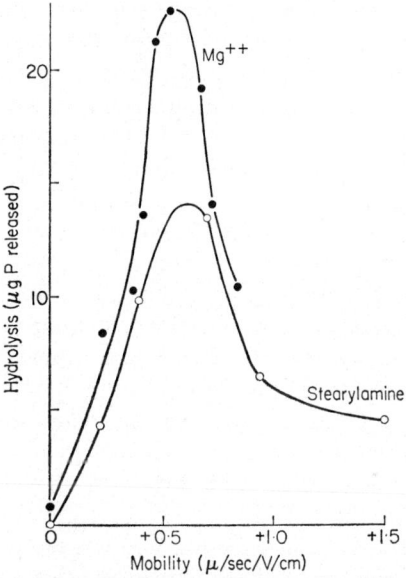

FIG. 6. A comparison of the enzymic hydrolysis of lecithin particles and their electrophoretic mobility when the latter is varied either by adding stearylamine to the phospholipid or magnesium ions to the bulk phase.

an anionic amphipath into positively-charged lipid particles, or ferricyanide $(Fe_3CN_6)^{3+}$ ions into the aqueous phase, completely inhibited the hydrolysis if sufficient was added to cause charge reversal of the ζ potential.

Phospholipase C has an isoelectric point in the region of pH 5·5 and hydrolysis of positively-charged lecithin particles only took place above this pH, i.e. when the protein has a net negative charge. However, if the lecithin particles were made negatively-charged by the introduction of an anionic amphipath, no hydrolysis of lecithin occurred below pH 5·5, i.e. when the enzyme has a net positive charge. This result suggests that the sign of the ζ potential on the substrate's surface is of supreme importance and that initiation of enzyme activity was not caused by a simple heterocoagulation of enzyme and substrate. Of course other interpretations of this result are possible such as the lack of ionization of certain ionogenic groups on the active centre of the enzyme below pH 5·5.

C. TRIPHOSPHOINOSITIDE PHOSPHOMONOESTERASE AND PHOSPHODIESTERASE*

These water-soluble enzymes hydrolyse triphosphoinositide; the phosphomonoesterase successively dephosphorylating the phospholipid into diphosphoinositide and monophosphoinositide while the phosphodiesterase cleaves it into diglyceride and inositol triphosphate. Triphosphoinositide-water interfaces are strongly negatively-charged due to the large number of negative charges (five possible) on each phospholipid molecule. In fact at a bulk pH of 5·5, the lowering of the surface pH was found to be sufficient to partially suppress ionization of some of the phosphate groups. If the surface charge was diluted with a zwitterionic phospholipid, the charged phosphate groups of triphosphoinositide were fully ionized at a bulk pH of 5·5.

At the usual substrate concentrations sodium triphosphoinositide is "water-soluble" and on electron microscopy after negative staining, the phospholipid was seen to exist as spherical micelles about 50 Å in diameter. When the partially purified enzymes were incubated with the pure substrate, a small amount of hydrolysis occurred but the reaction then ceased, possibly because a layer of denatured protein on the triphosphoinositide micelle blocked further formation of the enzyme-substrate complex. (This may not be the enzymes themselves since these would be negatively charged at the pH of the incubation medium). If the incubation was carried out in the presence of any agent that neutralized at least part of the negative-charge at the triphosphoinositide-water interface (e.g. long chain cations, divalent metal ions or even very basic proteins such as protamine which are stable to denaturation), the hydrolysis continued. Possibly this reduction in the magnitude of the negative ζ potential at the lipid-water interface prevents denaturation of the protein. In fact if a long chain cation was added to the triphosphoinositide-enzyme system after the hydrolysis had ceased, the reaction recommenced and continued at a rate equivalent to that occurring if the cationic amphipath was present in the system at the commencement of incubation (Fig. 7). It seems possible that the desorption of denatured protein produced by the reduction in the charge at the interface allows further formation of enzyme-substrate complex and hydrolysis of the triphosphoinositide.

* Dawson and Thompson, 1964; Thompson and Dawson, 1964; Hauser and Dawson, 1967

When anionic amphipathic substances like sodium hexadecyl sulphate and ganglioside were added to a triphosphoinositide-cetyltrimethylammonium bromide system, they acted as potent inhibitors of the enzymic hydrolysis; presumably these again increased the negative charge at the lipid-water interface and thus produced unfavourable conditions for the hydrolysis.

All of the substances which maintain the hydrolysis of triphosphoinositide are those which tend to combine preferentially with the phosphate groups on

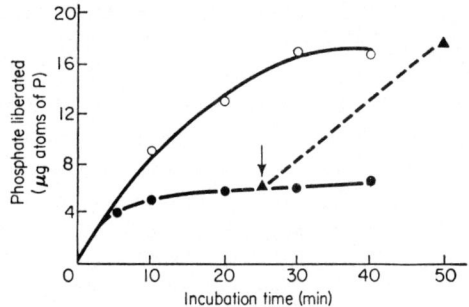

FIG. 7. Hydrolysis of triphosphoinositide by its phosphodiesterase. Basic incubation medium (0·43 μmole triphosphoinositide) (●); basic medium with 0·62 μmole cetyltrimethylammonium bromide added at zero incubation time (○) and after 25 min (▲).

the substrate and decrease or neutralize its excess negative charge. The result of this combination builds up an organized structure of substrate and activator which can be detected under the electron microscope or seen as a visible precipitate. It would also result in a reduction of the ζ potential at the lipid-water interface although activation was produced when the particles were still very negatively charged. However, any connection between the enzymic activation and ζ potential is certainly not so well defined as with the two enzymes previously discussed. Thus the particles formed from triphosphoinositide and sufficient stearylamine to cause optimal activity of the phosphodiesterase had an electrophoretic mobility of -18 μ/sec/V/cm while with Mg^{2+} as activator the equivalent activation was produced at $-8·9$ μ/sec/V/cm. Moreover although the electrophoretic mobilities of stearylamine/triphosphoinositide particles were almost identical with those of palmitylamine/triphosphoinositide particles, stearylamine acted as an activator while palmitylamine did not. Presumably this sensitivity to chain length may reflect a precise requirement for the stereo-chemical orientation of enzyme and lipid on the formation of the enzyme-substrate complex.

H*

D. PHOSPHOLIPASE D*

Phospholipase D from cabbage brings about the hydrolysis of lecithin into phosphatidic acid and free choline but only in the presence of a high concentration of calcium. It can also catalyse the transfer of the phosphatidyl unit of lecithin to water-soluble aliphatic primary alcohols (e.g. propanol) forming the corresponding phospholipid (phosphatidyl propanol).

Coarse lecithin particles prepared by shaking dry lecithin with water were hydrolysed only very slowly by the purified enzyme even in the presence of adequate Ca^{2+}. When certain amphipathic substances were introduced into the lecithin particle, e.g. dodecylsulphate, phosphatidic acid, triphosphoinositide, the reaction was markedly stimulated while other acidic lipids were comparatively inactive (Fig. 8). No evidence could be obtained from turbidity measurements that the activators were stimulating the reaction by merely dispersing the substrate; in fact, the high concentration of calcium present effectively prevented the activators acting as detergents (Fig. 8). The activation was reduced if the anionic amphipathic activators were added in excess

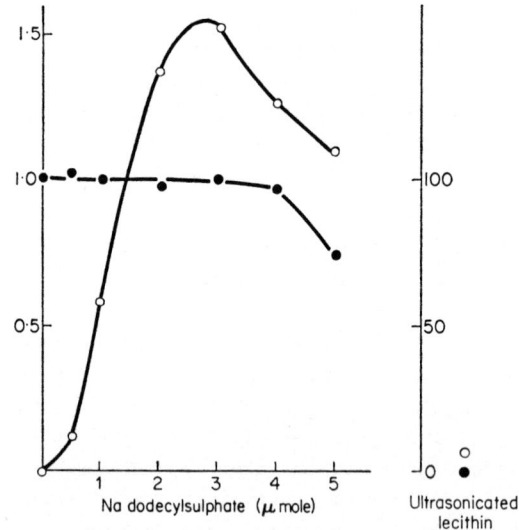

FIG. 8. The effect of Na dodecylsulphate on the dispersion of lecithin (with Ca^{2+}) and its hydrolysis by phospholipase D. The points on the right show that although ultrasonication of the lecithin produces far more dispersion, the rate at which it is hydrolysed is considerably less than in the presence of dodecylsulphate. Solid circle = O.D. as % of original lecithin dispersion, open circle = choline liberated $\triangle$ O.D. 365 mμ).

* Dawson and Hemington, 1967; Dawson, 1967.

and also by the addition of long chain cations. However these latter substances reversed the inhibition caused by using an excess of anionic amphipathic substance as activator.

At first sight these observations would suggest that the phospholipase D activity might be dependent on the ζ potential of the lecithin-water interface. However, measurements of the electrophoretic mobilities of the substrate particles clearly indicate this is not so. Thus although the activators in the absence of Ca^{2+} gave the lecithin particles a substantial anodic mobility, this was nullified and reversed by the counter ion binding of the calcium added. In certain instances the activated particle had precisely the same positive ζ potential as that possessed by a pure lecithin particle in the presence of the same concentration of calcium. Similarly, the progressive addition of long chain cations produced an inhibition of the system well before any change in the ζ potential could be detected.

The addition of activator did, however, promote binding of enzyme onto the lecithin particle. When the purified enzyme was exposed to a lecithin-dodecylsulphate particle, it was adsorbed and rapidly and irreversibly denatured on the highly negatively-charged lipid-water interface. In the presence of calcium, adsorption still occurred but the decrease in ζ potential produced by counter ion binding protected the enzyme molecule from complete unfolding. This could be shown by centrifuging down the lecithin/dodecylsulphate particle and washing it several times to remove excess enzyme. On incubation without further added enzyme, hydrolysis of lecithin took place. On the other hand, little enzyme adsorption occurred on a pure lecithin surface in the presence of calcium. This adsorption of the enzyme may be an essential prerequisite for the formation of the enzyme-substrate complex and may explain the mechanism of activation by certain anionic amphipathic substances.

Adsorption on lecithin + dodecylsulphate particles occurred almost as readily when Mg^{2+} was substituted for Ca^{2+}; Mg^{2+} also prevented the denaturation of enzyme. However, on incubation, no hydrolysis of the lecithin occurred until Ca^{2+} was added to the system; presumably after adsorption this cation plays a highly specific role in the further formation or breakdown of the enzyme-substrate complex.

The pH optimum of phospholipase D was 5·4 and at lower H^+ concentrations the activity rapidly fell so that little choline liberation from lecithin was observed at pH 6. From electrophoretic studies the iso-electric point of the enzyme appeared to be about 5·5 and, because of the calcium ion requirement which would make the protein molecule more net positive, it can be concluded that the enzyme only reacts with the substrate when it has a net positive charge. Since the activation by long chain anions was produced when the lecithin-water interface had a positive ζ potential, the electrostatic conditions in the diffuse double layer would not favour the approach of an enzyme molecule on the basis of its net charge. Possibly the enzyme approaches the interface with its net negatively charged regions orientated towards the fixed double layer and in this it is probably assisted by the shrinkage of the electrical double layer caused by the calcium present. After overcoming any electrical barrier, the enzyme molecule probably locks onto the anionic sites on the lipid surface either through divalent metal ion bridges or by direct interaction of its NH_3+ groups, forming a tightly bound complex. With the second alternative the protein must compete with calcium ions bound to fixed anionic sites on the surface. In this respect it is known that Ca^{2+} ions in the "fixed double" layer around anionic lipids are still in dynamic equilibrium with the bulk aqueous phase (Hauser and Dawson, 1967). This mechanism might help to explain why basic proteins such as histone are potent inhibitors of the enzyme reaction since they presumably would compete with the enzyme for the anionic sites on the surface of the lipid mixture.

E. PHOSPHOLIPASE A (*NAJA NAJA* VENOM)*

Phospholipase A (lecithin → lysolecithin + fatty acid) caused very little hydrolysis of coarse lecithin particles both in the absence or presence of calcium. No activation was produced when either cationic or anionic amphipathic substances were introduced into the lecithin particle. Clearly, therefore, the initiation of hydrolysis does not appear to be helped by the introduction of charged groups into the lecithin surface (as occurs with phospholipase D) or by an alteration of the ζ potential (as occurs with phospholipases C and D). On the other hand, the enzymic hydrolysis was greatly stimulated by saturating the aqueous phase with diethyl ether.

The ether penetrates into the lipid particle and produces two physico-chemical consequences both of which could have a bearing on the stimulation of the reaction. One is that the lecithin molecules orientated at the lipid-water interface are spaced out by the ether molecules and this probably allows easier access of the enzyme to the susceptible acyl ester bond. The second consequence is that the ether encourages the removal from the surface of the lecithin particle of fatty acids produced in the reaction and their replacement with fresh substrate molecules from the interior thus preventing product inhibition of the hydrolysis.

A small amount of calcium was necessary for optimal enzymic hydrolysis in the presence of ether but the concentration required had virtually no effect on the ζ potential; neither could it be replaced by Mg^{2+} which produces about the same ζ potential on a lecithin surface as Ca^{2+} (Bangham and Dawson, 1962, Dawson and Hemington, 1967).

* Dawson, 1963a, 1966.

Phosphatidyl ethanolamine particles were rapidly hydrolysed by the enzyme even in the absence of ether. It is possible that the net negative charge on the polar head groups of this phospholipid at the pH used (produced by suppression of ionization of the amino group) may in some way help the access of the active centre of the enzyme to the susceptible acyl ester bond. In this connection it was observed that the rate of hydrolysis of phosphatidyl ethanolamine increased as the pH was increased from 6 to 9·3 while that of lecithin was comparatively unchanged. At the same time the ζ potential of phosphatidyl ethanolamine particles became more negative with increasing pH while that of the lecithin particles remained isoelectric.

A study of the digestion of unimolecular films of ^{32}P-lecithin and ^{32}P-phosphatidyl ethanolamine by cobra venom phospholipase A indicated that the pure phospholipids were hydrolysed without the necessity for any activating agent and that the lysophospholipid released by the reaction largely left the film and entered the bulk phase. With lecithin, the hydrolysis of high pressure films was slow but if the film was expanded the rate appreciably increased at a surface pressure of about 30·dynes/cm. This corresponds to the pressure at which protein can penetrate into lipid films (Matalon and Schulman, 1949). On the other hand, the hydrolysis of unimolecular films of phosphatidyl ethanolamine occurred rapidly at pressures near collapse.

V. The Formation of Lipoprotein Membranes

In the preceding sections consideration has been given to the way in which protein can combine with lipids associated in an organized microstructure, i.e. bimolecular layers or orientated monolayers at the oil-water or air-water interface. Such combinations, of course, must presumably occur when multi-enzyme systems inactivated by solvent extraction are reactivated by reaction with micellar phospholipids (e.g. Green and Fleischer, 1963b; Tzagoloff and MacLennon, 1965; Mazanowska et al., 1966, The results can only act as a successful model for understanding the formation of a natural lipoprotein membrane if such a combination does indeed occur in the cell as the component parts (lipid and protein) are synthesized during growth or normal dynamic turnover of membranes. Assuming for the present that such is the case, the approach of the protein to the lipid surface will initially be governed by the sign and magnitude of the electrostatic field in the diffuse double layer around the lipid-water interface and the net charge on the whole globular protein molecule or possibly a certain specific region of its structure. Such an effect can be seen in the experiments of Matalon and Schulman (1949) on the adsorption of haemoglobin and albumin on lipid monolayers or the specific electrostatic conditions at the phospholipid-water interface required before *P. notatum* phospholipase B or *Cl. perfringins* phospholipase C can degrade lecithin. Intermolecular attractions other than electrostatic will only become significant when the lipid and protein are in close contact (< 6 Å) (Salem, 1962).

This of course does not mean that a protein cannot react ionically or through non-ionic bonding with a lipid-water interface whose ζ potential is virtually zero, e.g. lecithin. It is apparent that phospholipase A can hydrolyse lecithin and it is probable that the other phospholipases can react with a lecithin surface although at a rate very much slower than that when a favourable electrical diffuse double layer surrounds the substrate. In this connection recent observations have shown that many phospholipases which hardly attack coarse lecithin particles can hydrolyse ultrasonically dispersed lecithin at an appreciable rate (Dawson, 1963b; Kates *et al.*, 1965; Dawson and Hemington, 1967). Although this could be due to the enormously greater lecithin surface increasing the statistical chance of the enzyme reacting, it is now clear that changes in the molecular microstructure can also occur when coarse lecithin particles are ultrasonicated (Fig. 8) (Attwood and Saunders, 1965). Recently it has been shown that differences in the asymmetry of particles of synthetic lecithin produced by ultrasonic dispersion can have a marked influence on their hydrolysis by phospholipase A (Attwood *et al.*, 1965). Similarly one of the mitochondrial enzymes, β hydroxybutyrate dehydrogenase (EC 1.1.1.30), has an absolute requirement for lecithin (Jurtshuk *et al.*, 1963). The insoluble apoenzyme will not react with coarse lecithin particles, but it does when these particles are subjected to ultrasonic disintegration or dispersed by other means. A stable lipoprotein complex is formed which can be purified and shows enzymic activity. It has been shown recently by de Pury and Collins (1966) that the rate of reaction of sonicated lecithin with the structural protein of acetone-extracted mitochondria is dependent on the nature of the fatty acids in the lecithin. The combination shows a rapid initial uptake of lecithin (also seen with non-acetone extracted mitochondria) followed by a slower constant binding, the rate of which is probably governed by the rearrangement of the organized structure of the lecithin molecules to form predominantly hydrophobic bonds with the structural protein (Green and Fleischer, 1963 a, b).

In general when proteins are injected below pure lecithin films having surface pressures above 16 dynes/cm, little adsorption can be measured in terms of an increase in surface pressure (Doty and Schulman, 1949). However, such films are readily digested by enzymes up to pressures of about 30 dynes/cm (Bangham and Dawson, 1960; Dawson, 1966). Eley and Hedge (1956) found that there was a pressure increase when insulin or serum albumin were injected under synthetic lecithin films at an initial surface pressure of 2 dynes/cm but with such expanded films there is a possibility that whole protein molecules penetrate into the monolayer and denature at the air-water interface.

From the lipid analysis of natural membranes (see Chapter 2), it is apparent that the ionogenic lipids are mainly the phospholipids although gangliosides and cerebroside sulphate occur to a lesser extent. The phospholipids present can be divided at physiological pH values into the zwitterionic lipids (lecithin, sphingomyelin, choline plasmalogen), those which will have a small net negative charge (phosphatidyl ethanolamine, ethanolamine plasmalogen) due to partial ionization of their amino groups, and those which will have between one to five net negative charges per molecule (phosphatidyl serine, cardiolipin, phosphatidyl inositol, phosphatidic acid, diphosphoinositide, triphosphoinositide). It is apparent that as all membrane lipids are without exception either isoelectric or negatively-charged at physiological pH values, the apoprotein of the lipoprotein being formed must approach an organized lipid structure with a surplus of negative sites on its surface unless, as seems unlikely, the individual lipids themselves are organized into separate structures. The diffuse double layer will also have a negative potential unless it is reversed by a high concentration of divalent or multivalent cations in the fixed ionic environment immediately surrounding the lipid structure.

Although the relative signs and magnitudes of the electrostatic fields surrounding the protein and lipid interfaces must influence the statistical chance of a protein molecule coming into close enough proximity to bind with the surface and form a more or less stable complex, it cannot be assumed that this will not happen unless the net charge on the protein is opposite to that in the diffuse double layer. It is seen, for example, that phospholipase D below its isoelectric point (positively charged) can approach and be adsorbed on a surface composed of lecithin plus an anionic amphipathic substance where the ζ potential has suffered charge reversal by the addition of a high concentration of calcium or magnesium. Possibly in such cases the electrostatic field in the diffuse double layer can attract and orientate negative sites and areas on the protein molecule even though the total net charge of the whole is positive. This could be helped by the divalent ion shrinking the effective thickness of the diffuse electrical double layer around the lipid to below the diameter of the globular protein molecule. The final result of this initial attraction therefore, would be that the net positive protein would be held to the lipid surface by a divalent metal ion bridge.

After the initial approach of the protein molecule through the diffuse double layer it is likely that an electrostatic interaction will occur directly between ionogenic groups on the protein and the lipid surface. Even in the presence of salts in the bulk phase, the counter ions present in the "fixed" double layer around the lipid would still be in dynamic equilibrium with those in the bulk phase. On ionization of the metal-lipid salt linkage, $NH_3 +$ groups on the protein would be free to react in a competitive way with the exposed anionic groups on the lipid surface. This attachment could be reinforced by the proteins carboxyl groups interacting with the basic groups of zwitterionic phospholipids and possibly also by hydrogen bonding and hydrophobic bonding of the side chains of the protein penetrating into the spaces opened up between the lipid molecules. It is important to realize that for protein side chains of lengths 4–5 Å to penetrate into a phospholipid sheet (head groups 8–10 Å) and reach the hydrophobic region of the fatty acid chains, some unfolding of the polypeptide chain would have to occur if the lipid molecules were fairly closely spaced (Haydon and Taylor, 1963). This unfolding and the separation of the fatty acid chains (the gap being partially filled with water molecules) would require energy, which would probably have to be provided by the electrostatic attractions involved. This is possibly why the action of phospholipase B and C on high pressure lecithin monolayers >30 dynes/cm only takes place when the appropriate electrostatic conditions exist at the lipid-water interface. Of course Van der Waals attraction with limited or no penetration could still occur between, for example, the methylene groups of the ethanolamine and choline in the head groups of phospholipids and those in the protein side chains but the forces involved would be minimal compared with those of Coulombic binding (Salem, 1962).

Thus, depending on the stereochemical organization of the ionogenic groups and the distribution of its hydrophobic regions, the protein would form a tightly bound complex with the lipid stabilized by multi-point and multi-type bonding. Of course, the stability of the complex would still be to a greater or lesser degree sensitive to the salt concentration in the aqueous phase, depending again on the relative importance of ionic and non-ionic bonding in the completed complex.

The initial electrostatic bonding and subsequent non-ionic bonding will cause a certain deformation of the protein molecule and we have seen that in certain model systems with highly charged lipid-water interfaces, the forces involved can be sufficient to denature the protein. However, it is abundantly apparent that in natural lipoproteins the proteins are not all in the completely denatured form. Thus lipoproteins often possess enzymic activity and active water-soluble enzymes can be extracted from lipoproteins by various agents which disrupt the lipid-protein association. None of the protein of erythrocyte ghosts has the β- or completely unfolded conformation when examined by infra-red spectroscopy and, on isolation and examination by optical rotatory dispersion, it appears to be a globular protein with 17% in the α-helical form and the remainder probably random chain (Maddy and Malcolm, 1965). However, the forces of the interaction must result, to a greater or lesser degree, in changes in the spacial arrangement of the polypeptide chains from that typical of the native protein to a more disordered arrangement, a process which on completion leads to denaturation. There is a great need to know to what extent such conformational changes modify the protein in natural lipoprotein combination both biologically and chemically. In such experiments it would be essential to compare the properties of the protein both free and in lipoprotein combination by a whole battery of chemical, physical and biological tests which can be used for deducing conformational changes in proteins (Kauzmann, 1959). It is to be expected that such changes would in the initial stages be wholly or partially reversible; as an analogy, it is known, for example, that serum albumin is unfolded in salt-free acid or strong urea solutions but on removal of the acid or urea the protein returns to a form which is completely water-soluble at the isoelectric point. Even the heat denaturation of enzymes is sometimes reversible as, for example, the nucleotide pyrophosphatase of *Proteus vulgaris* which is completely inactivated at 70° but becomes reactivated on cooling to 37° (Swartz *et al.*, 1958). Recently it has been shown that the lipid-protein complexes formed between serum albumin and dodecyl-sulphate (Cockbain, 1953) result in an exposure of two disulphide groups on the albumin to reduction with β-mercaptoethanol and the complexing also protects the albumin from urea denaturation (Habeeb, 1966).

The picture which has been presented of the way in which proteins could react with organized lipid structures is inevitably over simplified because of the overall lack of knowledge we have about the process. In membranes the ionic lipids are, of course, combined with neutral lipids such as cholesterol and its esters. These could be present in the organized lipid structure before its reaction with protein or they could be attracted by the hydrophobic matrix of the initial ionic lipid-protein complex. Furthermore, metal ions may be retained as an intricate and essential part of the complex. Examples of this are probably the necessity of calcium for the phospholipase A and D reactions and the same metal for the formation of a stable biologically active complex between the blood-clotting factor X and anionic phospholipids (Papahadjopoulos *et al.*, 1964). If Ca^{2+} merely had a non-specific neutralizing effect on anionic groups in such interactions, it is to be expected that Mg^{2+} with its almost identical hydrated ion diameter would have a similar effect as it does in the activation of phospholipase C or the protection of phospholipase D from denaturation on negatively-charged surfaces.

Finally another possible mechanism for the formation of lipoprotein membranes must be briefly discussed. This is based on the well known property of differing lipoproteins exchanging their lipid moieties in colloidal solution. Thus individual serum lipoproteins containing ^{32}P-labelled phospholipids have been isolated by physical means and incubated in an aqueous medium with inactive serum lipoproteins of differing structure. On re-isolation of the individual lipoproteins, a rapid exchange of phospholipids between the two types of structure had taken place (Eder et al., 1954; Kunkel and Bearn, 1954). Similar lipid exchange reactions also occur between serum lipoproteins and other types of lipoprotein including the membrane of erythrocytes (Nichols and Coggiola, 1966; Sakagami et al., 1965). It seems likely that prior to exchange, collision of the two lipoprotein colloidal structures must occur resulting in coalescence and the formation of a collision complex (Gurd, 1960a, b,). Within the complex, the phospholipid molecules must overcome the energy barriers holding them to the individual apoproteins and other lipids and exchange freely. Subsequently the complex dissociates into two individual lipoprotein moieties and so the phospholipid arrangement has occured without any free phospholipid molecules being released into the aqueous phase. Cholesterol, which is presumably held in lipoprotein complex by weaker non-ionic forces, appears to exchange very rapidly even with membrane lipoproteins (Porte and Havel, 1961).

Now it is possible that the lipids which are synthesized during the growth of the cell and the maintenance of its structure during normal dynamic turnover may at the final stage of synthesis exist as a product-enzyme complex. Two alternatives are then possible. The complex could dissociate into the enzyme and newly-synthesized lipid and the individual molecules of the latter, being unable to fit into a water matrix, would immediately form a micro-structure, e.g. micelles or bimolecular leaflets. Such a lipid structure could then react with protein in the manner that has been discussed previously. An alternative possible fate of the lipid-enzyme complex is that it could pass the newly synthesized lipid molecule directly to an apoprotein or lipoprotein complex releasing the enzyme. The apoprotein could be of similar nature to that recently described in plasma by Roheim et al. (1965) and which reacts with liver lipids forming the plasma lipoproteins. The resultant lipoprotein could then react with another lipoprotein complex and exchange its lipid molecule in an analogous way to that which we have seen occurs in the collision between two colloidal lipoprotein particles. The whole network of lipoprotein membranes in the cell could be envisaged to exist as an intergrated continuum so that newly formed lipid molecules would, in fact, never leave the lipoprotein complex but would move by a process of exchange diffusion within the framework of the continuous lipoprotein cytoskeleton. The whole lipoprotein cytoskeleton would be in dynamic equilibrium so that any lipid molecules used at a given part of the cell would result in a net flow of lipid molecules to the same area. It is not to be expected that such a movement would necessarily form lipoprotein membranes of a constant composition throughout the cell. The final equilibrium composition of the lipid mixture in the membrane would depend on such environmental factors as, for example, the composition of the structural protein or proteins in a given membrane.

References

Attwood, D. and Saunders, L. (1965). *Biochim biophys. Acta* **98**, 344.
Attwood, D., Saunders, L., Gammack, D. B., de Haas, G. H. and van Deenen, L. L. M. (1965). *Biochim. biophys. Acta* **102**, 301.
Bangham, A. D. and Dawson, R. M. C. (1959). *Biochem. J.* **72**, 486.
Bangham, A. D. and Dawson, R. M. C. (1960). *Biochem. J.* **75**, 133.
Bangham, A. D. and Dawson, R. M. C. (1962). *Biochim. biophys. Acta* **59**, 103.
Chapman, D., Kamat, V. B., de Gier, J. and Penkett, S. A. (1967). *Nature, Lond.* **213**, 74.
Cockbain, E. G. (1953). *Trans. Faraday Soc.* **49**, 104.
Das, M. L. and Crane, F. L. (1964). *Biochemistry* **3**, 696.
Das, M. L., Myers, D. E. and Crane, F. L. (1964). *Biochim biophys. Acta* **84**, 618.
Das, M. L., Haah, E. D. and Crane, F. L. (1964). *Biochemistry* **4**, 859.
Dawson, R. M. C. (1967). *Biochem. J.* **102**, 205.
Dawson, R. M. C. (1958a). *Biochem. J.* **70**, 559.
Dawson, R. M. C. (1958b). *Biochem. J.* **68**, 352.
Dawson, R. M. C. (1963a). *Biochem. J.* **88**, 414.
Dawson, R. M. C. (1963b). *Biochem. biophys. Acta* **70**, 697.
Dawson, R. M. C. (1966). *Biochem. J.* **98**, 35C.
Dawson, R. M. C. and Hauser, H. (1967). *Biochim. biophys. Acta* **137**, 518.
Dawson, R. M. C. and Hemington, N. (1967). *Biochem. J.* **102**, 76.
Dawson, R. M. C. and Thompson, W. (1964). *Biochem. J.* **91**, 244.
dePury, G. G. and Collins, F. D. (1966). *Chem. Phys. Lipids* **1**, 1.
Dervichian, D. G. (1949). *Discuss. Faraday Soc.* **6**, 7.
Doty, P. and Schulman, J. H. (1949). *Discuss. Faraday Soc.* **6**, 21.
Eder, H. A., Bragdon, J. I. and Boyle, E. (1954). *Circulation* **10**, 603.
Eley, D. D. and Hedge, D. G. (1956). *J. Colloid Sci.* **11**, 445.
Eley, D. D. and Hedge, D. G. (1957a). *Discuss. Faraday Soc.* **21**, 221.
Eley, D. D. and Hedge, D. G. (1957b). *J. Colloid Sci,* **12**, 419.
Elkes, J. J., Frazer, A. C., Schulman, J. H. and Stewart, H. C. (1945). *Proc. R. Soc.* A **184**, 102.
Frazer, M. J. and Schulman, J. H. (1956). *J. Colloid Sci.* **11**, 451.
Frazer, M. J., Kaplan, J. G. and Schulman, J. H. (1955). *Discuss. Faraday Soc.* **20**, 44.
Green, D. E. and Fleischer, S. (1963a). *In* "Metabolism and Physiological Significance of Lipids" (R. M. C. Dawson and D. N. Rhodes, ed.) p. 581, John Wiley Limited, London.
Green, D. E. and Fleischer, S. (1963b). *Biochim biophys. Acta* **70**, 554.
Gurd, F. R. N. (1960a). "Association of Lipids with Proteins". *In* "Lipide Chemistry", (D. J. Hanachan; ed.) John Wiley and Sons, Inc., New York.
Gurd, F. R. N. (1960b). "Some Naturally Occurring Lipoprotein Systems". *In* "Lipide Chemistry" (D. J. Hanachan, ed.) John Wiley and Sons, Inc. New York.
Habeeb, A. F. S. A. (1966). *Biochim. biophys. Acta* **115**, 440.
Hauser, H. and Dawson, R. M. C. (1967). *European J. Biochem.* **1**, 61.
Haydon, D. A. and Taylor, J. L., (1963). *J Theoret. Biol.* **4**, 281.
Jurtshuk, P. Jr, Sekuzu, I. and Green, D. E. (1963). *J. biol. Chem.* **238**, 3595.
Kates, M., Madeley, J. R. and Beare, J. L. (1965). *Biochim. biophys. Acta* **106**, 630.
Kavanau, J. L. (1965). *In* "Structure and Function in Biological Membranes", Vol. 1 p. 56, Holden-Day, Inc., San Francisco.

Kauzmann, W. (1959). *Adv. Protein Chem.* **14,** 1.
Kruyt, H. R. (1949). *Colloid Sci.* **2,** Ch. 9. Elsevier, Amsterdam.
Kunkel, H. G. and Bearn, A. G. (1954). *Proc. Soc. exp. Biol. Med.* **86,** 887.
Lovern, J. A. (1954). "The Chemistry of Lipids of Biochemical Significance". Methuen and Co. Ltd., London.
Maddy, A. H. and Malcolm, B. R. (1965). *Science, N.Y.* **150,** 1616.
Matalon, R. and Schulman, J. H. (1949). *Discuss. Faraday Soc.* **6,** 27.
Mazanowska, A. M., Neuberger, A. and Tait, G. H. (1966). *Biochem. J.* **98,** 117.
Nichols, A. V. and Coggiola, E. L. (1966). *J. Lipid Res.* **7,** 215.
Palmer, K. J. and Schmitt, F. O. (1941). *J. cell. comp. Physiol.* **17,** 385.
Palmer, K. J., Schmitt, F. O., Chargaff, E. (1941). *J. cell. comp. Physiol.* **18,** 43.
Papahadjopoulos, D., Yin, E. T. and Hanahan, D. J. (1964). *Biochemistry* **3,** 1931.
Pethica, B. A. (1965). *Soc. Chem. Ind. Monogr.* **19,** 85.
Porte, D. and Havel, R. J. (1961). *J. Lipid Res.* **2,** 357.
Reich, M. and Wainio, W. W. (1961). *J. biol. Chem.* **236,** 3058.
Roheim, P. S., Miller, L. and Eder, H. A. (1965). *J. biol. Chem.* **240,** 2994.
Salem, L. (1962). *Can. J. Physiol. Biochem.* **40,** 1287.
Sakagami, T., Minari, O. and Orii, T. (1965). *Biochim. biophys. Acta* **98,** 111, 356.
Swartz, M. N., Kaplan, N. O. and Lamborg, M. F. (1958). *J. biol. Chem.* **232,** 1051.
Thompson, W. and Dawson, R. M. C. (1964). *Biochem. J.* **91,** 237.
Tzagoloff, A. and MacLennon, D. H. (1965). *Biochim. biophys. Acta* **99,** 476.

Chapter 6

Theoretical and Experimental Models for Biological Membranes

J. A. Lucy

Royal Free Hospital School of Medicine, University of London, England

I.	INTRODUCTION	233
II.	BIMOLECULAR LEAFLET MODELS	234
	A. General Considerations	234
	B. Some Recent Developments	237
III.	MICELLAR MODELS	244
	A. Structure	244
	B. Permeability	253
	C. Membrane-bound Enzymes	255
	D. Steroid Hormones	256
IV.	PHASE CHANGES WITHIN MEMBRANES	261
V.	EXPERIMENTAL MODELS FOR MEMBRANES	265
	A. Molecular Interactions in Monolayers	265
	B. Monolayers as Model Membranes	267
	C. Macro-models for Membranes	270
	D. Bilayers as Model Membranes	273
	E. Dispersions of Phospholipids	277
VI.	CONCLUSIONS	284
	REFERENCES	285

I. Introduction

One of the underlying themes of this book is the absorbing question of the relationships between the structures of different membranes and their various biochemical functions. This is by no means a new field of study but, for approximately thirty years, scientific thought about membranes was focussed almost exclusively on the concept of the bimolecular leaflet and on the Danielli model for the structure of membranes (Danielli and Davson, 1934–35). The situation in 1962 was summarized by Danielli in his concluding remarks at a symposium on the structure and function of membranes. Referring to the problem of the basic structure of the plasma membrane, Danielli said, "It now seems to be agreed that its basic structure is that which

I suggested in 1934, and it is also highly probable that the same structure is present in many other intercellular membranes" (Danielli, 1963). The introduction within recent years of new and improved techniques for the study of both model membrane systems and biological membranes has, however, raised many questions with regard to membrane structure. As a result, a number of theories and new models concerned with various aspects of the structure and function of lipoprotein membranes have been proposed and the scene is one of renewed interest.

It has not proved possible to discuss in this chapter more than a quite small proportion of recent papers that are directly or indirectly relevant to membrane models and some readers may, unfortunately, look in vain for specific topics in which they are particularly interested. The chapter is mainly concerned with some, but by no means all, of the recent speculative theories relating to the basic structure of the lipid moiety of biological membranes and it includes considerations of certain current experimental models for membranes. In addition, original suggestions are put forward concerning the interactions of steroid hormones with membranes and an interpretation of the properties of phospholipid particles dispersed in an aqueous environment is suggested that is relevant to the differential permeabilities of cations and anions through the plasma membrane of the red cell.

With regard to the micellar arrangement of lipids previously suggested for the structure of biological membranes by the writer (Lucy, 1964a), particular attention has been paid in this chapter to some of the implications of the dynamic aspects of the original model in which it was envisaged that the micellar configuration would be in equilibrium with the bimolecular leaflet structure and it is emphasized that the properties of natural membranes appear to indicate the presence of both bimolecular leaflet and micellar configurations.

II. Bimolecular Leaflet Models

A. GENERAL CONSIDERATIONS

Davson (1962) has summarized the growth of the concept of the paucimolecular membrane and the Danielli-Davson bimolecular leaflet model for the structure of cell membranes is so well known that no extensive discussion of this model will be attempted here (Danielli and Davson, 1934–35; Davson and Danielli, 1943). Recent experimental work in support of the model has been summarized by Stoeckenius (1966).

The concept of the bimolecular leaflet arose, in the first place, from studies such as those of Gorter and Grendel (1925) who used the monolayer technique to determine the thickness of the lipid layer in cell membranes. Gorter and Grendel concluded that the erythrocyte is bounded by a bimolecular layer of lipid (Fig. 1), although it now appears that they used too small a value for the surface area of erythrocytes in their calculations and that this error was compensated for by incomplete acetone-extraction of lipids from the cells (Bar *et al.*, 1966). A similar model for the plasma membrane was proposed by Danielli (Danielli and Davson, 1934–35) since the peculiar permeability relations of living cells could be explained if the typical surface

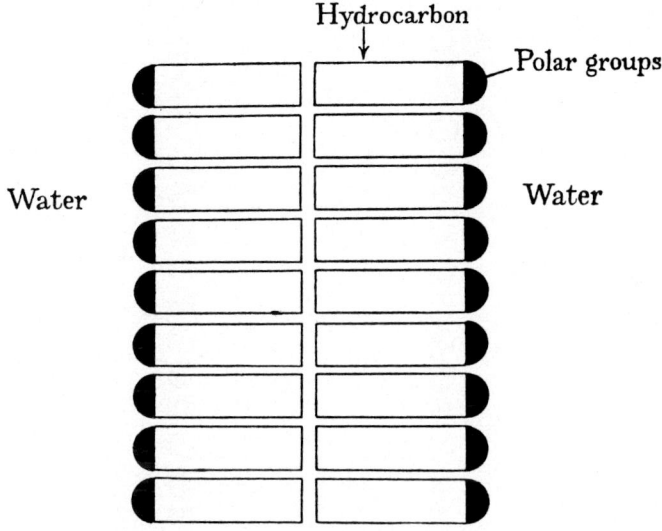

FIG. 1. Model for the structure of the red cell membrane (after Gorter and Grendel, 1925) (from Davson and Danielli, 1943 with permission of Cambridge University Press).

(plasma membrane) of the cell were lipoidal and paucimolecular. In view of the low values of interfacial tension observed at the surfaces of living cells, it was assumed in the model that a layer of protein molecules is adsorbed on the lipid. This assumption was based on work by Danielli and Harvey (1934–35) which showed that an adsorbed layer of protein is present at the interface between the aqueous part of the contents of mackerel eggs and the egg oil and that this layer is responsible for the low interfacial tension. Figure 2 reproduces the Danielli model for the structure of the plasma membrane which was later published by Davson and Danielli (1943). Within recent years it has become apparent that the phospholipid components of membranes can themselves also provide very low values of interfacial tension at the surfaces of membranes (cf. Huang *et al.*, 1964; Section V-D below) but there is, nevertheless, no doubt that proteins are important constituents of biological membranes.

A simple bimolecular leaflet model would appear to be too impermeable to water, water-soluble substances and small ions by comparison with the properties of biological membranes and it was therefore proposed later that pores of some kind extend through the lipid layer, from the aqueous phase on one side of the membrane to that on the other. Thus, Stein and Danielli

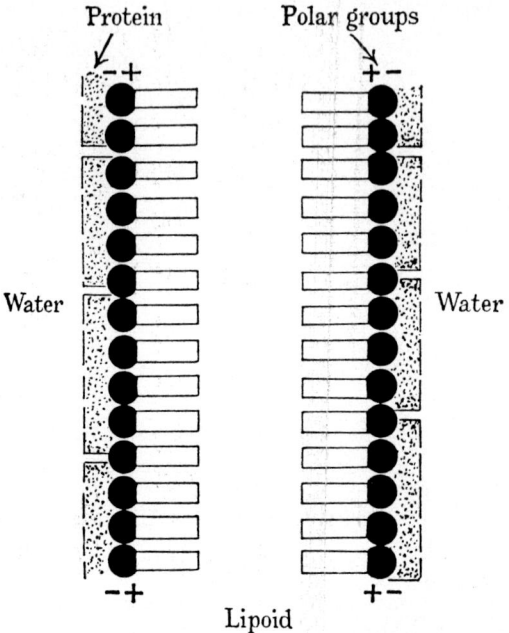

FIG. 2. Structure of cell membrane (Danielli) showing the proposed general pattern (from Davson and Danielli, 1943 with permission of Cambridge University Press).

(1956) obtained results from permeability studies which they interpreted as being compatible with the view that facilitated diffusion of a substance such as glucose occurs through a hydrogen-bonding component which extends through the thickness of the membrane. They suggested that the specific hydrogen-bonding space could be provided by protein lamellae extending through the lipid layer (Fig. 3).

The bimolecular leaflet model appears, in general terms, to be consistent with the known electrical properties of cells. The resistance of cell membranes is in the range of 10^3 to 10^6 ohms/cm^2 of membrane (Cole, 1940; Davson and Danielli, 1943), while their electrical capacity is rather less than 1 μF/cm^2 (cf. Table 4 in Whittam, 1964). If a value of 3 is assumed for the dielectric constant of such a membrane, it can be shown that its thickness lies between 30 and 50 Å. The electrical characteristics of the cell membrane are, therefore, similar to those of a lipid layer of about 40 Å in thickness and they thus support the concept that the membrane is indeed paucimolecular. A recent comparison of the electrical properties of cell membranes with those of artificial "bilayers" has indicated, however, that the lipids of cell membranes may not be in a completely continuous bimolecular leaflet and that "polar pores" may be present (see Hanai et al., 1965b; Section V-D below).

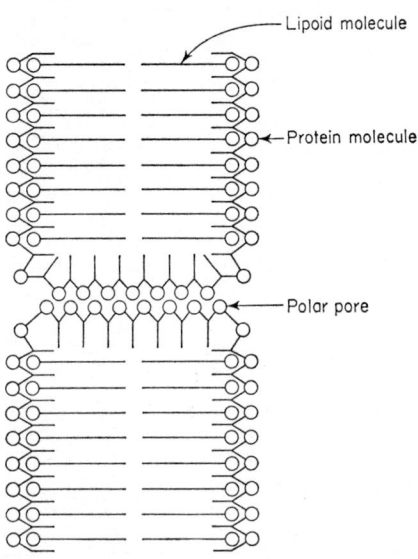

FIG. 3. Diagrammatic cross section of a pore composed by two polypeptide lamellae in a membrane. The basic structure of the membrane is a bimolecular lipid leaflet stabilized by adsorbed protein monolayers (from Stein and Danielli, 1956 with permission of Aberdeen University Press).

Further support for the bimolecular leaflet model appeared to be forthcoming with the development of electron microscopy, although electron micrographs of cell membranes have also recently indicated that this structure may not be universally present throughout all membranes at all times. This matter is referred to below (Section III-A), but for an extended consideration of the electron microscopy of membranes the reader is referred to Chapter 4.

B. SOME RECENT DEVELOPMENTS

With regard to more detailed models for the configuration of individual lipid molecules within the lipid layer in the bimolecular leaflet model, Finean (1953) has proposed a model for the structural arrangement of a specific phospholipid-cholesterol complex that he has suggested may occur in myelin. In this model, the polar end of the phospholipid is bent round so that it projects back towards the interior of the leaflet and participates in a general ionic interaction with the hydroxyl group of cholesterol. At the same time, the model facilitates the close association of the hydrophobic portions of the molecules in the complex (Fig. 4). This "walking-stick" configuration was suggested as an attempt to explain the observed fact that, in the lipid layer of myelin, the longest lipid molecules do not contribute their fully extended

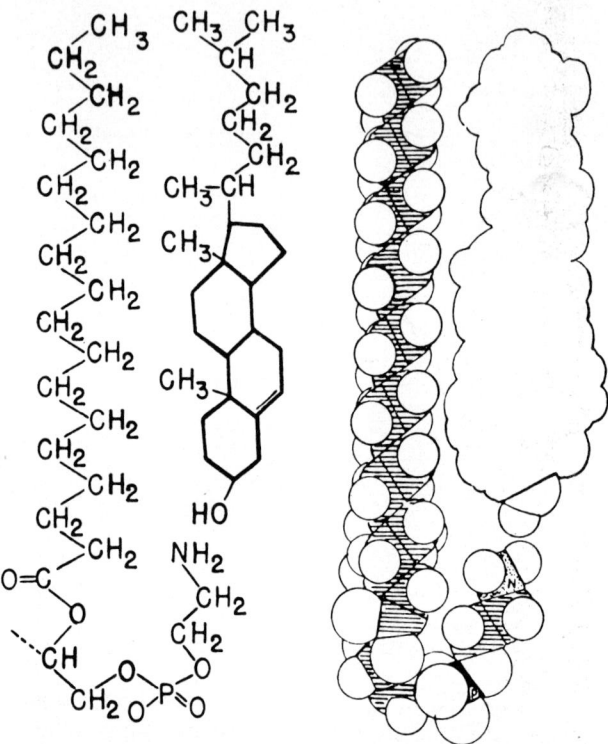

FIG. 4. A possible arrangement for the complex between cholesterol and phospholipid in myelin (from Finean, 1953 with permission of Birkhaeuser).

length of the thickness of the leaflet. Finean's speculations concerning the possible structure of the phospholipid-cholesterol complex have recently been extensively criticised by Vandenheuvel (1963) on the basis of work with molecular models which indicates that the walking-stick configuration is both sterically and energetically improbable. Vandenheuvel has made orthogonal projections of Dreiding stereo-models of the lipid molecules present in the myelin sheath of nerve. Using these models, and with maximum van der Waals interaction as a guiding principle, he has put forward a new model for the lecithin-cholesterol complex and also suggested a structure for the arrangement of the lipids within the bimolecular leaflet of the myelin sheath membrane. In the model for the lecithin-cholesterol complex, cholesterol fits between the choline moiety of lecithin and the unsaturated chain of the phospholipid which curves round to interact with the side chain of cholesterol (Fig. 5).

Pethica (1964) has commented that neither the structure suggested by Finean, nor an arrangement similar to that proposed by Vandenheuvel, is likely to be present in a mixed monolayer of cephalin and cholesterol at an air-water interface since both of these models would give large values for the surface dipole moment and this is not found experimentally (see Section V-B). It is interesting, however, in relation to the emphasis placed by Vandenheuvel on the possible van der Waals interactions of cholesterol with other lipid molecules, that Chapman and Penkett (1966) have recently made observations in nuclear magnetic resonance studies which indicate that an interaction may occur between the hydrocarbon chain of lecithin and the cholesterol molecule. Chapman *et al.* (1966) have also investigated the properties at the air-water interface of monolayers of phospholipids containing *trans* double bonds. It was found that replacement of a *cis* by a *trans* double bond causes the monomolecular films to be appreciably more condensed in character. In addition, it was observed that an interaction occurs between cholesterol and phospholipids containing a double bond in the 9, 10 position of the fatty acid chain, provided that the phospholipid film is expanded. The more condensed film of a *trans* phospholipid, therefore, can apparently interact with cholesterol only over a relatively narrow range of surface pressures.

Hechter (1965) has recently published a discussion of the possible roles of water in the organization of cell membranes and he has also put forward a hypothesis relating to the protein layers of the bimolecular leaflet model for the structure of cell membranes. This theory is, in turn, a development of Warner's theoretical hypothesis concerning a hexagonal conformation for polypeptides (Warner, 1961). In the membrane model suggested by Hechter, each protein layer is composed of two parallel layers of flat, interlocked discs. The discs are hexagonal and each disc is composed of two plate-like subunits associated in such a way that the surfaces of the discs are hydrophilic. Hechter suggests that the discs are held in position by water layers having an "ice-like" arrangement and that the discs form a precisely ordered lattice system (Fig. 6). The layers of discs are so constructed that aqueous channels run between them at right angles to the plane of the protein layers. These channels through the protein are lined with hydroxyl groups and either positively or negatively charged side-chains. Hechter has discussed questions of membrane permeability and depolarization in terms of this model. He suggests that a resting membrane is a highly ordered structure with an ordered lipid bilayer and that water is present in a hexagonal ice-like structure between the protein units. In contrast, in a depolarized membrane the protein subunits have

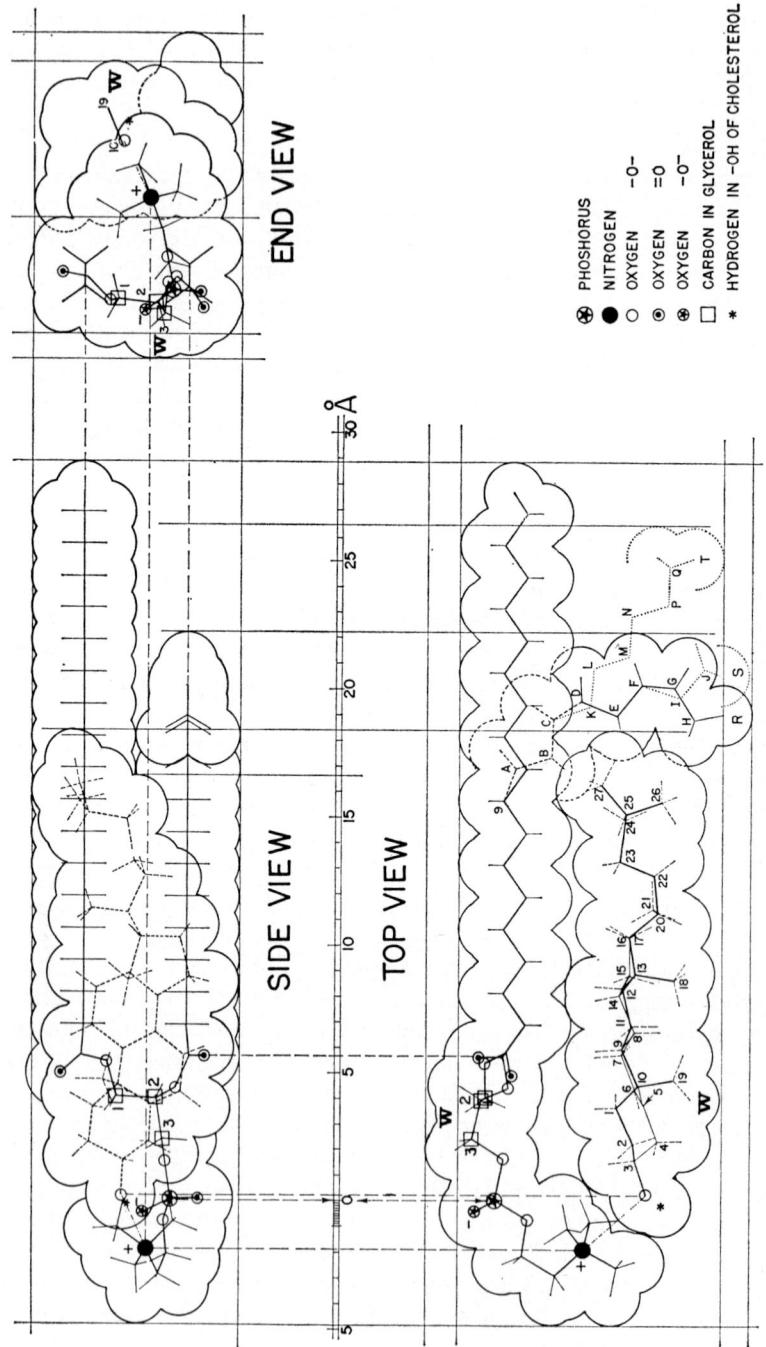

FIG. 5. Projection diagrams of molecular models in the cholesterol-lecithin complex suggested by Vandenheuvel (from Vandenheuvel, 1963 with permission of The American Oil Chemists' Society).

changed from hexagonal units to a more globular form, while the lipid and water molecules have become disordered. Mobile water molecules are

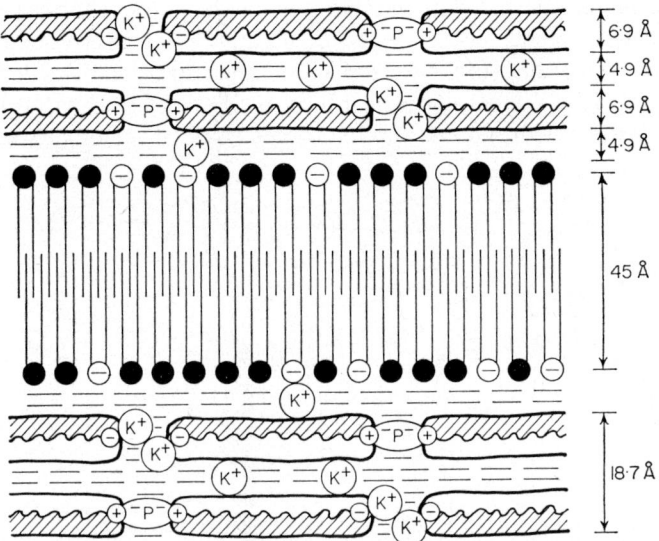

FIG. 6. Schematic representation, proposed by Hechter (1965), of the resting membrane where the basic features of the unit-membrane concept are retained and the protein layers are represented as a system of interlocked hexagonal discs cemented together by water layers in an ice-like arrangement to form a precisely ordered lattice system. The individual hexagonal subunits are shown interlocked through hydrophobic surfaces to form disc units held together by two layers of water in an ice-like state, this type of water being represented as (- -) (from Hechter, 1965 with permission of The Federation of American Societies for Experimental Biology).

envisaged as providing aqueous channels through the disordered hydrocarbon region of the membrane lipids; these channels permit relatively free diffusion of cations through the membrane with the electrochemical gradient (Fig. 7).

Hechter's model is valuable because it draws attention to the importance of water in the structure and function of membranes but several aspects of this model appear to be unsatisfactory. For example, there seems to be little basis for supposing that water molecules can penetrate through a barrier of virtually liquid hydrocarbon in the interior of the membrane in the way he suggests. Hechter refers to the lipids in the depolarized membrane as being in a random micellar arrangement but in the drawing of the bimolecular leaflet structure, the hydrocarbon chains simply have flexibility (Fig. 7). It was in an attempt to explain the permeability properties of membranes to water and ions that the concept of pores in the lipids of membranes was introduced, as for example in the model of Stein and Danielli (1956) in which the water in the pores is not in contact with the hydrocarbon chains of the lipids in the interior of the membrane (Fig. 3). Hechter's model seems to be a less acceptable alternative than models involving pores. A further difficulty with Hechter's model concerns the configuration of the protein of the membrane. According to Dickerson (1965), the hexagonal models which were proposed by Warner and which form the basis of Hechter's model do not have distinct hydrophobic and hydrophilic sides if the models are made

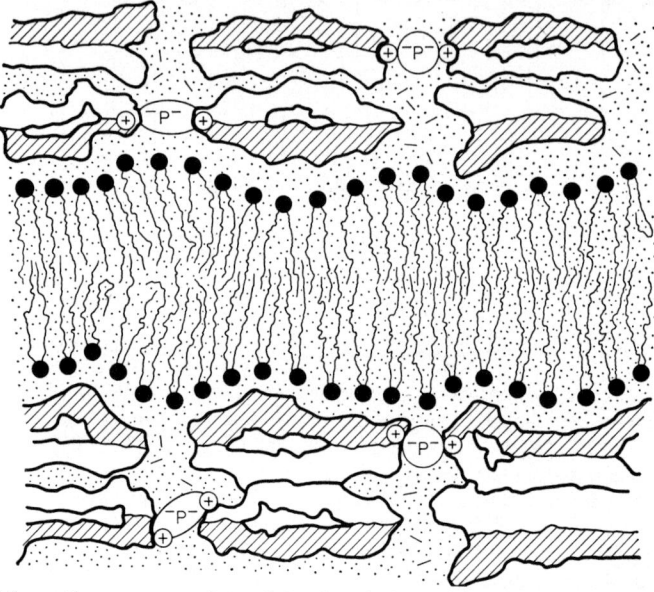

FIG. 7. Schematic representation of the depolarized membrane, where the arrangement of protein subunits, lipid and of water no longer provide a precisely ordered lattice (Hechter, 1965). In exaggerated form, the diagram illustrates that the protein subunits have changed from hexagonal discs to a more globular helical form, the lipid bilayer to a more random micellar arrangement and the ice-like water structures to less ordered water structures. Mobile water molecules are represented as small dots (·) (from Hechter, 1965 with permission of The Federation of American Societies for Experimental Biology).

with planar amides and they do not have carbonyl oxygens at the corners of a hexagonal mesh, 4·8 Å on a side. Moreover, the strain involved in twisting the amides out of planarity is apparently so great that even the most favourable hydrogen bonding and hydrophobic interactions cannot make these structures stable. Dickerson, therefore, maintains that the association by hydrophobic bonds of pairs of Warner's plate-like hexagonal structures, to make hexagonal discs with hydrophilic surfaces that are located in a hydrogen-bonding environment as suggested by Hechter in his membrane model, does not produce stable structures. Dickerson has concluded that it is probably reasonable to say that any peptide model with nonplanar amides can be regarded from the outset as extremely improbable, but if it proves useful for certain purposes to hypothesize models of this type, then their instability must be kept clearly in mind and at some point a quantitative numerical justification for such a distorted structure must be made.

An interesting discussion of pharmacological receptors and membrane permeability has been published by Watkins (1965). He has drawn attention to the similarities in structure and charge distribution that are seen when acetylcholine, γ-aminobutyric acid and glutamic acid are compared with the polar groups of lecithin, phosphatidylethanolamine and phosphatidylserine, respectively, and he has suggested that these relationships may provide a basis on which to explain the pharmacological actions of the first three substances. Watkins proposes that membranes normally contain complexes between the three lipids and membrane proteins (or other macromolecular substances) and that the pharmacological actions of acetylcholine, γ-aminobutyric acid and glutamic acid might result from dissociation of these complexes and the induction, thereby, of changes in membrane permeability in the "polar discontinuities" that may occur in a bimolecular leaflet type of membrane. For example, because of the similarity of structure between acetylcholine and the cationic polar groups of lecithin, some competition could be expected between these two entities for association with the complexing sites of a membrane protein. Watkins has suggested that lecithin may be attached to the protein by three different bonds, one of which is a linkage, via a divalent metal ion, between the phosphate of the lecithin and an anionic site on the protein (Fig. 8). Conformational changes in the protein and dis-

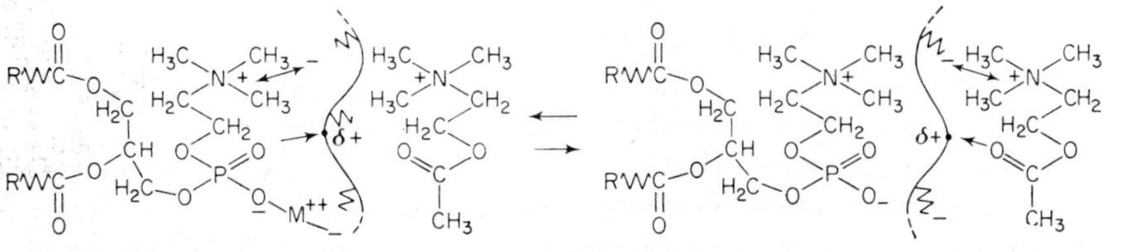

FIG. 8. Dissociation of a hypothetical lecithin-protein complex by acetylcholine (Watkins, 1965). The conformational change in the protein, induced by the interaction with acetylcholine, weakens the binding of the divalent metal ion and facilitates its displacement by sodium and/or potassium ions in the environmental fluid (from Watkins, 1965 with permission of Academic Press).

placement of the divalent cation by monovalent cations that are present in excess may occur when the protein binds acetylcholine and releases lecithin. These changes may then open up cation-selective channels in the membrane and increase the permeability of certain regions of membrane to sodium and potassium ions, hence producing the observed physiological effect of acetylcholine at the amphibian end plate.

Davies (1965) has proposed a theory concerning the quality of odours which also involves perturbation of the structure of a bimolecular lipid leaflet. He suggests that an odorant molecule first penetrates the cell membrane of an olfactory nerve cell. When the molecule subsequently leaves the membrane either by passing into the interior of the cell or into the mucus outside, sodium and potassium ions are thought to leak through the hole thus produced which initially has an area equal to that of the cross-sectional area of the orientated odorant molecule. If the hole heals relatively slowly, ionic exchange will occur through the hole and a nervous impulse will be initiated; when the hole heals rapidly behind the desorbing molecule, the particular nerve cell concerned will not be stimulated. Davies suggests that different nerve cells have different physical properties, depending on their constituent lipid molecules and that different odorant molecules, therefore, stimulate specific cells.

To conclude this section, it may perhaps be suggested that there is little doubt that the ability of cell membranes to act as barriers to free diffusion rests mainly on the properties of the bimolecular leaflet of lipid molecules. This structural arrangement may, therefore, be responsible for the very existence of cells and intracellular particles. However, membranes are not only structural components of cells, they are also highly important functional components and the many different functions of biological membranes do not all seem to be compatible with a simple bimolecular leaflet model. In this connection it is interesting that, although the theories of Hechter (1965), Watkins (1965) and Davies (1965) on membrane depolarization, the action of acetylcholine and the mechanisms of olfaction, are respectively all based on the bimolecular leaflet model, each theory involves some modification, discontinuity or perturbation in the lipid leaflet. This brings us to the point of considering that the bimolecular leaflet arrangement may, indeed, not be universal and that some of the varied properties of biological membranes may be based on other macromolecular organizations of lipid molecules, such as the micellar arrangements that are discussed in the following section.

III. Micellar Models

A. STRUCTURE

1. *Globular Micelles*

On the basis of observations made with the electron microscope on macromolecular lipid complexes that are apparently composed of globular micelles, it has been suggested that it is not necessary to consider structures containing a high proportion of phospholipids solely in terms of bimolecular leaflets (Lucy and Glauert, 1964b). Biological structures that might be regarded as representing bimolecular leaflets may, in certain circumstances, be formed by the association of globular micelles of lipid which function as biological building blocks in a manner similar to that in which protein subunits are assembled in complex structures such as viruses and bacterial flagella. The observations that complex lipid assemblies can apparently be constructed from globular micelles of lipid also led to the suggestion being made at a symposium held in Modena in April, 1963 that similar micelles may, under appropriate circumstances, also be present in cell membranes (Lucy and Glauert, 1964a). Towards the end of the same year, Sjöstrand (1963a,b,c) published electron micrographs of cell membranes which indicated the existence of an organized sub-structure, within the plane of the lipids of membranes, which he discussed in terms of lipid micelles.

As a development of ideas concerning the assembly of globular micelles of lipid into macromolecular structures, the writer has put forward a theoretical micellar model for the lipids of biological membranes and also proposed that globular micelles of lipid might be in dynamic equilibrium with the bimolecular leaflet structure within membranes (Lucy, 1964a). Figure 9 is a diagrammatic illustration of a cross-sectional view of the suggested arrangement of globular lipid micelles in a biological membrane.

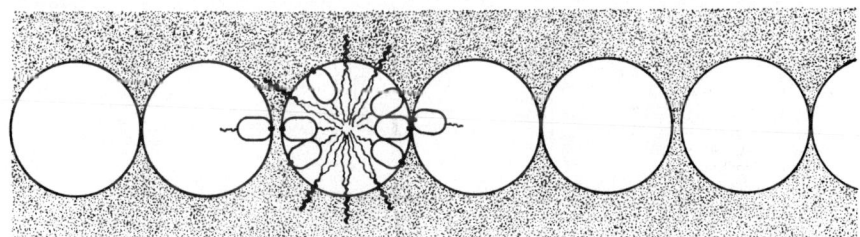

FIG. 9. A cross sectional view of a model for a biological membrane in which the lipids are in small globular micelles. The micelles are arranged in a plane and a layer of protein and/or glycoprotein is shown on each side of this plane of micelles. The organization of lipid molecules (phospholipid and cholesterol) within the micelles is illustrated diagramatically, but not stoichiometrically, for one of the micelles in the row (from Lucy, 1964a withpermission of Academic Press).

For comparison, Figure 10 shows a similar cross-sectional diagram of the bimolecular leaflet arrangement. Cross sections of biological membranes studied in the electron microscope may not distinguish between these two different arrangements of lipids since all except the very thinnest of sections would contain a number of superimposed layers of micelles in an area of micellar membrane and, as a result, the subunits might not be detectable (cf. however, Sjöstrand, 1963a).

I

A diagrammatic illustration of the surface view of the lipid micelles in the suggested model is shown in Fig. 11. For simplicity, the drawing shows the globular units in a regular, hexagonal array. It was proposed however, that the micelles would probably be in continuous random movement about their

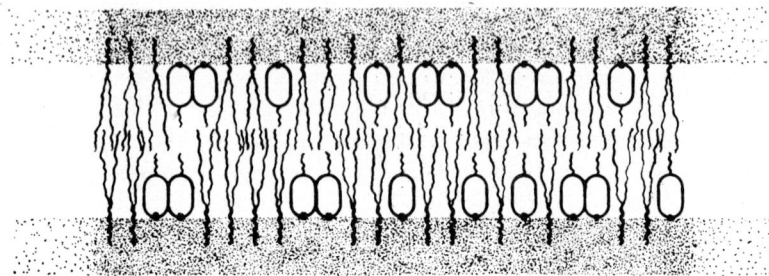

FIG. 10. A cross sectional view of a model for a biological membrane in which the lipids are arranged in the form of a bimolecular leaflet. A layer or protein and/or glycoprotein is shown on each side of the lipid leaflet (from Lucy, 1964a with permission from Academic Press).

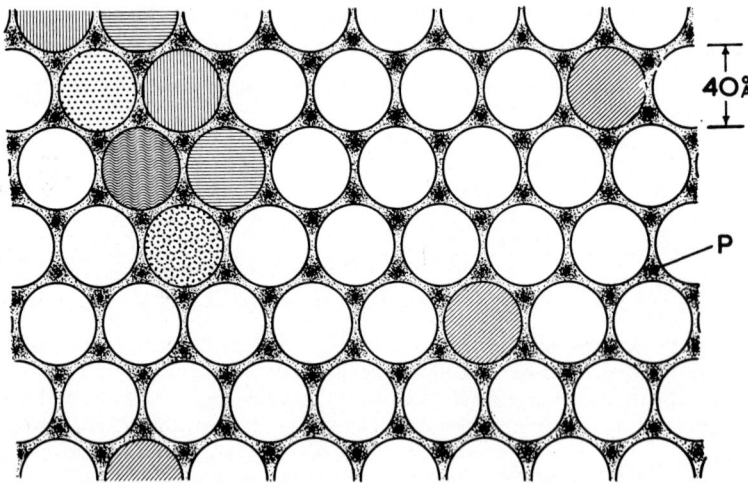

FIG. 11. A surface view of the lipid moiety in a micellar model for a biological membrane (Lucy, 1964a). The structure is to be regarded as flexible, with the individual micelles in continuous slight random movement. The shaded units are intended to represent globular proteins with enzymic or hormonal properties that have replaced the lipid micelles at certain points in the plane of the lattice. Water-filled pores (P), approximately 4 Å in radius, penetrate through the membrane between the micelles; the pores are lined by the polar groups of the lipids comprising the micelles and, in certain areas, by the polar groups of functional, globular molecules of protein (from Lucy, 1964a with permission from Academic Press).

mean positions (cf. Clunie *et al.*, 1965). It might perhaps even be suggested that Bernal's structural concept of a liquid as a "heap" of molecules (Bernal, 1964), if it would be applied to the arrangement of micelles in the relatively restricted—but flexible—plane of a membrane, may offer a suitable model with which to describe some of the properties of a micellar membrane composed of globular subunits. The globular micelles illustrated in Figs 9 and 11 have lipophilic cores of about 40 Å in diameter but they would be rather larger if the proportion of C_{18} fatty acids in the phospholipids were high. The overall diameters of the micelles would be about 50 Å if the polar groups of the phospholipid molecules are arranged tangentially to the surface of the globular units (cf. Pethica, 1964; Section IV-B below; Hanai *et al.*, 1965a and Section IV-D below). Ajacent micelles might be held together by hydrogen bonds, as suggested for the macromolecular lipid assemblies studied with the electron microscope (Lucy and Glauert, 1964b), while electrostatic interactions would be expected to play a more or less important part according to the nature of the phospholipid molecules comprising the micelles.

The stability of such a flat sheet of micelles may be dependent upon interactions with protein, or perhaps glycoprotein molecules, on the surface of the lipid layer. The question of the interactions between the lipid and protein moieties of membranes is, however, far from clear. If hydrophobic bonding is involved between the lipids and the structural proteins of membranes, as is suggested for example by the work of Fleischer *et al.* (1963) on the structural protein of mitochondria, this may occur both with lipids in the bimolecular leaflet configuration and with globular micelles of lipid. In the latter case, short loops of protein consisting predominantly of hydrophobic amino acid residues could perhaps penetrate into the liquid, hydrocarbon interiors of the micelles. These considerations lead to the possibility, which is discussed below in Section III-C, that certain membranes may be composed of globular subunits each of which is primarily a lipoprotein unit, rather than a globular micelle of phospholipid that is loosely associated with protein. Gent *et al.* (1964) have reported relevant studies on the solubilization of myelin by lysolecithin. A single component was observed in the soluble material by electrophoresis and sedimentation and it was suggested that myelin is composed of relatively stable lipid-protein units (estimated dimensions 60 × 60 × 40 Å) which are removed by the action of lysolecithin to form a single species of a complex micelle composed of protein-lipid-lysolecithin. It has been pointed out, however (Glauert and Lucy, 1968) that lysolecithin is known to break down macromolecular assemblies of lecithin (Bangham and Horne, 1964) and the observations of Gent and his colleagues may alternatively be interpreted as a further demonstration of the breakdown of phospholipid structures by lysolecithin.

2. Experimental Data

(a) *Negative-staining of lipids.* As has already been mentioned, indirect evidence for the possible occurrence of globular micelles in biological structures has come from studies on the macromolecular structures of lipids dispersed *in vitro* (Lucy and Glauert, 1964b). In these experiments, negatively-stained preparations of various mixtures of lecithin and cholesterol, dispersed in water with and without the haemolytic agent saponin, were investigated by electron microscopy. Lamellar, tubular, hexagonal and helical structures were observed in these preparations and, after consideration of a large number of possible alternative hypotheses, it was concluded that many of the structures observed could only be interpreted satisfactorily in terms of globular micelles of lipid. Figure 12 is an electron micrograph of a preparation of purified ovolecithin and cholesterol, negatively-stained with potassium phosphotungstate. In the past, myelin figures have been interpreted in terms of arrays, or concentric tubes, of bimolecular leaflets.

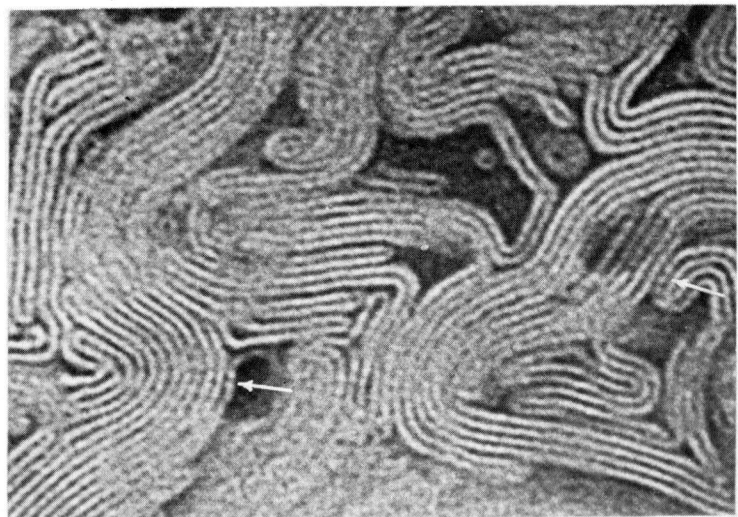

FIG. 12. Electron micrograph of a negatively-stained preparation of purified ovolecithin and cholesterol (from Lucy and Glauert, 1964b with permission of Academic Press). Micro-myelin figures of lecithin and cholesterol; in places the light (lipid) layers appear to contain globular subunits (arrows). × 300,000.

However, the micro-myelin figures seen in the electron micrograph are apparently not entirely consistent with the presence of bimolecular leaflets since parts of these structures appear to contain discrete globular units with diameters of about 40 Å. Indeed, the appearance of the structures in Fig. 13 would seem to be explicable only in terms of subunits assembled to form tubular structures (cf. bacterial flagella; Kerridge et al., 1962). End-views of short lengths of these tubes are seen in Fig. 14. Application of the photographic rotation technique of Markham et al. (1963) to the end-view of the tubes indicated that the tubes have five-fold symmetry in cross section, i.e. the tubes are apparently composed of five parallel rows of subunits.

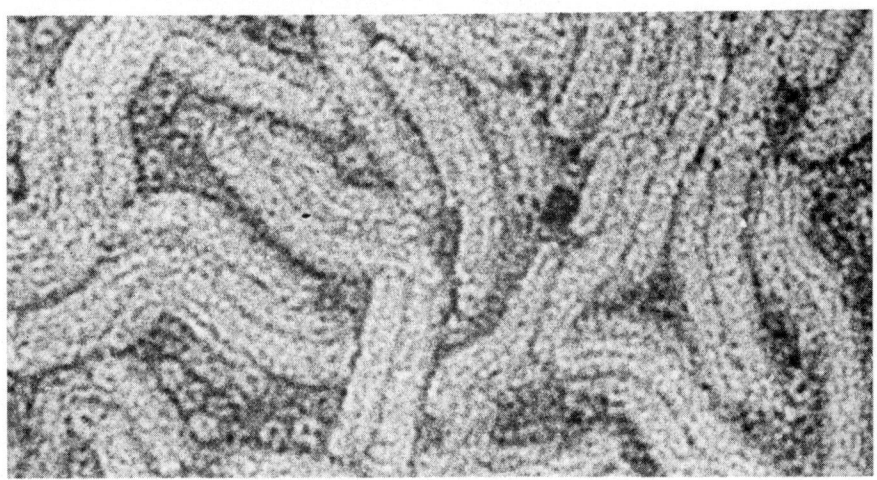

FIG. 13. Electron micrograph of a negatively-stained preparation of purified ovolecithin and cholesterol (from Lucy and Glauert, 1964 with permission from Academic Press). Granular, banded structures of low contrast in a mixture of lecithin and cholesterol are probably tubes into which the negative stain has failed to penetrate. The granular subunits appear to be arranged in rows parallel to the lengths of the bands. × 300,000.

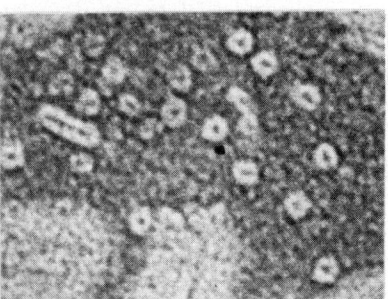

FIG. 14. Electron micrograph of a negatively-stained preparation of purified ovolecithin and cholesterol (from Lucy and Glauert, 1964b with permission from Academic Press). End-view of hollow tubes observed in negatively-stained preparations of lecithin and cholesterol. × 300,000.

The validity of the image obtained by negative-staining of lipid-containing materials has been questioned by Finean and Rumsby (1963) since it is known from X-ray diffraction studies that dehydration of lipoproteins generally leads to structural re-organization. These is considerable evidence nevertheless, which has recently been discussed by Glauert and Lucy (1968), that the presence of the negative stain prevents any radical alteration in the structure of lipid systems during drying (Horne et al., 1963). With regard to the tubular structures that are observed in negatively-stained preparations of lecithin and cholesterol, similar structures have also been observed in preparations of lecithin and cholesterol that have been dried, in the presence of the negative stain, five minutes after the addition of osmium tetroxide (1 %) to the aqueous dispersion of lipids (Glauert and Lucy, 1968). It therefore seems unlikely that the structures observed are artefacts produced by drying the lipids and it appears possible, as suggested by Horne and Whittaker (1962), that the stain sets in the form of an electron-dense glass and preserves the lipid structures as though they were in the presence of water.

A further criticism of the use of potassium or sodium phosphotungstate as negative stains for the study of lipids, which has been raised by Bangham and Horne (1964), is that a considerable electrostatic stress may develop at the lipid-water interface due to the difference between the valencies of the phosphotungstate and potassium ions. This stress may cause bimolecular leaflets of phospholipid to break up into aggregates of radially orientated molecules (Haydon and Taylor, 1963). Lecithin dispersions have been shown to acquire a net positive charge in the presence of bi- or ter-valent salts of univalent acids, and the ferricyanide ion has the opposite effect (Bangham and Dawson, 1962). Bangham and Horne consider that, although a 1 % solution of potassium phosphotungstate might not impose an intolerable electrostatic force of repulsion between lipid head groups, it must be remembered that the concentration will rise considerably before effective "freezing" occurs. They suggested that bi- or ter-valent salts of phosphotungstic acid might give more valid pictures of lipid structures. In studies with the helical assemblies of globular subunits that are formed in preparations of lecithin, cholesterol and saponin, it has been found, however, that the primary effect of the divalent calcium ion (as calcium phosphotungstate) is to retard the rate at which the helices are formed (Glauert and Lucy, 1966). Further experiments with potassium phosphotungstate and ammonium molybdate have shown that each of these salts also reduces, in differing degrees, the rate of formation of the helical structures, as compared with the rate observed in a saponin solution that is free from negative stain. It is conceivable that the cations of the negative stains may compete for anionic binding sites on the surfaces of the micelles that would otherwise be concerned in the formation of the macromolecular lipid assemblies.

Although it would not seem possible at this stage to exclude the possibility that the micelles which appear to be present in the negatively-stained preparations of lipid may owe their origin, at least in part, to the ionic properties of the negative stain, this does not affect the relevance of the electron microscopic studies on isolated lipid systems to the possible presence of globular micelles in biological membranes. As Bangham and Horne (1964) have pointed out, electrostatically-induced instabilities in the bimolecular leaflet, due to transient variations in local pH or bivalent ion concentration, might perhaps occur in excitable membranes and such instabilities might be concerned in the regulation of membrane permeability.

(b) *Electron microscopy of membranes.* Until quite recently, most electron micrographs of preparations of biological membranes and thin sections of fixed lipids could be interpreted in terms of the bimolecular leaflet model for the structure of membranes and the related "unit membrane" concept of Robertson (see Chapter 4). Within the past few years, however, a number of studies have indicated the existence of an organized sub-structure within the plane of the lipids of membranes, while investigations on fixed lipids have revealed the presence of both hexagonal and lamellar phases (Stoeckenius, 1962; Luzzati and Husson, 1962). Sjöstrand (1963a,b,c) and Sjöstrand and Elfvin (1964), in studies on potassium-permanganate fixed mouse kidney and freeze-dried pancreas tissue, have observed that some mitochondrial membranes and smooth cytomembranes consist of globular-components having diameters of about 50 Å and separated by stained septa, 10 to 20 Å thick. On the basis of these observations, Sjöstrand has proposed that these membranes consist of globular particles of lipid in the form of small micelles and that the stained material is protein which covers the surfaces of the lipid particles and thus prevents their fusing to form either a bimolecular leaflet or larger droplets. This proposal is similar to, but not identical with, the micellar model for the lipids of biological membranes suggested by the writer on the basis of investigations on isolated lipids.

Comparable globular units, but some having rather different dimensions, have been observed in a variety of other tissues by several different workers: for example, in the membrane elements of the retina of the tadpole and the frog (Nilsson, 1964a,b, 1965; see also Blasie *et al.*, 1965, referred to below in the section on X-ray diffraction); structural units with a repeat period of 85–95 Å in synaptic discs from goldfish (Robertson, 1963); in the mitochondria of mouse pancreas and nervous tissues studied by Malhotra (1966); by an electron cytochemical method in rat liver mitochondrial membranes (Rosa and Tsou, 1965); and in a number of investigations on chloroplast membranes (e.g. Weier *et al.*, 1966). Branton (1966) has made a study of the fracture planes revealed by freeze-etching of onion root tip cells. He considers that frozen membranes are split on fracturing and thus expose an internal membrane face. In the light of his experiments, Branton suggests that biological membranes are organized in part as an extended bilayer of lipid and in part as globular subunits. The relative proportion of the membrane which exists in these two configurations appears to vary with different cell organelles (cf. phase changes within membranes, Section IV below).

It has been argued that the appearance of a globular sub-structure within the plane of the lipids of membranes can be explained as an artefact of preparation. It is perhaps needless to say that this view-point is not held by many of the workers who have reported the presence of such a globular sub-structure in electron micrographs of membranes.

(c) *X-ray diffraction.* A recent paper by Clunie et al. (1965) on the structure of lyotrophic mesomorphic phases is indirectly relevant to the possible presence of globular micelles of lipid in biological membranes. These workers have made a comparative optical, density and X-ray diffraction study of the liquid crystalline phases occuring at room temperature in a semi-polar, surface-active agent-solvent system, with H_2O and D_2O as solvents. The surface active agent, *N,N,N*-trimethylamino dodecanoimide, was used to test the theory that lyotropic mesomorphic phase structures represent different ways of packing micellar units. Clunie et al. confirmed previously accepted structures for the neat and viscous isotropic (or cubic) phases (Luzzati and Husson, 1962) but proposed a more detailed structure for the middle phase. To account for their observations that the transition from viscous istropic to middle phase takes places without any dicontinuity in density or long spacing, they consider that it is necessary to postulate that there is little change in structure in passing from one phase to the other. These investigators have therefore suggested that a rigid "string of spheres" of unspecified length appears to be a more reasonable model for the elongated units of the middle phase than the indefinitely long cylinders suggested by Luzzatti and Husson (1962). It is postulated that the "string of spheres" is a linear aggregate of spherical micelles; the average length of the aggregates was estimated as 450 Å, i.e. 12 spherical micelles. Clunie et al. conclude that, for specific volume concentrations of surface active agent of less than 0·74, the primary structural units are spherical micelles and that the various phases, other than neat phase (a parallel array of bimolecular sheets), merely represent different aggregation states of spherical micelles having an essentially liquid interior (Hartley, 1955).

Indications of a micellar sub-structure do not appear to have been obtained in studies by X-ray diffraction of phospholipid myelin figures. In this connection, attention has been drawn to the possibility that arrays of micelles would form very flexible structures and, in an aqueous environment, any constituent micelles would be in continual movement about their mean positions (Lucy and Glauert, 1964b). Clunie et al. failed to observe any discreet X-ray reflections corresponding to the subperiod within the linear aggregates of the middle phase of the N,N,N-trimethylamino dodecanoimide-water system and they ascribed this to the presence of one-dimensional disorder, i.e. the strings of spherical micelles are randomly displaced parallel to their neighbours while retaining strict lattice order in cross-section. The very weak diffuse X-ray scattering expected from this one-dimensional disorder may remain undetected when superimposed on the strong sharp pattern due to the ($hk0$) planes. However, certain of the interpretations put forward by Clunie *et al.* have recently been criticized by Luzzati and Reiss-Husson (1966), see Chapter 3.

In a recent paper, Blasie *et al.* (1965) have reported data obtained by electron microscope and low-angle diffraction studies on outer segment membranes from the retina of the frog. They observed arrays of particles, with an average diameter of 40 Å, in surface views of membranes which were either negatively-stained with sodium phosphotungstate or fixed with potassium permanganate. The unit cell of the array of particles appeared to be square, with an average dimension of about 70 Å. It was presumed that the observed diameters of the particles corresponded to the non-polar cores of the units, so that the total diameter of each particle could approach 70 Å, a limit imposed by the square array in which the particles occurred. Blassie *et al.* commented that the composition of the observed globular particles is unknown, but that their dimensions are consistent with the globular micelles of lipid observed by Lucy and Glauert (1964b) in negatively-stained preparations of purified lipids. However, on the basis of size alone, each globular particle could be a rhodopsin molecule. Low-angle X-ray diffraction patterns obtained from centrifuged pellets of partially-dried (unstained and unfixed), isolated, outer segment membranes gave reflections consistent with a square array of particles having a unit cell side of 70 Å. The intensity ratios of the reflections were consistent with a spherical particle of 40 Å diameter. From combined X-ray diffraction and birefringence studies, Blasie *et al.* concluded that it can be argued that globular particles may be present in these membranes *in vivo* but that their observations do not distinguish between whether the particles occur on the surface of the membrane or within it. They also concluded that a modification of the uniformly layered, structural model for biological membranes is necessary to account for the structure of the outer segment membranes of the retinal receptor.

B. PERMEABILITY

Since a bimolecular leaflet is a relatively impermeable structure, models for the cell membrane based on such a leaflet incorporate pores to account for the observed permeability properties (see Fig. 3) (Stein and Danielli,

1956). In contrast, the micellar model illustrated in Fig. 11 contains pores which are an inherent feature of its structure although the membrane remains, nevertheless, essentially lipid in character. In this model the polar groups of the phospholipids, and their associated counter ions, line the pores each of which has an effective radius of approximately 4 Å at its narrowest point. This value compares well with the experimental figures that have been obtained for the equivalent pore radius of various membranes. For example, values of 3·5 Å and 4·2 Å have been obtained for red cell membranes by Paganelli and Solomon (1957–58) and Goldstein and Solomon (1960–61), respectively. The value of 4·0 Å has been obtained for the pore radius of the luminal surface of intestinal mucosal cells (Lindemann and Soloman, 1961–62) and 4·25 Å for the resting axolemma in the giant axon of the squid (Villegas and Barnola, 1960–61).

It is suggested below (Section V-E) that the relative impermeability of biological membranes to cations, as compared with anions, may be due to the pores having a positively-charged lining that results from the non-coplanar arrangement of the phospholipid polar groups.

With regard to the differential permeability of neural membranes to sodium and potassium ions, Gammack (1966) has drawn attention to the fact that monolayer and "bilayer" membranes do not generally show selectivity between these two ions. In contrast, titration at acid pH of micellar dispersions of phosphatidylserine by sodium and potassium salts has been observed to exhibit a differential pH change that is claimed to be well beyond experimental error; a reasonable assumption to explain the observed results being that there is a preferential binding of sodium over potassium (Abramson et al., 1964). Gammack has therefore suggested that ion selectivity in neural membranes may thus depend on the specific orientation and spacing of the polar groups of lipid molecules as might occur in the lining of membrane pores or, in the light of the greater selectivity of lipids when in micellar form, it may depend on a micellar organization of lipids in membranes such as that suggested by the writer.

The permeability of the micellar model illustrated in Fig. 11 to water and small ions probably exceeds the requirements of most biological membranes. It would appear, indeed, that less than 1% of the erythrocyte membrane needs to have a micellar configuration in order to account for its measured permeability. Myelin may perhaps have little or none of its lipid molecules in a micellar arrangement and, as Sjöstrand (1963b) has commented, although it has become commonly accepted that any membrane structure which is not in harmony with the myelin sheath is questionable, we have nevertheless every reason to believe that cellular membranes in general play a more active role than that of an electrical insulator. An alternative possibility is that, instead of the proportion of micelles being quite low in relatively impermeable membranes, the pores may effectively be blocked, for example by an ice-like crystalline structure for the water within the pores (cf. Hays and Leaf, 1961–62; Fernandez-Moran, 1959; Hechter, 1965, Section II-B above).

In connection with their observations on the hexagonal liquid-crystalline phase of phospholipids in water, Luzzati and Husson (1962) commented that if similar structures exist *in vivo* they would have remarkable permeability properties resulting from the long and narrow water channels which are covered by the polar groups of the lipid molecules. The hexagonal structure of the phospholipid-water is not, however, directly applicable to the configuration of phospholipids in membranes since, in the liquid-crystalline array, the hexagonally arranged cylinders of water extend indefinitely in one dimension while the hydrocarbon of the lipids fill the gaps between the cylinders.

C. MEMBRANE-BOUND ENZYMES

A micellar structure for certain areas of membranes might provide a means of enabling enzyme molecules, or molecules of certain protein hormones, to be structurally incorporated into the plane of the lipid layer. It was therefore suggested that one, or a number of lipid micelles in an organized array, could be replaced by globular protein molecules of the same order of size as the lipid micelles without disturbing the basic micellar patterns, Fig. 11, (Lucy, 1964a). Relatively large protein molecules could be inserted, provided that the local disorder in the array of globular units and the consequent increase in size of the aqueous pores did not become a limiting factor in the function of the membrane. Thus the micellar configuration offers a means by which the functional groups and active sites of enzymes may become an integral part of the structure of a membrane, while partial rotation of individual enzymes within the flexible lattice of lipid micelles might be concerned in the transfer of molecules and ions from one side of a membrane to the other in processes of active transport and metabolism (cf. Whittam, 1964; Mitchell, 1963). Sjöstrand (1963a) has suggested that the enzymes of the electron transfer chain in mitochondrial membranes may be located between the globular lipid particles that he has observed in biological membranes and that the enzymes would thus be in contact with both the lipid phase and with the aqueous phase of the mitochondrial matrix. Since the stained septa between the globular areas have a width of only about 10 Å, however, it would seem unlikely that the bulky enzyme complexes of the electron transfer chain are located between the lipid micelles.

A mixed array of globular enzymes and globular micelles of lipid might be particularly suitable for membranes whose primary function is to provide a locus for complex enzyme reactions rather than to function as a permeability barrier and it was originally suggested that the micellar configuration might therefore apply especially to the inner membrane of the mitochondrion (Lucy, 1964a). However, recent work indicates that the penetration of this membrane by anions probably does not occur via pores, since phosphate ions (H_2PO_4—, 5·2 Å; HPO_4—, 6·4 Å) penetrate readily while Cl^-, Br^- and NO_3^- (2·3–2·7 Å hydrated diameter) do not (Chappell, 1966).

McConnell *et al.* (1966) have reported interesting observations on negatively-stained preparations of cytochrome oxidase examined with the electron microscope. They observed that, on removal of bile acids from a purified preparation of cytochrome oxidase, the subunits (50 and 100 Å in diameter) which were present in the preparation organized themselves spontaneously into vesicular structures. These structures appeared to be bounded by a membrane composed of particles. Cytochrome oxidase from which phospholipid had been removed was unable to form vesicles but this capacity was restored by adding back mitochondrial phospholipid to the lipid-depleted particles. These findings may provide support for the idea that a mosaic of globular lipid micelles and enzyme molecules might occur in certain membranes. Alternatively, the observations may be interpreted as favouring the hypothesis that membranes are composed of lipoprotein subunits. Green and Tzagoloff (1966) have recently published a paper on the role of lipids in the structure and function of biological membranes in which it is considered that membranes are composed of repeating units that are composite macromolecules containing both protein and lipid. This concept is based on studies with mitochondria and the mitochondrion is assumed to be a model for membranes generally. In the view of the writer, it may be unwise to assume that any one system, whether it be the mitochondrion, the myelin sheath or the red blood cell, can be regarded as a universal model except in quite general terms. Benson (1966), in a discussion of the orientation of lipids in chloroplasts and cell membranes, has also recently proposed that membranes are formed by the association of globular lipoprotein subunits. It may perhaps be suggested that some membranes contain globular micelles of lipid and separate globular molecules of protein in the plane of the lipid micelles, while other membranes are composed of globular lipoprotein units and yet others contain all three species.

D. STEROID HORMONES

Willmer (1961) has proposed the stimulating hypothesis that there may be a correlation between the type of physiological action caused by any given steriod hormone and the groups at the ends of the steroid molecule, particularly the groups at C_3 and C_{17}, as a result of the packing of the steroid into lipoprotein membranes. He suggested that a steroid will pack with its long axis parallel to the hydrocarbon chain of the phospholipids in a bimolecular leaflet and that its physiological action would then be determined mainly by the polar groups thus exposed at the hydrophilic surface of the membrane. On this hypothesis, the ability of a steroid molecule to act on any particular cell will depend on the capacity of the steroid to enter and remain within the plasma membrane. This, in turn, will be governed chiefly by the molecular shape of the hormone, its affinity for particular phospholipids already in the membrane and on the action which any proteins associated with the hydrophilic surface of the lipid might have upon it. Willmer noted that these considerations would apply not only to the plasma membrane of cells but also to the membranes of organelles within cells. Recent work on the mechanism of action of oestrogens may perhaps be consistent with an effect of oestrogens on the nuclear membrane. For example, Mueller (1966) has found that RNA polymerase activity is affected by protein synthesis and he has suggested that oestrogens may activate a process by which protein, which is synthesized at a site that is distant from the genome, is made available to the nuclear sites of RNA synthesis.

In putting forward his thesis, Willmer assumed, as a starting point, that steroid molecules do not insert themselves into a phospholipid membrane at random but that the end groups are incorporated in a definite order of preference. Thus, it was assumed that the phenolic hydroxyl group at C_3 in the aromatic ring A of oestradiol would be at the polar surface of the phospholipid membrane, while the hydroxyl group at C_{13} would be associated with the hydrocarbon chains of the phospholipids in the interior of the membrane. It was also suggested that, since all the active steroids are much shorter than cholesterol, four molecules of steroid may be packed end-to-end across the width of a bimolecular leaflet membrane. Willmer commented that, if this occurred, then as many as three hydrophilic pools could develop at different levels within the membrane (Fig. 15). It seems to the writer than an inherent difficulty with both of these concepts is that they result in one or more polar groups being located in the hydrocarbon region of the membrane; on energetic grounds this would seem to be an unlikely occurrence. Furthermore, location of some of the polar groups of steroid hormones within the interior of the membrane would appear to preclude participation of these groups in complexes involving, for example, hydrogen bonds with the polar groups of other molecules in the aqueous environment of the membrane. These difficulties can be avoided, however, by considering the possible interactions of steroid hormones with a membrane in which the lipids can be induced to form local areas of micellar configuration. Before attempting to do this, it is necessary to discuss the experimental evidence that is available on the interfacial orientation of steroid hormones.

Munck (1957) has studied the behaviour of a number of steroid hormones at the water-heptane interface. With tightly packed molecules of testosterone at the interface, the experimentally determined area per molecule could only be explained by supposing that the testosterone molecules form a monomolecular layer and that the molecules lie more or less flat in this layer. Within the accuracy to which the dimensions could be determined, these

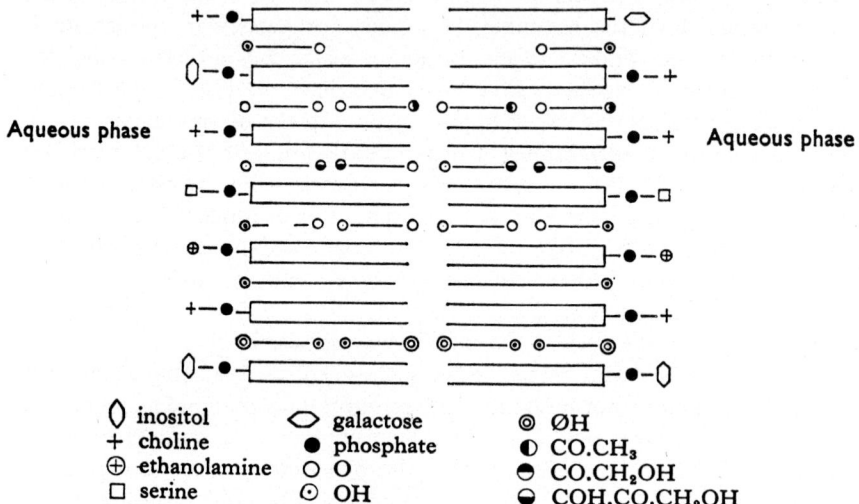

FIG. 15. The main phospholipids and steroids as they may pack in a bimolecular layer system, according to the suggestions of Willmer (from Willmer, 1961 with permission of Cambridge University Press).

conclusions applied equally to progesterone and to deoxycorticosterone, while hydrocortisone appeared to lie "on edge" in the interface. With diethylstilboestrol, the planes of the aromatic rings were apparently parallel to the interface. Munck concluded that, since almost all known steroid hormones and synthetic oestrogens have a large hydrocarbon nucleus with polar groups at each end and sometimes additional polar groups at other positions, it would be anticipated that all these molecules would adsorb at interfaces between polar and non-polar media and that the molecules would lie flat or on edge in the interface.

With regard to orientation at the air-water interface, molecules which have only one polar group, like cholesterol, pack vertically in monolayers. In contrast, as was shown by the experiments of Danielli et al. (1933), a steroid molecule will lie flat at the interface if there is a water-attractive group at each end of the molecule. Taylor and Haydon (1965) have recently investigated the interaction of progesterone with lipid films at the air-water interface. They observed that close-packed monolayers of cholesterol and L-α-(dipalmityl)-lecithin can absorb a small amount of progesterone. The amount however, was so small that it was not feasible to determine whether the progesterone molecules were orientated parallel to or at right angles to the interface. Nevertheless, since the results suggested that progesterone may be absorbed from an aqueous phase into a bimolecular lipid leaflet they were regarded as being consistent with Willmer's hypothesis.

It must be recognized that to proceed by analogy from the orientation of individual steroids in monolayers at the oil-water or air-water interfaces to their possible orientations in biological membranes is a considerable extrapolation. Specific complexes may occur in membranes and thereby invalidate such arguments. Nevertheless, it would seem probable that the orientation in membranes of steroid hormones with two or more polar groups would, in general, be more closely related to the behaviour of these same compounds in monolayers than to the behaviour of a steroid like cholesterol that has only one polar group. It is therefore suggested that steroid hormones that are physiologically active may be orientated with their lengths parallel to the lipid-water interface in biological membranes and not at right angles to it. Orientation in this way would maintain maximum contact between all the polar groups of the steroid molecules and the aqueous phase and would avoid burying polar groups within the hydrocarbon interior of the membrane. In connection with this suggestion, it is of interest that recent studies have shown that nearly all the polar side-chains of horse oxyhaemoglobin are in contact with water, either on the surface of the molecule or in the internal cavity. The large non-polar side chains lie either in the interior of the individual subunits, or in superficial crevices so designed as to minimize contact of these side-chains with water, or at the point of contact between unlike subunits (Perutz, 1965).

The orientation of steroid hormones parallel to the interface between hydrocarbon and water in a bimolecular leaflet model for biological membranes would not appear to provide any mechanism of action for steroid hormones in the control of membrane permeability. In contrast, location of hormone molecules, either flat or on edge, at the hydrophobic-hydrophilic interface of a globular micelle of phospholipid in an area of micellar membrane may enable the polar groups of the lipid-soluble steroids to interact with ions and polar molecules passing through the pores between the phospholipid micelles (Fig. 16). The presence of steroid molecules would be expected to favour the local formation of an area of micellar membrane (see Section IV) (Lucy, 1964a; Haydon and Taylor, 1963). Thus, the hormone molecules would re-orientate the "cytoskeleton", in accordance with Sir Rudolph Peter's original hypothesis on the mechanism of action of hormones

(Peters, 1956). It is also interesting to note that the lengths of steroid hormone molecules are such that it seems possible that the polar groups at each end of a hormone may project into two adjacent pores, as depicted in Fig. 16. An implicit requirement for this model, which the writer suggested at a conference on hormones held in Princeton in February, 1966 (Lucy, 1967), would seem to be that the phospholipid micelles have an essentially liquid hydrocarbon interior in order to facilitate the packing of the steroid molecules at the interface.

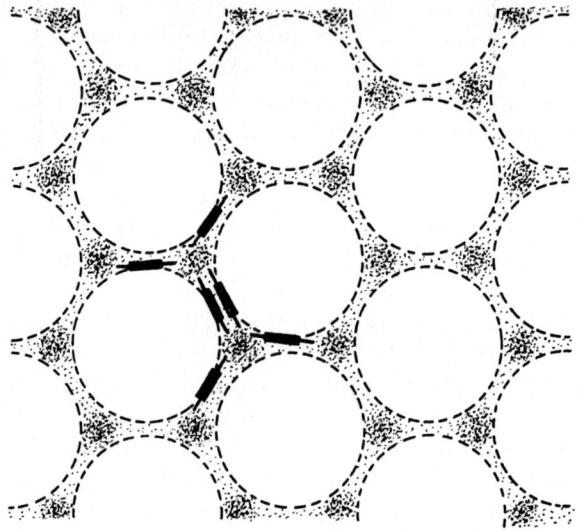

Fig. 16. A diagram to illustrate a possible mode of action of steroid hormones in an area of micellar membrane (Lucy, 1967) with permission of the New York Academy of Sciences). Each hormone molecule lies flat or on edge at the hydrophobic-hydrophilic interface of a phospholipid micelle. The polar groups at the ends of the steroid molecules are associated with the aqueous pores that penetrate through the thickness of the lipid moiety of the membrane between the micelles of phospholipid. These polar groups interact with ions and molecules passing through the pores. In this way, the permeability of membranes for polar substances may be controlled by the functional groups of lipid-soluble steroid hormones.

The molecular specificity, which is such an intriguing feature of the actions of steroid hormones and which does appear to be associated in some way with the polar groups of the different steroid molecules, may perhaps depend on the occurrence of specific ternary complexes between the polar groups of hormones and polar substances passing through the pores. Thus, if A and B represent the polar groups at each end of one particular steroid and substance X (having more than one binding site) is travelling through a pore, three different ternary interactions are possible between two molecules of the steroid and one molecule of substance X, viz, $A-X-A$; $B-X-B$ and $A-X-B$. These interactions may either facilitate or hinder the passage of substance X through the pores, and hence through the lipid moiety of the membrane, according to the sterochemistry and the dissociation constants of the complexes formed. Molecules of other steroid hormones, in which one or both of the polar groups are different from those in the original hormone, would be unable to form the specific complexes necessary to control the passage of substance X but would interact with other substrates.

IV. Phase Changes Within Membranes

Many investigators have been attracted by the idea of transitions in lipid structures occurring within membranes. If the bimolecular leaflet structure is the configuration that is present for most of the time in most membranes, phase transitions of varying degrees of reversibility may result in the formation of other configurations of lipid that have varying degrees of stability It is possible, for example, that a transition of a fairly permanent nature may be responsible for the properties of the junctional membranes in epithelial cells (salivary gland, renal, urinary bladder, sensory and liver cells) which have been shown to be highly permeable structures (Loewenstein, 1966). The conductance of such junctions is, in all cases, at least four orders of magnitude greater than that of the non-junctional surface membranes.

Luzzatti and Husson (1962) pointed out that highly developed polymorphism is a common feature of lipid-water systems. They also speculated that, in a cell membrane, the hydrocarbon chains are normally liquid but that physiological conditions may be near the borderline between the liquid crystalline structure and a coagel. When one parameter (concentration, temperature, electric potential) is altered, the hydrocarbon chains of the phospholipids may then crystallize and so block some physiological activity of the lipid such as membrane permeability.

It has been proposed by the writer that the adaptability and versatility of biological membranes may result from the utilization of both globular micelles and bimolecular leaflet arrangements of lipid molecules *in vivo* (Lucy, 1964a). The differences between one cell type and another, and the variations in properties of different areas of plasma membrane within the same cell, may depend upon the differences between the properties of the micellar membranes and the bimolecular leaflet. For example, in addition to the properties of the micellar structure discussed above, membrane flexibility may be associated with micelles and rigidity with the bimolecular leaflet. This general concept would appear to be supported by the recent observations of Branton (1966), which are referred to above in Section III-A on the structure of membranes studied by the freeze-etching technique.

On theoretical grounds, Haydon and Taylor (1963) have proposed that, as the proportion of non-phospholipid molecules is increased within an array of phospholipids, the bimolecular leaflet structure will ultimately become unstable. It has therefore been suggested that substances like vitamin A or steroid hormones, on incorporation into the lipids of biological membranes, may induce transitions from the lamellar to the micellar configuration (Lucy, 1964a). Small variations in molecular size, shape, hydrophilic–hydrophobic balance, lipid solubility and electrostatic charge in a series of closely related molecules, such as the steroids, might be expected to produce relatively large effects in a membrane that possesses the ability to change from a lamellar structure to a micellar structure.

Pethica (1967), in a recent discussion of the question of the degree of liquidity of the hydrocarbon chains in liquid-crystalline and micellar systems (cf. Luzzati and Husson, 1962; Segerman, 1965), has concluded that the balance of evidence is against the presence of randomly coiled chains in membranes. He considers that the chains are probably semi-solid and rather straight and that they may exhibit a high degree of ordered packing while, at the same time, being sufficiently liquid to allow of chain rotation. In his view, sub-structure transformations depend to a large degree on the packing configurations of the hydrocarbon moieties of the mixed lipids. Pethica also suggests, as an extension of the writer's speculations, that the bimolecular leaflet structure may correspond to a more solid interior, and the micellar form to a more liquid interior for membranes, thus associating the hypothetical lamellar-micelle transition with a change in the structural organization of the hydrocarbon chains of the phospholipids.

Kavanau (1963) has proposed that biological membranes may be visualized as protein-lipoprotein-lipid complexes capable of existing in, and readily transforming between, several equilibrium states. One state, the "closed" configuration, consists of the bimolecular lipid leaflet, 50 to 60 Å thick, sandwiched between protein monolayers. Another state, the "open" configuration, consists of a regular hexagonal arrangement of cylindrical micelles (80×180–200 Å) at a centre-to-centre spacing of about 150 Å. It was proposed that transformations between these two states result in cytoplasmic streaming and that this theoretical model might be the basis for protoplasmic streaming. There would, however, seem to be little directly relevant evidence, either from studies on biological membranes or from investigations on model, lipid-water systems, in support of the presence of cylindrical micelles of lipid, 80 Å in diameter and 180 to 200 Å long, occurring in areas of membrane having a cross sectional thickness of approximately 200 Å. In addition as previously remarked (Lucy, 1964a), contrary to Kavanau's interpretation, the findings of Dourmashkin et al. (1962) on saponin-treated membranes also appear not to provide support for his hypothesis concerning the structure of membranes.

An interesting theory has been proposed by Beament (1964) in relation to the problem of the uptake of water by insects from sub-saturated air. New lipid is continuously secreted by the cockroach and insects are, in general, able to repair their surface by secreting lipid (Beament, 1955). It seems reasonable to suppose that the newly synthesized lipids are transported to the surface of an insect in the form of micelles dispersed in water. Filaments, 60 to 130 Å in diameter, have been observed in the epicuticle of *Calpodes ethlius* by means of the electron microscope (Locke, 1965) and, since several pieces of evidence point to the presence of pores in the lipoprotein substrate, Beament suggests that these structures may be pores that contain water and dispersed micelles of newly synthesized lipid. At the mouth of each pore, there would be an orientated monolayer of lipid spread across the surface of the water and this monolayer would be continuous with the fixed monolayer of lipid on the surface of the insect cuticle. When the insect is in dry air, the pores will normally be full of water but, as there is a monolayer of lipid at the air-water interface, evaporation of water from the insect will be low. In contrast, if a drop of water is placed on the cuticle, a continuous water path may be developed between the external water and the water in the pores; water may thus enter the insect if internal suction is applied (Fig. 17). The formation of the continuous pathway of water (which necessarily results in a discontinuity in the surface lipid) is postulated to result from, firstly, the formation of an orientated bilayer of lipid on the surface due to the presence of the external water and, secondly, the equilibrium between this bimolecular leaflet and the globular micelles of lipid present in the water which fills the pores. However, as Beament remarks, it remains difficult to see why a cockroach that is actively taking in water does not take in ions and small water-soluble particles through the "open" pore and he raises the questions of whether emulsion micelles act as filters and whether the charge around the lip of the pore due to the polar groups of the lipids is sufficient to keep ions out. Furthermore, the mechanism by which the insect sucks water into the pores remains obscure, although this may be a further manifestation of capillary phenomena.

Locke (1965), himself, considers that the filaments that he has observed in the epicuticle of insects are filaments of wax. He suggests that the material in the filaments is liquid-crystalline lipid in the middle phase (Luzzati and Husson, 1962), and that newly synthesized polar lipids are continuously added to the inner ends of the wax filaments as the surface wax is utilized.

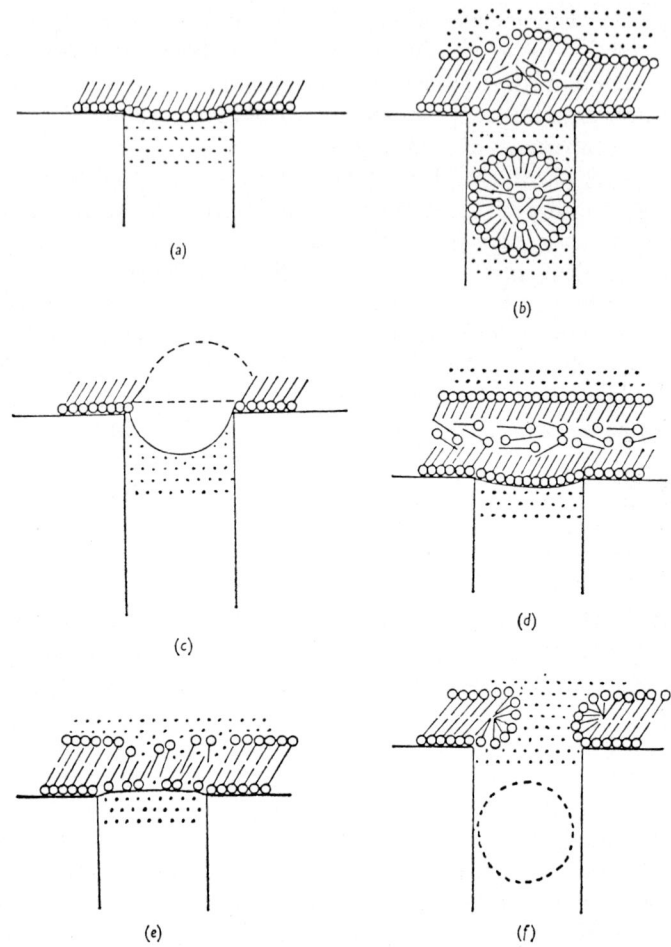

FIG. 17. Models of molecular behaviour at the mouth of a pore, about 200 Å in diameter, present at the surface of insect cuticle. (a) Pore filled with water, covered by a monolayer continuous with that on solid cuticle. (b) Water over the grease, forming an inverted monolayer, squeezing out grease, and forming a double layer over solid cuticle; a potential micelle over the pore. A micelle of grease in the pore. (c) Diagram indicating the range of meniscus shapes available for very small changes in level along the molecules of lipid lining the lip of the tube. (d) Water over thicker grease; an earlier stage than (b). (e) The double layer over the pore becoming unstable rather than forming a micelle. (f) An open pore indicating the possible form of lipid round the lip. The polar groups would create a strong electrical field at the orifice (from Beament, 1964 with permission of Academic Press).

Locke also proposes that the variable permeability of insect cuticle to water under differing conditions may be due to phase changes in the liquid crystals occurring within the 60–130 Å channels leading to the surface.

V. Experimental Models for Membranes

A. MOLECULAR INTERACTIONS IN MONOLAYERS

Monolayers of lipid molecules at the air-water or oil-water interface provide a valuable means of investigating molecular interactions in an orientated system. Monolayers have therefore been used extensively to study the properties of membrane-active substances (Lucy, 1964b) which, when present in only microgram quantities *in vivo*, have far reaching biological effects that result from interactions with lipoprotein membranes. Thus, Skou (1958) found that there is a close parallel between the ability of both tertiary amine anaesthetics and various alcohols to block nervous conduction and the capacity of these different substances to penetrate into, and increase the pressure of, a monolayer of lipids from nerve tissue. Skou's work stimulated a number of comparable studies. For example, Bangham *et al.* (1962) have investigated the penetration of lipid films by compounds preventing liver necrosis in rats treated with carbon tetrachloride. They observed a correlation between the protective activity of a given compound and its interaction with a lecithin/cholesterol mixture at both the air-water and lipid-water interfaces. In addition, three simple long-chain amphipathic cations, cetyltrimethylammonium bromide, docosanyl pyridinium bromide and stearyl amine, which interacted strongly with the lipid films were highly effective in preventing leakage of hepatic β-glucuronidase into the sera of rats treated with carbon tetrachloride.

The penetration of monolayers by compounds in the vitamin A series has been investigated by Bangham *et al.* (1964) in the light of the actions of retinol on erythrocytes, lysosomes, mitochondria and other membrane-bound organelles (Dingle and Lucy, 1965). Retinol was the only compound of those tested that was able to cause both a large increase in surface pressure at constant area and also a large increase in the area of a lecithin-cholesterol film at a constant pressure of 30 dynes/cm. In addition, it was found that retinol interacts more strongly with lecithin than with cholesterol. It was proposed that the penetration of lipoprotein membranes may be an initial step in certain of the actions of the vitamin in biological systems and that the molecular specificity seen in the model system would appear to go a considerable way towards providing an explanation for the molecular specificity of vitamin A observed in its action in animals, in organ culture, on erythrocytes and fibroblasts *in vitro* and on isolated subcellular particles.

It is of interest to compare these studies on vitamin A with investigations made by Demel et al. (1965) on the penetration of lipid monolayers by polyene antibiotics. Filipin and nystatin readily penetrate monolayers of cholesterol and ergosterol at initial surface pressures greater than the collapse pressure of the antibiotics. Under the same conditions, there was essentially no interaction with a variety of pure synthetic phospholipids unless a sterol was present; this result contrasts with the above observations made with vitamin A. It is significant, furthermore, that vitamin A causes lysis of protoplasts of *Bacillus megaterium* (Kinsky, 1963; K. McQuillen, A. M. Glauert and J. A. Lucy, unpublished observations), while polyene antibiotics are without effect. Thus, while retinol appears to require phospholipids for its initial interaction with membranes, polyene antibiotics apparently interact specifically with sterols (but see Sessa, 1966; Section IV-E below). Demel et al. have suggested that the polyene antibiotics produce a reorientation of the sterol molecules in membranes and a consequent alteration in cell permeability; the haemolytic action of polyene antibiotics may therefore be similar to that of saponin.

Investigations of the interactions in monolayers of membrane lipids with haemolytic agents reveal not only the value of the monolayer technique but also its limitations. Rideal and Schulman (1939) demonstrated that it is possible to analyse the various factors that govern adsorption by, or the penetration into, a monolayer of a given structure. In addition, as Rideal and Schulman pointed out, considerable molecular specificity results from the principle that both polar and non-polar interactions apparently participate in molecular adlineation in monolayers. There is no doubt that a positive interaction between a lytic substance and a monolayer of membrane lipids, like that which was observed at the air-water interface between cholesterol and saponin by Schulman and Rideal (1937), provides strong evidence that a direct molecular action of the lytic agent on the erythrocyte membrane is responsible for the haemolysis and that other more indirect mechanisms, such as metabolic changes, are unlikely to be involved. Nevertheless, twenty-five years were to elapse before electron microscopy of saponin-treated erythrocytes (Dourmashkin et al., 1962) led to more detailed information on the macromolecular changes that are associated with the action of saponin on the lipids of membranes (Glauert et al., 1962; Lucy and Glauert, 1964b; Bangham and Horne, 1962; Husson and Luzzati, 1963).

As yet, despite the data obtained from the monolayer experiments, the detailed mechanisms by which vitamin A and the polyene antibiotics cause haemolysis remain obscure. With regard to the former, it is not known whether retinol is haemolytic as a result of configurational changes which it may produce in the organization of the lipids of the red cell membrane (such as the formation of extensive areas of micellar membrane; Lucy, 1964a), or whether the membrane is damaged by reactions associated with autoxidation of retinol that may occur within the membrane (Lucy and Dingle, 1964; Lucy, 1965; Lucy, 1966). Similar considerations would seem to apply to the use of monolayers in other studies. For example, van Deenen and Demel (1965) have recently made the interesting observation that there is a very strong interaction between monolayers of gangliosides and the psychoactive drugs, orphenadrine and chlorpromazine but not between these drugs and lecithin, phosphatidylethanolamine or sphingomyelin. This finding may well prove to be very important in the determination of the mechanism of action of these drugs but further investigations will no doubt need to be undertaken with a variety of techniques before the nature and possible functions of the complexes which the drugs form with gangliosides becomes apparent. It seems conceivable, as has been suggested in connection with the physiological actions of vitamin A (Dingle and Lucy, 1965), that penetration into membranes is simply the initial step in the action of membrane-active drugs and that the drugs subsequently participate in specific reactions occurring at the sites of the membranes concerned.

Following the studies of Skou (1958) on the interactions of local anaesthetics with lipid monolayers, the effects of narcotic gases on the surface of lipoprotein-covered water have been investigated by Clements and Wilson (1962) who observed a significant interaction. These workers were, however, unable to conclude whether the interactions occurred primarily between lipid film and narcotic or between water and narcotic facilitated by the presence of the lipid. Balasubramanian and Wetlaufer (1966) have recently studied reversible alterations in the structure of globular proteins induced by general anaesthetic gases. They investigated changes in optical rotation in aqueous solutions of β-lactoglobulin and bovine plasma albumin and obtained results which indicated a correlation between the potency of a general anaesthetic and its effect on protein structure. These workers consider that their findings are significant in relation to the mechanism of general anaesthesia, as well as in the production of side effects and they believe that the observations of Clements and Wilson should be re-evaluated to admit that an aqueous interface is not an essential condition for significant interactions.

Haydon and Taylor (1963) have considered the relevance of monolayer penetration studies to adsorption into bimolecular lipid leaflets from a general thermodynamic viewpoint and they have concluded that the air-water monolayer is not entirely satisfactory as a model, although the deficiencies are of a quantitative rather than a qualitative nature.

B. MONOLAYERS AS MODEL MEMBRANES

Although monolayers of lipid molecules at an air-water interface, or at an oil-water interface, can be used to considerable advantage for the evaluation of the chemical and physical interactions between molecules, such films are less satisfactory as structural models for biological membranes. Some of the reasons for this are obvious enough: a monolayer consists of only one layer of lipid molecules which lies at the interface between water and air, or between oil and water, and it does not separate two aqueous phases. Hence, factors that affect the stability of the lipid film in relation to the macromolecular arrangement of its constituent molecules are probably quite different in the monolayer from what they are in natural membranes.

The difficulty of extrapolating from monolayers to membranes may be illustrated by a consideration of the permeability of natural membranes and of monolayers to water. Thus, it is sometimes suggested that the bimolecular leaflet of lipid as a theoretical model for the structure of biological membranes does not need to be modified by the introduction of pores to account for the permeability to water, in view of the fact that water is known to pass through a monolayer of lipid at an air-water interface. However, Shanes (1963) has pointed out that the ratio of the penetration rate of water in monolayers to that in the squid giant axon membrane is approximately 1000:1. The lack of correspondence between the behaviour of water passing through monolayers and the permeability properties of natural membranes was analysed concisely by Danielli (1962) during one of the discussions at a symposium on the plasma membrane. Danielli pointed out that a major potential energy jump is involved when a water molecule passes from the surface of the water into the gas phase. A monolayer of material which is spread at the interface will not have any effect on the rate of evaporation unless the potential energy change involved in passing from water, through the monolayer, into the gaseous phase is significantly greater than that in passing directly from the surface of the water into the gas phase. Thus, only if the monolayer is exceptionally impermeable to water, and water finds it difficult to get into the monolayer, will there be any effect of the monolayer on evaporation. This situation is not equivalent to the impedance to movement when there is water on both sides of a thin lipid layer, since what is then measured is the resistance to diffusion at the rate-limiting position, which may be at the lipid-water interface or in the interior of the membrane.

For a detailed consideration of the retardation of the evaporation of water by monolayers, the reader is referred to the recent review by LaMer and Healy (1965).

Pethica (1964) and his colleagues have made surface potential and force-area measurements on monolayers of saturated cephalins at the air-water interface and have used the data obtained in this way in an analysis of the orientation of the polar groups of lecithins and cephalins at the interface. Their observations indicated that the positive and negative ionic charges of the polar groups of the phospholipid molecules lie in a plane, which is parallel to the surface of the monolayer and normal to the direction of the close-packed hydrocarbon chains. This conclusion is consistent with that reached by Hanai *et al.* (1965a) in studies on the orientation of the polar groups in a bilayer of lecithin (see Section V-D). It was estimated by Pethica that the normal displacement of the positive and negative charges in the phospholipid zwitterions from the plane may be considered to be very small in a spread monolayer (less than 0·1 Å on a time average). Calculations for a sheet of phospholipid molecules indicated that coulombic interactions between the zwitterions is a major factor in stabilizing the sheet, while cohesion between sheets arranged back-to-back to form bimolecular leaflets results from the energetically unfavourable interaction of the paraffin chains with water. With regard to interactions between phospholipid and cholesterol, Pethica suggests that the introduction of cholesterol into a phospholipid sheet will, in general, reduce the coulombic cohesion of the sheet and that it may also reduce van der Waal's stability. As judged by the monolayer experiments of Pethica's group, only those phospholipids containing double bonds interact positively with cholesterol. van Deenen *et al.* (1962) have also observed that the effect of cholesterol in reducing the area per molecule occupied by lecithins is much greater when cholesterol is mixed with unsaturated lecithin molecules than with saturated lecithins. Pethica therefore proposes that the interaction of phospholipids with cholesterol is not principally due to van der Waal's interactions, as suggested by Vandenheuvel (1963) for his model of the lecithin-cholesterol complex (see Section II-B above), but that unsaturated bonds may be particularly involved. Further investigations on the role of double bonds in these systems is clearly required.

It is interesting that Pethica and his colleagues found that the surface potentials of mixed cholesterol-cephalin monolayers at the air-water interface were close to the porportionate mean of the surface potentials of the monolayers of the pure components. This result implies that the phospholipid zwitterions and the dipoles of the cholesterol are not mutually distorted in the mixed monolayer. Pethica comments that this is strong evidence against the presence in the mixed monolayer of either Finean's "walking stick" configuration for cephalin-cholesterol, Fig. 4, (Finean, 1953) or the cholesterol-phospholipid arrangement suggested by Vandenheuvel (1963), Fig. 5 (see Section II-B). Both the Finean and Vandenheuvel models would give large surface dipole moments since these models require that the cationic and anionic centres of the phospholipid zwitterions are not tangentially coplanar.

In the Danielli-Davson model, it was suggested that the layer of protein immediately adjacent to the bimolecular layer of lipid might be in an extended configuration in view of the fact that protein denaturation occurs at an oil-water interface (Davson and Danielli, 1943). Eley and Hedge (1956) obtained results in experiments with the Langmuir trough on the interaction between lipid monolayers of stearic acid and cholesterol, and proteins (plasma albumin, fibrinogen, lysozyme and insulin) which supported this suggestion. They found that, in each system, the first sublayer of protein apparently consisted of denatured protein while there were indications that the second sublayer was native protein. The first layer of protein may not be in an extended form, however, if the protein molecules are in contact with the polar groups of phospholipid molecules. Investigations by Malcolm (1962) have indicated that the α-helix may be the stable configuration of proteins at the air-water interface and, by implication, at the oil-water interface. Malcolm suggests that the unfolding observed in proteins at interfaces is often simply a loss of tertiary structure. He has investigated the possible presence of the α-helix in monolayers of synthetic polypeptides at an air-water interface by following deuterium exchange in N-deuterated polypeptides. In solution, peptide deuterium exchanges for hydrogen rapidly when the peptide groups form hydrogen bonds to solvent molecules but exchange is very slow when the polypeptide is in the α-helical conformation. The polymers studied were poly-D-α-amino-n-butyric acid, poly-DL-leucine, poly-γ-methyl-L-glutamate, poly-γ-ethyl-L-glutamate and poly-γ-benzyl-L-glutamate which were spread as monolayers on either N/100 hydrochloric acid or N/100 sodium hydroxide and subsequently swept off for spectroscopic examination. Malcolm reported that the infra-red spectra of the films were consistent with the general presence of the α-helix in the monolayers; the α-helix was apparently less stable on an alkaline than on an acid substrate. These experiments are of considerable intrinsic interest but there are many differences between the system studied and the situation in natural membranes. Nevertheless, as Malcolm points out, the possibility that the α-helix may be stable at the lipid-water interface should be seriously considered in relation to the questions of specific interactions between the lipids and proteins of membranes.

C. MACRO-MODELS FOR MEMBRANES

The value of monomolecular layers of lipids as models for membranes derives mainly from their extreme thinness but, as we have seen, they have the disadvantage of not separating two aqueous phases. Many investigators have attempted to devise model membranes which do act as a barrier between two aqueous phases and membranes made from a variety of "unnatural" materials, such as collodion, have been used in these studies. Until the development of the bilayer membranes of phospholipid, which are considered in the next section (V-D), membranes separating two aqueous phases had the disadvantage of being very thick as compared with natural membranes. Such systems might perhaps be described as macro-models for membranes. Despite the disadvatange of relatively great thickness, however, much useful information relevant to the properties of natural membranes may be obtained from a study of this type of model, as is illustrated in this section.

Tobias *et al.* (1961–62) have described a cephalin-cholesterol model membrane, the electrical resistance of which can be reversibly raised by calcium chloride and lowered by potassium or sodium chloride. The lipid barrier used in their experiments was made by dipping a Millipore filter disc (cellulose nitrate-cellulose acetate material), having a pore size of 100 Å ± 20 Å and a thickness of 0·15 mm, into benzene containing lipid and then allowing the benzene to evaporate. Usually, 1·0 to 1·8 mg of lipid was deposited per cm². The lipid barrier was clamped between two plastic vessels which contained the solutions and electrodes required for measuring the electrical properties of the barrier. Experiments with this model at controlled pH values indicated that the effects of cations on the resistance of the membrane were dependent on combination of the cations with the acidic groups of the phospholipids. In addition, the changes in resistance were apparently correlated with decreased hydration of the membrane in the presence of calcium ions and increased hydration in solutions of sodium or potassium ions.

Nash and Tobias (1964) have published further studies on this experimental model which have illustrated the potential importance of the phosphatidylserine of membranes in relation to calcium ions. They found that the equivalent d.c. resistance of the Millipore membrane impregnated with phospholipid increased with exposure to increasing concentrations of calcium ions when the phospholipid was "animal cephalin" or phosphatidylserine but not when the membrane contained phosphatidylethanolamine or lecithin. Further experiments on the limiting conductivity of the membrane with decreasing ionic strength supported the suggestion that the membrane containing phosphatidylserine behaves as a cation exchanger for calcium. Nash and Tobias suggest that their work indicates that the phosphatidylserine present in cell membranes may provide negatively-charged sites for cation exchange, that preferential uptake of calcium by this phospholipid could account for the high concentration of the ion usually attributed to biological membranes and that certain physiological properties of the cell membrane previously ascribed to phospholipids in general may now be more accurately attributed to phosphatidylserine in particular. They have emphasized, however, that the membrane model is very thick relative to biological membranes. The absence of differential mobilities for sodium and potassium ions, and the lack of a non-linear or threshold response to the passage of current, also indicates that the model is not a perfect biological analogue and that it can only indicate some of the possibilities of how phospholipids might act in cell surfaces.

In the light of these and earlier experiments, Tobias (1964) has put forward a hypothetical molecular mechanism for nerve excitation. He has proposed that calcium ions, which may normally be associated with the polar groups of the phosphatidyl serine in the axon membrane, are displaced by potassium ions and that this displacement leads to increased hydration of the membrane and deformation of membrane proteins. These molecular changes are postulated to underlie the reversible change in membrane structure which expresses itself as a fall in resistance and increased permeability to sodium and potassium ions.

Studies of interfacial phenomena at the oil-water interface may perhaps be regarded as approaching one step nearer to biological membranes than the air-water interface in which one phase is gaseous. Colacicco (1965a) has recently re-investigated electrical potentials at the oil-water interface by means of an experimental system which may be regarded, in essence, as two interfacial monolayers arranged back-to-back and separated by a relatively thick layer of oil. This double-monolayer is presumably inherently much more stable than the bilayer system that is discussed below in Section V-D but the 10 mm layer of oil-phase prevents this system from being regarded as a satisfactory experimental model for the structure of cell membranes. Nevertheless, this disadvantage does not preclude its use in the study of interfacial phenomena (Colacicco, 1965b). In his experiments, Colacicco used a layer of water-saturated pentanol to separate a lower from an upper electrolyte solution. Changes of potential at a given interface were studied on the injection of quantities of surfactant at that interface without disturbing the other. A sudden change of potential ensued on injection and the new potential decayed at a rate depending largely on the surfactant's solubility and disappearance from the interface. It was found that the sign of the potential was determined by the sign of the fixed head charges of the amphipathic molecules as they appeared at the oil-water interface. The change of potential was smallest in 1 M potassium chloride and increased with decreasing electrolyte concentration; the largest potentials were obtained when charges of opposite sign lined the opposing interfaces in the absence of electrolytes. With sodium dodecyl sulphate at one interface and cetyltrimethylammonium bromide at the other, ΔV reached the value of 370 mV in the absence of salts. In contrast, a chromatographically homogenous sample of egg lecithin produced a zero change of potential. This latter finding is consistent with the observations of Anderson and Pethica (1956) on monolayers of synthetic lecithin, at the air-water interface, which demonstrated that lecithin is isoelectric over a wide range of pH.

D. BILAYERS AS MODEL MEMBRANES

A few years ago, Mueller et al. (1962a,b) reported experiments on the formation in vitro of very thin lipid membranes, up to 10 mm² in area, from the lipids of white matter. By forming such a membrane across a small hole in a barrier between two compartments filled with saline, its transverse electrical properties could be measured and controlled chemical investigations undertaken. Mueller et al. reported that the membranes studied were grossly manipulable, resilient, self-sealing to puncture, liquid in the plane of the layer, stainable with osmium tetroxide and 60 to 90 Å thick. The electrical capacity of the membrane was about 1 μF/cm²; its resistance was regularly found to be 10^7 and often greater than 10^8 ohm cm². Below pH 5·0, and above pH 9·0, the membrane was unstable. Mueller et al. assumed that this type of membrane has the structure of a bimolecular lipid leaflet.

The bilayer membrane system is, for many purposes, a considerably better experimental model for biological membranes than a monolayer of lipid molecules. Membranes of the type described by Mueller and his colleagues not only separate two aqueous phases but in addition the dimensions of bilayer membranes, unlike those of the macro-models discussed above, are closely similar to those of biological membranes. The attractiveness of the bilayer system, despite its reputation for requiring some manipulative expertise, had led to such films being studied extensively by several investigators. It is to be expected that an increasing number of workers will turn to this technique in the hope of gaining additional information on specific aspects of membrane structure and function. The modified technique described by van den Berg (1965) allows the preparation of films of up to 9 mm in diameter and also simplifies some of the experimental procedures involved in the study of these very thin membranes.

Huang et al. (1964) subsequently undertook a study of the properties of bilayer membranes which, unlike those originally described by Mueller et al., were prepared from defined materials of relatively simple constitution. Huang et al. used purified phospholipid preparations of known analytical composition which were dissolved in chloroform-methanol containing n-tetradecane. It was found impossible to prepare membranes from either tetradecane alone, or phospholipid alone, and the concentration of these two components proved to be critical. However, the actual composition of the very thin membrane that is actually formed may be quite different from that of the original mixture which is brushed over the hole. The membranes produced were apparently 61 $\pm$ 10 Å in thickness, as determined by optical techniques (Huang and Thompson, 1965); they had a resistance of about 10^6 ohm cm² and a surface tension of 0·5 dyne cm⁻¹ for each face. Huang *et al.* concluded that the lipid membranes studied were probably bimolecular leaflets although other arrangement of lipid molecules might not be inconsistent with the available evidence. The values obtained for the thickness of the bilayer membrane by Huang and Thompson (1965) have been queried by Tien (1966a) who has questioned an equation employed by them in the optical determination of membrane thickness. Tien has obtained a thickness of 69 Å, with an estimated error of $\pm 15\%$, for a similar type of membrane. Thompson and Huang (1966) have agreed that an erroneous equation was used in their original studies and that the value for the thickness of their membranes is 72 ± 10 Å.

It should perhaps be remarked at this point that, although there is a marked similarity between the properties of the bilayer membrane and biological membranes, it does not follow, as is sometimes implied, that either natural membranes or the artificial films have the structure of a bimolecular leaflet simply because one resembles the other.

Closely similar experiments to those of Huang *et al.* (1964) have been made by Hanai, Haydon and Taylor (1964) on the electrical properties of membranes containing purified lecithin and n-decane. They found that the thickness of the hydrocarbon part of these films was 48 ± 1 Å. The capacitance of the membranes was accurately reproducible in different experiments (0.38 ± 0.01 $\mu F/cm^2$) but the values obtained for the d.c. conductance, which ranged from about 2×10^{-10} to about 1×10^{-8} mho, were not. Hanai *et al.* considered that this latter finding indicates that there are leaks of varying sizes in different films. They suggested that micro-roughness in the machined teflon used in their experiments, in conjunction with contact angle phenomena, could easily cause small aqueous channels (of the order of considerably lesss than 1% of the total area of the membrane) to be present around the edges of the films. The lowest conductance observed was approximately 10^{-11} mho and it was suggested that it is probably reasonable to take the lower rather than the higher values of the conductance as more truly representative of the properties of the hydrocarbon films. This point has recently been investigated further (Hanai *et al.*, 1965c) in experiments in which the conductance of films having different areas could be determined by means of bulging the film under hydrostatic pressure. Hanai *et al.* subsequently concluded that conductance leakages at the edges of the films frequently completely mask the true film conductances. A maximum value for the conductance of the film of 1.38×10^{-9} mho/cm^2 was obtained.

The permeability to water of lipid bilayers has occupied the attention of both Huang and Thompson (1966) and Hanai *et al.* (1966) and both groups are concerned by the discrepancies between the values obtained for water permeability by radio-tracer and osmotic methods. The former method yields values in the region of 3×10^{-4} cm sec^{-1} and the osmotic method gives values of the order of ten-fold larger. Huang and Thompson consider that the difference between the two values cannot be attributed to stagnant layer effects or to a chemical isotope effect. They suggest that the mechanisms of water movement across the membrane may be different in the two types of experiment and that the inequality of the coefficients could reflect this basic difference. Earlier investigators have proposed that the net water flux observed when an osmotic gradient is applied across a membrane consists of two components: one, the result of diffusion through pores in the membrane and the other, the result of filtration. Huang and Thompson have commented that the simplest explanation of the different permeability coefficients is apparently the existence of pores but that such pores or their equivalents must be transient structures occupying a total area that is only a small fraction of the whole membrane. In contrast, Hanai et al. (1966), although they also consider an isotope effect unlikely to be the cause of the observed discrepancy, have concluded that the existence of water-filled pores in the membrane is improbable, chiefly because of the high electrical resistance of the membranes. Hanai et al. have suggested that the most probable explanation is the existence of relatively thick, unstirred layers of aqueous solution adjacent to the lipid membranes. The resolution of this problem is naturally awaited with interest.

Hanai et al. (1965a) have studied the question of the orientation of the polar groups in the bilayer leaflets of lecithin. They compared measured values for the electrokinetic potentials of lecithin, which were positive and very low, with those anticipated on theoretical grounds. The positive nature of the potentials was expected but the very low values in 0·1 M NaCl were not consistent with the phosphate and trimethylammonium ions being in an extended configuration. It was concluded, therefore, that the polar groups are nearly in one plane parallel to the leaflet (cf. Pethica, 1964; Section V-B above).

Hanai et al. (1965b) have also investigated possible reasons for the difference between the capacitances of cell membranes, which have an average value of 1 $\mu F/cm^2$, and the capacitance (approximately 0·38 $\mu F/cm^2$) of their artificial lecithin/n-decane membrane system. It was observed that inclusion of increasing quantities of cholesterol in the solution from which the bilayer membranes were prepared led to an increased capacitance that probably resulted from an increased dielectric constant of the lipid mixture and which levelled off at a value of about 0·6 $\mu F/cm^2$. The adsorption of proteins, in contrast, had little effect. If the average number of carbon atoms in the fatty acid chains of the phospholipids of cell membranes is less than eighteen, the cell membrane would be thinner than the artificial bilayers and hence have an even greater capacitance than the bilayers containing cholesterol and egg lecithin. Alternatively, Hanai et al. point out that if "polar pores" are present in cell membranes these pores would constitute a capacitance in parallel with and additional to the capacitance of the lipid. They conclude that it is not difficult to account for the basic electrical properties of cell membranes in terms of a lipid bilayer containing a small number of conducting channels. Since only small areas of the membrane (about 2%) would need to be polar (cf. Lucy, 1964a), the overall resistance of the membrane would be high but considerably lower than the true value of the resistance of the experimental bilayer membrane.

Tien (1966b) has most recently reported a modified technique for the preparation of bilayer membranes and using this procedure he has found that, although freshly recrystallized cholesterol does not give stable membranes, highly stable black membranes are easily formed from cholesterol that has been oxidized by molecular oxygen. In addition, 7-dehydrocholesterol was observed to form stable black membranes. These membranes were estimated to be less than 50 Å thick. Thus, phospholipids are apparently not indispensable for the formation of stable bilayer membranes. Tien comments that the finding that a cholesterol derivative alone can form stable bilayer membranes may necessitate a modification of bimolecular leaflet models which assume membranes to be composed of cholesterol and/or phospholipids. It might perhaps be added, however, that there is no *a priori* reason why the formation of bimolecular leaflets should be restricted to phospholipids. Any substance, or complex of substances, in which the hydrophobic and hydrophilic moieties are appropriately balanced with regard both to size and location, might behave similarly. It is relevant that Dervichian (1946) has remarked that lecithin swells, although it does not dissolve readily in water, because its two insoluble hydrocarbon chains counter-balance the hydrosoluble character of the phosphorylated choline group. He has reported that myelin figures can be formed not only by lecithin but also by cholesterol in mixtures with lysolecithin, sodium cetylsulphate or stearylcholine hydrochloride.

The use of bilayer membranes as a technique for the study of specific problems is illustrated by the experiments of Seufert (1965) on induced permeability changes. He has studied the effects of anionic, cationic and non-ionic detergents on the electrical properties of a membrane similar to that used by Mueller *et al.* (1962a) in the presence and absence of a salt gradient across the membrane. All three types of detergent decreased the membrane resistance within a few minutes by at least a factor of a hundred and also produced potential differences across the membrane in the absence of a salt gradient. In the presence of a salt gradient, anionic and non-ionic detergents generated a potential but only with anionics did a steady potential difference persist across the membrane for more than an hour. Seufert suggests that the action of the detergents could be explained by the occurrence of a phase transition from a bimolecular leaflet to a localized micellar arrangement. Localized micelles with a negative charge thus produced may form a fixed charge, cation-selective pore system.

The potentialities of the bilayer technique are further illustrated by the interesting experiments of Castillo et al. (1966) who have used the films in experiments with antigen-antibody and enzyme-substrate reactions. Using membranes, the composition of which is critical but unfortunately not defined, these workers observed that the electrical impedance of the films decreased markedly and reversibly when interactions between antigen and antibody, or enzyme and substrate, appeared to occur on the surfaces of the lipid membranes. van Zutphen et al. (1966) have recently studied the effects of polyene antibiotics on bilayer lipid membranes. They found that filipin and nystatin did not affect the stability of bilayers containing lecithin but no cholesterol. However, films prepared from an equimolar solution of lecithin and cholesterol in decane were disrupted shortly after addition of the antibiotics. These observations are consistent with those made earlier by Demel et al. (1965) on the interactions in monolayers (see Section V-A above). In recent studies on interactions between a protein from beef erythrocyte ghosts and bilayer membranes, Maddy et al. (1966) have observed effects of the protein on the physical properties of the membranes which are not given by a number of other proteins.

E. DISPERSIONS OF PHOSPHOLIPIDS

Electron microscopy of phospholipids dispersed in an aqueous environment has already been discussed above in Section III-A.

A number of interesting papers have recently been published by Bangham and his colleagues on the diffusion of ions from within swollen phospholipid structures to an external aqueous environment. These workers have employed an experimental system consisting of dispersed particles of lecithin containing up to 15% of dicetyl phosphoric acid (Bangham et al., 1965b). Birefringence data indicated that ovolecithin in 0·145 M KCl swells to form aqueous compartments which increase in thickness as a direct function of the surface charge imparted by the long-chain anion. The presence of a net charge, whether positive or negative, appears to cause a separation of adjacent layers of phospholipid. The amount of a cation, e.g. K^+, that remained associated with the swollen phospholipid structures, even after prolonged dialysis, was finite for pure lecithin and increased directly as a function of the surface charge, whether positive or negative. Bangham et al. (1965b) suggest that this indicates that the cations taking part in the original swelling process are restricted in their subsequent diffusion out of the lipid, i.e. they are "captured", and that the amount captured is proportional to a physical property of the lipid structure rather than to any chemical ion-pairing or binding. The diffusion rate for cations leaving the phospholipid structures decreased as the negative charge on the lipid structures was decreased. Cation leakage occurred down to zero surface charge, i.e. pure lecithin, but with the addition of as little as 5 moles % of a long-chain cation, leakage was completely prevented. No significant differences in exchange diffusion rate were detected for the cation series Li^+, Na^+, K^+, Rb^+ and choline. Anions diffused much more rapidly than cations and, suprisingly, anions appeared to be relatively free to diffuse whether fixed charges of either sign were present or not. The exchange diffusion rate of Cl^- and I^- was greater than that for F^-, NO_3^-, SO_4^{--} and HPO_4^{--} and it has been suggested by Bangham et al. (1965b) that this may be partially explained on the basis of the hydrated radii of the ions concerned. Water appeared to exchange as fast as, if not faster than, chloride.

Bangham et al. (1965b) have concluded that if the properties of the dispersions have been interpreted correctly then simple bimolecular leaflets of lecithin have been shown to present a diffusion barrier to cations several orders of magnitude greater than to anions. Furthermore, they suggest that the combination of such a property with the extreme degree of permeability to water, and the ensuing osmotic sensitivity, makes the analogy between the behaviour of these particles and that of certain biological membranes (e.g. erythrocytes) irresistible. Bangham et al. (1965b) also proposed that transient foci of micelles, initiated by local concentrations of the long-chain ions, may be responsible for the increased diffusion of cations with increased negative charge on the lipid structures, and that similar phenomena may also be involved in the functioning of excitable membranes.

There is no doubt that the experimental system studied by Bangham et al. (1965b) is of considerable interest and that it can be used, as Bangham and his colleagues have themselves shown in the papers discussed below, in investigations that are relevant to the actions of steroids, drugs and other membrane-active substances on biological membranes. The interpretation of their studies on the phospholipid dispersions is, however, a matter of some debate. There would appear to be no direct evidence for the assumption that the phospholipid particles, which form when dry lipids disperse in aqueous salt solutions, are completely closed structures and that there is actually a complete separation of the outer-most part of each structure from the continuous aqueous phase in which it is suspended. It remains possible that the interiors of the structures may be directly connnected to the exterior environment. This possibility need not entail contact between hydrocarbon and water, since the lipids could be curved around at points where the lipid layer is faulty, as at the edges of a membrane pore. The question of whether or not the lipid structures are closed is intimately connected with the role of dicetyl phosphate, which probably involves more than just the separation of adjacent layers of phospholipids. Bangham et al. (1965b) have remarked that the formation of spherical micelles, induced by the presence of the long-chain surface active ions, would breach a bimolecular leaflet structure completely and permit comparatively unrestricted diffusion to occur. It seems possible, however, that the dicetyl phosphoric acid which is presumably associated with the phospholipid during the whole course of the experiments, may induce the formation of micelles that are of a more permanent nature than the transient foci envisaged by Bangham and his colleagues.

The following proposals represent an attempt by the writer to interpret some of the above findings. It is suggested that globular micelles of lecithin-dicetyl phosphoric acid are formed when the lipid molecules are initially dispersed in the aqueous environment, as well as bimolecular leaflets of lecithin in which the local concentration of surface active agent is lower. It is additionally proposed that the larger particles present in the dispersions are not concentric spheres of continuous bimolecular leaflets but are relatively imperfect structures that are constructed from bimolecular leaflets and globular micelles; these structures also have relatively large discontinuities or faults in the layering and are therefore open to the exterior (Fig. 18).

It would appear from the findings of Bangham et al. (1965b) on the relative diffusion rates of anions of different sizes, that anions leave the phospholipid particles through quite small holes. However, both anions and water diffuse out of the particles very rapidly. It is therefore suggested that anions and water molecules may leak primarily through a large number of small pores, about 8 Å in diameter, which run between the globular micelles of phospholipid (Fig. 18) and which are similar to those discussed above in connection with micellar models for biological membranes (see Sections III-A and III-B). The pores would appear to be lined by positive charges, even in the presence of long-chain anions, since the leakage of chloride ions is not prevented by dicetyl phosphoric acid. The persistence of positive charges may perhaps be explained by supposing that in globular micelles of phospholipid the positive and negative polar groups deviate from the coplanar arrangement that may normally exist in a bimolecular leaflet of lecithin (cf. Sections V-B and V-D). It should be borne in mind, however, that arrangements of phospholipids in which the positive and negative charges are not coplanar will be energetically unfavourable unless compensated for by the binding of other ions, interdigitation of the zwitterions, etc. (Pethica, 1964). If the polar groups of the phospholipid do adopt a somewhat extended configuration around the pore, the pore will be lined by positively charged choline groups and the concentration of dicetyl phosphoric acid molecules may not be sufficient to reduce this local charge to zero. Assuming that the number of micelles, and hence the number of pores leading to the exterior, is considerable owing to the presence of the dicetyl phosphate, the leakage rate of chloride ions would be rapid. The average distance travelled by the internally-trapped anions to the exterior, via the pores in the multiple lipid layers, will be quite short (Fig. 18) while, in accordance with the experimental findings, large anions would be expected to negotiate the small pores less easily than small anions.

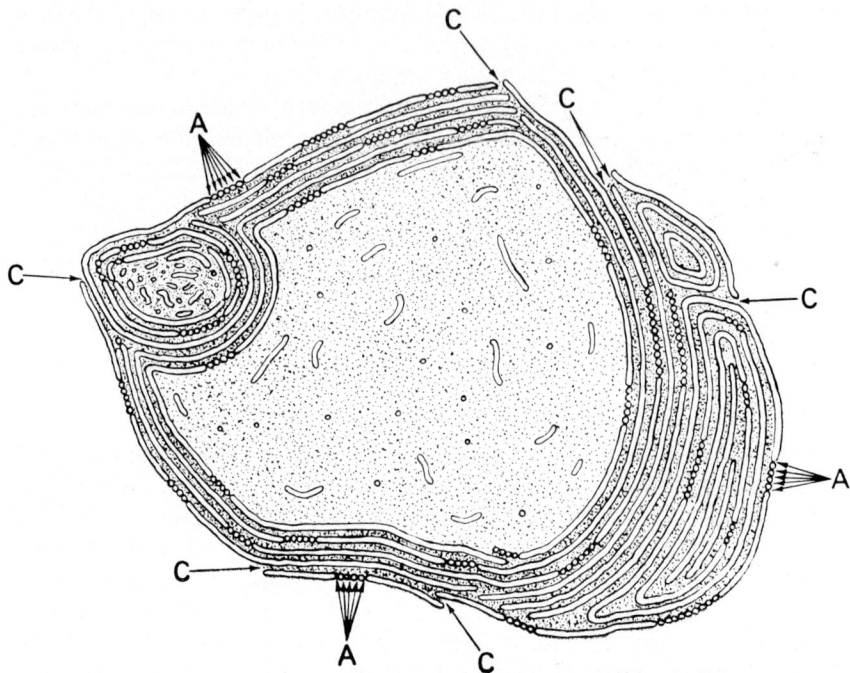

FIG. 18. A hypothetical model proposed by the writer for the cross sectional structure of a particle of lipid, composed of lecithin and dicetyl phosphoric acid, in an aqueous environment. The light layers represent lipid which is in bimolecular leaflets in some places and in globular micelles in others. The shaded layers represent the aqueous channels, about 20–25 Å wide, between the lipid layers. Anions, A, enter and leave the particle and also pass through the internal lipid layers by travelling in pores (about 8 Å in diameter) between the globular micelles of lipid. The pores are positively charged and cations pass through only with difficulty. Cations, C, enter and leave at the points indicated, where the layering is faulty and the particle is not completely closed; cations also diffuse between the lipid layers inside the particle. In the interior of the particle, various aggregates of lipid micelles are dispersed in the aqueous phase; these aggregates enable the layered structure of the particle to grow by accretion.

Cations could presumably escape only with difficulty, or not at all, through the small pores since the pore linings are supposed to be always at least slightly positive. It is suggested that an additional leakage of cations may occur at relatively few, larger openings where the lipid structures may not be fully closed (Fig. 18). Anions and water molecules could also leak away at these points. The exchange diffusion rate of cations leaving the particles via the larger openings would be mainly dependent on the rate at which the trapped cations move towards the surface by travelling in the internal maze between the lipid layers which may be approximately, but not perfectly, concentric. Cations from the exterior would enter the particles and travel inward by the same routes. Unless a one-way system is set up, in which cations enter and leave the particles by different channels, most of the channels would contain a two-way traffic of cations moving away from, and towards, the openings. Under these conditions, the observed increase in the leakage rate of cations with increased negative charge on the lipid might well be anticipated since a more rapid internal diffusion of cations may occur on increasing the distance between the lipid layers within the particles.

The surfaces of adjacent lipid layers on phospholipid dispersions are from 20 to 26 Å apart (Lucy and Glauert, 1964b; Bangham and Horne, 1964) in the presence of potassium phosphotungstate which, in the light of the proposals of Bangham and Horne concerning electrostatic stress, may induce maximum separation of the layers. The coplanar, isolectric polar groups of the lecithin might therefore be separated by, at most, only 15 to 20 Å of aqueous phase in the layered structures. It is of interest to speculate what effect the presence of the positive charges of added long-chain cations would have on the diffusion of potassium ions between the layers of polar groups in channels of this size. Repulsion of the diffusing cations by the positively-charged walls of the aqueous compartments may conceivably so reduce the width of the channel available to cations that their two-way passage is effectively prevented and the main pathway for cation exchange thus blocked. In these circumstances, the total leakage of cations from the particles would fall to zero since, in the presence of long chain cations, the positive charge on the linings of the small anion-pores would be increased, thus reducing still further the already small leakage of cations through the pores. In contrast, exchange diffusion of anions would be much less affected by the excess negative charges of added dicetyl phosphoric acid because the mean distances travelled by the trapped anions, via the numerous pores to the exterior, is postulated to be short and, as discussed above, the anion-pores may never become negatively charged.

The suggestion that the lecithin-dicetyl phosphoric acid particles possess relatively large openings to the exterior that are absent from fully closed structures, such as biological cells or sub-cellular particles, would appear to be supported by the fact that cations of differing sizes show equal rates of exchange diffusion into the artificial lecithin particles. If the present interpretation is correct, leakage of cations from particles that are completely closed by a single lipid membrane, which is partly in the form of globular micelles and partly in the form of a bimolecular leaflet, will be considerably less than the observed leakage of cations from the lecithin-dicetyl phosphoric acid particles studied by Bangham et al. (1965b). Furthermore, the relative impermeability to cations as compared with anions which these particles do, nevertheless, exhibit may not be a property of bimolecular leaflets of phospholipid as is suggested by Bangham and his colleagues. The phenomenon may, instead, be a reflection of the fundamental chemical structure of phospholipids in which the positive charge is remote from the hydrophobic, fatty acid chains; this may result in the pores of micellar areas of membrane having a positive charge.

The half time for the efflux of chloride ions from red cells (0·2 sec) is approximately half a million times less than the half time for potassium exchange (30 hr), despite the similarity in the hydrated radii of the two ions (Whittam, 1964). Anions and water may penetrate the erythrocyte membrane via pores which are present in localized micellar regions of the membrane that constitute somewhat less than 1 % of the total surface area of membrane (cf. Section III-B) and cations may fail to diffuse rapidly through the red cell membrane because of positive charges in the linings of the pores. By analogy with the interpretation offered here for the properties of the lecithin particles, the positive charges might be provided by the membrane phospholipids if the polar groups of these molecules deviate from a planar configuration when the molecules are oganized in globular micelles.

The action of steriods and streptolysin S on the permeability of the lipid structures (composed of lecithin, cholesterol and dicetyl phosphoric acid) to cations has been investigated by Bangham et al. (1965c). It was observed that a number of substances including diethylstilboestrol, etiocholanolone, desoxycorticosterone, progesterone, pregnanolone and streptolysin S increased the release of both sodium and potassium ions from the lipid particles. There was a correlation between the action of the steroids on these lipid particles and their ability to release acid phosphatase from lysosomes. Furthermore, as in other membrane systems, cortisone, cortisol and cortisone acetate modified the actions of the lytic steroids. Bangham et al. (1965c) have pointed out that their findings tend to support the hypothesis advanced by Willmer (1961) and others that steroids may exert their pharmacological actions by regulating the permeability of cells and their organelles (Section III-D above), since these experiments apparently demonstrate a direct interaction between the lipid particles and the added steroids. The difficulty remains, however, that the concentration of steroids necessary for the disruption or stabilization of membranous structures *in vitro* is far in excess of the physiological. As far as the mechanism of action of the steroids on the lipid structures is concerned, no mechanism in molecular terms was suggested.

It would perhaps seem that further clarification of the actual structure of the phospholipid-cholesterol-dicetyl phosphoric acid particles is required before this is attempted.

In related studies, Weissman et al. (1965) have observed that glucose can be sequestered within particles of lecithin-cholesterol-dicetyl phosphoric acid and that membrane-active steroids modify the permeability to glucose in a manner that is similar to their effects on the permeability of cations. The polyene antibiotics, filipin and etruscomycin facilitate release of glucose from lipid particles whether or not they contain cholesterol, while amphotericin, nystatin and pimaricin are apparently active only when cholesterol is present (Sessa, 1966).

Bangham et al. (1965a) have studied the cation permeability of lecithin-dicetyl phosphoric acid particles under the influence of narcotic substances. They investigated the actions of n-alkyl alcohols, chloroform and ether and also the local anaesthetics characterized by the presence of secondary or tertiary amino groups, or both. It was observed that, at constant thermodynamic activity, alcohols of the n-alkyl series increased the diffusion rate of potassium; the effect increasing as the alcohol chain-length decreases. Ether and chloroform at comparable concentrations also increased the diffusion rate of potassium ions. Bangham et al. (1965a) consider that it is not unreasonable to conclude that narcosis induced by these types of molecule could result from the increased permeability of membranes to cations. Further, these substances may induce the formation of micelles in bimolecular leaflets of phospholipid (cf. Haydon and Taylor, 1963). The local anaesthetics studied, tetracaine, nupercaine, cocaine and procaine, also affected the leakage of potassium ions but in this case, the leakage was decreased. All four of the compounds reduced the zeta potential of a negatively-charged membrane system. Since the diffusion rate of cations from the lipid particles depends on the sign and magnitude of the surface charge on the lipids (see above), the reduced leakage caused by the local anaesthetics was no surprise. Thus, although both groups of narcotics cause changes in cation permeability of the lipid particles, the changes were in opposite directions and were apparently dependent upon different physico-chemical mechanisms. Nevertheless, the action of the latter group of local anaesthetics in this model system is paralleled by the effects of tetrodotoxin, "Xylcaine" and prochlorperazine ethane disulphonate on the sciatic nerve of Rana pipiens (Hille, 1966). Since these three substances reversibly eliminated nerve conduction they may be said to have a local anaesthetic action. Hille observed that the effect of these three compounds on the current-clamped node was to raise the threshold, slow the rate of rise and decrease the amplitude of the action potential without affecting the resting membrane potential. It appeared that these agents act primarily by reducing the maximum available sodium conductance. It was suggested that the sodium channels of the nerve are closed or clogged individually by the binding of the anaesthetic molecule, in cation form, to a complementary structure in the channel. The binding of the molecule was depicted as a clogging of the sodium channel because anaesthesia seems to decrease the available number of channels without affecting the kinetic properties of those remaining.

VI. Conclusions

There seems to be little doubt that interest in the structure and function of membranes, and in membrane models, is at least as great today as at any other time in the history of the study of membranes. In many ways, however, we appear to be only at the beginning of the understanding of biological membranes and it is to be expected that research in this field will be intensified in the future. For example, it may be anticipated that, as interests shift increasingly from the mechanisms of protein synthesis to their control, membranes may receive even more attention. It may also perhaps be expected that further research on the biological roles of relatively weak chemical interactions—such as hydrogen bonding, "hydrophobic bonding" and charge-transfer—may proceed hand-in-hand with an improved knowledge of the functions of membranes in complex processes like photosynthesis. An additional stimulus to the study of membrane systems is the recognition that membranes are of more than academic interest. Thus model bilayer membranes are currently being studied by the electronic industry in the hope that phenomena occurring at the molecular level in liquid systems may eventually be utilized in the processing and storage of information (Hawkins, 1966).

With regard to the value of theoretical models in this field, it is sometimes suggested that theoretical models for the structure and function of membranes should not nowadays be proposed unless they are accompanied by a comprehensive thermodynamic analysis which demonstrates that the suggested structure or mechanism is energetically feasible. The complexities of natural membranes lead one to think that such analyses would, unfortunately, have a very "open-end" character. Furthermore, while no one would deny that increased attention to the thermodynamics of membranes is desirable, it is to be hoped that this suggestion will not deter those who are working on other aspects of biological membranes from putting forward theoretical models.

Finally, as regards the question of which arrangement of lipid molecules provides the correct model for biological membranes, in the writer's view different models cannot be regarded as mutually exclusive because no single model seems capable of accounting satisfactorily for the properties of such differing systems as, for example, nerve cells and mitochondria. Observations which indicate the presence of micellar arrangements in specific membranes are accumulating, as has been shown in this chapter. Nevertheless, it remains possible that certain properties of membranes, for example the electrical properties, may eventually be found to be quite incompatible with the concept of a micellar configuration for the lipids. Most investigators would probably agree, however, that much more detailed models than those which we have at present are probably required to explain the complex functions of biological membranes. Some of these models will no doubt be variations on existing ideas, and some may perhaps be revolutionary in concept.

I am most grateful to Dame Honor Fell, F.R.S., Dr. J. T. Dingle, Miss Audrey M. Glauert and Dr. F. U. Lichti for reading the manuscript of this chapter and for their helpful suggestions. I thank Mr. R. A. Parker for drawing Fig. 18.

References

Abramson, M. B., Katzman, R. and Gregor, H. P. (1964). *J. biol. Chem.* **239**, 70.
Anderson, P. J. and Pethica, B. A. (1956). *In* "Biochemical Problems of Lipids". Proc. 2nd Int. Conf. on Lipids. (G. Popjak and E. le Breton, eds) p. 24. Butterworths, London.
Balasubramanian, D. and Wetlaufer, D. B. (1966). *Proc. natn. Acad. Sci., U.S.A.* **55**, 762.
Bangham, A. D. and Dawson, R. M. C. (1962). *Biochim. biophys. Acta* **59**, 103.
Bangham, A. D. and Horne, R. W. (1962). *Nature, Lond.* **196**, 952.
Bangham, A. D. and Horne, R. W. (1964). *J. molec. Biol.* **8**, 660.
Bangham, A. D., Rees, K. R. and Shotlander, V. (1962). *Nature, Lond.* **193**, 754.
Bangham, A. D., Dingle, J. T. and Lucy, J. A. (1964). *Biochem. J.* **90**, 133.
Bangham, A. D., Standish, M. M. and Miller, N. (1965a). *Nature, Lond.* **208**, 1295.
Bangham, A. D., Standish, M. M. and Watkins, J. C. (1965b). *J. molec. Biol.* **13**, 238.
Bangham, A. D., Standish, M. M. and Weissman, G. (1965c). *J. molec. Biol.* **13**, 253.
Bar, R. S., Deamer, D. W., and Cornwell, D. G. (1966). *Science, N.Y.* **153**, 1010.
Beament, J. W. L. (1955). *J. exp. Biol.* **32**, 514.
Beament, J. W. L. (1964). *Symp Soc. exp. Biol.* **19**, 273.
Benson, A. A. (1966). *J. Am. oil chem. Soc.* **43**, 265.
Bernal, J. D. (1964). *Proc. R. Soc.* A, **280**, 299.
Blasie, J. K., Dewey, M. M., Blaurock, A. E. and Worthington, C. R. (1965). *J. molec. Biol.* **14**, 143.
Branton, D. (1966). *Proc. natn. Acad. Sci., U.S.A.* **55**, 1048.
Castillo, J. del, Rodrigues, A., Romero, C. A. and Sanchez, V. (1966). *Science, N.Y.* **153**, 185.
Chapman, D. and Penkett, S. A. (1966). *Nature, Lond.* **211**, 1304.
Chapman, D., Owens, N. F. and Walker, D. A. (1966) *Biochim. biophys. Acta* **120**, 148.
Chappell, J. B. (1966). *Biochem. J.* **100**, 43P.

Clements, J. A. and Wilson, K. M. (1962). *Proc. natn. Acad. Sci., U.S.A.* **48**, 1008.
Clunie, J. S., Corkill, J. M. and Goodman, J. F. (1965). *Proc. R. Soc.* A, **285**, 520.
Colacicco, G. (1965a). *Nature, Lond.*, **207**, 936.
Colacicco, G. (1965b). *Nature, Lond.* **207**, 1045.
Cole, K. S. (1940). *Cold Spring Harb. Symp. quant. Biol.* **8**, 110.
Danielli, J. F. (1962). *In* "The Plasma Membrane". Suppl. to *Circulation* **26**, 1175.
Danielli, J. F. (1963). *In* "The Structure and Function of the Membranes and Surfaces of Cells". *Biochem. Soc. Symp.* **22**, 172.
Danielli, J. F. and Davson, H. (1934–35). *J. cell. comp. Physiol.* **5**, 495.
Danielli, J. F. and Harvey, E. N. (1934–35). *J. cell. comp. Physiol.* **5**, 483.
Danielli, J. F., Marrian, G. F. and Haslewood, G. A. D. (1933). *Biochem. J.* **27**, 311.
Davies, J. T. (1965). *J. theoret. Biol.* **8**, 1.
Davson, H. (1962). *In* "The Plasma Membrane". Suppl. to *Circulation* **26**, 1022.
Davson, H. and Danielli, J. F. (1943). "The Permeability of Natural Membranes". Cambridge University Press, Cambridge.
Demel, R. A., van Deenen, L. L. M. and Kinsky, S. C. (1965). *J. biol. Chem.* **240**, 2749.
Dervichian, D. G. (1946). *Discuss. Faraday Soc.* **42B**, 180.
Dickerson, R. E. (1965). *Nature, Lond.* **208**, 139.
Dingle, J. T. and Lucy, J. A. (1965). *Biol. Rev.* **40**, 422.
Dourmashkin, R. R., Dougherty, R. M. and Harris, R. J. C. (1962). *Nature, Lond.* **194**, 1116.
Eley, D. D. and Hedge, D. G. (1956). *Discuss. Faraday Soc.* **21**, 221.
Fernández-Moràn, H. (1959). *In* "Biophysical Science—a Study Program". (J. L. Oncley, ed.) pp. 319–330. Wiley, New York.
Finean, J. B. (1953). *Experientia*, **9**, 17.
Finean, J. B. and Rumsby, M. G. (1963). *Nature, Lond.* **197**, 1326.
Fleischer, S., Richardson, S., Chapman, A., Fleischer, B. and Hultin, H. (1963). *Fedn Proc. Fedn Am. Socs. exp. Biol.* **22**, 526.
Gammack, D. B. (1966). *Biochem. J.* **98**, 18P.
Gent, W. L. G., Gregson, N. A., Gammack, D. B. and Raper, J. H. (1964). *Nature, Lond.* **204**, 553.
Glauert, A. M. and Lucy, J. A. (1968). *In* "The Membranes" (A. J. Dalton and F. H. Haguenau, eds). In Press. Academic Press, New York.
Glauert, A. M., Dingle, J. T. and Lucy, J. A. (1962). *Nature, Lond.* **196**, 953.
Goldstein, D. A. and Solomon, A. K. (1960–61). *J. gen. Physiol.* **44**, 1.
Gorter, E. and Grendel, F. (1925). *J. exp. Med.* **41**, 439.
Green, D. E. and Tzagoloff, A. (1966). *J. lipid Res.* **7**, 587.
Hanai, T., Haydon, D. A. and Taylor, J. (1964). *Proc. R. Soc.* A, **281**, 377.
Hanai, T., Haydon, D. A. and Taylor, J. (1965a). *J. theoret. Biol.* **9**, 278.
Hanai, T., Haydon, D. A. and Taylor, J. (1965b). *J. theoret. Biol.* **9**, 422.
Hanai, T., Haydon, D. A. and Taylor, J. (1965c). *J. theoret. Biol.* **9**, 433.
Hanai, T., Haydon, D. A. and Redwood, W. R. (1966). *Ann. N.Y. Acad. Sci.* **137**, 731.
Hartley, G. S. (1955). *In* "Progress in the Chemistry of Fats and Other Lipids". Vol. 3, pp. 19–55. (R. T. Holman, W. O. Lundberg, and T. Malkin, eds) Pergamon Press, Oxford.
Hawkins, J. K. (1966). *Science Jl.* **2**, 64.

Haydon, D. A. and Taylor, J. (1963). *J. theoret. Biol.* **4**, 281.
Hays, R. M. and Leaf, A. (1961–62). *J. gen. Physiol.* **45**, 933.
Hechter, O. (1965). *Fedn Proc. Fedn Am. Socs exp. Biol.* **24**(2), S–91 to S–102.
Hille, B. (1966). *Nature, Lond.* **210**, 1220.
Horne, R. W. and Whittaker, V. P. (1962). *Z. Zellforsch. mikrosk. Anat.* **58**, 1.
Horne, R. W., Bangham, A. D. and Whittaker, V. P. (1963). *Nature, Lond.* **200**, 1340.
Huang, C. and Thompson, T. E. (1965). *J. molec. Biol.* **13**, 183.
Huang, C. and Thompson, T. E. (1966). *J. molec. Biol.* **15**, 539.
Huang, C., Wheeldon, L. and Thompson, T. E. (1964). *J. molec. Biol.* **8**, 148.
Husson, F. and Luzzati, V. (1963). *Nature, Lond.* **197**, 822.
Kavanau, J. L. (1963). *Nature, Lond.* **198**, 525.
Kerridge, D., Horne, R. W. and Glauert, A. M. (1962). *J. molec. Biol.* **4**, 227.
Kinsky, S. C. (1963). *Archs Biochem. Biophys.* **102**, 180.
La Mer, V. K. and Healy, T. W. (1965). *Science, N.Y.* **148**, 36.
Lindemann, B. and Solomon, A. K. (1961–62). *J. gen. Physiol.* **45**, 801.
Locke, M. (1965). *Science, N.Y.* **147**, 295.
Loewenstein, W. R. (1966). *Ann. N.Y. Acad. Sci.* **137**, 441.
Lucy, J. A. (1964a). *J. theoret. Biol.* **7**, 360.
Lucy, J. A. (1964b). *Natn. Cancer Inst. Monogr.* **13**, 93.
Lucy, J. A. (1965). *Biochem. J.* **96**, 12P.
Lucy, J. A. (1966). *Biochem. J.* **99**, 57P.
Lucy, J. A. (1967). 4th Conf. on Cellular Dynamics. (L. D. Peachey, ed.), N.Y. Acad. Sci. Interdisciplinary Communications Program, 147.
Lucy, J. A. and Dingle, J. T. (1964). *Nature, Lond.* **204**, 156.
Lucy, J. A. and Glauert, A. M. (1964a). In "Symposium on Electron Microscopy; Modena, 1963" (P. Buffa, ed.), p. 233. C.N.R., Rome.
Lucy, J. A. and Glauert, A. M. (1964b). *J. molec. Biol.* **8**, 727.
Luzzati, V. and Husson, F. (1962). *J. biochem. biophys. Cytol.* **12**, 207.
Luzzati, V. and Reiss-Husson, F. (1966). *Nature, Lond.* **210**, 1351.
Maddy, A. H., Huang, C. and Thompson, T. E. (1966). *Fedn Proc. Fedn Am. Socs exp. Biol.* **25**, 933.
Malcolm, B. R. (1962). *Nature, Lond.* **195**, 901.
Malhotra, S. K. (1966). *J. Ultrastruct. Res.* **15**, 14.
Markham, R., Frey, S. and Hills, G. J. (1963). *Virology* **20**, 88.
McConnell, D. G., Tzagoloff, A., MacLennan, D. H. and Green, D. E. (1966). *J. biol. Chem.* **241**, 2373.
Mitchell, P. (1963). In "The Structure and Function of Membranes and Surfaces of Cells". *Biochem. Soc. Symp.* **22**, 142.
Mueller, G. C. (1966). Proc. 6th Pan-Amer. Congr. on Endocrinology, Mexico City, 1965. *Excerpta Med. Int. Congr. Series* **112**, 27.
Mueller, P., Rudin, D. O., Tien, H. T. and Wescott, W. C. (1962a). *Nature, Lond.* **194**, 979.
Mueller, P., Rudin, D. O., Tien, H. T. and Wescott, W. C. (1962b). In "The Plasma Membrane". Suppl. to *Circulation* **26**, 1167.
Munck, A. (1957). *Biochim. biophys. Acta* **24**, 507.
Nash, H. A. and Tobias, J. M. (1964). *Proc. natn Acad. Sci., U.S.A.* **51**, 476.
Nilsson, S. E. G. (1964a). *Nature, Lond.* **202**, 509.
Nilsson, S. E. G. (1964b). *J. Ultrastruct. Res.* **11**, 581.
Nilsson, S. E. G. (1965). *J. Ultrastruct. Res.* **12**, 207.

Paganelli, C. V. and Solomon, A. K. (1957–58). *J. gen. Physiol.* **41**, 259.
Perutz, M. F. (1965). *J. molec. Biol.* **13**, 646.
Peters, R. A. (1956). *Nature, Lond.* **177**, 436.
Pethica, B. A. (1964). In "Symposium on Surface Activity and Microbial Cell". *Soc. Chem. Ind. London*, **19**, 85.
Pethica, B. A. (1967). In "Symposium on Biophysics and Physiology of Biological Transport". Frascati, Rome, 1965. *Protoplasma* **63**, 147.
Rideal, E. K. and Schulman, J. H. (1939). *Nature, Lond.* **144**, 100.
Robertson, J. D. (1963). *J. biophys. biochem. Cytol.* **19**, 201.
Rosa, C. G. and Tsou, K-C. (1965). *Nature, Lond.* **206**, 103.
Schulman, J. H. and Rideal, E. K. (1937). *Proc. R. Soc.* B, **122**, 29.
Segerman, E. (1965). In "Surface Chemistry". 2nd Scand. Symp. on Surface Activity. 1965. (V. Runnström-Reio, ed.). p. 157. Munksgaard, Copenhagen.
Sessa, G. (1966). *Fedn Proc. Fedn Am. Socs exp. Biol.* **25**, 358.
Seufert, W. D. (1965). *Nature, Lond.* **207**, 174.
Shanes, A. M. (1963). *Science, N.Y.* **140**, 824.
Sjöstrand, F. S. (1963a). *Nature, Lond.* **199**, 1262.
Sjöstrand, F. S. (1963b). *J. Ultrastruct. Res.* **9**, 340.
Sjöstrand, F. S. (1963c). *J. Ultrastruct. Res.* **9**, 561.
Sjöstrand, F. S. and Elfvin, L-G. (1964). *J. Ultrastruct. Res.* **10**, 263.
Skou, J. C. (1958). *Biochim. biophys. Acta* **30**, 625.
Stein, W. D. and Danielli, J. F. (1956). *Discuss. Faraday Soc.* **21**, 238.
Stoeckenius, W. (1962). *J. biophys. biochem. Cytol.* **12**, 221.
Stoeckenius, W. (1966). *Ann. N.Y. Acad. Sci.* **137**, 641.
Taylor, J. L. and Haydon, D. A. (1965). *Biochim. biophys. Acta* **94**, 488.
Thompson, T. E. and Huang, C. H. (1966). *J. molec. Biol.* **16**, 576.
Tien, H. T. (1966a). *J. molec. Biol.* **16**, 577.
Tien, H. T. (1966b). "Formation of 'Black' Lipid Membranes by Oxidation Products of Cholesterol". 10th Ann. meeting, Biophysical Society, Boston, 1966.
Tobias, J. M. (1964). *Nature, Lond.* **203**, 13.
Tobias, J. M., Agin, D. and Pawlowski, R. (1961–62). *J. gen. Physiol.* **45**, 989.
van Deenen, L. L. M. and Demel, R. A. (1965). *Biochim. biophys. Acta* **94**, 314.
van Deenen, L. L. M., Houtsmuller, U. M. T., de Haas, G. H. and Mulder, E. (1962). *J. Pharm Pharmac.* **14**, 429.
van den Berg, H. J. (1965). *J. molec. Biol.* **12**, 290.
Vandenheuvel, F. A. (1963). *J. Am. oil chem. Soc.* **40**, 455.
van Zutphen, H., van Deenen, L. L. M. and Kinsky, S. C. (1966). *Biochem. biophys. Res. Commun.* **22**, 393.
Villegas, R. and Barnola, F. V. (1960–61). *J. gen. Physiol.* **44**, 963.
Warner, D. T. (1961). *Nature, Lond.* **190**, 120.
Watkins, J. C. (1965). *J. theoret. Biol.* **9**, 37.
Weier, T. E., Bisalputra, T. and Harrison, A. (1966). *J. Ultrastruct. Res.* **15**, 38.
Weissman, G., Sessa, G. and Weissman, S. (1965). *Nature, Lond.* **208**, 649.
Whittam, R. (1964). "Transport and Diffusion in Red Blood Cells". E. Arnold, London.
Willmer, E. N. (1961). *Biol. Rev.* **36**, 368.

Chapter 7

Membranes and Bioenergetics

R. B. LESLIE

Molecular Biophysics Unit, Unilever Research Laboratory, Welwyn, Herts, England.

I.	INTRODUCTION	290
II.	STRUCTURAL ASPECTS AND GROSS MORPHOLOGY OF LAMELLAR SYSTEMS INVOLVED IN BIOENERGETICS	291
	A. General	291
	B. Outline of Optical Techniques used to Study Lamellar Structures	292
	C. The Structure of Rod Outer Segment Membranes . . .	293
	D. Mitochondrial Membranes (Cristae)	295
	E. Photosynthetic Lamellae	296
	F. The Arrangement of Pigments Within the Photosynthetic Lamella	301
	G. Summary	304
III.	FUNCTIONAL ASPECTS OF MEMBRANES AND BIOENERGETICS . . .	305
	A. Chemical Composition	305
	B. Some Molecular Structures Involved in Membrane Bioenergetics	307
	C. Some Functional Similarities	311
IV.	THE EVOLUTION OF PHYSICAL BIOENERGETICS	312
V.	PHYSICAL BIOENERGETICS	314
	A. Elementary Concepts	314
	B. Energy Transfer by Resonance Mechanisms . . .	316
	C. Possible Roles of the Membrane in Energy Transfer . .	320
	D. Energy Transfer Coupled with Discrete Charge Migration .	321
	E. Donor–Acceptor Interaction	322
	F. The Role of Lamellar Organization in Charge Separation and Electron Mobility	323
VI.	ENERGY AND ELECTRON MIGRATION IN BIOENERGETIC LAMELLAR SYSTEMS.	325
	A. Rod Outer Segments	325
	B. Mitochondrial Membranes	328
	C. Photosynthetic Systems	333
VII.	SOME MODEL SYSTEMS	336
	A. General	336
	B. Monolayers at Interfaces	337
	C. Bilayers Separating Aqueous Chambers	338
	D. Micellar Models	339
VIII.	CONCLUSION AND OUTSTANDING PROBLEMS	340
IX.	ACKNOWLEDGEMENTS	342
	REFERENCES	342

I. Introduction

Membranes, or lipoprotein lamellar formations, have two major functions in biological systems; (1) as barriers with specific permeability properties and (2) as structural frameworks for processes intimately associated with absorption, stabilization, transfer and utilization of various forms of energy. This chapter considers the role of the membrane in association with bioenergetic processes, which constitutes much of the subject matter of the field of physical bioenergetics (Augenstine, 1960).

Recent research, for example the work of Lehninger (1962) and Packer (1966) in the metabolically linked swelling, contraction and ion uptake phenomena in mitochondria and chloroplasts, indicates conclusively that the membrane, or lamellar structure, is inseparably bifunctional. However, we have separated membrane function into two arbitrary areas, those of permeability and function as a "structural framework", to limit the areas under consideration.

Van Niel (1960), Szent-Györgyi (1964) and many others have posed the question: "what is the connection between the ubiquitous lamellar structures seen, for example, in chloroplasts and mitochondrial cristae, bioenergetics and, indeed, life?".

The structural organization and orientation of molecules, such as chlorophyll, quinones, carotenoids, cytochromes and enzymes, in largely lipid protein lamellar formations now seems to be unequivocally accepted. These molecules constitute, in large measure, the lowest common molecular denominator of the basic bioenergetic processes, photosynthesis, mitochondrial oxidative phosphorylation and vision. The occurence, in relatively high concentrations, of these "energy transduction" molecules (Green, 1960; Green and Fleischer, 1962) into highly ordered quasi-crystalline, quasi-solid state lamellar systems may generate conditions which are favourable to physical energy transfer mechanisms such as semiconduction, photoconduction, "exciton" migration, resonance transfer and charge transfer interaction. These phenomena, some of which certainly have relevance to bioenergetics, will be examined in the context of the known occurrence of the relevant molecules predominantly in lamellar formations. The lack of simple and meaningful model systems of bioenergetic processes has greatly hindered the investigation of the intimate relation between structure and function. Section VII attempts to illustrate this thesis and several model systems will be discussed and evaluated. In this chapter, we shall consider the relationship between structure and function in membrane systems only as they occur in three types of organelle: the chloroplast, the mitochondrion and retinal outer segments. Much discussion will be restricted specifically to the chloroplast lamella to avoid repetition. Results obtained in the last few years tend to show a unity of structure and function which would not have been considered possible earlier.

II. Structural Aspects and Gross Morphology of Lamellar Systems Involved in Bioenergetics

A. General

This section will illustrate the structural and morphological aspects of the three organelles chosen to demonstrate the association of lipoprotein lamellar organization and bioenergetics—i.e. the chloroplast, the mitochondrial christae, the rod outer segments. The function of the latter (referred to from now on as ROS membranes) is to convert electromagnetic energy into a visual pattern. The mitochondrial cristae (MC) and the photosynthetic lamella (PL), whether individual or multiple, represent, on the other hand, the two extremes of the energy cycle of life; energy capture and storage from the sun and energy utilization by the cell.

In all three processes, energy changes from one form to another, often with considerable efficiency. The processes can be separated into two phases; the initial biophysical processes concerned in the energy conversion act and the secondary biochemical utilization of the products of the first phase.

This *arbitrary* division into biophysical and biochemical aspects can be clearly seen in the photosynthetic process, where the concepts of light and dark reactions are now well established. The former, which involves the absorption of a quantum of light, the generation of a strong oxidant and strong reductant (probably electrons and "holes"), physically separated in space through the mediation of chlorophyll in an electronically excited state, is clearly a biophysical process. The utilization of the strong oxidizing and reducing power to reduce carbon dioxide to the level of carbohydrate, in the series of dark reactions known as the Calvin cycle, is clearly the result of well established biochemical enzymology (Calvin and Bassham, 1957). The emphasis throughout this chapter will be largely devoted to the initial biophysical aspects of membranes and bioenergetics.

New electron microscope techniques, such as freeze-etching and negative staining, combined with improved biochemical separation and analytical procedures have within the last few years shed considerable doubt on the correctness and utility of the simple Danielli–Davson membrane model in bioenergetically important systems (Sjöstrand, 1962, 1963; Moor and Mühlethaler, 1963). The cell membrane and a chloroplast membrane, are designed by nature to perform quite different functions. Up to 20%, on a weight basis, of non-lipid molecules, such as pigments in PL and ROS membranes and enzyme complexes in MC membranes, are known to be associated with bioenergetically important membranes. On these considerations alone, it would be more realistic to expect the two types of membrane to be different at the ultrastructural level. Most of the earlier electron microscopy (pre-1960) is collected and discussed for PL, ROS, MC membrane in a comparative way by Sjöstrand (1959, 1960), Fernandez-Moran (1959) and Hodge (1959).

Questions of significance are whether membranes, such as the plasma membrane or membranes involved in bioenergetics, are best described as bimolecular lipid bilayers with attached or partially incorporated functional molecules or as two-dimensional arrays of globular or micellar subunits. In this latter case, the subunits, probably lipid–protein or pigment–lipid–protein assemblies, must possess both built-in structural organization and functional integrity. It may well emerge that both descriptions are correct in appropriate circumstances and that the bilayer–micellar transition will have far reaching functional potentialities (see Chapter 6 by J. A. Lucy).

B. OUTLINE OF OPTICAL TECHNIQUES USED TO STUDY LAMELLAR SYSTEMS

Much of the experimental evidence on which this chapter is based must inevitably be provided by electron microscopy but some limitations of electron microscopy can be avoided by using refined optical techniques. The type of information which polarized light techniques can yield will be described but for details of the physical and theoretical basis of these studies, the papers referred to must be consulted. Polarization microscopy is described fully by Hartshorne and Stuart (1960) and Oster (1955).

The anisotropy giving rise to the related phenomena of birefringence and dichroism has two principal sources. The *intrinsic anisotropy* of molecules (due to the fact the electron clouds may be more easily distorted in some directions than others) will give rise to intrinsic birefringence or dichroism. Alternatively oriented sub-structures of several molecules, or particles which may or may not be intrinsically anisotropic, can give rise to an overall anisotropy if they differ in refractive index from the medium in which they are immersed. This is called *form* anisotropy and leads to form birefringence and form dichroism, respectively. If the subunits of the overall structure are arranged as long cylindrical rods around a central axis, this will exhibit positive form birefringence if the system is observed with light incident perpendicular to the axis. If the subunits are arranged as thin parallel disc-like platelets, negative form birefringence will be observed if light incident perpendicular to the axis is used. Intrinsic and form anisotropy can be induced, or enhanced, in certain cases by mechanical or electrical alignment of the material under study. This leads to flow birefringence (dichroism) and electric birefringence (dichroism), respectively. Intrinsic and form anisotropy can be separated by disrupting the composite structure or balancing the refractive index of the subunits and the medium.

Birefringence and dichroism studies are thus valuable to reveal and characterize the presence of an oriented, organized substructure within a lamellar system and have the advantage that measurements may be made on living, aqueous systems in a functionally active condition.

Molecules which are excited into the fluorescent state* with plane polarized light will emit plane polarized fluorescent light, providing that the lifetime of the excited state is less than the rotational relaxational time. Even if unpolarized incident light is used to elicit fluorescence from oriented, relatively fixed molecules, polarized fluorescent light will be omitted. In the former instance, however, if the fluorescent quantum is emitted by a molecule other than that which absorbed the light, then partial depolarization of the emitted fluorescent light may occur, particularly if the absorbing and emitting molecules, although relatively fixed, have different orientations (see Section V).

C. THE STRUCTURE OF ROD OUTER SEGMENT MEMBRANES

Eyes, in general, have two types of lamellar pigment-containing receptor sites, the rods and cones. Rods are generally used for seeing in very low intensity light and cones for colour and high light intensity vision. Only the rod outer segment (ROS) will be considered in detail, since it is here that the primary biophysical quantum conversion act takes place. The rod outer segments have a lamellar stacked disc structure, a discovery due to the careful observation of osmium tetroxide fixed rod segments by Schultze almost exactly one hundred years ago (Schultze, 1866).

Application of polarized light and spectroscopic techniques enabled the ultrastructure of rod outer segments to be further elucidated. This work, largely due to W. J. Schmidt (1935 a,b,c) in the case of ROS membranes, clearly tried to bridge that most difficult of gaps—the relationship between what is observed in the microscope (or electron microscope) and the dimensions and chemical peculiarities of the individual molecules.

This early work has been evaluated in the recent review by Moody (1964). It seems now to be well established that rods show a birefringence made up of two components, reflecting two types of molecular organizations or orientations within the rod. There appears to be a large positive birefringence, attributed to oriented lipid, probably phospholipid, molecules and a smaller negative birefringence, due to some other type of organized substructure. Further birefringence studies, both after lipid extraction and in media of

*Those unfamiliar with fluorescence terminology may find it helpful to refer to Section V.

differing refractive indices, suggested the probable origin of the two types of birefringence. The smaller component is probably form-birefringence, arising from small oriented substructures suggested by Schmidt to be a pile of parallel platelets. The larger, and normally dominating, positive birefringence arises from the presence of lipid in the outer segment membranes with the hydrocarbon chains parallel to the rod (long) axis. In his model, Schmidt visualized the lipid as being arranged as a stack of bimolecular lipid leaflets, closed off at each end and separated possibly by orientated protein-containing regions.

Concurrently, dichroism was yielding valuable structural information (Moody, 1964). These studies also strongly suggested a high degree of structural organization or orientation of rhodopsin within the rods. Maximum light absorption was found by Schmidt to occur when the electric vector was perpendicular to the long axis of the rod, i.e. parallel to the plane of the bimolecular lipid leaflets suggested by the birefringence studies. However, as Moody pointed out, the observed dichroism could be consistent with the transition dipole moments of the rhodopsin chromophores lying approximately parallel to the plane of the bimolecular leaflets, corresponding to intrinsic dichroism, or rhodopsin molecules being concentrated in localized disc regions, leading to form dichroism.

Early electron microscopical studies, largely the work of Sjöstrand (1949), seemed to confirm and extend previously accepted thoughts about the structure and ultrastructure of ROS membranes. Until recently, electron microscopy seemed to suggest that each disc was the result of two unit membranes coming together, back-to-back, to give a composite structure. It is also possible that discs may arise from infoldings of the surface (unit) membrane. Semi-quantitative calculations, combined with analytical data, indicated that the lipid in each half of a given disc probably forms a bimolecular layer but that the amount of lipid present may be insufficient to form a continuous bimolecular layer. The relevant suggestion is also made (*vide supra*) that rhodopsin may be a structural component of the disc membrane (Moody 1964).

In the papers of Blaisie et al. (1965) and Nilsson (1964, 1965) definite evidence for the occurrence of globular particulate substructure on, or within, the disc membranes is presented. In view of the probable occurrence of functional subunits in or on both photosynthetic lamellae and mitochondrial christae, these latest observations may have great significance Frog retinal rod photoreceptor membranes and unstained, moisture-containing samples were studied by combined optical, electron microscopical and low angle x-ray scattering techniques. The results of these two sets of investigations were consistent with the occurrence of largely nonpolar (lipidic?) particles of 40–70 Å diameter, arranged in a square array on the surface of, or within, the disc membrane. This phraseology brings to mind the statement of Park and Biggins (1964), that the surface of the quantasome appears to contain four or more subunits. It will be of great interest to see the results of electron microscopic examination of ROS membranes, using the relatively new freeze-etching technique. The protein moiety of the rhodopsin complex (opsin) may also prove to have a functional role in stabilizing the substructure of the ROS membrane in the same way as the structural protein of mitochondrial christae.

D. MITOCHONDRIAL MEMBRANES (CHRISTAE)

The major function of mitochondrial membranes (christae) is to provide a locus for the efficient extraction of biologically useful energy from the substrates of the Krebs citric acid cycle by controlled sequential oxidative processes. Electrons derived from substrates pass unidirectionally along a series of molecules of differing redox potential to the terminal acceptor, oxygen. During their journey along the chain of complexes, the electrons give up their energy, which is used to generate molecules of ATP, and hence provide energy in a readily usable and transportable form for the multitudinous purposes of the cell. The combined processes of electron transfer and coupled phosphorylation, which constitute mitochondrial respiration, take place in association with, and are completely dependent upon, the organized lipoprotein lamellar structure of the mitochondrial christae. In medium resolution electron microscopy and the older preparative techniques, mitochondria appear as sausage-shaped objects with a surrounding double membrane and a characteristic inner transverse system of invaginations called cristae which are continuous with the outer membrane structure (cf. ROS membranes).

Within the last few years, there have been great advances in the understanding of mitochondrial oxidative phosphorylation much of the credit for which is due to D. E. Green and his various associates at the Enzyme Institute in Wisconsin. However, rapid progress leads inevitably to considerable conflict in interpretation of results and this is particularly evident in the field of mitochondrial membrane structure. On the other hand, it is perhaps in the case of cristae that the most convincing evidence has been brought forward to seriously challenge the pauci-molecular (or Danielli–Davson) unit membrane model. Sjöstrand (1963) presented e.m. evidence which indicated that mitochondrial membranes were composed of small globular lipid micelles separated by protein molecules. These protein molecules could be both the enzymatic proteins of the electron transfer chain and the largely hydrophobic structural protein. The structural protein would stabilize the two-dimensional micellar array. This type of model would be consistent with the previously somewhat puzzling electron microscope studies of Fleischer *et al.* (1962), which indicated that the overall structure of mitochondrial membranes was not significantly changed by extraction of over 85% of the phospholipids.

The recent review by Green and Tzagoloff (1966) illustrates the rapidly emerging viewpoint that membrane systems, particularly those involved in bioenergetics, are composed of repeating functional, micellar subunits, rather than unit membranes with attached, or partially embedded, functional subunits. Though Green and Tzagoloff draw evidence largely from mitochondrial studies, they extrapolate their views probably legitimately to membrane systems generally.

These authors visualize the mitochondrial membrane as "... a continuum, one particle thick, made up of fused or nesting repeating particles". The repeating particles are functionally active lipoprotein macromolecules. In oxidative phosphorylation, the transfer of electrons from the final electron donors (NADH and succinate) to molecular oxygen is brought about by the presence in the correct stoichiometric proportions and position of four clearly definable complexes.

Green and Tzagoloff present evidence that these functionally active complexes (which, combined together with phospholipids and structural proteins, form the elementary particles) are themselves the "membrane forming elements". The repeating units of the inner mitochondrial membrane have the built-in ability, moreover, of spontaneously aligning themselves to form membranes. This ability requires the presence of phospholipid which, it is suggested, somehow restricts the alignment of the repeating units to the required two-dimensional array. Evidence from lipid removal and reinsertion studies previously available (Green and Fleischer, 1963; Green and Lester, 1959; Lester and Fleischer, 1961) is still valuable to show that lipids and ubiquinones may have a functional role in oxidative phosphorylation.

The organized structure of the mitochondrial inner membrane is nature's way of bringing about electron transfer by the coupled redox processes in the most efficient way. As discussed by Lehninger (1959), if the individual enzyme complexes of the electron transfer chain were not confined by membrane location to a relatively fixed array, the rate of respiration would be extremely small. Organization of the complexes by, and within, the cristae must be very efficient to account for the observed rate of respiration.

E. PHOTOSYNTHETIC LAMELLA

In an analogous way to that described for ROS membranes and at roughly the same time, similar optical studies were being performed on leaf sections to localize the position and structure of the pigments involved in the photosynthetic process. This work is now largely of historical interest but we may single out the contributions of Frey-Wyssling (1937), Menke (1938) and Frey-Wyssling and Steinmann (1948), who, on the basis of birefringence and dichroism studies, proposed that the pigments, chlorophyll, proteins and lipids were arranged in ultra-thin lamellae.

The advent of electron microscopy enabled this proposal to be easily checked and the association of the "light" reactions with the pigment containing lamellar regions is now universally accepted. The early electron microscopy seemed to suggest that compound lamellae (or grana) might be built up from a pile of structurally asymmetric unit membranes of which chlorophyll and, perhaps carotenoids, would be integral structural components (Hodge, 1959).

Recent electron microscopy has led to the seemingly inescapable conclusion that chloroplast lamellae have a pronounced subunit substructure (Weier and Benson, 1966). Figure 1 shows an electron micrograph of a chloroplast from maize showing the highly developed grana lamellae and the interconnecting stroma lamellae. Figure 2 shows the membrane within a single granum of a *vicia faba* chloroplast, at much higher magnification, where the particulate nature of the membranes is clearly apparent. The work of Park and Pon (1963) and Park and Biggins (1964) demonstrated by chemical means that the light reactions of photosynthesis were associated with pigment-containing lamellar structures, the individual lamellae being made up apparently of two layers of subunits. The later papers are concerned with a more precise chemical and structural characterization of these subunits which were termed quantasomes (see also Lichtenthaler and Calvin, 1964). The analytical data and other considerations suggested that the quantasome could represent the photosynthetic unit (*vide supra* Section VI-C). Park and Biggins also suggest, on the basis of the size and composition, that the lamellae consist of proteins embedded in a lipid matrix.

The newer technique of freeze-etching, introduced and exploited particularly by Mühlethaler and his colleagues, has also revealed a particulate substructure in chloroplast lamellae (Mühlethaler *et al.*, 1965; Mühlethaler, 1966). This model visualizes a chloroplast membrane as being composed of a central lipid layer (40 Å) with protein particles (60 Å) on both sides. The dimensions of the structure are such that the protein particles are in direct contact through the layer. The model for the chloroplast grana, developed by Mühlethaler and his colleagues, is reproduced in Fig. 3. On the basis of comparative studies, using both the relatively mild freeze-etching technique and the more usual chemically stained and fixed preparations, Mühlethaler demonstrates how fixation procedures can lead to the protein uncoiling and to the transformation of the correct membrane with attached or

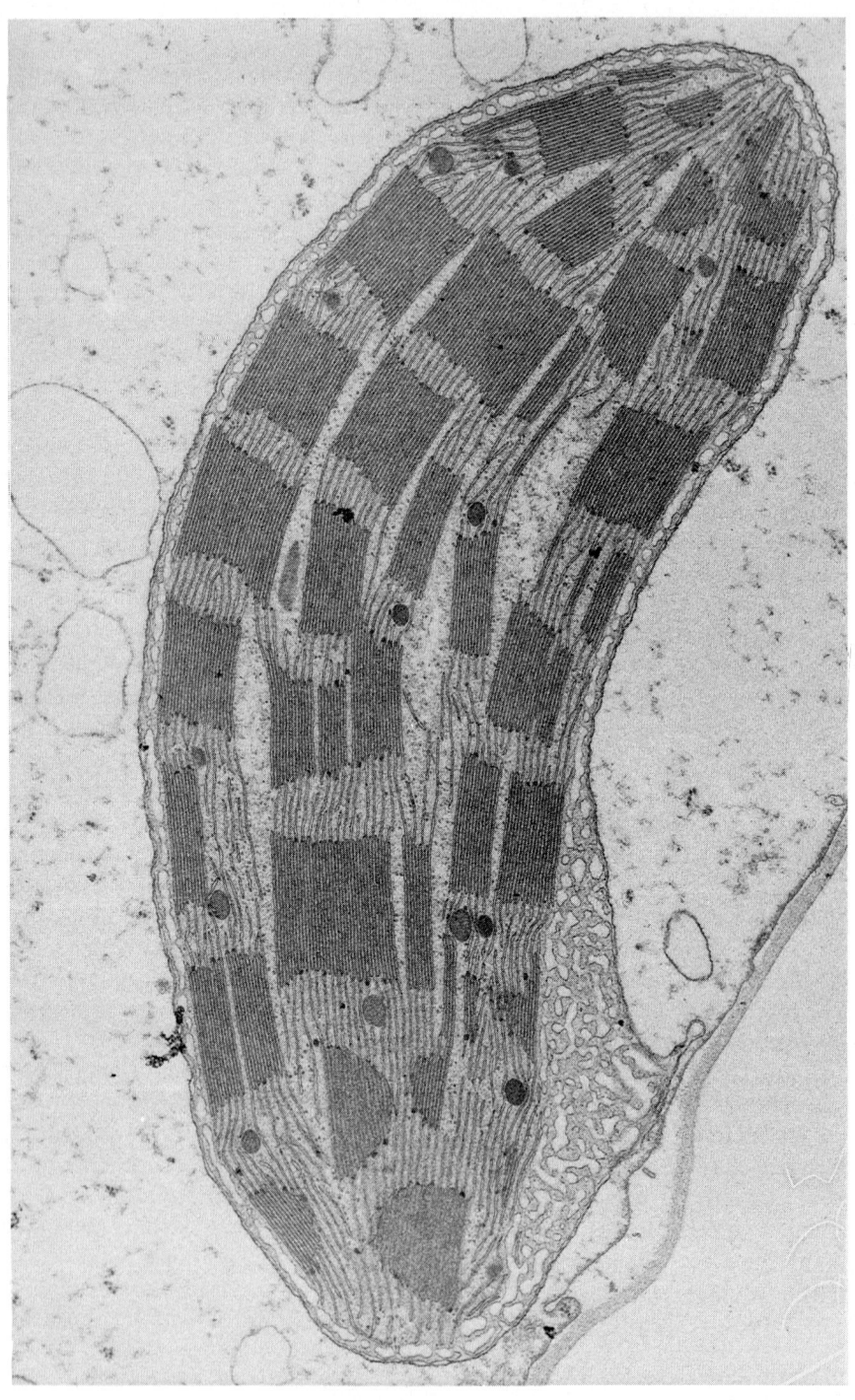

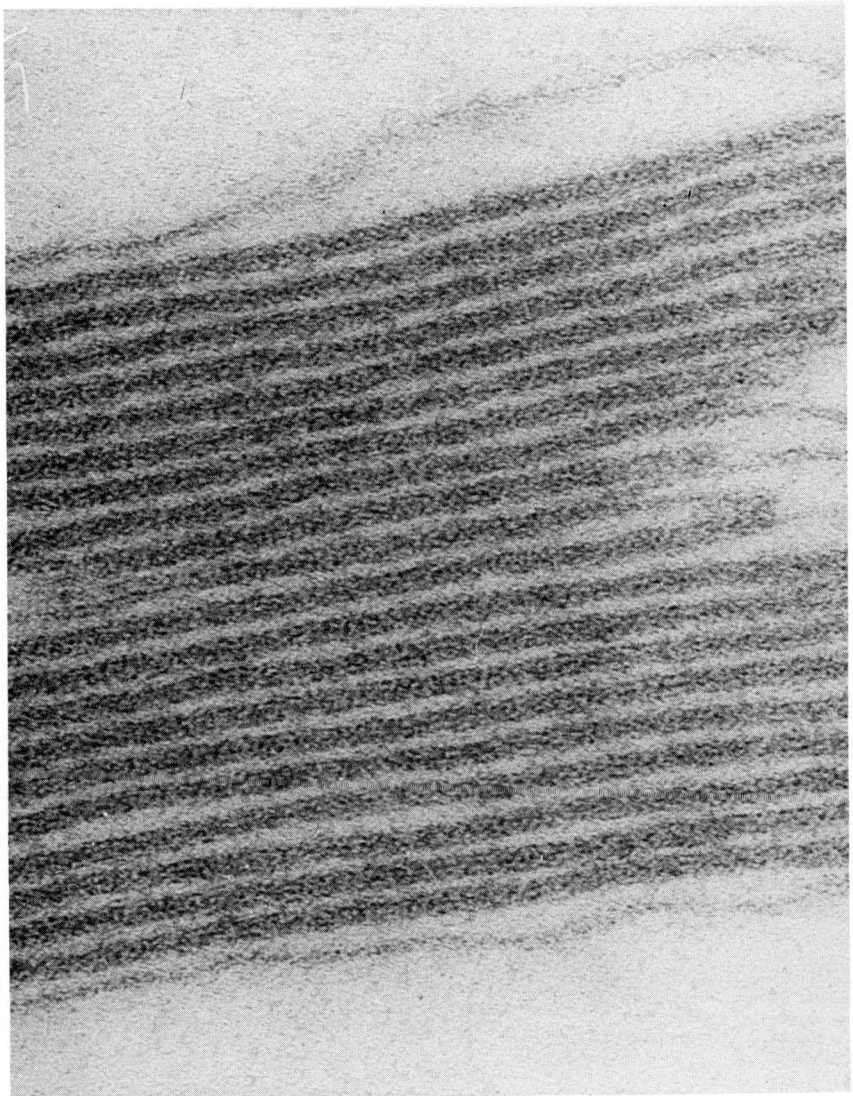

FIG. 2. Electron micrograph showing the membranes within a single granum in an isolated *Vicia faba* chloroplast. Notice the particulate appearance in the print which may correspond to membrane subunits. Magnification × 428,000. Micrograph taken by Dr. L. K. Shumway and reproduced through the courtesy of Prof. T. E. Weier.

FIG. 1. (Facing) Electron micrograph of a chloroplast from a mature *Zea mays* plastid, showing the highly developed grana and stroma lamella system. Magnification × 37,000. Micrograph taken by Dr. L. K. Shumway and reproduced through the courtesy of Prof. T. E. Weier.

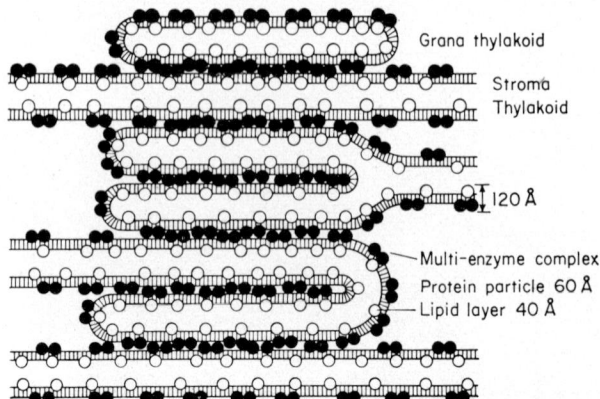

Fig. 3. The structure of a chloroplast granum as inferred from electron microscope studies using the relatively mild freeze-etching technique. Membrane seems to be composed of a central lipid layer covered in both sides by attached or partially embedded protein or lipoprotein particles. Reproduced by kind permission of Prof. K. Mühlethaler.

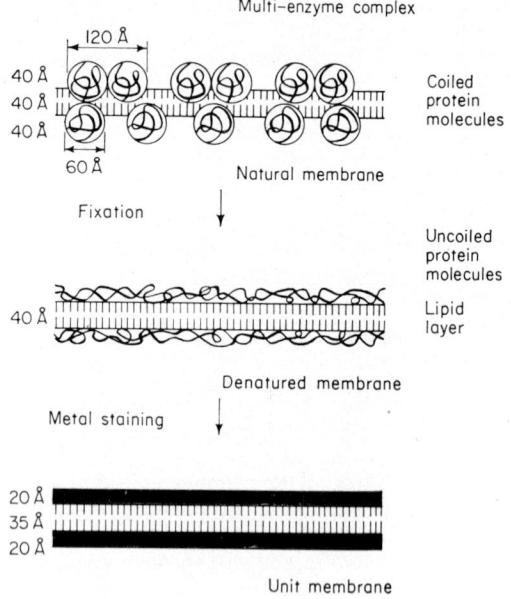

Fig. 4. Diagrammatic representation showing how a globular subunit type of chloroplast membrane may be changed by preparative procedures used in electron microscopy to give the appearance of a "unit membrane" of the Danielli-Davson type. Reproduced by kind permission of Professor K. Mühlethaler.

embedded particle structure into typical unit membrane-like structures (Fig. 4).

Other aspects and confirmatory evidence of the subunit structure of chloroplast lamellae were discussed by various contributors to the recent Aberystwyth meeting on chloroplast biochemistry (Goodwin, 1966 and Benson, 1966).

F. THE ARRANGEMENT OF PIGMENTS WITHIN THE PHOTOSYNTHETIC LAMELLA

Before the recent acceptance of the corpuscular nature of the chloroplast lamellae, most of the evidence seemed to point to the chlorophyll being arranged roughly as a monomolecular layer over almost all of the available lamella surface area (Wolken, 1962; Jacobs et al., 1957; Rabinowitch, 1959; Hodge, 1959). In this view, the chlorophyll molecule would be arranged so that the long phytol tail (about 20 Å long) would fit into the layer system and perhaps contribute some stability to the structure. The porphyrin ring system, approximately planar, of area ~ 200 Å^2 per molecule and thickness across the plane of 3·5–4 Å, was then visualized as forming amorphous monolayers at the hydrophilic protein lipid interface. Variations on this theme are quite numerous (see Calvin, 1959 and Lumry and Spikes, 1960). The essential features of such a model are shown in Fig. 5. In this illustration, some of the more recently discovered and characterized molecules have been included. The model is also intended to illustrate the concept of a reaction centre (chlorophyll a) and the directions of possible energetic processes, such as energy and charge migration, which are considered in later sections.

Though the chloroplast lamella may, in fact, be particulate, the type of model depicted in Fig. 5 may still represent the local environment and interrelationships of the various pigments, enzymes, etc. quite adequately for the present purpose. Furthermore, this type of model provides a useful starting point in the consideration of the primary process of the photosynthetic process. The actual proximity, mutual orientation and relative position of the various types of chlorophyll and other molecules (quinones, ferredoxin, carotenoids, etc.), if known accurately, could be decisive from the standpoint of energetic interaction and energy transfer processes. As pointed out by Calvin (1965), the real problem which must be solved concerns the ordering of the chlorophyll and other molecules in the lamellae, or lamellar repeating subunits, and bridging the gap between the molecular structure of the chlorophyll, carotenoids, etc. and the structures seen in the electron microscope.

Studies of polarized fluorescence, dichroism and birefringence, although

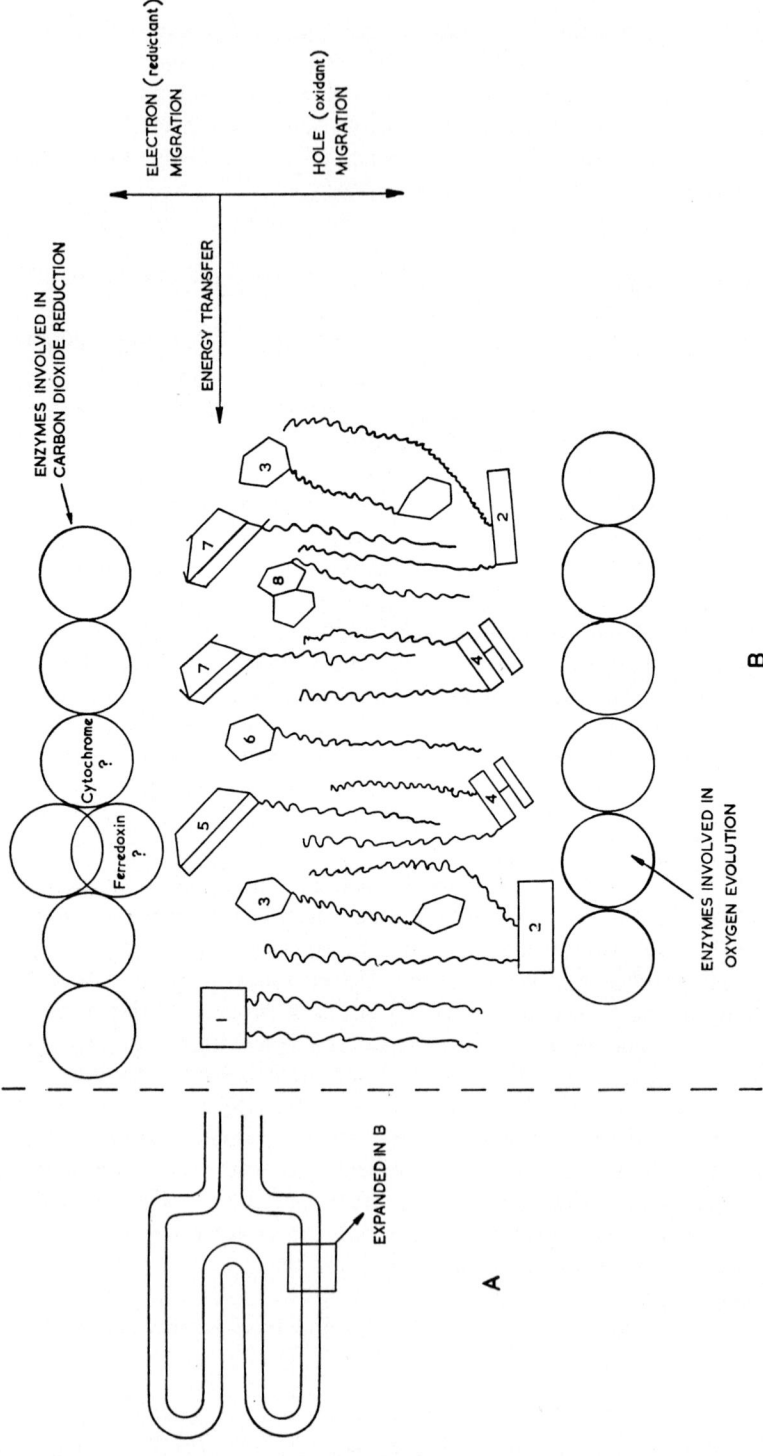

FIG. 5. Highly schematic representation of possible molecular arrangement in a chloroplast lamella near to the reaction centre. Greatly modified and updated version of a diagram originally given by Calvin (1959). 1. galactolipid; 2. phosphatidylglycerol; 3. β-carotene; 4. digalactolipid; 5. chlorophyll a; 6. plastoquinone; 7. chlorophyll b; 8. vitamin K_1. Possible direction of energy transfer and electron or hole migration also indicated. Detailed molecular structures are also given in Figs 6 and 7.

technically very difficult, offer some hope of a successful answer. Once again the problem of the two types of birefringence (or dichroism) makes the interpretation of any observed effects difficult to interpret unambiguously.

In his early studies of birefringence and dichroism, working with very favourable samples of large chloroplasts, Goedheer (1955) showed that the lamellae seemed to contain both lipid and proteinaceous layers and also a very thin, probably monomolecular, layer of pigments. Moreover, the porphyrin head groups of the chlorophyll molecules seemed to have a component parallel to the plane of the lamellae, whilst the lipid molecules appeared oriented perpendicular to the plane. Thus, at best, these studies of dichroism and birefringence indicated only a relatively small degree of orientation of the chlorophyll molecules and studies using fluorescence polarization techniques have since supported this idea. A little later, Goedheer also suggested that the relatively small degree of porphyrin orientation observed could be due to chlorophyll molecules being bound peripherally to spherical lipoprotein macromolecules which formed part of the lamellae (see Weier and Benson, 1966 and Rabinowitch, 1959). The fluorescence depolarization studies of Arnold and Meek (1956) were also interpreted as indicating largely random orientation of the porphyrin planes. Later, more refined measurements extended to longer wavelengths ~ 700 mμ by Olson and his colleagues definitely indicated that a fraction of the chlorophyll a present in *Euglena* chloroplasts was aligned parallel to the plane of the lamellae (Olson *et al.* 1961, 1962; Olson 1963). Studies of electric dichroism and electric birefringence by Sauer and Calvin (1962), exhibited by particulate subunits isolated from spinach chloroplasts, indicated that about 5% of the chlorophyll molecules in a quantasome are oriented parallel to the lamellar structure, while the remainder and the carotenoids are probably relatively unoriented. Spectral studies on model systems also suggest that the degree of order or crystallinity in the chlorophyll-containing layers is probably not very high (Rabinowitch, 1959).

A very detailed model of the molecular arrangement of the chloroplast lamella has been proposed by Weier and Benson (1966) and its more novel features extended to membrane systems generally by Benson (1966). Salient features of the new model are that it takes account of the detailed chemical properties of the molecules which are known to be associated with the chloroplast lamella, for example, of the surfactant nature of the lipids, the structure of the paraffinic chains of the chloroplasts and quinones, etc. Furthermore, it is a globular subunit type of model but, interestingly enough, the "partitions" formed by two layers of asymmetric lipoprotein subunits are separated by a highly hydrophobic region where the chlorophyll porphyrin planes, the carotenoids and energetically important benzoquinone end of plastoquinone are embedded and probably oriented in an as yet unspecified way.

G. SUMMARY

Views concerning the membraneous structures which are obviously of critical importance in bioenergetics have undergone considerable revision in the last few years.

The intimate details of how the various molecules involved in bioenergetic processes are actually located within the lamellar structures are, as yet, not known within any degree of precision. The paucity of this information makes the choice between various mechanisms of energy transfer and charge migration phenomena very difficult. The main conclusion of the work discussed in this section suggests that the lamellar systems involved in bioenergetics are probably made up of more corpuscular subunits than hitherto believed. Furthermore, it seems extremely likely that the actual particulate subunits, which link together spontaneously to give two-dimensional arrays and the appearance of the characteristic lamellar structures, are themselves also the functionally active subunits. This thesis seems to be most clearly illustrated by the elementary particle and quantasome concepts. A given subunit appears to require several constituents to be both functionally and structurally efficient. Specific lipid composition, both the nature of the polar end groups whether glycolipids or phospholipids and the nature of the hydrocarbon chains (degree, type and extent of the unsaturation), seems highly variable yet specific. No less important are those special molecules involved in the bioenergetics of the process (quinones, porphyrins, cytochromes, carotenoids, etc.). Finally, lamellar or structural protein whose largely hydrophobic nature, lack of redox groups and spontaneous tendency to polymerize (Green and Tzagoloff, 1966 and Benson, 1966) must indicate that their role is mainly to provide the structural matrix for the bioenergetic efficiency of the process to be maximized (probably by hydrophobic interaction with specific lipids).

A great gap in our knowledge at the present time is clearly a lack of real understanding of the various types of interactions involved in the formation of lipid–protein, lipid–pigment, pigment–protein subunit complexes.

For the purpose of this chapter, it seems only possible to consider bioenergetics from the following standpoint. The molecules involved occur in quite high concentrations ($\sim 10^{-2}$ mole/l for chlorophyll in the chloroplast and $\sim 10^{-3}$ M for rhodopsin in ROS membranes) and are highly localized in regions of the organelle under consideration. However, the degree of organization or orientation, although considerable, is far from justifying the term "crystalline array".

III. Functional Aspects of Membrane and Bioenergetics
A. CHEMICAL COMPOSITION

1. *Retinal Rods*

Rod outer segments appear simpler in chemical composition than cristae or photosynthetic lamellae. The main constituents of the retinal photo-receptor membranes are proteins, lipids (particularly phospholipids) and the pigment retinene, possibly small amounts of cholesterol and an oxidizing component. The oxidizing component was described as an ubiquinone, located in the lipid plane of the outer disc and ascribed an electronic function by Pearse (1965). The identity of the redox component has recently, however, been challenged by Fleischer and McConnell (1966). The most striking feature of the chemical composition of the retinal membranes is the extremely high lipid content which may be 60% on a weight basis. Of the lipid fraction, some 90% is phospholipid, mainly phosphatidylethanolamine, lecithin and acidic phospholipids such as serines (Fleischer and McConnell, 1966). Earlier analyses (quoted from Wolken, 1962) indicated that the outer segments consisted of 40–50% protein, 20–40% phospholipids and 4–10% retinene. The pigment retinene (shown in Fig. 6) is, of course, intimately associated both structurally and functionally with the protein opsin and it seems very likely that rhodopsin itself must contain a very high percentage of phospholipid ($\sim 50\%$). Largely on analytical grounds, it was suggested that rhodopsin could form 35% of the weight of frog outer segments and that this rhodopsin is probably a structural subunit of the ROS membrane (Moody, 1964).

It is particularly interesting that the globular particles, observed by Blaisie et al. (1965) in or on the membranes, were suggested by these authors to possibly be, rhodopsin molecules on the basis of their size.

2. *Photosynthetic Lamellae*

The chemical composition of photosynthetic lamellae is extremely variable and complex (Goodwin, 1966). Roughly, photosynthetic lamellae are 40–50% proteinaceous and the remainder may be characterized as soluble in lipid solvents. The various functional chemical classes seem to be both structural and enzymatic proteins, lipid (phospholipids, glycolipids), pigments, chlorophyll, carotenoids and quinone type compounds and inorganic materials (e.g. non-haem iron, copper, manganese).

The most accurate analytical data on the composition of functionally active chloroplast lamella preparations (quantasomes) are those of Park and Pon (1963), Lichtenthaler and Calvin (1964) and Park and Biggins (1964). From data of Park and Biggins, the quantasome composition may be expressed as in Table I. The lipid soluble contribution may be further subdivided as in Table II.

TABLE I
Quantasome composition

Total molecular weight	1,920,000
Protein contribution to total molecular weight	930,000
Lipid soluble contributions to total molecular weight	990,000

Data taken from Park and Biggins (1964).

TABLE II
Composition of lipid soluble fraction of spinach quantasomes, taken from data of Park and Biggins (1964)

Total "lipid soluble" fraction	990,000
Chlorophyll *a, b*	206,400
Carotenoids	27,000
Quinones	31,800
Phospholipids	90,800
Glycolipids	402,000
Sulpholipids	41,000
Unidentified lipids	190,000

3. *Mitochondrial Cristae (or Elementary Particles)*

Michondrial cristae, elementary particles, and also the four primary complexes into which the elementary particle has been fractionated, all contain about 30% of lipid. Of this lipid fraction more than 90% is phospholipids. The main phospholipids are lecithin, cephalins and cardiolipin, the balance of the lipid content being made up predominantly by quinones (ubiquinones) and cholesterol (Green and Fleischer, 1963; Fleischer *et al.*, 1961).

Mitochondrial lipids show significant poly-unsaturation, an average of 3·2 double bonds per atom of phosphorus being found by Fleischer *et al.* (1961). The protein complement of mitochondrial preparations is made up of the oxidation-reduction proteins involved energetically in the electron transfer process, cytochromes, flavoproteins, metalloproteins etc. ($\sim 20\%$), soluble proteins involved in other mitochondrial functions (20%), the remaining 60% being structural protein (Criddle *et al.*, 1962). The structural protein appears to combine with phospholipids to form the membraneous matrix on or in which the elementary particles are bound.

In this chapter, however, we shall merely accept that, because of the structural protein–phospholipid interaction, the functional groups are arranged for the most efficient energetic interaction. The essential role of lipids, particularly phospholipids, in electron transfer has been recognized for several years (Green and Lester, 1959; Green and Fleischer, 1962, 1963).

The structures of various lipids which are found in the lamellar systems under discussion are shown in Fig. 6.

B. SOME MOLECULAR STRUCTURES INVOLVED IN MEMBRANE BIOENERGETICS

A unifying feature of the function of retinal rod photoreceptors, electron transfer particles and photosynthetic lamellae is the overall similarity of the detailed structures involved, which we believe extends down to the various molecular species involved (Fig. 7).

1. Long Chain Quinones

Long chain quinones have been shown to perform an energetic role in both mitochondrial function and photosynthesis. In Section III-A-1 above, the position with regard to the occurrence of ubiquinones in the retina was mentioned. The significant feature, as far as the ROS system is concerned, is that a compound with redox potentiality, which may be an ubiquinone, does exist (Fleischer and McConnell, 1966). The quinones under discussion are ubiquinones (coenzyme Q_n) in mitochondrial function and plastoquinones and Vitamin K in photosynthetic function. In beef heart mitochondria, ubiquinone (CoQ_{10}) was suggested to be on the main path of electron transfer in oxidative phosphorylation, mediating electron flow between complexes I, and II and complex III, but this is disputed by Redfearn (1966).

Plastoquinones, of which several have now been characterized, are composed of a 2,3 dimethyl benzoquinone nucleus with a long hydrocarbon chain attached (Crane et al., 1966). The most commonly occurring side chain in chloroplasts seems to have nine isoprenoid units (PQA45) and is thus of some considerable length. A plastoquinone-like compound also undergoes oxidation-reduction during photosynthesis and has been suggested to be a component of the electron transfer pathway between photochemical systems 1 and 2 (Amesz, 1964).

Before the discovery of the widespread occurrence and probable absolute requirement for ubiquinones and plastoquinones, a role for other long chain quinone derivatives, the K vitamins, had been demonstrated in electron transfer in bacteria (Brodie et al., 1957; Brodie, 1961) in redox processes involving pyridine nucleotides and cytochromes (Wosilait, 1961) and in photosynthesis (Bishop, 1958, 1959; Arnon, 1961). The structures of the quinones under discussion are shown in Fig. 7.

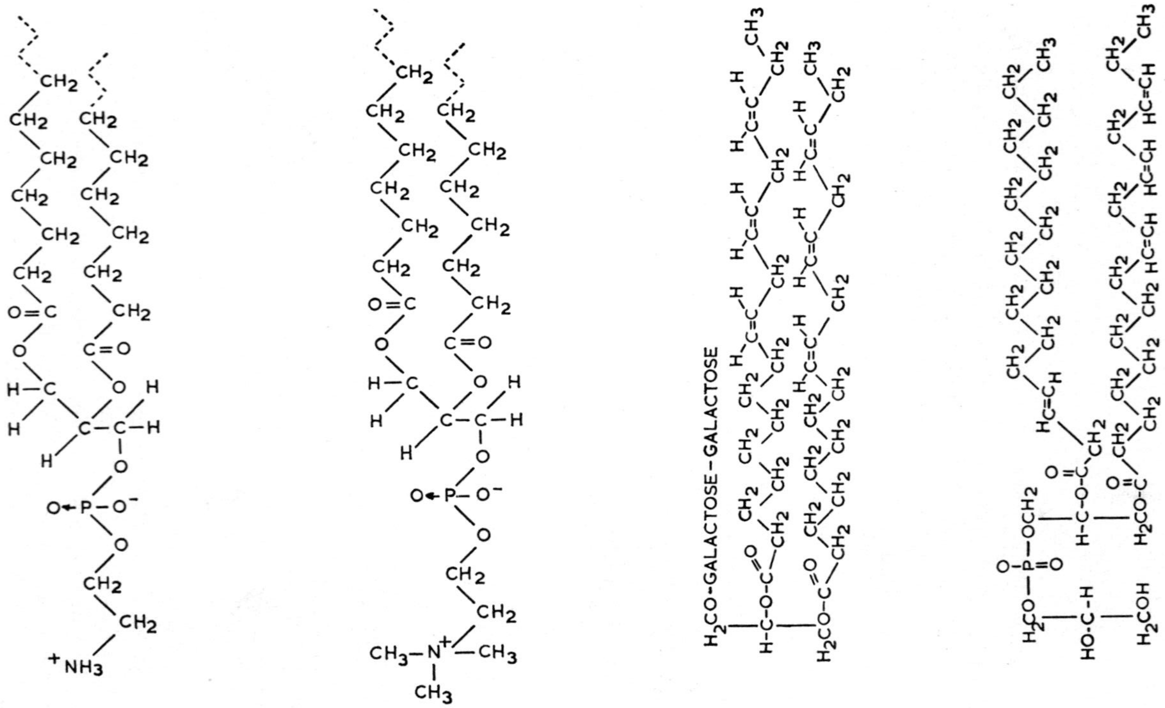

Phosphatidyl ethanolamine (cephalin)

Phosphatidyl choline (lecithin)

Digalactosyl lipid

Phosphatidyl glycerol

FIG. 6. Typical molecular structures of lipids involved in membrane formation in mitochondrial cristae, photosynthetic lamellae and retinal rods.

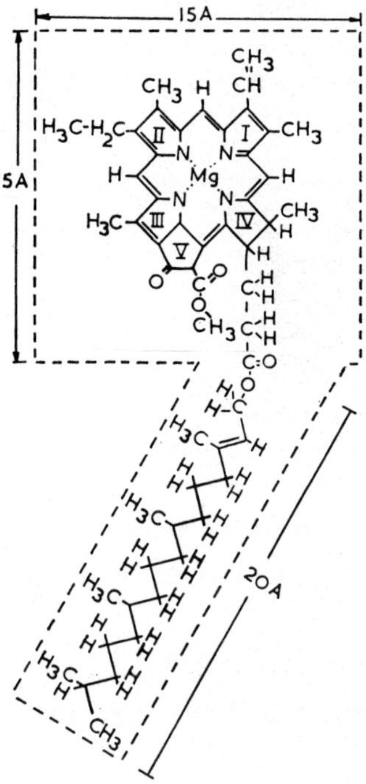

Chlorophyll *a* 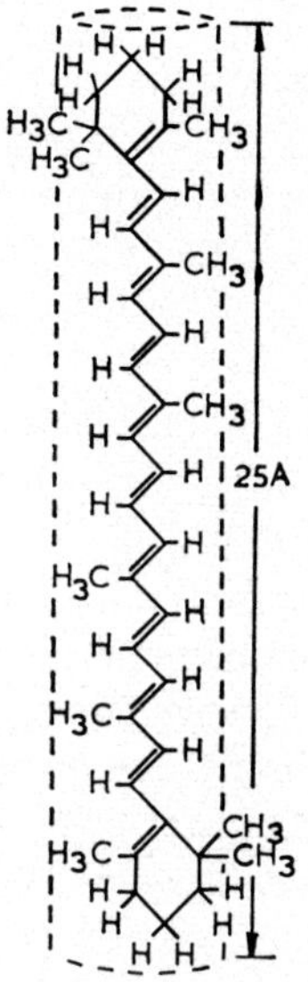   β-carotene 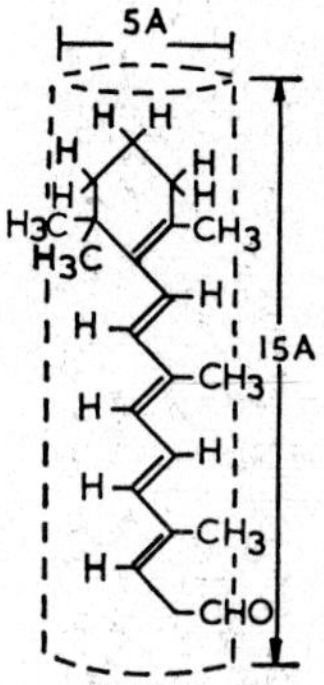   Retinene₁ (retinaldehyde)

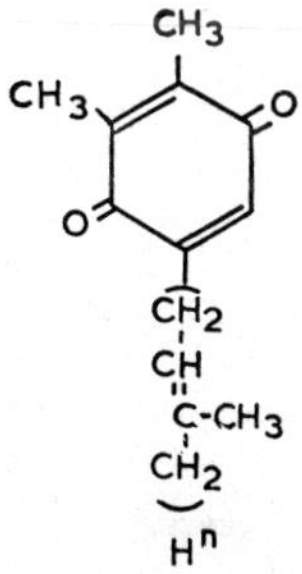

Plastoquinone 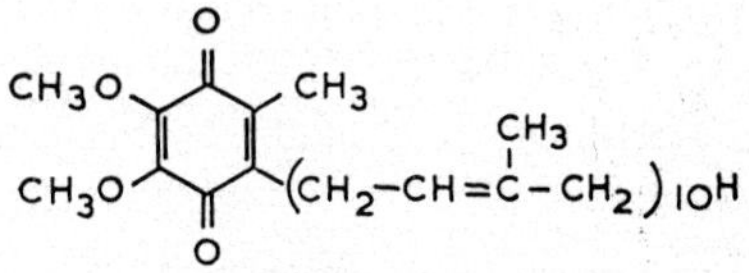   Ubiquinone $C_0 Q_{10}$

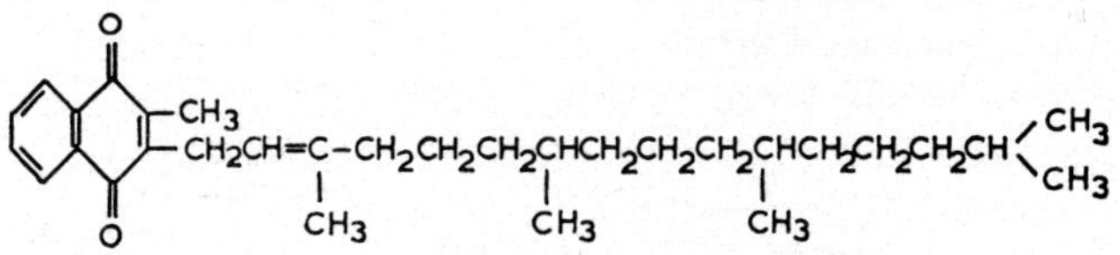

Vitamin K_1

FIG. 7. Molecular structures of some pigments involved in bioenergetics.

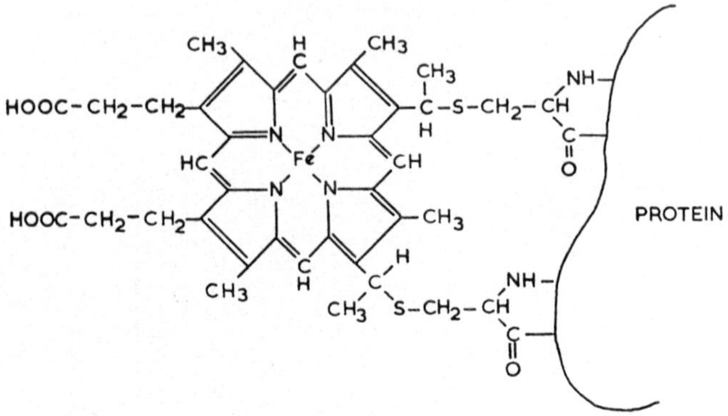

Prosthetic group of cytochrome *c*
FIG. 7. (continued).

2. Carotenoids and Carotenoid Derivatives

The occurrence and functional role of carotenoid pigments is largely confined to the rod outer segment and the photosynthetic lamella.

The structure of β-carotene and retinene, the initial light receptor pigments in the ROS membrane, are shown in Fig. 7.

The complex chemistry of retinene and its derivatives, worked out largely by Wald and his colleagues (Wald, 1965; Kropf and Hubbard, 1958) involved in the visual process, does not concern us here since the "only" action of the light (i.e. the quantum conversion act) is to cause the isomerization of 11-*cis* retinaldehyde to the all *trans* configuration and, in some way, initiate the visual impulse. However, it is clear that changing the hydrocarbon chain of retinene from a planar to a kinked configuration which may trigger the visual impulse must be related in some way to the problem of the conformation of rhodopsin or the rod outer segment membrane structure.

In chloroplast lamellae, carotenoids of the β-carotene type occur ubiquitously but only to the extent of a few per cent (Wolken, 1962; Menke, 1966). Their role is thought to be as protective agents against destructive photosensitized oxidations but from time to time a more active role has been proposed (Lumry and Spikes, 1960). For example, Calvin (1958) suggests that electronic effects involving carotenoids may play a role in the primary electron-hole separation and Lynch and French (1957), on the basis of solvent extraction studies, allotted β-carotene a definite but unspecified role in chloroplast function. Certainly, it is well established that carotenoids may absorb energy which can be transferred to chlorophyll and contribute to the overall efficiency of photosynthesis (Duysens, 1964). The question of carotene function in photosynthesis is, however, kept open by the recent observation of Lundegardh (1966) which suggests that carotenoids may be involved in the transfer of electrons to ferredoxin, one of the very early stages of the photosynthetic process.

3. Cytochromes

The cytochromes, together with their attached proteins, are largely involved only with mitochondrial function and photosynthesis. In mitochondrial function, the cytochromes involved are cytochrome c, c_1, b, a, a_3, whereas in both plant and bacterial photosynthesis, cytochrome f, cytochrome b_6 have also been characterized. Purely for comparative purposes, Fig. 7 contains the prosthetic group of cytochrome c.

4. Chlorophylls

As retinene is the heart of the visual process, so chlorophyll in its many guises is the heart of the energy absorption and conversion act in photosynthesis.

The many functionally important types of chlorophyll are discussed in the recent review by Duysens (1964) and in connection with some other biophysical problems by Clayton (1965). Chlorophyll a seems to occur in all higher plant chloroplasts and probably it, or a percentage of it, constitutes the photochemically active reaction centre. The approximate dimensions and the chemical structure of chlorophyll a are shown in Fig. 7.

5. Other Molecules

Other molecules, which are present and definitely play a role in bioenergetics, are pyridine nucleotides and trace metals, e.g. non-haem, copper and manganese. In the specific case of photosynthetic quantum conversion, unique low molecular weight iron-containing proteins, ferredoxins, are currently of great interest because of their exceptionally low negative redox potential ($E_0' = -423$ mV at pH 7·55) (Arnon, 1965). Ferredoxins appear to be the first clearly defined chemical molecules which are produced by the primary quantum conversion act.

The approximate dimensions of the various molecules involved in bioenergetics may be estimated by comparison with chlorophyll and β-carotene shown in Figs 6 and 7.

C. SOME FUNCTIONAL SIMILARITIES

Clearly, the molecular structures of some of the molecules involved in bioenergetics will predetermine not only their localization but also their orientation into lamellar formations. In the case of the quinones and chlorophylls, particularly, the molecules will be oriented by both their slightly hydrophilic and bioenergetically important end groups and by hydrophobic interaction between their long chains and hydrophobic amino acid side chains or lipid hydrocarbon chains. This contribution to the orientation of the end group (quinone or porphyrin) by hydrophobic interactions involving the "tail" will probably prove to be much more important than is generally recognized.

At the present time, the visual process seems to function differently from either the photosynthetic quantum conversion process or mitochondrial oxidative phosphorylation. Energy and electron transfer processes, as exemplified by quantasome and elementary particle composition and function, show very marked similarities (Park and Pon, 1963). Both quantasomes and the elementary complexes of mitochondria contain cytochromes, quinones, pyridine nucleotides and trace metallic elements. In both systems, initially large quanta of energy are provided at the beginning of a chain of similar molecular complexes which have specific redox potentialities. These electron transferring molecules, by sequentially coupling redox reactions, enable the initially large energy quanta to be broken up into much smaller units, each unit being sufficient to bring about the coupled production of ATP from ADP and inorganic phosphate. In many ways, mitochondrial function and photosynthetic energy conversion can be visualized as the same process, running in opposite directions. In elementary particles, the direction of electron transfer is from reduced pyridine nucleotides to oxygen, an oxidative exergonic process, leading to the production of water. On the other hand, photosynthesis is an endergonic process. In non-cyclic photophosphorylation, water is split, to give separated oxidizing and reducing entities, oxygen is evolved and ATP is produced. The electrons travel via chlorophyll and ferredoxin and thence to pyridine nucleotides, finally leading to the reduction of carbon dioxide (Arnon, 1965, 1966).

Cyclic photophosphorylation in chloroplasts also shows marked similarities to mitochondrial electron transfer. In this process, an electron is raised to an excited level of the chlorophyll molecule by absorbed radiant energy and returns to the chlorophyll ground state through a chain of carriers (plastoquinone, cytochrome b_6, cytochrome f) with concomitant coupled production of, and energy storage in, ATP. Both elementary particles and chloroplasts contain structural proteins and both have a high lipid content of about 30%.

IV. The Evolution of Physical Bioenergetics

The need for a bioenergetic approach to mitochondrial function, photosynthesis and vision using physical concepts has been apparent for many years (Augenstine, 1960). The problem of energy transfer in photosynthesis was first clearly hinted at in the well known and classical experiments of Emerson and Arnold (1931, 1932) who were studying photosynthesis in flashing light. The work of Emerson and Arnold and its interpretation by Gaffron and Wohl (1936) as indicating a "photosynthetic unit" in which several hundred chlorophyll molecules or pigments were energetically coupled to one particular "reaction centre", has now been completely verified (Clayton, 1965; Bay and Pearlstein, 1963). The photosynthetic unit in chloroplasts seems to contain about 300 chlorophyll molecules coupled to one photochemically active reaction centre.

It is worthwhile to bear in mind that the primary biophysical aspects of the three processes under discussion may be concerned with the interrelated phenomena of energy migration and also the migration of separated primary oxidants and reductants. In the limit, the primary reductants and oxidants are probably electrons and "positive holes". These two different aspects of bioenergetics, energy migration and charge separation by the uncoupled movement of discrete electrons and holes, are seen most clearly to occur concurrently in the photosynthetic process. Here, it is well established that the radiant energy absorbed by any one of a number of different pigments is quickly and efficiently transferred to the photochemically active reaction centre (Clayton, 1965; Bay and Pearlstein, 1963). On arrival at the reaction centre or its particular chloroplast environment, the peculiar properties of the special chlorophyll molecules allows this excitation energy to be converted into a stabilized and utilizable separation of oxidizing and reducing power as holes and electrons. This separation and stabilization of the primary oxidants and reductants, which then migrate independently and perform their functions individually, is the central problem of the quantum conversion and energy storage act in photosynthesis (Calvin, 1961; Arnold, 1965). This remarkable feat, the conversion of 30–40 Kcals of energy into discrete electrons and holes, separated or stabilized in space so as not to back react, results from the organization of chlorophyll and other molecules into lamellar formations.

Energy transfer *per se* from the site of initial absorption to the reaction centre is now fairly well understood as a result of both theoretical and experimental studies (Bay and Pearlstein, 1963; Duysens, 1964). However, lack of understanding of the chlorophyll-mediated initial charge separation and subsequent redox reactions represents the greatest gap in the understanding of the photosynthetic process at the present time. The first clear statement that the primary quantum conversion act was the production of a transient oxidant and a transient reductant, is usually considered to be the comparative biochemical studies by van Niel (1941) on a range of living organisms. This thesis has apparently guided the thoughts and experimental approach of a great many eminent workers of whom we may cite Calvin (1961) and Arnold (1965). The replacing of van Niel's (1941) conception of the oxidant and reductant from the original (OH) and (H), respectively, into the language of physics as "holes" and "electrons" is due to Katz (1949) and Bradley and Calvin (1955).

In mitochondria, the discrete electron transfer through the medium of sequential coupled redox reactions is of great significance in the context of membranes and bioenergetics. A similar requirement for electron transfer is involved in the linking of systems I and II of the photosynthetic process and during the process of cyclic and non-cyclic photophosphorylation (Clayton, 1965). In the specific case of the components of the electron transfer chain, whether one considers the elementary particles or the isolated complexes themselves, the problem is: how do the electrons actually travel from prosthetic group to prosthetic group with the necessary specificity and particularly flux to account for the high rate of respiration observed, bearing in mind the actual physical dimensions of the molecules involved (Lehninger, 1959).

In retinal rods, the bioenergetic process represents the conversion of electro-magnetic radiation into a characteristic nerve impulse.

Because of the great efficiency and sensitivity of the visual process, shown by the fact that absorption of a quantum of green light by any one or perhaps million of rhodopsin molecules has a 30% chance of generating a nervous impulse (Hecht et al., 1942), there seems to be a need for efficient energy migration in the rods (Hagins and Jennings, 1959). The rod outer segment is a good example of a system whose highly organized and quasi-crystalline structure suggests that typical solid state energy transfer mechanisms may take place (Brown et al., 1963; Wald, 1965).

As discussed above, there is no doubt that the processes of energy and electron transfer are intimately connected with lamellar organization and orientation of specific molecules with particular combinations of bioenergetic properties. We will now consider some physical ideas encountered in bioenergetics which are necessary prerequisites to any attempt to answer the following questions: (1) what is the role of the membrane in bioenergetics, (2) which particular bioenergetic mechanisms will be facilitated by membrane location of specific molecules and (3) how can membranes separate oxidizing and reducing power?

V. Physical Bioenergetics

A. ELEMENTARY CONCEPTS

Many of the molecules of importance in bioenergetics are highly coloured. As colour reflects the absorption of electromagnetic energy and excitation of electrons between characteristic energy levels, it is reasonable to infer that these specific energy levels play a fundamental role in bioenergetics.

A convenient energy level diagram to explore this role is shown in Fig. 8 which depicts the $(n\pi^*)$ excited states as being of lower energy than the corresponding (π, π^*) excited states. This is a situation probably appropriate to many, but not all, biomolecules. In singlet states, represented by $^1\Gamma\,(n\pi^*)$, the electron spins are still paired. In the triplet state, represented by $^3\Gamma\,\pi, \pi^*$, the excited electron loses a little excess energy and reverses its direction of spin. In Fig. 8, if the electron moves between various energy levels and pho-

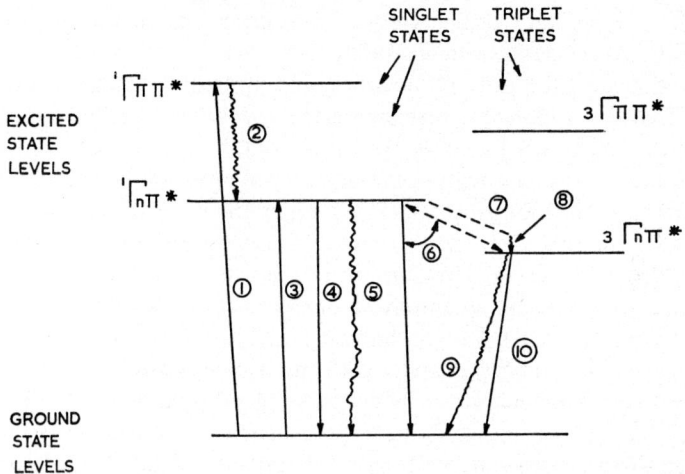

FIG. 8. Schematic energy level diagram. Upward pointing arrows indicate energy absorption, downward arrows energy emission or dissipation; solid lines represent radiative transitions; wavy lines radiationless transitions; dashed lines indicate radiationless intersystem crossing. See text for classification of various transitions.

tons are absorbed or emitted, this is a radiative transition and is depicted by a solid line. Wavy lines represent electronic transitions which are not accompanied by emission or absorption of radiation (radiationless transition) because the energy exchanges with the environment, largely thermally.

The absorption of energy, leading to population of excited singlet states $^1\Gamma\pi, \pi^*$, $^1\Gamma n\pi^*$ is represented by processes 1,3. The remaining processes are as follows:

2,8—extremely rapid radiationless internal conversion.

5,9—relatively unimportant slow radiationless deactivation of states $^1\Gamma n\pi^*$, $^3\Gamma n\pi^*$.

4—emission of a quantum of (prompt) fluorescence light (usually within 10^{-8}—10^{-9} sec after absorption).

10—this represents the return of an excited electron to the ground state from the triplet $^3\Gamma n\pi^*$ state with the emission of a quantum of phosphorescent light (10^{-3}—10^2 sec).

The remaining process, 6, represents correctly "delayed" fluorescence. This arises when some of the electrons in the triplet state gain a little energy and revert to the singlet state followed by emission of a quantum of light. The emitted "delayed" light is identical in spectral characteristics to a quantum of normal prompt fluorescence. The "delay", the longer lifetime and the fact that this type of "fluorescence" has a characteristic temperature dependence reflects the manner in which the triplet–singlet transition is induced, whether by further energy absorption or mutual triplet–triplet anihilation processes (Robinson, 1963).

To understand the fate of absorbed and potentially biologically useful energy, consider a molecule with an electron in the first excited singlet state. The energy may be dissipated in several ways: (a) the electron may return to the ground state and a quantum of fluorescence light will be emitted (process 4). In this case, the biological system (of which the molecule is a part) has not benefitted in any way. (b) The excess energy may be degraded into heat (5) or (c) after spin reversal, the electron will go into the triplet state (7). The final possibility (d) is that the quantum of energy may be transferred (usefully or wastefully) to another molecule. Should possibility (c) be operative and the electron goes into the triplet state, then the triplet excitation energy may be dissipated in several ways: (1) by emission of phosphorescence, (2) thermal degradation, (3) by energy transfer to another molecule or system, (4) reversion to the singlet state, followed by emission of delayed fluorescence (6).

B. ENERGY TRANSFER BY RESONANCE MECHANISMS

1. *The Förster Mechanism*

The particular mode of energy transfer will depend upon the following factors:

(1) if we are considering the transfer of energy from an excited singlet or excited triplet state.

(2) the degree of intermolecular coupling between the subunits which go to make up the aggregate. This applies particularly in the case of transfer from excited singlet states.

(3) is energy transfer being considered in an aggregate of similar molecular subunits or different molecular subunits?

If the system is relatively fluid, unordered and not very strongly coupled on a molecular scale, then energy transfer from molecules in the triplet state is not very probable. We may concentrate on transfer from the excited singlet state. Several mechanisms are possible.

(a) the trivial radiative mechanism. The quantum of excitation energy is emitted by the excited molecule and reabsorbed by an unexcited molecule. This may be repeated several times but the mechanism is only of minor importance.

(b) in non-radiative, intermolecular energy transfer, there are two important mechanisms:

(1) Energy transfer between molecules which come into close contact and (2) energy transfer between molecules which are relatively widely separated on a molecular scale, particularly between widely separated dissimilar molecules.

The latter condition will approximate, of course, to several mixed pigment bioenergetic systems of relatively small degree of orientation. Energy transfer under these circumstances is usually now termed the "Förster mechanism" or "resonance transfer" or "energy transfer by inductive resonance". The main role of the membrane systems involved in "Förster energy transfer" will be to bring about those conditions under which the efficiency of the transfer process is maximized or perhaps confined to a two-dimensional plane. In the Förster mechanism, there are always two molecules, the energy donor and the acceptor, which may be represented as in Fig. 9.

If the molecules D, A, are sufficiently strongly coupled, have suitable spectral overlap and, particularly, proximity, then there is a definite probability of energy transfer between D and A.

The detailed mathematical theory developed largely by Förster following earlier treatments by Perrin (Förster, 1960) shows that energy transfer by this mechanism of inductive resonance depends mainly upon:

(1) the degree of overlap of the fluorescence spectrum of the donor and the absorption spectra of the acceptor and

(2) the sixth power of the separation of the donor and acceptor molecules.

Resonance transfer is clearly demonstrated experimentally by the appearance of sensitized fluorescence. Energy absorbed by the donor is emitted as fluorescence characteristic of the acceptor (shown in Fig. 9). Resonance transfer between similar molecules can be detected experimentally by the phenomenon of concentration depolarization of the emitted fluorescent light. As examples of experimental studies, we may mention the sensitization of chlorophyll *a* fluorescence by chlorophyll *b* (Watson and Livingston, 1950; Duysens, 1951) and depolarization of chlorophyll fluorescence observed in grana by Arnold and Meek (1956).

2. *Energy Transfer by Excitons*

Turning to more highly ordered and strongly coupled systems, energy transfer is more complex. The problem of membrane function in bioenergetics largely revolves around the extent of the intermolecular coupling it introduces into the subunits. Thus, from a bioenergetic viewpoint, a different situation exists depending on whether the subunits which make up the molecular lamella are associated by weak van de Waal's forces or very crystalline arrays held by orbital overlap interactions.

A simplified picture of an "exciton" can be introduced as follows. In widely separated and non-coupled molecular aggregates, each molecule may be characterized by its own set of ground and excited state energy levels. As the degree of molecular ordering and intermolecular coupling increases, the

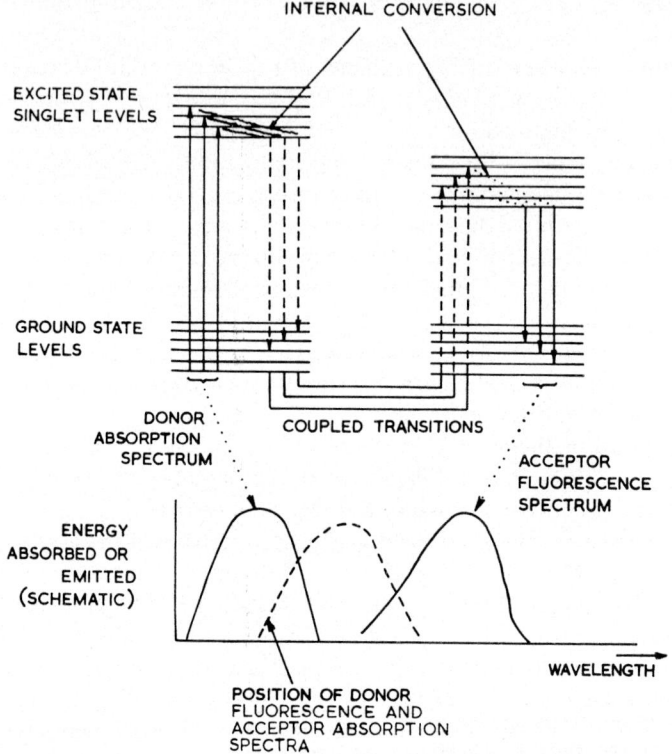

FIG. 9. Schematic diagram to illustrate energy transfer by the Förster-type mechanism. In the energy level diagram heavy lines represent electronic levels, light lines are vibrational levels; solid lines show energy absorption by donor and fluorescence emission by acceptor; dashed lines represent corresponding coupled transitions in donor and acceptor, respectively. Schematic below shows relative position on wavelength scale of various transitions. Diagram indicates 100% efficient energy transfer between donor and acceptor.

individual energy levels interact and fuse into closely spaced bands of energy levels. These energy levels are now more truly characteristic of the molecular aggregates as a whole. Such energy levels, belonging to the whole structure, are called "exciton bands".

The formation of a set of exciton bands, from the individual molecular energy levels, is shown schematically in Fig. 10.

In Fig. 10 the formation of singlet exciton levels is indicated; of course, there can also be triplet exciton levels, probably of considerable importance in bioenergetics. Assume that we have such a coupled aggregate as represented on the right band side of Fig. 10. In the resting state, all the lower valence band energy levels will be filled with paired electrons.

Absorption of a quantum of energy will raise an electron from the valence band into the singlet exciton level. However, because the spacing of the individual singlet exciton levels is extremely small compared to thermal quanta, the excited electron may wander relatively freely from molecule to molecule in the aggregate.

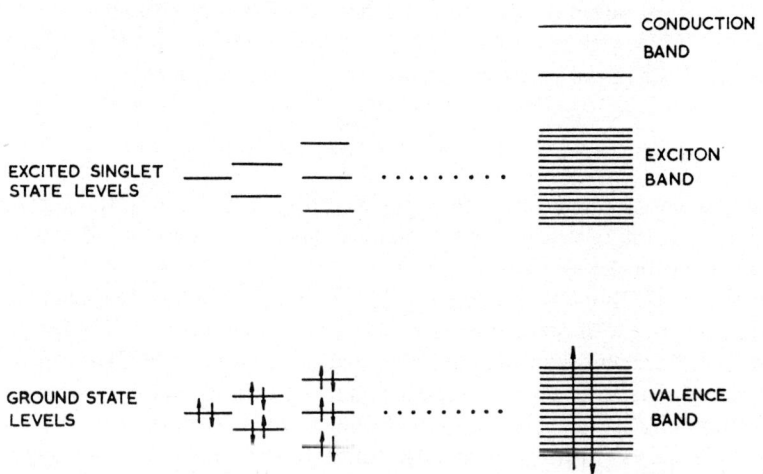

FIG. 10. Schematic illustration of the formation of a band of singlet "exciton" energy levels in a molecular aggregate from the interaction of a large number of individual molecular energy levels. Conduction band also shown. Arrows represent direction of electron spin.

Thus we think of the excited state as being characteristic of the whole array of molecules and not localized on any particular molecule in the array. However, under certain circumstances, if we had "instantaneous" time resolution the excitation energy could be visualized as being localized in a particular area of the array or even on a particular molecule. The important point is that we must regard the excitation energy in general as not being localized, but rather as being free to wander or migrate throughout the whole array of molecules. This excited state still represents, however, an electron in an excited level and its associated vacancy (a "positive hole") in the valence band. This associated "electron–hole" pair held together by coulomb forces is called an exciton. Just as the excitation energy was free to wander throughout the array of molecules forming the aggregate, so, too, must the electron–hole pair (exciton) have free mobility throughout the aggregate, since, in reality, the mobile excited "state" and the "exciton" are merely cause and effect.

Since an "exciton" represents an associated electron–hole pair, it is clear that exciton migration *per se* cannot lead to charge separation or independent mobility of the electron or hole and hence to electrical conductivity.

Exciton theory in its more advanced presentations is divided into two categories; (1) the free (strong coupling) fast exciton case, where all molecular individuality is complete lost and (2) the weak, slow exciton transfer, where the "excitons" may be regarded as jumping from molecule to molecule.

In the strong coupling exciton case, the rate of energy transfer is extremely rapid ($\sim 10^{15}$ sec^{-1}), much faster than the rate of molecular rearrangements. The rate (k_1) is given by an expression of the following form

$$k_1 \sim f\frac{(M)}{R^3}$$

where f is the oscillator strength of the electronic transition in the individual subunit, R is the separation of the dipoles and M is a function related to the arrangement of the dipoles.

In the weak coupling exciton case, a similar $[(m)/R^3]$ dependence is observed together with other factors—transfer rates here are $\sim 10^{12}$ sec^{-1}.

In the Förster mechanism, the rate equation contains the factor $[(m)/R]^3$ squared and therefore there is a sixth power dependence upon the separation of the subunits. Transfer rates by the Förster mechanism are generally slower than by either exciton mechanisms. Rates in the range 10^6–10^{10} sec^{-1} are predicted by detailed theory comparable with the rate of prompt fluorescence. It is because the rate of the Förster process approximates to natural fluorescence lifetimes that energy transfer by this mechanism can compete with fluorescence as a means of deactivating the excited state.

C. POSSIBLE ROLES OF THE MEMBRANE IN ENERGY TRANSFER

A high degree of order and a high local concentration of pigments in lamellar systems will lead to the following effects:

(1) The most efficient absorption or capture of energy.

(2) In strongly coupled aggregates, the excitation energy will be led away from the site of absorption extremely rapidly, possibly to a more biologically useful site.

(3) Distance effects. This will be controlled by localization of the pigment into the membrane and will influence the rate of energy transfer processes (d^6 for Förster energy transfer, d^3 for exciton transfer).

(4) Orientational effects. The relation of one chromophoric group to its neighbour in the molecular array, which may be controlled by the localization of the non-chromophoric group in the lamellar formation, will influence the exciton band structure and processes which are related to the band structure (cf. Hochstrasser and Kasha, 1964).

(5) A particularly important effect of pigment aggregation in lamellae may be to lead to enhanced population of the relatively long-lived triplet state by an increased probability of intersystem crossing from the excited singlet state. This has been pointed out and developed by McRae and Kasha (1958) and the possible biological implications outlined by Kasha (1959). If lamellar aggregation leads to enhanced triplet state population, this provides the cell or organism with a mechanism of temporary storage of the excitation energy some million times longer than the singlet state lifetime. For this reason alone, energy transfer via triplet state excitons have many potential advantages over singlet state excitons.

Examples of energy transfer in biological systems by resonance methods are given in a later section.

D. ENERGY TRANSFER COUPLED WITH DISCRETE CHARGE MIGRATION

A feature of many bioenergetic systems is that energy transfer probably precedes, or is accompanied by, the uncoupled migration of individual holes and electrons. Two processes of electron or hole migration may be important in bioenergetics. (a), photoconductivity and (b), electronic semiconductivity.

Lying somewhat above the lowest exciton levels shown in Fig. 10 is another band of energy levels called the conduction band. If an electron should find itself in a conduction band level, it is no longer associated by any type of interaction with the positive hole with which it was formerly associated in the ground state. This electron and the positive hole, therefore, now have a free independent existence. Should they migrate independently under the influence of an applied electric or bioelectric field, they will confer the property of electrical conductivity on the material. In the most general way, if the free electrons and holes are generated by the absorption of light, the material is classed as a photoconductor. If electrons are raised into the conduction band by thermal quanta alone, then this represents semiconductivity.

At the present time in biologically relevant organic molecular aggregates there is confusion as to the charge carrier generation process, whether by light or by thermal energy. This confusion stems largely from deficiencies in the knowledge of the relative positioning of the conduction band energy levels, relative to the singlet and triplet exciton levels, and uncertainties as to how these energy level positions may be influenced by aggregation, defects, impurities or imperfections in the aggregate. Finally, little is known about the ionization of excitons, whether singlet or triplet, by additional quanta (light or thermal) or trapping and this could well be an important stage in the production of independently free and mobile electrons and holes.

The existence of free electrons and holes in bioenergetic lamellar systems has been studied in many ways, such as by electron spin resonance, semiconduction, thermoluminscence, photoconductivity and paramagnetism.

Semiconduction may be detected experimentally by measuring the resistance of a sample as a function of temperature, while photoconductivity is detected by resistance measurements as a function of temperature and light energy. For an intrinsic semiconductor, the resistance R varies with temperature T as shown in the equation,

$$R = Ro \exp \frac{\Delta E}{2KT}$$

The conduction activation energy, ΔE, is biologically very relevant since it determines the energy required to generate the independent electrons and holes required for conductivity. Most semiconduction studies on biologically relevant materials have, unfortunately, been concerned solely with determination of the activation energy ΔE. This has often been at the expense of measurements and interpretation of the pre-exponential factor Ro, which could theoretically be informative as to the mobility of the charge carriers.

In photoconduction studies, experiments on spectral and temperature dependence of the photocurrent are revealing as to the nature of the charge carrier generation process.

We are not concerned here with the details of semiconduction and photoconduction theory but interested readers are referred to reviews by Kearns (1964) and Eley (1962) and to the papers by Kearns (1960) and Tollin (1960).

The generation of charge carriers has been discussed, so far, from the viewpoint that promotion of an electron to an excited level (whether singlet, triplet or exciton) precedes the actual electron–hole separation process. For completeness, we should mention that charge separation in aggregates has been related to the properties of the individual molecules by a more individualistic molecular picture, a treatment originally due to Lyons (1957; Kearns, 1964). In this theory, essentially classical chemical concepts, such as ionization potential, electron affinity and the polarization energy associated with the individual molecules in the aggregate, are used to compute the conduction activation energy, ΔE.

E. DONOR–ACCEPTOR INTERACTION

In the present context, donor–acceptor interaction may be equated to charge transfer complex formation.

In the formation of charge-transfer complexes, two molecules, each capable of independent existence, are involved. One of these molecules must be an electron donor (D) and one an electron acceptor (A). Though some molecules of biological importance (e.g. flavins) may function both as donors and acceptors and form self charge transfer complexes, we shall regard D, A as different molecules (Pullman and Pullman, 1958).

The crux of charge transfer complex formation is that partial or complete transfer of a *single* electron from an orbital of the donor to an orbital of the acceptor, either spontaneously or photo-induced, leads to a resonance interaction which imparts stability to the complex compared to the isolated partners. This resonance interaction may be regarded as the charge transfer force.

The quantum mechanical basis of charge transfer forces was formulated by Mulliken (1952) and accounts with biological implications are given by McGlynn (1960) and Szent-Gyorgyi (1960).

Charge transfer complex formation may be represented as below:

$$D + A \leftrightarrows [DA] \leftrightarrows [D^+ A^-] \leftrightarrows D^+_{(Sol.)} + A^-_{(Sol.)}$$

The interaction in the "no bond" state represented by $[DA]$ may arise from any of the normal chemical intermolecular forces such as dipole–dipole, H-bonding, dispersion, etc. The "dative" state represented by $[D^+ A^-]$ has extra stabilizing forces, both electrovalent and covalent, which arise from the transfer of an electron from $D \rightarrow A$.

The stabilization of the charge transfer complex arises from the resonance interaction between the "no bond" $[DA]$ and "dative" states $[D^+ A^-]$ and the charge transfer force represents this resonance interaction. Under certain conditions, using very strong donors and acceptors and polar solvents, the complex may most accurately be represented by the dative structure $[D^+ A^-]$; such complexes have a strong ionic character and may even dissociate into the two (solvated) free radical ions in polar media. Such strong charge transfer is often accompanied by the production of the appropriate free radicals (Isenberg and Baird, 1962; Isenberg and Szent-Gyorgyi, 1958; Forster and Thomson, 1962).

Charge transfer interaction is usually accompanied by the appearance of an intense new absorption band in the region of the spectrum (usually red-shifted) where neither of the isolated components absorb.

F. THE ROLE OF LAMELLAR ORGANIZATION IN CHARGE SEPARATION AND ELECTRON MOBILITY

The formation of charge transfer complexes requires several conditions to be filled. These conditions, which indicate the probable role of the lamellar system in complex formation, are:

(1) very close approach and significant orbital overlap,
(2) favourable energetic relationship (donor–acceptor ability) between D, A,
(3) favourable dielectric or ionic environment and
(4) correct symmetry relations between the overlapping orbitals.

In strong, or even moderate, charge transfer complexes, there is partial uncoupling of electron spins and the transfer of an electron from a filled level of the donor to an unfilled level of the acceptor. Such complexes might intuitively be expected to show enhanced electronic semiconductivity with significantly reduced energy gaps (Szent-Gyorgyi, 1960).

Although the relationship between the energy of the characteristic charge transfer absorption band and the conduction band is unclear, this enhanced electronic conduction is almost universally observed (Eley, 1962; Kearns, 1964).

At the outset, it must be admitted that any possible function of the membrane in electron mobility by crystal semiconduction mechanisms, particularly by formation of conduction bands, is unlikely. The reason lies in the nature of the lamellar systems themselves. A characteristic of a lamellar system, even of a pure pigment lamella, undiluted by lipid, is its two-dimensional nature and its extreme mono- and bimolecular thickness. This at once precludes the development of conventional conduction bands as in inorganic crystals which require strong overlap and coupling in three dimensions. It might of course be argued, as originally proposed by Katz (1949) for the chloroplast lamella, that a two-dimensional semiconductor is possible in which the electrons and holes may have independent mobility in the plane of the lamella. Although there is no theoretical objecton to a two-dimensional semiconductor, the evident weak coupling in biological lamellar systems would require electron or hole migration to proceed by a hopping mechanism.

The perversity of nature may be such that the interfaces, which separate the lamelae, provide, in some way, traps selective for holes or electrons. If this were so and, say, electrons were selectively trapped, then holes might have mobility throughout the rest of the aggregate.

Those conditions which militate against conventional conduction bands may enhance conduction by hole or electron mobility after charge transfer complex formation.

The arrangement of molecules, such as quinones and chlorophyll discussed above, into lipid formations will almost certainly be such that the long fatty chains will be associated with the hydrophobic lipid regions. The bioenergetically important end groups will then be associated together in the more polar aqueous regions.

In this way it may be said that the membrane, because of chain packing considerations, arranged for the "end" groups to: (1) come together in high concentration, (2) possibly come together with the correct orbital symmetry for maximum interaction and (3) arranged, perhaps, that the $D-A$ interaction will proceed in the region favouring maximum charge transfer (micropolarity and micro-dielectric constant).

The fact that so many bioenergetic systems are made up of several components suggests that if semiconduction plays a role, it must be rather indirect and as a result of charge transfer interaction or electron injection (Eley et al., 1966), possibly by an enzymatic reaction. Since the enzymatic processes always follow the initial biophysical charge separation process, the semiconduction role will be a secondary one.

VI. Energy and Charge Migration in Bioenergetic Lamellar Systems

In this section, the principles developed in Sections IV and V will be applied to energy transfer and charge migration* in vision, respiration and photosynthesis. Although the sequence of events *in vivo* may be energy transfer, charge separation and, finally, charge migration or simultaneous energy transfer and charge migration, no attempt will be made to delineate the time dependence of the various processes.

A. ROD OUTER SEGMENTS

1. *Energy Transfer by Resonance Mechanisms*

According to Brown et al. (1963), the *a priori* prospects of energy transfer by resonance mechanisms in ROS membranes are poor. This opinion follows from a consideration of the efficiency of the energy transfer as a function of the separation of the interacting retinene pigments and the size of the rhodopsin molecule itself. If, as pointed out by these authors and by Wald (1965), the rhodopsin molecules are organized by the lamellar system so that the retinene chromophores form local clusters, the efficiency of the energy transfer within the cluster will be increased but resonance transfer from one cluster to the next will be correpondingly reduced and may be completely prevented.

There have been few experimental studies relevant to resonance transfer of energy in the ROS systems. Hagins and Jennings (1959) could detect no photodichroism of retinal rods and this points to an efficient, intermolecular, radiationless energy transfer. These authors, after applying the Förster theory and comparing the result with the size and concentration of rhodopsin in the lamellar structure (cf. Wald (1965) above), concluded that energy transfer

---

*To avoid confusion with the "charge transfer" concept as outlined in connection with the formation of molecular complexes in the preceding section, we shall refer to the independent movement of electron or holes in an aggregate (semiconduction) as charge migration.

by resonance was very unlikely. They preferred restricted molecular rotation, photoconduction or electrolytic conduction to explain the absence of photodichroism.

2. Semi- and Photoconduction in Visual Function

As discussed in Section II-C, the lipoprotein membranes which form the matrix on, or in which, the rhodopsin molecules are located, may be derived from invaginations of the surface membrane in much the same way as are mitochondrial cristae (cf. Moody, 1964). A conclusion of Section II-C was that rhodopsin, a lipid–pigment–protein complex, probably forms an integral part of each disc membrane; since the final impulse transmitted to the brain almost certainly proceeds by an action potential mechanism, this continuity and partial equivalence of the light receptor membrane and the visual impulse transmitting membrane is probably significant. Such continuity makes ti easier to visualize a change-over from electron–hole migration in the primary stages to the ionic, electrolytic changes in the action potential generating and transmission stages. The actual change-over mechanism might conceivably involve the exchange of electrons through proteins or the carotenoids and ions via pores or active transport across a membrane according to the theories developed by Jahn (1962) and Cope (1963).

It may be unnecessary to invoke such a distinct change-over requirement. Perhaps photoisomerization of the rhodopsin molecules, which form an integral part of the membrane, could lead to local permeability and hence conductance changes directly in the membrane.

If ubiquinone, or an ubiquinone-like redox compound, (Section III-A-1) is present in ROS membranes, it seems reasonable to ascribe some function to it. Pearse (1965) suggests that such a redox compound is well situated to function as an electron sink and to couple, electronically, metabolic processes in the mitochondria-rich inner segment with electron liberation by photoisomerization of retinene in the outer segments. Pearse visualizes the photoreceptor, containing reduced ubiquinone, as a "miniature capacitor".

In connection with visual excitation, Dingle and Lucy (1965) suggest that photoisomerization of the kinked chain all-*cis* retinal to the relatively straight chain all-*trans* retinal yields a molecule in which increased orbital overlap and electron delocalization makes the latter molecule a much more effective electron transfer agent and may provide a molecular pathway for the transmission of electrons associated with the visual impulse.

Recalling that not only the degree but also the type (i.e. *cis* or *trans*) of unsaturation present in the acyl chains of phospholipids can lead to monolayers with different fluidity and, hence, presumably different permeability properties (Chapman *et al.*, 1966), we may perhaps take Dingle and Lucy's suggestion one stage further. It may be that photoinduced *cis-trans* isomerization of the retinene, localized within a characteristic phospholipid environment, may lead to local molecular rearrangements within a membrane and hence to localized ionic permeability changes. If this were so, the "changeover" requirement referred to above would be largely inapplicable and it is particularly interesting in this connection that Brown et al. (1963) suggest that the site of light absorption and the excitation process must be quite close on a molecular scale.

Several experimental studies of the semi- and photoconductive properties of carotenoids have been made, both with single crystals and with relatively thick films deposited on quartz slides. Before such studies may be said to indicate, much less demonstrate, a role of semi- or photoconduction in the visual process, much more work is required. Specifically, the conduction parameters, photoconduction activation energy, carrier mobility, etc. must bear comparison with the physiological kinetics. To illustrate this, Brown et al. (1963) believe that the photoconduction observed by Rosenberg et al. (1961) in dried sheep's rods was much too slow to be of significance in the excitation process.

Environmental factors which must also be taken into account include the much lower degree of "crystallinity" in the rod, the presence of strong donors or acceptors, hydration effects and, possibly, surface effects due to absorbed oxygen.

Thus, Rosenberg (1961) found that oxygen increased both the photocurrent and the dark current in carotenoids, whereas only the dark current activation energy was decreased. Chapman and Cherry (1964) observed that oxygen increased the bulk conductivity of well defined single crystals of β-carotene by several orders of magnitude. These authors also observed a particularly interesting effect; the appearance of current pulses similar in several aspects to the spike discharge observed under illumination of single optic nerve fibres of Limulus under a constant applied potential difference.

The effects of hydration, and the presence of strong donors or acceptors, on the conduction parameters of biomolecules are very striking but relatively little understood.

These effects have been studied in most detail with proteins, such as serum albumin and haemoglobin; the latter may approximate in size to rhodopsin. Davis et al. (1960) found that when small amounts of strong electron acceptors such as chloranil, a tetrachlorinated quinone (cf. Fig. 11), were incorporated into proteins, the conductivity was increased by a factor of 10^6. The explanation, since substantiated, was that the chloranil accepted electrons from the protein leaving unpaired holes free to migrate in the electric field in a protein conduction band. The effect of adsorbed water on protein conductivity is no less marked. For example, haemoglobin and other proteins show an increase of approximately 10^8 in conductivity for 10% adsorbed water, the charge carriers still being predominantly holes or electrons (Cardew and Eley, 1959; Eley and Leslie, 1963, 1964; Rosenberg, 1962; Maricic et al., 1964).

Rosenberg et al. (1961) observed semiconduction in sheep's rods, dried by water and sucrose washing. Only in the latter preparations was photoconduction observed. Their observations were believed to be due to the retention

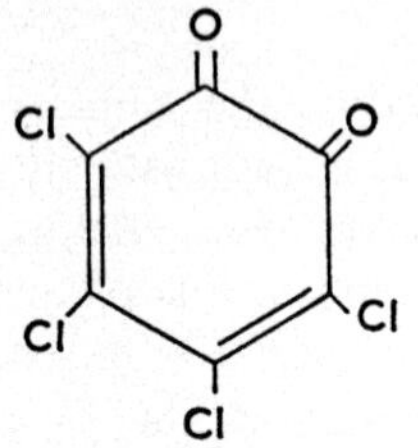

ORTHO-CHLORANIL
A

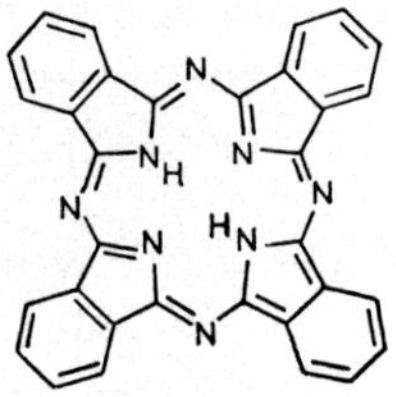

METAL-FREE
PHTHALOCYANINE
B

Fig. 11. Molecular structures of o-chloranil and metal-free phthalocyanine.

of water in the structure or to the prevention of osmotic destruction of the rod structure by sucrose washing prior to drying. As yet, nothing is known about the potentially important area of membrane hydration.

B. MITOCHONDRIAL MEMBRANES

1. *Resonance Energy Transfer*

Energy transfer by resonance mechanisms along the electron transfer chain in mitochondria is unlikely because most resonance transfer mechanisms, such as the Förster mechanism (Section V-B), require that the molecule be first excited into an electronically excited singlet state. Although many of the respiratory carriers have a characteristic fluorescence spectrum and absorb in the visible region, the required initial excitation energy is unlikely to be available in mitochondria.

On the other hand, transfer of excitation energy by the Förster mechanism has been demonstrated from one side of a protein molecule to the other (Teale and Weber, 1957; Styrer, 1960).

Karreman et al. (1958) demonstrated that in the sequence: aromatic amino acids $\longrightarrow$ reduced diphosphopyridine nucleotides $\longrightarrow$ oxidized riboflavin $\longrightarrow$ cytochromes (cf. mitochondrial chain), the overlap of the absorption and fluorescence spectra is such that a path for the efficient resonance transfer of excitation energy should exist. Karreman et al. also point out that the energy flow would be undirectional and proceed in the same direction as electron flow during respiration. The mitochondrial membrane would then maintain the proper donor–acceptor sequence and electronic oscillator alignment.

However, because the singlet excitation energy for amino acids lies in the u.v. and because discrete electrons must pass along the chain, it seems necessary to invoke other energy and charge migration mechanisms to explain mitohcondrial function.

2. Charge Migration in Mitochondrial Function

In mitochondrial function, electrons flow from reduced pyridine nucleotides to molecular oxygen through a sequence of molecules, each with specific redox capabilities. Discrete electron flow is demonstrated by the fact that actively respiring mitochondria exhibit characteristic electron spin resonance signals (Commoner and Hollocher, 1960) and definite u.v. spectra as a function of the redox state of the components (Chance and Williams, 1956).

Two types of possible electron transferring mechanisms may be distinguished and involve, respectively, considerable molecular motion, such as rotation and translation of the redox molecules, or electron movement *through* relatively fixed molecules. Proponents of the first type of mechanism envisage that restricted rotation of the molecules within the lipoprotein matrix is possible (Chance and Williams, 1956). Such restricted rotation permits collision of the prosthetic groups and electron exchange by classical redox concepts. In an analogous way, Green and colleagues (Green, 1960; Green and Lester, 1959) postulated that the lipid matrix of the electron transfer chain permitted small molecules, such as coenzyme Q (ubiquinone) or cytochrome c, to have translational freedom. These small molecules would thus act as reversible and mobile electron carriers, ubiquinone linking flavoprotein and cytochrome c_1 and cytochrome c linking cytochrome c_1 and cytochrome a. Clearly, this is not a complete answer, because cytochrome

c_1 would be required to interact sequentially with two different components localized in different compartments of the matrix.

The finding that cytochrome c_1 probably exists in mitochondria in a polymeric form suggested a partial solution. Green (1960) suggested that in the polymeric cytochrome c_1, one of the haem groups was spatially located to react with ubiquinone and a different haem group was spatially located to react with cytochrome c. The electrons would then flow through a static cytochrome c_1 by ". . . internal links in the polymeric cytochrome c_1". Here we see the requirement of electron flow through a lipoprotein complex emerging.

Green and Fleischer (1962, 1964) suggest that the oxidation-reduction groups of each complex are sufficiently close, because of attachment to their respective fixed proteins by "flexible molecular arms", to permit interaction of the redox groups by rotation or oscillation. Alternatively, the redox groups may be brought sufficiently close together by membrane location so that the components are ". . . hybridized to form a single oxidation–reduction system"; this again would seem to point to an electronic mechanism.

The second broad class of electron transferring mechanisms is submolecular in origin and starts from the assumption that the redox molecules are relatively firmly and specifically located in lipoprotein membrane frameworks. Organization within or by the lipoprotein matrix brings the prosthetic groups, the proteins themselves, and possibly the unsaturated phospholipids, into sufficiently close apposition so that electron, and possibly proton, transfer through these molecules collectively is possible.

The transfer of electrons along the electron transfer chain by a straightforward semiconduction mechanism is very unlikely. A comparison of the rate of transfer of electrons along the chain during respiration with the rate of electron flow (or current) through a molecule on the main pathway *in vitro* under an appropriate voltage gradient was very unfavourable to a semi-conduction mechanism (Cardew and Eley, 1959).

The micro-heterogeneity of the respiratory chain in molecular terms also militates against normal semiconduction. On the other hand, the likely organization of molecules such as ubiquinones suggests that indirect semiconduction via charge transfer interaction may be important. Accepting that semiconduction via charge transfer interaction may be present, the mandatory presence of lipid for efficient mitochondrial function would be expected. Specific phospholipid involvement could be a very efficient way to fulfil the conditions necessary for strong interaction.

The semiconduction and photoconduction properties of many charge transfer complexes have been studied in the last few years and, almost invariably, complex formation leads to greatly increased conductivity and decreased conduction activation energies (Eley, 1962; Labes *et al.*, 1960; Kearns, 1964).

Thus whereas dry proteins, such as haem proteins, have typical specific resistivities of $10^{18} \Omega$—cm and activation energies of 2–3 eV, charge transfer complexes quite closely approximating to the biological situation have specific resistivities reduced by perhaps 10^{10} and conduction activation energies of perhaps $\frac{1}{10}$ of the haem protein values quoted above.

Activation energies of this magnitude (a few Kcal per mole) should certainly make charge carrier generation by metabolic processes quite feasible. When corrected for lamella thickness, the specific resistances of the more highly conducting complexes should also fall into a biologically useful range. It is interesting that some of the earliest work on the conduction properties of charge transfer complexes was concerned with complexes formed between aromatic amines, substituted quinones, aromatic heterocycles, phthalocyanines, carotenes and iodine; all molecules which are characterized by lipid lamellar solubility and relevance to bioenergetics.

Throughout the early part of the chapter the occurrence of bioenergetic molecules in various aggregated lamellar formation was discussed. It was pointed out that, although the systems exhibited a highly developed ultrastructural organization, this was also characterized by pronounced molecular heterogeneity. The amphipathic properties of lipids also provided the molecular means of linking or creating hydrophilic and hydrophobic phases. For these reasons (and also because the possession of long hydrophobic chains in ubiquinone, plastoquinone, carotenoids and chlorophyll will bring the energetically important end groups into close and probably stereospecific proximity in a polar region) charge transfer interaction and related conduction phenomena are likely to be of the greatest importance in bioenergetics. Perhaps the formation of charge transfer complexes furnishes the most elegant example of the way in which nature uses molecular structure, of both the bioenergetic molecules, and lipids to bring about conditions best suited to bioenergetic function.

To illustrate the possible role of charge transfer complexes in mitochondrial function we shall choose only a few examples which seem to approximate to the electron transfer chain. The formation of a complex between protein and the isoalloxazine moiety of FMN due to charge-transfer interactions was demonstrated by Harbury and Foley (1958). Similarly Ciliento and Giusti (1959) suggested charge transfer complex formation between pyridine nucleotides and π electron containing amino acids (indole or tryptophan) found in proteins.

Isenberg and Szent-Gyorgyi (1958) also studied similar complexes between flavins, amino acids and pyridine nucleotides. Free radical formation, charge transfer complex interaction and enzyme function was discussed in a later review by Isenberg (1964).

In the earlier paper, Isenberg and Szent-Gyorgyi speculate that transfer of an electron from the protein to the flavin may lead to hole conductivity in a protein conduction band. Refilling of the protein conduction band by electrons derived from reduced pyridine nucleotides would provide a mechanism of electron transport *through* the protein molecule.

Snart (1964) formed complexes between cytochrome c and ubiquinone (coenzyme Q_{10}, see Fig. 7) and triple complexes formed from this haem protein, ubiquinone and mitochondrial lipid. The complexes were, from the optical evidence and the weak e.s.r. signal observed, possibly formed by donor–acceptor interaction between the haem or aromatic amino acid side chains of the protein and the quinone. However, Snart suggested that the quinone is somehow complexed with the hydrogen bond system of the protein. In this interaction positive holes are created in the protein conduction band associated with the hydrogen bond peptide network by electron transfer from the protein to the electron accepting quinone.

All the complexes, and whole mitochondria, showed very much lower conduction activation energies and specific resistivities roughly a million-fold lower than the components. Indeed, the conduction parameters according to this study are compatible with electron transfer and oxidative phosphorylation occurring at least partially by this modified semiconduction mechanism. If the greatly enhanced conduction is a result of stereospecific donor–acceptor interaction, this again points strongly to the specific role of lamellar organization.

A new model of electron transport along chains of protein–prosthetic group molecules, such as those found in mitochondria and perhaps photosynthetic systems, has recently been developed by Urry and Eyring (1963, 1965 a,b).

The starting point for the development of the "imidazole pump model" of electron transport due to these authors is the remarkably rapid rate of *in vivo* mitochondrial electron transport compared to *in vitro* model systems. This rapidity results from the electronic properties of the molecules involved and particularly from the integrity of the mitochondrial membrane. It is implied that the membrane brings about the most favourable steric arrangement and juxtaposition of the molecules so that orbital overlap and energetic interaction is maximized, leading to a rapid electron flow.

In essence, the model requires the central iron atoms of two adjacent porphyrin systems of the various cytochromes to be linked by imidazole molecules, possibly derived from histidine side chains from protein.

The adjacent porphyrin and imidazole molecules are, however, spatially oriented, presumably by the membrane, so that contiguity of orbitals belonging to the ferrous or ferric iron and the nitrogen atoms of imidazole is obtained. By sequential coordination first to one iron atom and then to the other, electrons may be pumped from electron donors interacting with the first iron atom to electron acceptors interacting with the second iron atom.

This electronic bridging of the two cytochrome systems will be particularly rapid because of the symmetry of the imidazole molecule, providing the various groups involved are favourably oriented by the membrane.

In a recent study, charge transfer complexes were formed between quinones and isoalloxazine derivatives, molecules relevant both to mitochondrial and photosynthetic function (Ray et al., 1965). The complexes, which gave suitable single crystals, were studied by semiconduction, photoconduction and spin resonance techniques. The results clearly indicate that strong charge transfer interaction is involved. The specific resistance of the complexes was greatly decreased compared to the isolated components but the activation energies for the riboflavin-hydroquinone complex at least, did not seem unduly different from the value quoted for Riboflavin or FMN by Eley and Leslie (1963).

Ray et al., however, draw attention to the importance of imperfections and trapping centres in free carrier generation. They conclude that donor–acceptor interaction between biologically relevant molecules, in the appropriate structural environment, can lead to complete electron delocalization throughout an aggregate and to useful biological consequences.

C. PHOTOSYNTHETIC SYSTEMS

1. *Energy Transfer Within the Photosynthetic Unit*

In photosynthetic systems, energy transfer in the photosynthetic unit leads to the localization and trapping of the excitation energy at those special protein–pigment complexes which constitute the reaction centre (cf. Fig. 5). At the reaction centre, the excitation energy is converted into separated oxidizing and reducing power, probably initially a hole and an electron. The hole and electron must then be considered as individuals and free to go their separate ways.

Energy transfer in the photosynthetic unit is fairly well understood. It has been demonstrated between the molecules which go to make up the unit both *in vivo* and in appropriate model systems *in vitro* (Duysens, 1964; Clayton, 1963, 1965; Lumry and Spikes, 1960; Teale, 1959).

In the experiment reported by Teale, transfer efficiencies from the aromatic amino acid residues of grana proteins to chlorophyll approached 100% in isolated chloroplasts; disruption of the grana structure abolished the transfer with a concomitant increase in amino acid fluorescence.

Energy transfer, both between the various spectroscopically distinct chlorophylls and the accessory pigments, is by the Förster inductive resonance mechanism via singlet excited states. In certain cases, the energy absorbed by the accessory pigments (e.g. carotenoids) can be transferred to the reaction centres with an efficiency approaching 100% (Duysens, 1964).

It will be recalled that a high degree of crystallinity or molecular homogeneity was not required for efficient energy transfer by the Förster mechanism but spectral overlap and, particularly, closeness of approach was required. These characteristics are just those which seem most consistent with the localization of the photosynthetic pigments within the chlorophyll lamella or quantasome (Section II-E,F).

In view of the latest morphological evidence, which indicates that a quantasome or other globular substructure may represent the photosynthetic unit, the calculations of Bay and Pearlstein (1963) are of considerable interest. These authors developed a very satisfactory mathematical model to describe energy transfer within the photosynthetic unit, using this morphological data and the weak coupling Förster resonance transfer mechanism. Their calculations indicate that this mechanism predicts that the excitation energy will be delivered quite efficiently to the reaction centres well within the singlet excitation lifetime of the pigments involved. Making the likely assumption that, once the excitation energy reaches the reaction centre it is converted with high efficiency into the separated oxidant and reductant, resonance transfer of the excitation energy to this centre and photosynthetic utilization will compete very effectively with the wasteful fluorescence emission.

The efficiency of energy transfer by the Förster mechanism depends upon dipolar interactions with nearest neighbours and Bay and Pearlstein show that if the photosynthetic unit is globular, rather than a two-dimensional array, this will lead to a significant increase in efficiency.

2. *Semi- and Photoconduction in Photosynthetic Function*

The existence and, particularly, the use of conduction bands encompassing many molecules or photoconduction phenomena cannot yet be unequivocally included or excluded in theories of the primary generation of redox potentiality. Considerable confusion exists in the literature pertaining to this problem, much of it generated by our ignorance of the detailed micro-environment of the photochemically active molecules at the reaction centre.

The importance of purely electronic effects in the primary photosynthetic processes is suggested by the following observations.

(1) Photoinduced optical density changes, the result of redox processes, have been detected in photochemically active systems at $1°K$. At these temperatures, ordinary molecular chemistry is impossible (Clayton, 1965).

(2) Chloroplasts emit luminescence (delayed light) after prior illumination. This seems to require the energy to be trapped in metastable energy levels. The delayed light may reflect the recombination of electrons and holes (Arnold, 1965).

(3) Photoinduced e.s.r. signals also occur at very low temperatures ($77°K$). This seems inescapable proof that free electrons are involved.

The confusion regarding the relevance of conduction phenomena has been increased by very suggestive results obtained using chloroplasts, chloroplast pigments and model complexes (Arnold and Sherwood, 1957, 1959; Nelson, 1957; Eley and Snart, 1965; Terenin et al., 1959; Rosenberg and Camiscoli, 1961; McCree, 1965; Kearns, 1964).

These workers have demonstrated that chloroplasts, and pigments relevant to photosynthesis in various states of aggregation, show semiconduction and photoconduction properties. Perhaps the most significant obervations are the activation energy for semiconduction, which for "chlorophyll" is about $1\cdot1 \to 1\cdot4$ eV, is considerably less than the energy content of a quantum of red light (680 m$\mu \approx 1\cdot 8$ eV ≈ 42 Kcal/mole). Thus, if the conduction activation energy truly represents the energy required to generate independetly mobile electrons and/or holes, this should be easily feasible by those quanta active in photosynthesis.

The question of photoconduction activation energy also raises some problems. Early measurements (Rosenberg and Camiscoli, 1961) indicated a significant photocurrent activation energy; this led Rabinowitch (1959) to question the relevance of photoconduction since the primary photophysical process is essentially temperature independent. However Eley and Snart (1965) report a much smaller photocurrent activation energy and McCree (1965) found that plant materials are not photoconductive to any biologically useful degree so the question remains unsettled. Terenin (Rabinowitch, 1959) found photoconductivity in "crystalline" but not amorphous chlorophyll layers but crystallinity is, according to some authors, not an absolute requirement for photoconduction (Rosenberg, 1959; Jacobs, 1959).

In view of the morphological evidence discussed at the beginning of this chapter and the molecular heterogeneity to which reference has been made, the presence of extensive conduction bands on a molecular scale is unlikely. The danger of dismissing extensive conduction bands on incomplete morphological data must be realized, however. Perhaps conduction bands could develop in the approximately 5% highly oriented chlorophyll in quantasomes (Sauer and Calvin, 1962; Butler, 1961). Since such oriented chlorophyl absorbs at longer wavelengths, it is probably associated with the quantum conversion process.

How then is the high quantum efficiency, definite involvement of electrons and holes, essentiality for intact lamellar structure and molecular heterogeneity to be reconciled? In the grana of higher plants evolutionary adaptation in surely highly developed and molecular heterogeneity and specific lipid involvement point to the involvement of electron and hole migration via donor-acceptor charge transfer interaction or trapping phenomena.

If, for example, an exciton is dissociated and the electron is given by charge transfer interaction to ferredoxin, say, the "hole" (Chl^+) should be able to move freely through the array of chlorophyll molecules (cf. Fig. 5).

In some experiments in Calvin's laboratory lamellar systems, approximating to the chloroplast, were formed by evaporating strong electron acceptors (e.g. o-chloranil) on to layers of phthalocyanine (see Fig. 11) (Tollin, 1960; Kearns, 1960, 1964). These were examined for semiconduction, photoconduction and light-induced electron spin resonance signals. Excellent correlations were observed. Strong interactions were demonstrated by the low activation energies observed and the very pronounced falls in specific resistivity of the composite layers, compared to those of the isolated components. For example, a 10^7-fold decrease in dark resistivity and a 10^5-fold decrease in steady state light resistivity were seen.

The general explanation was that, because of the intimate contact between the layers, electrons are transferred from the donor to the acceptor layer, leaving holes free to migrate in the donor layer.

Extrapolation of these results to the biological environment unguardedly is not justified. However, the results indicate what may be expected by close apposition of donor and acceptor systems across interfaces.

In a recent presentation exciton migration, charge migration and charge transfer interaction have been implicated as playing a role in the primary photophysical process of photosynthesis (Calvin, 1965).

VII. Some Model Systems

A. GENERAL

Bioenergetic systems are associated with lipoprotein lamellar formations. The lipoprotein membrane may bring about conditions favourable to energy transfer and possibly also to charge migration. It is no less important that the interface, possibly consisting of lipid polar groups, separates the interior of the hydrophobic membrane from the surrounding hydrophilic medium. The initial biophysical aspects of bioenergetics take place in association with the hydrophobic regions and the biochemical, enzymological aspects in association with the hydrophilic regions. The membrane, or the interface, must, paradoxically, both separate and link these biophysical and biochemical processes.

In this final section some model systems will be outlined, the study of which would further the understanding of membrane bioenergetics. The model must have hydrophobic and hydrophilic regions separated by an interface to be of direct biological relevance. Preferably, amphipathic lipids or lipoproteins should form the interface. Three models which fulfil some or all of these requirements, but which have so far been little used in bioenergetic studies, are discussed.

B. MONOLAYERS AT INTERFACES

Monolayer models should be useful in studies of the energetic properties of pigment and mixed pigment systems, and, in particular, in obtaining valuable information from lipid-pigment-protein monolayers. If lipid-enzyme monolayers could be made with the enzyme still retaining activity, the door would be opened to the study of the linking of the biophysical and biochemical aspects discussed. These are difficult experiments but the fact that the membranes involved in bioenergetic systems may be composed of functional lipoprotein subunits, rather than relatively discrete lipid-protein layers as in the Danielli-Davson model, suggests that such experiments should be possible. For example, in the case of the enzyme complexes of the mitochondrial electron transfer chain, Green and Tzagoloff (1966) discuss the way in which the complexes have, besides their enzymic properties, the capacity to organize themselves into membranes, the enzyme molecules themselves actually forming the membrane subunits. It is to be noted that membrane formation from the active subunits is dependent upon, and controlled by, the presence of mitochondrial phospholipids.

Few experimental attempts to make lipoprotein monolayers have been reported. Eley and Hedge (1956) made useful preliminary observations but concluded that some protein denaturation took place when they injected protein under a lipid monolayer. This denaturation counts against retention of enzyme activity. Fraser *et al.* (1955) studied the association of enzyme molecules (catalase) with oil-water interfaces. The extent of surface denaturation of the enzyme was assessed by the loss in activity when it was adsorbed on to lipid stabilized oil-water emulsions. In view of the rapidly emerging evidence that many enzyme systems actually require the presence of lipid for efficient functioning (Green and Tzagoloff, 1966), renewed effort in this direction would seem well worthwhile.

Hyono *et al.* (1962) have made a beginning with an active bioenergetic system. They studied monolayers of rhodopsin at the air-water interface and at low surface pressures they found that illumination of the monolayer led to enhancement of the tendency of the pressure to increase spontaneously; this enhancement did not occur on illumination after bleaching.

Pure pigment and mixed lipid pigment monolayers have been used by several workers, mainly in connection with chlorophyll and photosynthetic lamellae. Many properties of monolayers of chlorophyll at the air-water and oil-water interfaces have been related and compared to the postulated monomolecular state of chlorophyll *in vivo*. Considering first spectroscopic studies, which presumably relate to the pigment orientation, degree of energetic interaction and organization, we may cite the following authors: Goedheer (1955) oriented chlorophyll *a* at an ammonium oleate interface and studied birefringence and dichroism; Jacobs et al. (1954, 1957) and particularly Trurnit and Colmano (1959) studied absorption spectra under various conditions.

This earlier work has been much refined and extended by Litvin et al. (1965) in the U.S.S.R. and the group at the General Electric Research Laboratory, U.S.A. (Bellamy et al., 1963; Tweet et al., 1964a,b,c; Gaines et al., 1964, 1965).

The Russian workers studied absorption and fluorescence spectra of monolayers and films of β-carotene, chlorophylls a and b and also mixed monolayers and films of these chloroplast components. They were able to demonstrate wavelength shifts in the absorption spectra on aggregation and energy transfer between them and, possibly, complex formation. The elegant work of the General Electric group deserves special consideration. These workers have developed spectrometers to measure both fluorescence and absorption spectra of individual monolayers and polarization and surface potential techniques to study the orientation of, for example, the porphyrin planes of chlorophyll. Only highlights can be mentioned here.

Absorption and fluorescence emission spectra have been recorded for chlorophyll monolayers and energy transfer studied as a function of chlorophyll concentration in the monolayer. In this particular type of study, the relative importance of energy transfer by the weak coupling (Förster) mechanism or the stronger coupling (localized or delocalized) exciton mechanisms may be assessed by studying fluorescence quenching for diluted and undiluted monolayers. For undiluted and non-crystalline monolayers, the situation which is probably appropriate to the chloroplast lamella, the General Electric workers demonstrated that rapid energy transfer did exist. In a more recent paper, mixed films containing vitamin K_1 and chlorophyll were studied, both by surface pressure, surface potential and fluorescence quenching techniques. This system is clearly one stage nearer the photosynthetic lamella and it is interesting to note that some evidence was brought forward to suggest some interaction between the porphyrin rings and the quinone moiety of the vitamin K_1.

If charge transfer interactions are important in bioenergetic systems, the monolayer technique, with its control of degree of interaction, should be of great value in their study.

The use of monolayer techniques in photoconduction studies by Arnold and Sherwood (1959) and McCree (1965) has been mentioned in Section VI.

C. BILAYERS SEPARATING AQUEOUS CHAMBERS

If, as has been suggested (cf. Calvin, 1959, 1961), the ubiquitous lamellar systems found in bioenergetics are used to spatially separate oxidizing and reducing power or discharge "intermediates" into non-connecting hydrophilic regions, a bilayer model might be useful.

Such phospholipid bilayers have become available in the last few years (see Chapter 6 by J. A. Lucy for a review) but their use so far has been confined almost entirely to the study of permeability and excitability properties of cell membranes.

In principle, there seems to be no reason why pigments, chlorophyll, long chain quinones or carotenoids should not be incorporated into such phospholipid bilayers. Alternatively, and perhaps concomitantly, enzymes might be associated with such phospholipid bilayers. Perhaps chloroplast lipids and enzyme complexes of the electron transfer chain themselves could be studied. Such composite phospholipid films containing incorporated pigments and attached enzymes might be excellent model systems with which to study simultaneously both the initial biophysical quantum conversion and subsequent biochemical enzymology. Some very preliminary attempts to incorporate pigments (β-carotene, ubiquinone, retinene, vitamin K_1) into phospholipid bilayers and to study their spectral properties were recently reported by Leslie and Chapman (1967).

D. MICELLAR MODELS

In micellar models we include, arbitrarily, dispersions, true micelles, lipoproteins solubilized by detergent action and liquid crystals. The use of micellar models, like those of all models, has both advantages and disadvantages. Since an important characteristic of all the bioenergetic systems being discussed is the possession of a lipoprotein interface immersed in an aqueous phase, it is to be expected that study of models which also contain this interface could contribute useful information. The main advantage is their experimental convenience since the great technical difficulties of studying monolayers and bilayers containing biologically relevant pigment concentrations may be easily overcome.

Teale dispersed various carotenoids and chlorophyll *a, b* into detergent micelles to give relatively high local pigment concentration (0·1 m). Although β-carotene could not be very successfully dispersed into micelles, other carotenoids such as fucoxanol were and gave energy transfer efficiencies (to chlorophyll *a*) approaching 100%. This correlates very well with the high efficiency found for these compounds in brown algae by Dutton *et al.* (1943).

As an example of a detergent solubilized pigment-lipoprotein system, we may cite chloroplastin and rhodopsin (Wolken, 1962). According to Wolken, the anionic detergent (digitonin) solubilizes the photoreceptor systems of the chloroplast and retinal rods to give colloidal solutions. In these colloidal solutions the pigments, lipids, proteins are somehow complexed to digitonin micelles. Thus chloroplastin contains chlorophyll, carotenoid, cytochrome, phospholipid and protein. Inasmuch as chloroplastin micelles show many of the properties of intact chloroplasts, oxygen evolution, photochemical activity in solution and the conversion of inorganic phosphate to ATP, they may be quite useful model systems.

If lipids, such as phospholipids, are added to water and the mixture agitated, various liquid crystalline phases are formed depending upon the relative proportion of lipid to water and many other factors (Dervichian, 1964).

According to Bangham et al. (1965), in salt solution these liquid crystals are composed of a series of concentric aqueous chambers, each limited by discrete bimolecular membranes. Various long chain species, both negatively and positively charged, may be incorporated into the liquid crystals. Such liquid crystals might also furnish useful model systems, if pigments concerned in bioenergetics may be incorporated into the limiting membranes and enzymes into the aqueous chambers. Such pigments may well be compatible with incorporation into the liquid crystals; with certain mixed pigment molecules charge-transfer interaction within the liquid crystals may be possible but preliminary attempts to demonstrate this have been inconclusive.

Apart from the demonstration by Menke (1966) that chloroplast lipids form both myelin figures and micelles, there appears to be no other relevant reports of the use of this type of model.

VIII. Conclusion and Outstanding Problems

Lamellar organization and bioenergetic efficiency are inseparable. Rapidly emerging evidence is suggesting that the lamellar structure involved in bioenergetics may have a more pronounced globular structure than hitherto believed. The molecules involved in bioenergetics and associated with lamellar systems are themselves probably an integral part of that lamellar structure and contribute to its stability.

The conservatism and skill of nature in developing both energetic and steric moieties within the same molecule is particularly striking. The molecules involved in a range of bioenergetic processes seem structurally designed to encompass two functions, an inbuilt tendency to associate themselves with membrane forming lipids and the correct electronic properties. The electronic properties seem to arise most frequently from the presence of a resonating aromatic system derived from relatively few molecular types. The steric moieties arise from the possession of long hydrophobic chains or from general lipid solubility.

The presence of a long hydrophobic chain does not significantly modify the electronic energy level structure of the individual molecules; this may be the explanation for the relatively scanty attention which has been focussed on the *total* molecular structure of bioenergetically important molecules. It is, however, this association of the bioenergetic molecules with lamellar systems, probably by means of hydrophobic interaction, which leads to the emergence of new energetic properties and potentialities of the aggregate compared to the isolated molecules.

Because in many of the molecules discussed in this chapter the energetic and hydrophobic lipid moieties occur within the same molecule, specific organization of the lipid moiety will lead inevitably to specific organization of the energetic moiety. In turn, specific organization of the energetic moiety will lead to energetic interactions and energetic potentialities in the organized systems quite different from those of the isolated molecules. In this sense, the hydrophobic lipid chains have a direct functional role in bringing about energetic interaction.

Furthermore, the wide range of possible hydrophobic interactions between bioenergetic molecules and membrane lipids or membrane proteins, which remains largely unexplored, may lead to a corresponding wide range of energetic interactions. Changes in membrane organization will be translated by the hydrophobic chain into changes in degree of organization and hence in the degree of coupling or interaction associated with the energetic moiety. This implies that membrane location can give a degree of energetic specificity in solution. This specificity may encompass directional effects or the type of energetic interaction, such as exciton transfer, donor-acceptor interaction or development of conduction bands, which develops.

Since most of the energy and electron transferring mechanism described in the chapter involved a degree of energetic interaction between several molecules, changes in their membrane environment will provide the possibility of changing from one type of mechanism to another.

However, rather than stress the importance of perfect organization which morphological evidence might imply, I would draw attention to the importance of overall molecular organization containing specific imperfections. Such localized imperfections in an otherwise perfect molecular array are likely to be of considerable importance. Collection sites (sinks) for energy gathered by an aggregate, or trapping centres involved in free electron or hole generation are the more obvious possibilities. Such localized, but important, imperfections would probably occur in such low concentration *in vivo* as to be undetectable by electron microscopy. They may also be transient imperfections which could be related to the state of the membrane in the vicinity of only a few molecules.

I would like to stress the importance of work on model systems for further advances in the understanding of the role of membranes in bioenergetics.

Monolayers, bilayers and micellar models could all contribute a great deal to the elucidation of the interrelationship between degree of order, or energetic coupling, and membrane bioenergetics. Membrane models also offer a reasonable hope of success in linking the initial biophysical quantum conversion act with the subsequent biochemical enzymology which seems to be an outstanding problem in membrane bioenergetics at the present time.

The study of enzyme reactions and redox reactions associated with these model membrane systems coupled with quantum conversion could well be a fruitful research area.

IX. Acknowledgements

Sincere thanks are due to the following: Professor D. D. Eley, Professor A. Szent-Györgyi and Professor Irvin Isenberg, whose collective enthusiasm for bioenergetics greatly stimulated the author.

I would like to thank Dr. D. Chapman for guiding me into the membrane area and his continued encouragement both in research and in the preparation of this chapter.

Last, but not least, I thank Mr. R. J. Cherry for many helpful discussions on various points and Mr. P. Brown for his considerable help in editing the manuscript.

References

Amesz, J. (1964). *Biochim. biophys. Acta* **79**, 257.
Arnold, W. and Meek, E. S. (1956). *Archs Biochem. Biophys.* **60**, 82.
Arnold, W. and Sherwood, H. K. (1957). *Proc. natn. Acad. Sci., U.S.A.* **43**, 105.
Arnold, W. and Sherwood, H. K. (1959). *Brookhaven Symp. Biol.* No. **11**, p. 1.
Arnold, W. (1965). *J. phys. Chem.* **69**, 788.
Arnon, D. I. (1961). *Fedn Proc. Fedn Am. Socs exp. Biol.* **20**, 1012.
Arnon, D. I. (1966). *Experimentia* **22**, 273.
Arnon, D. I., Tsujimoto, H. Y. and McSwain, B. D. (1965). *Nature, Lond.* **207**, 1367.
Augenstine, L. G. (ed.) (1960). Radiation Research Supplement 2, "Bioenergetics", Academic Press, New York.
Bangham, A. D., Standish, M. M. and Watkins, J. C. (1965). *J. molec. Biol.* **13**, 238.
Bay, Z. and Pearlstein, R. M. (1963). *Proc. Natn. Acad. Sci. U.S.A.* **50**, 1071.
Bellamy, W. D., Gaines, G. L. and Tweet, A. G. (1963). *J. chem. Phys.* **39**, 2528.
Benson, A. A. (1966). *J. Am. Oil Chem. Soc.* **43**, 265.
Bishop, N. I. (1958). *Proc. natn. Acad. Sci. U.S.A.* **44**, 501.
Bishop, N. I. (1959). *Proc. natn. Acad. Sci. U.S.A.* **45**, 1696.
Blaisie, T. K., Dewey, M. M., Blaurock, A. E. and Worthington, C. R. (1965). *J. molec. Biol.* **14**, 143.
Bradley, D. F. and Calvin, M. (1955). *Proc. natn. Acad. Sci. U.S.A.* **41**, 563.
Brodie, A. F. (1961). *Fedn Proc. Fedn Am. Socs exp. Biol.* **20** No. 4 p. 995.

Brodie, A. F., Weber, M. M., and Grey, C. T. (1957). *Biochim. biophys. Acta* **25**, 448.
Brown, P. K., Gibbons, I. R., and Wald, G. (1963). *J. biochem. biophys. Cytol.* **19**, 79.
Butler, W. H. (1961). *Archs biochem. Biophys.* **93**, 413.
Calvin, M. (1958). *Brookhaven Symp. Biol.* No. **11**, 160.
Calvin, M. (1959). *Rev. mod. Phys.* **31**, 147.
Calvin, M. (1961). *J. Theoret. Biol.* **2**, 258.
Calvin, M. (1965) *In* "Recent Progress in Photobiology" (E. J. Bowen, ed.) 225–258. Blackwell Scientific Publications, Oxford.
Calvin, M. and Bassham, J. A. (1957). "The Path of Carbon in Photosynthesis". Prentice Hall, Englewood Cliffs, New Jersey.
Cardew, M. H. and Eley, D. D. (1959). *Discuss. Faraday Soc.* **27**, 115.
Chance, B. and Williams, G. R. (1956). *Adv. Enzymol.* **17**, 65.
Chapman, D. and Cherry, R. J. (1964). *Nature, Lond.* **203**, 641.
Chapman, D., Owens, N. F. and Walker, D. A. (1966). *Biochim. biophys. Acta* **120,** 148.
Ciliento, G. and Guisti, P. (1959). *J. Am. Chem. Soc.* **81**, 3801.
Clayton, R. K. (1963). *A. Rev. Pl. Physiol.* **14**, 159.
Clayton, R. K. (1965). *Science, N.Y.* **149**, 1346.
Commoner, B. and Hollocher, J. C. (1960). *Proc. natn. Acad. Sci. U.S.A.* **46**, 405.
Cope, F. W. (1963). *Bull. math. Biophys.* **25**, 165.
Crane, F. L , Henninger, M. D., Wood, P. M. and Barr, R. (1966). *In* "Biochemistry of Chloroplasts" Vol. 1 (T. W. Goodwin, ed.) pp. 133–151, Academic Press, London and New York.
Criddle, R. S., Bock, R. M., Green, D. E. and Tisdale, H. (1962). *Biochemistry, N.Y.* **1**, 827.
Davis, K. M. C., Eley, D. D. and Snart, R. S. (1960). *Nature, Lond.* **188**, 724.
Dervichian, D. G. (1964). *Prog. Biophys. molec. Biol.* **14**, 265.
Dingle, J. T. and Lucy, J. A. (1965). *Biol. Rev.* **40**, 422.
Dutton, H. J., Manning, W. M., Duggar, B. M. (1943). *J. Am. Chem. Soc.* **47**, 308.
Duysens, L. N. M. (1951). *Nature, Lond.* **168**, 548.
Duysens, L. N. M. (1964). *A. Rev. Prog. Biophys.* **14**, 1.
Eley, D. D. (1962). *In* "Horizons in Biochemistry" (M. Kasha and B. Pullman, eds) 341–380. Academic Press, New York and London.
Eley, D. D. and Hedge, D. G. (1956). *Discuss. Faraday Soc.* No. **21**, 221.
Eley, D. D. and Leslie, R. B. (1963a). *Nature, Lond.* **197**, 898.
Eley, D. D. and Leslie, R. B. (1963b). *In* "Structure and Properties of Biomolecules" (J. Duchesne, ed.) pp. 238–258. Interscience, New York.
Eley, D. D. and Leslie, R. B. (1964). *In* "Electronic Aspects of Biochemistry" (B. Pullman, ed.) pp. 105–117, Academic Press, New York.
Eley, D. D. and Snart, R. S. (1965). *Biochim. biophys. Acta* **102**, 379.
Eley, D. D., Willis, M. R. and Jones, K. W. (1966). *Nature, Lond.* **212**, 72.
Emerson, R. and Arnold, W. (1931). *J. gen. Physiol.* **15**, 391.
Emerson, R. and Arnold, W. (1932). *J. gen. Physiol.* **16**, 191.
Fernanez-Moran, H. (1959). *In* "Biophysical Science" (J. L. Oncley, ed.) p. 319, John Wiley, New York.
Fleischer, S. and McConnell, D. G. (1966). *Nature, Lond.* **212**, 1366.
Fleischer, S., Klouwen, H. and Brierley, G. (1961). *J. biol. Chem.* **236**, 2936.
Fleischer, S., Brierley, G. P., Klouwen, H., Slauterback, D. B. (1962). *J. biol. Chem.* **237**, 3264.

Förster, Th. (1960). Radiation Research Suppl. 2, "Bioenergetics" pp. 326–339. Academic Press, New York.
Förster, R. and Thomson, T. J. (1962). *Trans. Faraday Soc.* **58**, 860.
Fraser, M. T., Kaplan, T. G. and Schulman, J. H. (1955). *Discuss. Faraday Soc.* No. **20**, 44.
Frey-Wyssling, A. (1937). *Protoplasma* **29**, 279.
Frey-Wyssling, A. and Steinmann, E. (1948). *Biochim. biophys. Acta* **2**, 254.
Gaffron, H. and Wohl, K. (1936). *Naturwissenschaften* **24**, 81.
Gaines, G. L., Bellamy, W. D. and Tweet, A. G. (1964). *J. chem. Phys.* **41**, 538.
Gaines, G. L., Tweet, A. G. and Bellamy, W. D. (1965). *J. chem. Phys.* **42**, 2193.
Goedheer, J. C. (1955). *Biochim. biophys. Acta.* **16**, 471.
Goodwin, T. W. (1966). "Biochemistry of Chloroplasts" Vol. I. Academic Press, London and New York.
Green, D. E. (1960). Radiation Research Supplement No. 2, "Bioenergetics", pp. 504–527. Academic Press, New York.
Green, D. E. and Fleischer, S. (1962). *In* "Horizons in Biochemistry". (M. Kasha and B. Pullman, eds) pp. 381–420. Academic Press, New York.
Green, D. E. and Fleischer, S. (1963). *Biochim. biophys. Acta* **70**, 554.
Green, D. E. and Fleischer, S. (1964). "Metabolism and Physical Significance of Lipids" (R. M. C. Dawson and D. C. Rhodes, eds) p. 582. J. Wiley, New York.
Green, D. E. and Lester, R. L. (1959). *Fedn Proc.* **18**, 987.
Green, D. E. and Tzagoloff, A. (1966). *J. Lipid Res.* **7**, 587.
Hagins, W. A. and Jennings, W. H. (1959). *Discuss. Faraday Soc.* No. **27**, 180.
Harbury, H. A. and Foley, K. A. (1958). *Proc. natn. Acad. Sci. U.S.A.* **44**, 662.
Hartshorne, N. H. and Stuart, A. (1960). "Crystals and the Polarising Microscope". 3rd Ed., 1960. Edward Arnold, London.
Hecht, S., Shlaer, S. and Pirenne, N. H. (1942). *J. gen. Physiol.* **25**, 819.
Hochstrasser, R. M. and Kasha, M. (1964). *Photochem. Photobiol.* **3**, 317.
Hodge, A. J. (1959). *In* "Biophysical Science" (J. L. Oncley, ed.) p. 331, John Wiley, New York.
Hyono, A., Kuriyama, S., Tsujo, K., Hosoya, Y. 1962). *Nature, Lond.* **193**, 679.
Isenberg, I. (1964). *Physiol. Rev.* **44**, 487.
Isenberg, I. and Baird, S. L. (1962). *J. Am. Chem. Soc.* **84**, 3803.
Isenberg, I. and Szent-Györgyi, A. (1958). *Proc. natn. Acad. Sci. U.S.A.* **44**, 857, 862.
Jacobs, E. E. (1959). *Brookhaven Symposium* No. 11, 32.
Jacobs, E. E., Vatter, A. E. and Holt, A. S. (1954). *Archs Biochem. Biophys.* **53**, 228.
Jacobs, E. E., Holt, A. S., Kromhout, R. and Rabinowitch, E. (1957). *Archs Biochem. Biophys.* **72**, 495.
Jahn, T. L. (1962). *J. Theoret. Biol.* **2**, 129.
Karreman, G., Steele, R. H. and Szent-Györgyi, A. (1958). *Proc. natn. Acad. Sci. U.S.A.* **44**, 140.
Kasha, M. (1959). *Rev. mod. Phys.* **31**, 162.
Katz, E. (1949). *In* "Photosynthesis in G Plants". (W. E. Loomis and J. Franck, eds). p. 291. Iowa State College Press, Ames, Iowa.
Kearns, D. R. (1960). Radiation Research Suppl. 2, "Bioenergetics" pp. 407–431. Academic Press, New York.
Kearns, D. R. (1964). *In* "Structure and Properties of Biomolecules" (J. Duchesne, ed.) pp. 282–338, Interscience, New York.
Kropf, A. and Hubbard, R. (1958). *Ann. N.Y. Acad. Sci.* **74**, 266.
Labes, M. M., Sehr, R. and Bose, M. (1960). *J. chem. Phys.* **33**, 868.

Lehninger, A. L. (1959). *Rev. mod. Phys.* **31**, 145.
Lehninger, A. L. (1962). *In* "Horizons in Biochemistry" (M. Kasha and B. Pullman, eds) pp. 421–433, Academic Press, New York and London.
Leslie, R. B. (1962). PhD. Thesis, Nottingham University.
Leslie, R. B. and Chapman, D. (1967). *J. phys. Chem. Lipids* **1**, 143.
Lester, R. L. and Fleischer, S. (1961). *Biochim. biphys. Acta* **47**, 358.
Lichtenthaler, H. K. and Calvin, M. (1964). *Biochim. biophys. Acta* **79**, 30.
Liebman, P. A. (1962). *Biophys. J.* **2**, 161.
Litvin, F. F., Gulyaev, B. A. and Sineshchekov, A. (1965). *Dokl., (Proc.) Acad-Sci. U.S.S.R.*, Biophysics Section, 160–162, 88.
Lumry, R. and Spikes, J. D. (1960). Radiation Research Suppl. 2, "Bioenergetics", pp. 539–577, Academic Press, New York.
Lundegårdh, H. (1966). *Nature, Lond.* **212**, 606.
Lynch, V. H. and French, C. S. (1957). *Archs Biochem. Biophys.* **70**, 382.
Lyons, L. E. (1957). *J. Chem. Soc.* 5001.
Maricic, S., Pifat, G. and Pravdic, G. (1964). *Biochim. biophys. Acta* **79**, 293.
McCree, K. J. (1965a). *Biochim. biophys. Acta* **102**, 90.
McCree, K. J. (1965b). *Biochim. biophys. Acta* **102**, 96.
McGlynn, S. P. (1960). Radiation Research Suppl. 2, "Bioenergetics". pp. 300–323, Academic Press, New York.
McRae, E. G. and Kasha, M. (1958). *J. chem. Phys.* **28**, 721.
Menke, W. (1938). *Kolloidzschrift* **85**, 256.
Menke, W. (1966). *In* "Biochemistry of Chloroplasts" (T. W. Goodwin, ed.) Vol. 1. pp. 3–18, Academic Press, London and New York.
Moody, M. F. (1964). *Biol. Rev.* **39**, 43.
Moor, H. and Mühlethaler, K. (1963). *J. biophys. biochem. Cytol.* **17**, 609.
Mühlethaler, K. (1966). *In* "Biochemistry of Chloroplasts" (T. W. Goodwin, ed.). Vol. 1, pp. 49–64, Academic Press, New York.
Mühlethaler, K., Moor, H. and Szarkowski, J. W. (1965). *Planta* **67**, 305.
Mulliken, R. S. (1952a) *J. Am. chem. Soc.* **74**, 811.
Mulliken, R. S. (1952b). *J. phys. Chem.* **56**, 801.
Nelson, R. C. (1957). *J. chem. Phys.* **27**, 864.
Nilsson, S. E. G. (1964). *J. Ultrastruct. Res.* **11**, 581.
Nilsson, S. E. G. (1965). *J. Ultrastruct. Res.* **12**, 207.
Olson, R. A. (1963). *In* "Photosynthetic Mechanisms of Green Plants". Publication No. 1145, National Academy of Science—National Research Council, Washington D.C.
Olson, R. A. Butler, W. L. and Jennings, W. H. (1961). *Biochim. biophys. Acta* **54**, 615.
Olson, R. A., Butler, W. L. and Jennings, W. H. (1962). *Biochim. biophys. Acta* **58**, 144.
Oster, G. (1955). *In* "Physical Techniques in Biological Research" (G. Oster and A. W. Pollister, eds) Vol. 1, pp. 439–460, Academic Press, New York.
Packer, L. (1966). *In* "Biochemistry of Chloroplasts" (T. W. Goodwin, ed.) pp. 233–242, Academic Press, London and New York.
Park, R. B. and Biggins, J. (1964). *Science, N.Y.* **144**, 1009.
Park, R. B. and Pon, N. G. (1963). *J. molec. Biol.* **6**, 105.
Pearse, A. G. (1965). *Nature, Lond.* **205**, 708.
Pullman, B. and Pullman, A. (1958). *Proc. natn. Acad. Science U.S.A.* **44**, 1197.
Rabinowitch, E. (1959). *Discuss. Faraday Soc.* No. **27**, 256.

Ray, A., Guzzo, A. V. and Tollin, G. (1965). *Biochim. biophys. Acta* **94**, 258.
Redfearn, E. R. (1966). *Vitams Horm.* **24**, 465.
Robinson, G. W. (1963). *Proc. natn. Acad. Science U.S.A.* **49**, 521.
Rosenberg, B. (1959). *Discuss. Faraday Soc.* No. **27**, 254.
Rosenberg, B. (1961). *J. chem. Phys.* **34**, 812.
Rosenberg, B. (1962). *J. chem. Phys.* **36**, 816.
Rosenberg, B. and Camiscoli, J. F. (1961). *J. chem. Phys.* **35**, 982.
Rosenberg, B., Orlando, R. A. and Orlando, J. M. (1961). *Archs Biochem. Biophys.* **93**, 395.
Sauer, K. and Calvin, M. (1962). *J. Theoret. Biol.* **4**, 451.
Schmidt, W. J. (1935a). *Z. Zellforsch. mikrosk. Anat.* **22**, 485.
Schmidt, W. J. (1935b). *Z. wiss. Mikrosk.* **52**, 8.
Schmidt, W. J. (1935c). *Zool. Anz.* **109**, 245.
Schultze, M. (1866). *Archs Mikrobiol. Anat.* **2**, 175.
Sjöstrand, F. S. (1949). *J. cell comp. Physiol.* **33**, 383.
Sjöstrand, F. S. (1959). "Biophysical Science" (J. L. Oncley, ed.) p. 301, John Wiley, New York.
Sjöstrand, F. S. (1960). Radiation Research Suppl. 2, "Bioenergetics" pp. 349–386, Academic Press, New York.
Sjöstrand, F. S. (1962). *Nature, Lond.* **199**, 1262.
Sjöstrand, F. S. (1963). *J. Ultrastruct. Res.* **9**, 340.
Snart, R. S. (1964). *Biochim. biophys. Acta* **88**, 502.
Styrer, L. (1960). Radiation Research Suppl. 2, "Bioenergetics". pp. 432–451, Academic Press, New York.
Szent-Györgyi, A. (1960). "An Introduction to Submolecular Biology". Academic Press, New York.
Szent-Györgyi, A. (1964). *In* "The Structure and Properties of Biomolecules" (J. Duchesne, ed.) p. XI (Introduction).
Teale, F. J. W. (1959). *Discuss. Faraday Soc.* No. **27**, 251.
Teale, F. J. W. and Weber, G. (1957). *Biochem. J.* **65**, 476.
Terenin, A. N., Putseiko, E. and Akimov, I. (1959). *Discuss. Faraday Soc.* No. **27**, 83.
Tollin, G. (1960). Radiation Research Supp. 2, "Bioenergetics", pp. 387–404, Academic Press, New York.
Trurnit, H. J. and Colmano, G. (1959). *Biochim. biophys. Acta* **31**, 434.
Tweet, A. G., Gaines, G. L. and Bellamy, W. D. (1964a). *J. chem. Phys.* **40**, 2596.
Tweet, A. G., Gaines, G. L. and Bellamy, W. D. (1964b). *J. chem. Phys.* **41**, 1008.
Tweet, A. G., Bellamy, W. D. and Gaines, G. L. (1964c). *J. chem. Phys.* **41**, 2068.
Urry, D. W. and Eyring, H. (1963). *Proc. natn. Acad. Sci. U.S.A.* **49**, 253.
Urry, D. W. and Eyring, H. (1965a). *J. Theoret. Biol.* **8**, 198.
Urry, D. A. and Eyring, H. (1965b). *J. Theroet. Biol.* **8**, p. 214.
van Niel, C. B. (1941). *Adv. Enzymol.* **1**, 263–328.
van Niel, C. B. (1960). *In* "Light and Life" (W. D. McElroy and B. Glass, eds) pp. 315–316. The Johns Hopkins Press, Baltimore, Maryland.
Wald, G. (1965). *In* "Recent Progress in Photobiology" (E. J. Bowen, ed.) pp. 133–144 (and refs therein). Blackwell Scientific Publications, Oxford.
Watson, W. F. and Livingston, R. (1950). *J. chem. Phys.* **18**, 802.
Weier, T. E. and Benson, A. A. (1966). *In* "Biochemistry of Chloroplasts" (T. W. Goodwin, ed.) pp. 91–113, Academic Press, London and New York.
Wolken, J. J. (1962) *J. Theoret. Biol.* **3**, 192.
Wosilait, W. D. (1961). *Fedn Proc. Fedn Am. Socs exp. Biol.* **20**, 1005.

Chapter 8

Nerve Excitability and Membrane Macromolecules

IRWIN SINGER AND ICHIJI TASAKI

Laboratory of Neurobiology,
National Institute of Mental Health, Bethesda, Maryland, U.S.A.

I.	INTRODUCTION	347
II.	MATERIALS AND METHODS	349
	A. Material and Dissection	349
	B. Double-cannulation and Perfusion	349
	C. Electric Stimululation of Axons	352
	D. Observation of Action Potential	352
	E. Measurement of Membrane Resistance by Pulse Method	353
	F. Measurement of Membrane Impedance	354
	G. Measurements of Membrane Currents under Voltage-clamp	354
	H. Radiotracer Measurement of Cation Flux	355
III.	RESULTS	356
	A. Anatomy of the Squid Giant Axon	356
	B. Biochemistry of the Squid Giant Axon	358
	C. Chemical Requirements for Excitability	362
	D. Physical Requirements for Excitability	385
	E. Characteristics of the Resting Membrane	387
	F. Characteristics of the Physiological Membrane in the Excited State	389
	G. The Flux-resistance Product	395
IV.	EXCITATION AS A PHYSICO-CHEMICAL PROCESS	396
	A. The Membrane as a Macromolecular Complex with Cation-exchanger Properties	396
	B. The "Two-Stable-States" Theory of Excitation	399
	C. Summary	405
REFERENCES		406
ADDENDUM		410

I. Introduction

Since the discovery of the electrical signal which indicates propagation of a nerve impulse (Du Boise-Reymond, 1849), many investigations of the mechanism of nerve excitation have been made from a purely electrical point of view. The electrical events which accompany the process of excitation were easily demonstrated by electrophysiologists and it was natural to interpret these events in terms of combinations of resistors, capacitors, rectifiers, electric dipoles and batteries; consequently, the nerve membrane was thought of in terms of an "equivalent electrical circuit." Some early models for the excitation process were based on the electrical characteristics of such circuits (e.g. Hill, 1932; Erlanger and Gasser, 1937, p. 88) and most current theories of excitation are based on elaborate models of this type (e.g. Cole and Curtis, 1939; Hodgkin and Huxley, 1952d; Hodgkin, 1964, p. 56; Cole, 1965).

Physico-chemical interpretations of nerve excitation were not attempted until more than forty years after Du Bois-Reymond's discovery. At that time physical chemists began to study the properties of colloids and developed new concepts to deal with the behavior of small colloidal particles. (This early work in "colloid science" has been comprehensively summarized by Kruyt (1949, 1952).) Shortly after the turn of the century, several biologists tried to apply these new physico-chemical concepts to the emerging studies of muscle and nerve physiology. These early attempts to provide a physico-chemical basis for the excitation process were summarized by Höber (1928, 1945), who proposed a "colloidal theory" of excitation.

Although refined electrical analyses of the excitation process have continued to dominate studies of the nerve membrane in recent years, several investigators have attempted to place the studies of muscle and nerve on a more physico-chemical basis (e.g. Segal, 1958; Ling, 1962; Tasaki and Singer, 1965, 1966). However, in order to describe membrane phenomena adequately from the latter point of view, it is necessary both to control the physico-chemical environment on both sides of the membrane and to have some knowledge or theory of the chemical structure of the membrane itself. The difficulty of meeting these requirements in the past probably accounts for the paucity of physico-chemical studies dealing with the problems of biological membranes. With the use of squid giant axons and the development of micro-techniques for biochemical analysis and intracellular perfusion, it has become possible to satisfy, qualitatively, the requirement for a physico-chemical approach to the excitation process in nerve membranes.

Electrophysiological investigation of the squid giant axon began with measurement of the "action potential" by means of electrodes placed in the external medium (Young, 1936a). While the neurophysiologist could easily control the external medium, adequate physico-chemical investigation could not begin until it also became possible to control the intracellular environment. This internal control was achieved in two laboratories by entirely different methods: the "roller technique" (Baker *et al.*, 1961) and the "double cannulation technique" (Oikawa *et al.*, 1961). Since the composition of the solutions on both sides of the membrane may be controlled precisely, the squid giant axon has become the classic experimental preparation for the study of the excitation process in nerve. Some familiarity with the material and methods used in this experimental system is essential for critical analysis of the data obtained and these subjects are described in considerable detail below.

II. Materials and Methods

A. MATERIAL AND DISSECTION

The most common varieties of squid used for neurophysiological investigation are *Loligo forbesi* (Plymouth, England), *Doryteuthis bleekeri* (Misaki, Japan), *Dosidicus gigas* (Viña del Mar, Chile) and *Loligo pealii* (Woods Hole, Massachusetts, U.S.A.). Most of the investigations in this laboratory were conducted with the North Atlantic squid (*Loligo pealii*) available at the Marine Biological Laboratory in Woods Hole. All of these varieties contain giant axons suitable for internal perfusion. Large specimens of *Loligo* contain axons ranging from 300 to 600 μ in diameter; large specimens of *Dosidicus* often contain axons greater than 1·0 mm in diameter. In addition, axons can be obtained from *Dosidicus* which are free from branches. However, these anatomical differences do not appear to produce any discrepancies in the results obtained by similar methods from different species. The double-cannulation methods used in this laboratory are described below for *Loligo pealii*. Essentially similar techniques are used for *Dosidicus* and *Doryteuthis*. The "roller technique" used in other laboratories has been described elsewhere in detail (Baker *et al.*, 1961).

Dissection of the giant axon is performed under cool, running (natural) sea water. Dark field illumination and a dissecting microscope are desirable for partial cleaning of the axon. For convenience in subsequent manipulations, each end of the axon is ligated with thread prior to excision. Some axons are extensively cleaned by carefully dissecting away the connective tissue and small nerve fibers which accompany the giant axon. This cleaning procedure is performed under a microscope; it reduces the unstirred layer at the external surface of the axon and is particularly desirable in radio-isotope experiments. (Axons used without such extensive cleaning are less likely to be injured during manipulation.) The axon is then mounted horizontally in a Lucite chamber containing natural sea water (Fig. 1).

B. DOUBLE-CANNULATION AND PERFUSION

The method of double-cannulation has been modified for a variety of experiments since its original description (Tasaki *et al.*, 1962). The method described below is typical of those currently used (Tasaki *et al.*, 1965a;

Tasaki et al., 1965b). A small incision is made in the axon membrane overlying one Lucite side-piece and a glass cannula (about 300 μ in diameter) is inserted through the incision by means of a micromanipulator (Fig. 1, bottom). This cannula is carefully advanced along the axis of the axon from the incision, across the air gap, to its initial position within the perfusion chamber. Gentle suction is applied to remove the axoplasm while the cannula

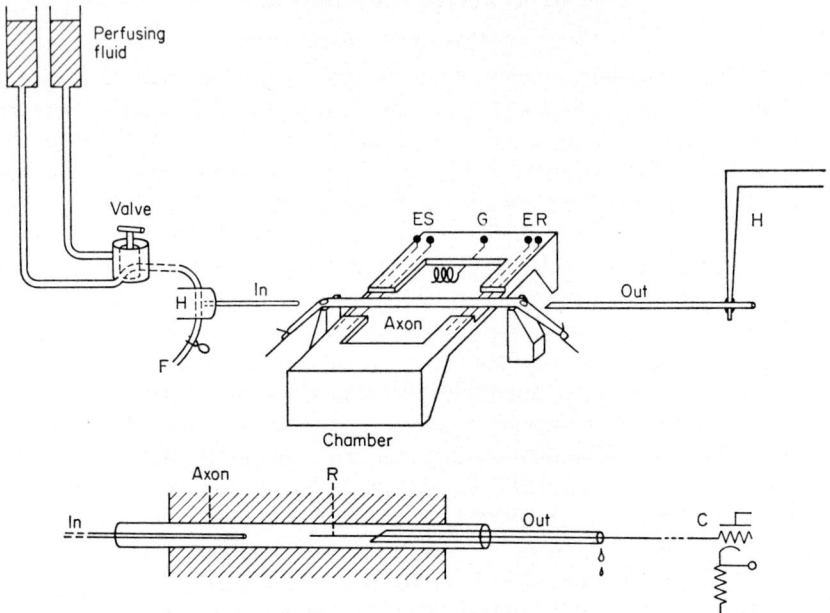

FIG. 1. Top: Experimental arrangement for double-cannulation and internal perfusion of *Loligo pealii* giant axons. The axon lies within the Lucite chamber, covered with natural sea water (not shown), supported by the two Lucite side-pieces and anchored by the ligatures at either end. The glass inlet (In) and outlet (Out) cannulae are supported in Lucite holders (H) connected to micromanipulators (not shown). The micromanipulators are used to insert both cannulae into the axon through incisions made above the sidepieces. The reservoirs for the perfusion fluid are connected to the inlet cannula through polystyrene tubing and a glass valve. The Lucite holder for the inlet cannula has a small T-shaped chamber within it which permits fluid to be delivered to the inlet cannula when the flushing outlet (F) is closed with the clip. The extracellular stimulating (ES) and recording (ER) platinum electrodes are built into the chamber and are exposed only within the lateral pools of the chamber; the platinum ground electrode (G) lies within the central pool. Bottom: Arrangement of cannulae and internal recording electrode after separation. The central pool, viewed from above, is indicated by the cross-hatching. The recording microelectrode (R) lies concentrically within the outlet cannula and the axon and is connected to the recording system through the cathode follower (C) (for details, see text).

is advanced. The cannula tip is brought to lie within the center pool without touching the membrane. This cannula serves as the outlet for the perfusion solution. Similarly, a second glass cannula (about 150 μ in diameter) is inserted into the other end of the axon to serve as the inlet. The inlet cannula is advanced to the position just concentrically within the outlet cannula. Then, the hydrostatic pressure in the inlet cannula is raised while the tips of the two cannulae overlap and more suction can be applied to remove the "core" of the axoplasm in the larger, outlet cannula; the perfusion fluid can also be used to flush the remaining axoplasm from the outlet cannula.

When internal recordings of action potential are desired, the appropriate recording electrode is introduced into the axon concentrically within the outlet cannula by means of a third micromanipulator and brought to lie 7 to 8 mm from the inlet cannula. The outlet cannula may now be withdrawn to its final position in the lateral pool, with the cannula tip close to the edge of the perfusion chamber; the resulting 12 to 20 mm separation between the inlet and outlet cannulae will be the "perfusion zone" when flow begins. The internal recording electrode thus remains in position in the center of the perfusion zone between the inlet and outlet cannulae (Fig. 1, bottom).

The inlet cannula is connected through a glass valve and polystyrene tubing to reservoirs for the perfusion fluid. The glass valve permits rapid alternation between two reservoirs containing different perfusion solutions. When solutions are alternated within the same axon, the dead space from which fluid cannot be flushed immediately is less than 0.5 mm^3. Flow is instituted by elevating the reservoirs; a complete change from one perfusion solution to another can be effected within one minute.

The entire perfusion system, from reservoirs to inlet, is flushed with 12 vol% glycerol solution before the introduction of any new perfusion fluid into the reservoirs. Flow rates of 10 to 30 mm^3/min are maintained when the reservoirs are raised to a height of 40 cm. Variation of the height of the reservoirs from 60 cm above the level of sea water in the chamber to 10 cm below the sea water level did not affect either the resting or action potential (Tasaki *et al.*, 1965a). When the axoplasm is removed by the roller technique (Baker *et al.*, 1962a) or by internal perfusion with mild proteases (Tasaki *et al.*, 1965a), axons are more sensitive to increased internal pressure.

The perfusion fluid passing through the axon may be collected directly from the outflow cannula into a small beaker for measuring the flow rate. Perfusion fluid is collected directly into planchets during isotope influx studies.

When isotope fluxes are measured it is necessary to isolate the radioactive external medium surrounding the perfused zone from the (non-radioactive) media surrounding the unperfused zones to either side. Vaseline partitions which isolate the lateral pools of external medium are erected in the Lucite chamber after double cannulation. If the external sea water in the lateral pools is replaced with an isohydric, isotonic solution of $MgSO_4$, the unperfused (lateral) portions of the axon can be made inexcitable without noticeably affecting events in the (central) perfusion zone. In order to measure isotope efflux, the fluid medium outside the perfusion zone is collected into planchets at periodic intervals and replaced with fresh solution.

C. ELECTRIC STIMULATION OF AXONS

1. *External Electrodes*

Propagated nerve impulses can be induced by applying electric shocks near the proximal end of the perfusion zone through a pair of external platinum electrodes built into the chamber (Fig. 1). In this case a stimulator with a stimulus-isolation unit is used. Stimuli of about 0·1 msec duration, and approximately twice the threshold level of intensity, are delivered at a frequency of 0·5 to 1·0 shock/sec for most studies. Higher frequencies (25 to 100 shock/sec) are used in isotope experiments and in studies of the effects of high frequency stimulation (e.g. Tasaki et al., 1965a).

2. *Internal Electrode*

When it is desirable to stimulate the perfusion zone directly with an internal stimulating electrode, an enamelled silver or platinum wire (50 μ in diameter) is inserted into the outlet cannula. This wire has a 10 to 20 mm bare portion which is brought to lie in the perfusion zone. (A second enamelled wire, which is used for intracellular recording, is twisted about this stimulating wire. The two wires are introduced into the axon simultaneously through the outlet cannula; see below.) Tektronix pulse generators may be used as sources of stimulating current. (A resistor of the order of $1M\Omega$ is connected between the internal wire and the pulse generator.)

D. OBSERVATION OF ACTION POTENTIAL

1. *Extracellular Recording*

Propagated diphasic action potentials may be recorded with a pair of platinum electrodes located beyond the perfusion zone, near the distal end of the axon. The potential variations may be displayed on an oscilloscope. The action potential may be monitored in this manner throughout the mounting procedure and subsequent manipulations.

2. *Intracellular Recording*

Several different kinds of internal microelectrode recordings are used. A glass pipette electrode filled with an isotonic solution of KCl is used most often in this laboratory. (No significant differences are observed when saturated KCl is used as the conducting medium.) The glass pipette electrode (about 80 μ in diameter) is advanced concentrically within the outlet cannula with a micromanipulator. The electrode is usually placed in the center of the perfusion zone; however, it may be moved within or beyond the perfusion zone if desired.

Other types of electrodes used frequently are: (1) a glass-coated, steel or platinum wire electrode (25 to 50 μ in diameter) and (2) an enamelled wire electrode (about 50 μ in diameter) with a 1 mm bare portion in the middle of the perfusion zone. The latter electrode is used primarily in conjunction with the internal stimulating electrode described above; the bare area of this recording electrode is positioned adjacent to the center of the bare area of the stimulating electrode. No significant differences are observed in the responses obtained by these various recordings electrodes.

Intracellular potential variations are monitored either through a Bak unity-gain cathode follower or through a Philbrick operational amplifier and are displayed on an oscilloscope. When the KCl-filled glass pipette is used for recording, a "chloridized" silver wire electrode is used to connect the input of the cathode follower with the KCl solution. In most cases, a large Ag–AgCl electrode is used in the external medium as a ground.

Although the variation of the membrane potential associated with excitation may be measured by any of these recording systems without ambiguity, the potential difference across the "resting" membrane can only be described operationally. The following procedure is used in this laboratory. The oscilloscope beam is brought to represent the potential level of the glass pipette electrode placed in the external medium. Then that microelectrode is withdrawn from the external medium and positioned within the axon in the center of the perfusion zone. The "resting potential" is defined as the difference between these two potential levels.

The necessity for an operational definition of the "resting potential" poses many problems for excitation theory, particularly for the "equivalent-circuit" model, and will be discussed in detail in Section III-E. Further discussion of the artifacts and ambiguities that can arise from the use of various electrodes and recording techniques, as well as a description of the procedures which can be used to reduce or eliminate them, can be found in other articles from this laboratory (Cavanaugh, 1956, Tasaki *et al.*, 1965a; Tasaki *et al.*, 1965c; Tasaki and Singer, 1966, 1968).

E. Measurement of Membrane Resistance by Pulse Method

For measurement of membrane resistance, the twisted pair of enamelled silver wire electrodes is used (see Fig. 9, top). Rectangular current pulses are obtained from a Tektronix pulse generator and delivered to the perfused zone of the axon through the internal wire electrode (the electrode with the 10 mm bare portion). These (inward-directed) current pulses are maintained at 0·5 to 1·0 μA in intensity and 10 msec in duration and are monitored on one oscilloscope beam. The amplitudes of the potential variations across the membrane are measured with the other enamelled wire electrode (the electrode with the 1 mm bare portion) and are monitored simultaneously by the other oscilloscope beam. The membrane resistance is determined from the simultaneous current and potential measurements.

F. MEASUREMENT OF MEMBRANE IMPEDANCE

For measurement of membrane impedance, a similar pair of enamelled platinum wire electrodes is used. In this case, the impedance electrode (current-carrying) has a bare portion of about 4 mm and the potential electrode has a bare portion of 2 mm. (The a.c. impedance bridge is shown diagrammatically in Fig. 10A. The fixed resistance (R) and capacitance (C) are 100 KΩ and 0·001 μF, respectively.) The output of an oscillator is attenuated so that the amplitude of the alternating current (16 kilocycles/sec) across the membrane of the axon does not exceed 5 to 10 mV, peak-to-peak. The impedance bridge output is led through a pair of high-pass filters to a low-level pre-amplifier and then through a band-pass filter (tuned to the bridge a.c.) to the oscilloscope. Simultaneous recording of the action potential and the impedance variation is obtained with this arrangement. (The electrode arrangement developed by Cole and Curtis (1939) can also be used for measuring the membrane impedance; details of their arrangement may be found in the article cited.)

G. MEASUREMENT OF MEMBRANE CURRENTS UNDER VOLTAGE-CLAMP

The voltage-clamp technique is a means for maintaining the potential difference across the membrane at a pre-selected constant value and measuring the membrane currents developed under various experimental conditions. A highly schematic diagram of the essential features of this arrangement is shown in Fig. 9, top. A Philbrick operational amplifier serves as the feedback control for the external current source. An internal pair of "platinized" platinum electrodes (about 50 μ in diameter) is used (Section II-E); the electrodes are inserted concentrically within the axon and outlet cannula in the same manner. One electrode has an uninsulated, platinized segment that is used for passing the voltage-clamping currents through the membrane. The other electrode has an uninsulated, platinized segment (about 1 mm in length) which is positioned adjacent to the bare portion of the current-passing electrode. A large coil of uninsulated platinum wire is used as the ground electrode in these experiments.

The voltage-clamp device used in these experiments is modified from that developed by Cole and Moore (1960) and further details may be found elsewhere (Tasaki and Singer, 1966; Tasaki et al., 1966b). Immediately after the onset of the clamping-potential pulse, the observed transmembrane current varies with time for the duration of the pulse. For a given clamping voltage the amplitude of this transmembrane current is usually measured at the moment when the inward-directed current reaches its peak value. In this manner, current-time and current-voltage relationships may be obtained. A representative study of this type is shown in Fig. 8 and will be discussed in Section III-C in more detail.

H. RADIOTRACER MEASUREMENT OF CATION FLUX

1. *Efflux Studies*

In order to study effluxes of internal cations, the appropriate radio-isotope is added to the internal perfusion fluid; perfusion is maintained for at least 5 to 10 min to ensure steady-state conditions. The two perfusion cannulae are then removed to the ends of the axon and the entire external medium is collected at regular 5 or 6 min intervals. Each collection is counted as a single sample. The following equation is used to determine the efflux, J_k, of cation species k:

$$J_k = \frac{n_k m_k}{n_k^0 S t}$$

where n_k represents the radioactivity (expressed in counts/min) in a single sample of external medium, transported through a membrane of area S (cm^2) during a time t (sec); m_k represents the amount of non-radioactive species k (pmoles) in a sample of the internal medium which has the radioactivity n_k^0 (counts/min). The cation efflux, J_k, is then expressed in pmoles $\times$ cm^{-2} $\times$ sec^{-1}.

Since the quantity of ions transferred across the axon membrane is very small, the isotope flux during a single action potential cannot be measured. In order to obtain the isotope efflux per impulse, it is necessary to stimulate the axon repetitively during a particular collection period and then to divide the total "extra" efflux by the number of impulses conducted. The "extra" efflux is defined as the total efflux during the period of stimulation, less the amount that would have appeared during a comparable period with no stimulation. Note that the limited time-resolution of the radio-isotope measurements prevents determination of the time-course of the ion-fluxes during a single cycle of excitation.

2. Influx Studies

Influxes of various external cations are measured by adding the appropriate radio-isotope to the external medium. Samples of the internal perfusion fluid are then collected from the outlet cannula at regular 5 or 6 min intervals. Since continuous flow of the internal perfusion solution is required for these samples to be collected, the two perfusion cannulae are not removed during the procedure. The amount of influx is determined, both at rest and during passage of an impulse, by the methods described for efflux measurements.

It is important to note that whenever isotope fluxes are used to trace either influx or efflux of a particular non-radioactive species, the radioactive and non-radioactive isotopes are present only on one side of the membrane. If the non-radioactive species were present on both sides of the membrane, the radio-isotope would not trace the movement of the non-radioactive species unambiguously. (It is assumed that there is no significant difference in ion mobility or selectivity between two different isotopes of the same chemical element within the membrane.)

The radioactive samples are dried in their planchets before counting. Radioactivity is determined with a low-background radiation counter. Further details of the tracer techniques used in this laboratory may be found elsewhere (Tasaki, 1963; Tasaki *et al.*, 1966b).

III. Results

A. ANATOMY OF THE SQUID GIANT AXON

Nerve fibers or axons are generally divided into two classes: myelinated and unmyelinated. Most vertebrates have both kinds of nerve fibers but the unmyelinated axons are much smaller in size. On the other hand, some invertebrates have very large unmyelinated axons; for example, the squid often contains giant axons as large as 1 mm in diameter (Young, 1936a,b). The multilayered lipid sheath (myelin) and its regular interruptions (nodes of Ranvier) are characteristic of vertebrate myelinated fibers but are absent in these large invertebrate nerve fibers. Such unmyelinated fibers consist of a core of protoplasm (axoplasm) and a "membrane" separating the axoplasm from the external environment. These fibers are capable of conducting nerve impulses for many hours, or sometimes for days, in the absence of the cell body and nucleus.

In addition to axoplasm and axon "membrane," the isolated squid giant axon includes a connective tissue sheath and a layer of Schwann cells. Since it is not possible to remove either the Schwann cells or all of the connective tissue, all physiological studies of the axon "membrane" include these components. Similarly, even in internally perfused giant axons, it is not possible to be certain that the axoplasm is removed completely; therefore, this material may also contribute properties to the "physiological membrane." The latter point is important since there is a concentration of mitochondria just beneath the anatomical "membrane" in some giant crustacean axons (Geren and Schmitt, 1954) which may supply energy to the "physiological membrane". However, the bulk of the axoplasm (including the neurofibrils and mitochondria) may be removed by the use of proteases. Excitability of such "empty" axons can be maintained if proper salt solutions are used for internal perfusion (see Takenaka and Yamagishi, 1966). Thus, the "physiological membrane" consists of any remaining internal axoplasm, the remaining external connective tissue, the Schwann cells and the "anatomical membrane."

Much of the information about the anatomical membrane has come from recent electron microscopy studies. The advantages and limitations of various techniques of fixation, staining, etc. for electron microscopy of cell membranes have been discussed in detail in recent reviews (e.g. Robertson, 1960). It is generally assumed that electron dense stains, such as OsO_4, are deposited on polar groups, such as proteins, which lie on either side of a bimolecular lipid leaflet which is not electron dense. With these techniques, the total thickness of the axon membrane has been found to be 70 to 100 Å; the two electron dense lines are 50 to 70 Å apart (Fig. 2A). (Recently, macromolecular concepts have been applied to the processes of membrane fixation and staining in electron microscopy; the underlying mechanisms are visualized in terms of "tricomplex" formation in a colloidal system (Elbers *et al.*, 1965).

While no substructure of the axon membrane has been revealed by these methods, the detailed relationship of the giant axon membrane to the Schwann cell investment has been studied in the squid (Geren and Schmitt, 1954) and in the earthworm (Hama, 1959). In the squid, electron microscopy reveals several complex layers of Schwann cell processes surrounding the axon with a space between the Schwann cell membrane and the axon membrane estimated to be about 150 Å by electron microscopy (Figs 2A and 2B). Although there are likely to be channels directly from the axon membrane to the external medium, the possible contribution of the Schwann cell to the various physiological functions of the cell cannot be ignored. It is generally agreed that the Schwann cell plays a role in myelin formation in myelinated axons (Geren, 1954) but its role in unmyelinated axons remains a subject for speculation (see Hodgkin, 1964, p. 26). Hence, the behavior of the "physiological membrane" may include properties contributed by small layers of axoplasm internally and connective tissue externally as well as by the Schwann cells and the "anatomical membrane".

B. BIOCHEMISTRY OF THE SQUID GIANT AXON

The composition of the intracellular axoplasm has been analyzed by several workers; the results are summarized in Table I for the major cations and anions (Steinbach, 1941; Steinbach and Spiegelman, 1943; Koechlin 1955; Keynes and Lewis, 1951, 1956; Caldwell, 1956 and 1960; Davison and Taylor, 1960; Deffner, 1961; Hodgkin, 1964). The potassium-containing axoplasm has a relatively low concentration of alkali earth metals; in addition to chloride, a variety of organic anions are present. The usual experimental

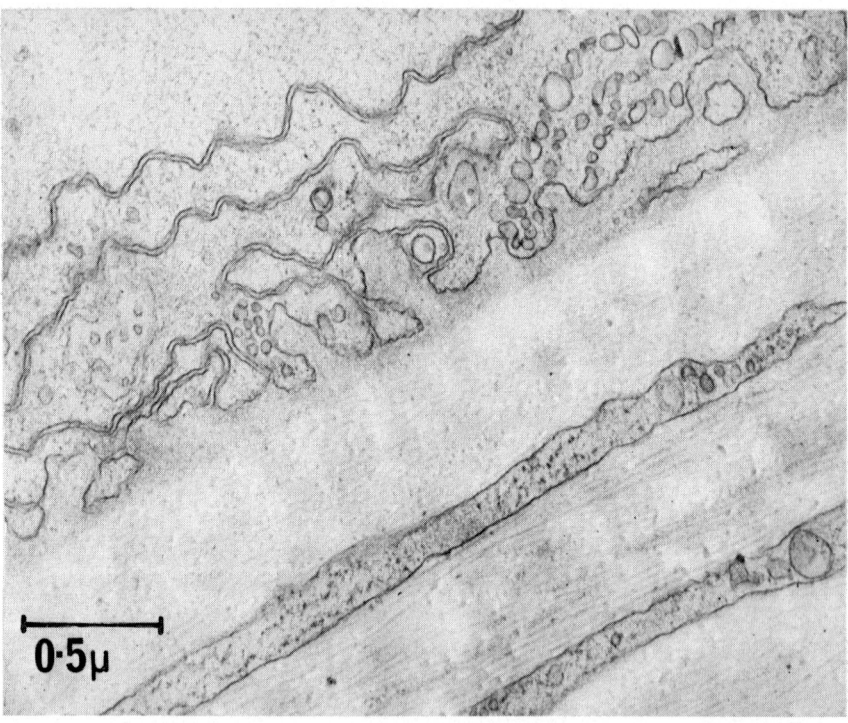

Fig. 2A. Electron micrograph of squid giant axon. The axon membrane is represented by the uppermost of the electron-dense lines in the upper left hand corner. The axoplasm is indicated by the clear zone above the "railroad tracks" marking the membrane. The electron-dense lines are stained with OsO_4 and are thought to represent the polar zones in the bimolecular leaflet. The 'complex region below the axon membrane represents multiple infoldings of processes which belong to the Schwann cell. (Note that the Schwann cell membrane also consists of a pair of electron-dense lines.) The clear areas below the Schwann cell (including the scale) are surrounding connective tissues.

external medium is natural sea water which has an ionic composition almost identical to squid blood (see Table I). In contrast to the internal medium, the usual sodium-containing external medium has a relatively high concentration of divalent cations; chloride is usually the only major anion present.

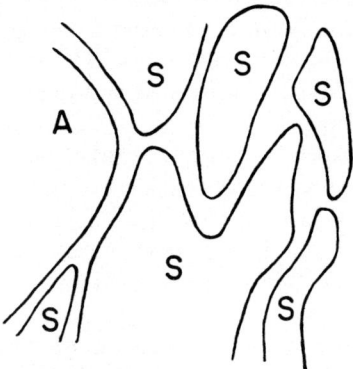

FIG. 2B. Diagrammatic relationship of Schwann cell processes and axon. The axon is indicated by A and the multiple processes of a Schwann cell by S. Each of the single lines in this figure corresponds to a pair of electron-dense lines in the electron micrograph. The intimate anatomical relationship of the Schwann cell processes to the axon membrane suggests that they are part of the "physiological membrane" of the axon (see text).

TABLE I
Concentration (mmoles/kg H_2O)

Component	Axoplasm	Squid blood	Natural sea water
K	400	20	10
Na	50	440	460
Ca	0·4	10	10
Mg	10	54	53
Cl	40–150	560	540
Phosphate	2·5–2·9	—	—
Aspartate	75	—	—
Glutamate	12	—	—
Isethionate	250	—	—

Since some axoplasm remains even in the internally perfused squid giant axon, various ions may be present in a form bound to axoplasm. However, it is extremely unlikely that any diffusible (unbound) ions or (water-soluble) metabolites remain within the axon after a few minutes of perfusion. Since the perfused axon can remain excitable for many hours with no overt sign of deterioration, any metabolite required for excitation must be either provided within the perfusion fluid or "stored" within the membrane.

Nerve and other cell membranes are generally assumed to contain proteins, lipids and phospholipids (Hodgkin, 1964, p. 29; Tasaki and Takenaka, 1964; Tasaki *et al.*, 1965a). In the classic picture of the "bimolecular leaflet" (Davson and Danielli, 1943), the lipids are arranged with their non-polar groups forming a central layer; their polar groups face inward (towards the cytoplasm) and outward (toward the external medium). The proteins are essentially confined to the two polar layers. This arrangement is consistent with the two electron dense lines separated by a relatively clear zone observed by electron microscopists.

However, the apparently symmetrical bimolecular leaflet visualized by electron microscopy is neither biochemically nor physiologically symmetrical. For example, proteases have no demonstrable effects on excitability when applied externally for hours (Tobias, 1955; Tasaki and Takenaka, 1964) but will rapidly suppress excitability when applied internally (Rojas and Luxoro, 1963; Tasaki and Takenaka, 1964; Tasaki *et al.*, 1965a). The effects of various neutral salts on excitability are also quite different when applied on either side of the membrane. For example, the internal surface of the physiological membrane is very sensitive to anion differences, whereas the external surface is relatively insensitive (Singer and Tasaki, 1965; Tasaki *et al.*, 1965a); in addition, cations which produce depolarization externally have no such effects when applied internally. Moreover, there are differences in the effects of pH, heavy metals and organic ions all of which reflect the dissimilarity of the two surfaces of the physiological membrane.

Some reasonable conceptions of the macromolecular structure of the physiological membrane can be derived from these data. Most (macromolecular) proteins and phospholipids are ampholytes, i.e. they carry both positively and negatively charged groups (White *et al.*, 1964, pp. 73–77, 122–125). Possible sources of positive charge include N-terminal groups of proteins, as well as amino (arginine, ethanolamine, lysine), amidino (asparagine, glutamine), guanidino (arginine, histidine), imidazo (tryptophan) and choline side-groups. Possible sources of negative charge include C-terminal groups and carboxylate side-groups of proteins and phosphate groups of phospholipids. Furthermore, it may be assumed that strong (electrochemical) interaction exists between adjacent charged groups in the membrane and between such charged groups and the physico-chemical environments on each side of the membrane (Schellman and Schellman, 1964, pp. 15, 48). At physiological pH, most biological macromolecules are on the alkaline side of their isoelectric points, hence at physiological pH they carry a net negative charge although both positive and negative groups may be ionized (Jirgensons and Straumanis, 1962, p. 152).

The existence of some of these groups in the axon has been assumed from biochemical analysis of other cell membranes; the presence of several particular groups has been confirmed by enzyme studies of the giant axon. For example, the fact that trypsin is very effective in eliminating the action potential implies that there are free amino groups adjacent to some of the peptide bonds in the membrane (Tasaki et al., 1965a). The ability of phospholipase C and D to block conduction confirms the presence of phospholipids in the physiological membrane (Tasaki and Takenaka, 1964). The presence of free C-terminal carboxylate and N-terminal amino groups in membrane proteins may be inferred from the similar actions of aminopeptidases and carboxypeptidases. (Details of these and other enzyme investigations may be found in the articles cited above.)

Physiological studies interpreted from a physico-chemical point of view also suggest the existence of free amino, carboxyl and phosphate groups in the membrane. The particular lyotropic cation and anion sequences found for the internal and external surfaces of the squid giant axon suggest that the external surface of the physiological membrane contains (predominately) carboxyl groups and a relatively high excess of fixed negative charge, whereas the internal surface contains (predominately) phosphate groups (Tasaki et al., 1965a). The relationship of the lyotropic series to the process of excitation will be discussed in detail in Sections III and IV.

In summary, the physiological membrane (which may include axoplasm, axon membrane, Schwann cells and connective tissue) may be visualized as a macromolecular complex of proteins and phospholipids (Fig. 3; see also

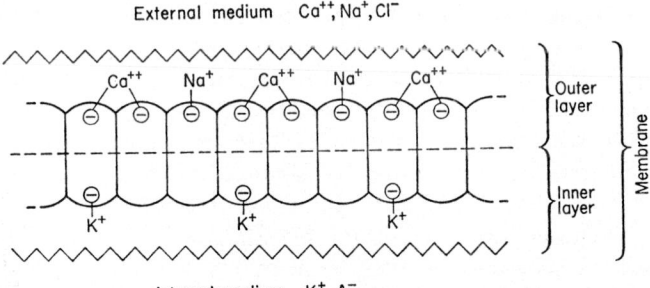

FIG. 3. Model for the macromolecular membrane. This highly schematic diagram was constructed solely on the basis of physiological data obtained from the squid giant axon. The external physiological layer contains a relatively large number of fixed negative charges which are occupied primarily by divalent cations derived from the external medium. The internal physiological layer has relatively few fixed anionic sites, occupied primarily by cations derived from the axon interior. The "physiological membrane" includes the ill-defined boundary between the two layers (dashed line) and the two boundaries between each layer and its surroundings (jagged lines). The cation-exchanger properties of the membrane are attributed to the external physiological layer (see Fig. 14). (Reprinted from Tasaki and Singer, 1966a).

Fig. 14, left). This structure is held together primarily by intra- and intermolecular salt linkages between adjacent oppositely-charged groups (Tasaki et al., 1965a). Some counter ions, particularly external divalent cations, may also contribute to maintaining the membrane structure by forming bridges between two negatively-charged side groups (Steinhardt and Beychok, 1964, p. 265). Other weak bonds, such as hydrogen bonds in polar regions and hydrophobic bonds in non-polar regions, may also be important (see also, Schellman and Schellman, 1964, p. 46). In addition, there is likely to be an excess of fixed negative charges, predominately in the layer near the external surface, which confers some cation-exchanger properties on the system (Teorell, 1953). These concepts have been developed in detail in recent articles from this laboratory and provide the basis of the "two-stable-states" theory of excitation (Tasaki, 1963; Tasaki et al., 1965a; Tasaki and Singer, 1965, 1966). The results of experiments with internally perfused squid giant axons to be discussed in the remainder of this article will be interpreted in terms of the interactions of such a complex macromolecular system with its physico-chemical environment.

C. CHEMICAL REQUIREMENTS FOR EXCITABILITY

1. *pH Effects*

The pH requirements for excitability are known to be different on the two sides of the membrane (Tasaki et al., 1965a). For most studies with squid giant axons the internal pH is maintained between 7·2 and 7·4 by a potassium phosphate buffer (less than 5 mM phosphate). The external pH is maintained at about 8·0 with a *tris*hydroxymethylaminomethane ("*tris*")-HCl buffer (less than 5 mM *tris*). (At greater concentrations, these buffers become significant in the ionic composition of their respective solutions.) The optimum internal pH for producing large action potentials for a long period of time is about 7·3. Action potentials of smaller amplitude can be produced for shorter times within the pH range 6·5 to 7·0. At pH values greater than 7·5, spontaneous repetitive discharge of impulses occurs with a loss of excitability which is often irreversible. To some extent, increasing the external divalent cation concentration can reduce repetitive firing (Tasaki et al., 1965a). At pH values less than 6·5 excitability is suppressed but can be restored by rapidly changing to a solution with a pH closer to the optimum. The optimum external pH is about 8·0, with an excitable range from 7·0 to 8·5. As with the internal medium, external pH values greater than 8·5 induce spontaneous repetitive firing of impulses, followed by irreversible loss of excitability and pH values lower than 7·0 produce a suppression of excitation that is often reversible.

The pH sensitivity of macromolecular systems is very well known (Overbeek and Bungenberg de Jong, 1949, p. 184; Kauzmann, 1959). This sensitivity is directly related to the role of pH in determining the degree of side-group ionization as well as the stability of peptide and other intra- and intermolecular bonds (Katchalski *et al.*, 1964, pp. 472 and 493). Small variations from the optimum pH can produce variations in the amplitude of the action potential by determining the number of ionized fixed negative charges available, thus controlling the ion-exchanger properties of the membrane. Extreme values of pH probably eliminate excitability by irreversible protein denaturation. The more reversible changes in protein structure produced by a mildly acid pH are characteristic of biological macromolecules. The fact that the pH optimum is more alkaline on the external side of the membrane supports the likelihood that there is a relatively greater number of ionized fixed negative charges on the external membrane surface than on the internal surface (see Steinhardt and Beychok, 1964, p. 261).

2. *Concentration Effects*
(a) *Dilution of the external medium.* Dilution of the electrolyte solutions on either side of the membrane produces very different results. Such dilution is usually accomplished with an isotonic sucrose solution externally and with either isotonic sucrose or glycerol solutions internally. When the ionic strength of an external medium of natural sea water is reduced in this manner, the magnitude of the resting potential of the membrane increases (i.e. the axon interior becomes more negative); after a ten-fold dilution, the axon becomes inexcitable. If the resting potential is plotted as a function of the log of the concentration, the slope is roughly 30 to 40 mV for a ten-fold dilution (I. Teorell and C. S. Spyropoulos, unpublished).

It must be noted that such dilution with isotonic non-electrolyte solution has several possible ambiguities. Even if the possible role of the non-electrolyte is ignored, not only are the concentrations of the individual ions reduced, but the total ionic strength is decreased as well. Furthermore, when there are fixed negative charges of a high density in the membrane and both univalent and divalent cations externally, a simple ten-fold dilution of the external medium has the same effect as a ten-fold increase in the external concentration of the divalent species without a change in the univalent cation concentration (Helfferich, 1962, p. 156; see also Tasaki and Singer, 1968).

The fact that after dilution the intracellular potential becomes more negative with respect to the external medium implies that it is dilution of the external cations (rather than the anions) that determines the change in resting potential. In the equivalent circuit model for excitation, it is often assumed that the resting potential of the membrane is determined solely by the distribution of potassium ions across the membrane, i.e. that the resting membrane essentially behaves as a potassium-sensitive glass electrode and is impermeable to sodium

(Hodgkin, 1964, pp. 30–31). In this case, the membrane potential (expressed in mV) should vary with the potassium concentration according to the Nernst relationship:

$$E_r = 58 \log \frac{[K]_e}{[K]_i} \qquad (1)$$

The resting potential, E_r, would then be expected by vary by 58 mV with a ten-fold change in the external concentration of potassium ion, $[K]_e$; the internal concentration of potassium ion, $[K]_i$, is assumed to be constant. One of the major pieces of evidence for this interpretation is that Eq. (1) is obeyed over a wide range of external potassium ion concentrations (Hodgkin, 1964, p. 31). It should be pointed out, however, that axons exposed to such high concentrations of potassium ion are not excitable and there is no assurance that the physiological membrane behaves in the same way under excitable and inexcitable conditions. The resting potential cannot be described by the equation above in those cases where no potassium ion exists in the external solution and complete elimination of potassium ion in the surrounding natural sea water does not alter the resting potential of the squid giant axon. There are many other difficulties with descriptions of the "resting" membrane potential and with the equivalent circuit model for excitation which will be discussed subsequently (Sections C, E, F).

(b) *Dilution of the internal medium.* The effects of dilution of the internal salt solution in the perfused squid giant axon have been analyzed in detail. Excitability can be maintained in the squid giant axon with solutions of very low ionic strength (Tasaki *et al.*, 1962; Baker *et al.*, 1962b; Narahashi, 1963; Baker *et al.*, 1964; Moore *et al.*, 1964; Tasaki *et al.*, 1965a). Ordinarily, twenty-fold dilution of the internal electrolyte will not abolish excitability. In addition, a variety of "unfavorable" perfusion media which produce conduction block in moderate or high concentrations have little or no harmful effect at low concentrations (Tasaki and Shimamura, 1962; Tasaki *et al.*, 1965a). Of particular interest is the fact that when excitation is suppressed by perfusion with some "unfavorable" salt solutions, conduction can be restored merely by dilution of the perfusion fluid with isotonic non-electrolyte solution (Tasaki *et al.*, 1965a).

An example of this phenomenon in an axon immersed in natural sea water is shown in Fig. 4. The initial perfusion fluid contained 500 mM KCl. Both the resting and action potentials remained almost constant for roughly 20 min. This plateau period was followed by a rapid fall in the amplitude of the action potential and a complete loss of excitability after an additional 10 min. The magnitudes of the resting potential and membrane resistance gradually decreased during the same time period. Within one minute after conduction was blocked, the internal perfusion solution was changed from 500 mM KCl to an isohydric, isotonic solution containing 200 mM KCl. This substitution resulted in an immediate restoration of excitability with conduction of action potentials of even larger amplitude than initially. In general, when more dilute restoring solutions were used, recovery was even more marked. If more than a few minutes elapsed after conduction was blocked, dilution produced incomplete or no recovery.

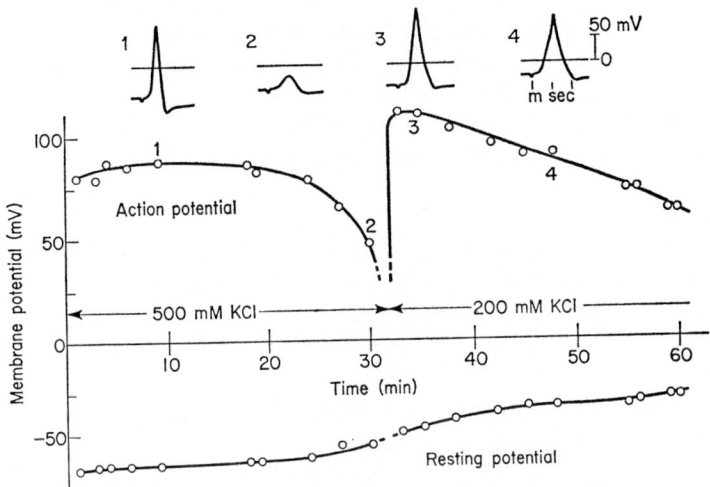

FIG. 4. Effect of dilution of internal electrolyte. The time-courses of the action potential "overshoot" and the resting potential are indicated by the upper and lower curves, respectively. Stimuli were delivered at 1 shock/sec throughout this experiment. The break in each curve represents the time during which conduction was blocked and during which the internal medium was changed from 500 mM to 200 mM KCl. Four representative oscilloscope records are shown at the top; the numbers refer to corresponding positions in time along the action potential curve. Note that action potentials were restored within one minute following dilution and that there was almost no change in the resting potential. (Reprinted from Tasaki et al., 1965a).

It should be noted that dilution restored the action potential without a significant change in the resting potential. This finding cannot be explained within the original formulation of the electrical equivalent circuit theory which required a change in the resting potential to precede any change in the action potential (Hodgkin and Huxley, 1952).

Furthermore, the resting potential does not obey the Nernst equation for potassium ion when the internal concentration of potassium is varied by dilution with isotonic sucrose. There is only a 10 mV change in the resting potential over the entire range of $[K]_i$ from 25 mM to 500 mM (Tasaki and Takenaka, 1964). The deviations of the resting membrane from ideal potassium electrode behavior according to Eq. (1) have been attributed to the existence of fixed negative charges near the internal surface of the membrane (Baker *et al.*, 1964). It is assumed that as the internal solution becomes more dilute, the electric field produced by these fixed charges establishes an electrical potential profile near the inner surface of the membrane such that the region immediately adjacent to the inner surface of the membrane becomes relatively negative compared to the remainder of the perfusion solution. Since the internal microelectrode records the potential level at some distance from the membrane, the "true" resting potential is thought to be much larger than the measured resting potential.

However, if there were such a field of electric potential in an electrolyte solution, the ions in the solution would redistribute themselves so that the electrochemical potential remained constant throughout the system; such an ionic double layer, and the inner surface of the membrane, must be in thermodynamic equilibrium with the immediately adjacent region in the solution. Therefore, there is effectively no change in electrochemical potential introduced by the assumption of fixed negative charges at the internal surface and the relative deviations from Eq. (1). (In deriving Eq. (1) for a potassium-sensitive membrane electrode, the effect of such a double layer has already been taken into consideration and consequently cannot make an additional contribution to the membrane potential.)

The effect of dilution of the internal perfusion fluid can be easily understood in terms of the macromolecular approach to the physiological membrane. The membrane proteins and phospholipids are assumed to form a compact complex structure because of coulombic and other molecular interactions. The relatively high resistance of the excitable, "resting" axon membrane is attributed to this close packing which makes a large proportion of the charged groups unavailable for transmembrane ion transport; i.e. in the resting state, the functional ion-exchange capacity of the excitable physiological membrane is relatively small (Winter, 1956). If some process disrupts inter- and intramolecular linkages, previously unavailable charged groups become available for transmembrane ion transport (see also, Fig. 14) and membrane resistance would be decreased. If this process continued to disrupt the membrane structure, excitability would eventually be lost. Such disruptions can be produced by the high concentrations of electrolytes usually used for internal perfusion.

In systems consisting of charged macromolecules and neutral salts, electrostatic interactions between macromolecular charged groups are reduced by the presence of the electrolyte; this effect is produced by "screening" of the electric charges on the macromolecules by the neutral salts (Overbeek and

Bungenberg de Jong, 1949, p. 226; Kauzmann, 1959, pp. 46 and 49). The magnitude of this effect is determined primarily by the concentration and valence of the ions derived from the neutral salts. Thus, dilution of the salt would tend to increase electrostatic interactions between oppositely charged groups in a macromolecular system such as the physiological membrane. In the present case, dilution of the intracellular salt would restore high membrane resistance and increase excitability.

The deviation of the change in resting potential from the Nernst slope during dilution experiments can be attributed to the continuous interdiffusion of cations in the resting state (Section III-E) and the multi-layer structure of the axon membrane. The influx of external cations is expected to create a diffusion potential across the axoplasm and (internal) unstirred layers; efflux of internal cations produces a similar effect across the Schwann cell and (external) unstirred layers. Since dilution would tend to increase the diffusion potentials across these unstirred layers, this mechanism may account for the deviations from the Nernst slope.

An alternative (or additional) mechanism may involve the formation of small numbers of "active patches" in the outer physiological membrane by the efflux of internal (depolarizing) cations in the resting state (see Section IV-B; Tasaki and Singer, 1966). Dilution would also reduce this effect and could account for the deviations from Eq. (1).

3. *Nature of Internal Cations and Anions*

There are differences in the physiological effects of electrolytes of the same concentration and valence (Singer and Tasaki, 1965; Tasaki *et al.*, 1965a). The anions were studied by comparing isotonic, isohydric salts of the same cation (usually potassium); cations were studied in a similar manner, using a common anion (usually chloride). Two different methods were employed for these comparisons. The first method consisted of determining the time required to block conduction during continuous intracellular perfusion with a particular salt; this time is defined as the "survival time." The second method consisted of determining the ability of one ion to restore excitability which has been depressed by another ion; this ability is defined as the "restoring ability". Both methods gave entirely consistent results.

Typical examples of both "survival time" and "restoring ability" methods are shown in Fig. 5 for anions (Tasaki *et al.*, 1965a). In this case the axon was perfused with a 400 mM KCl solution initially. The action potentials which could be elicited from the perfused zone of the axon remained at an almost constant level for 20 to 25 min. This plateau period was followed by a rapid decrease in amplitude and eventually by conduction block (at 38 min). The survival time of this axon is thus 38 min; for 18 axons perfused with 400 mM KCl solutions, the average survival time was 27·4 min. Survival times with solutions of the other halide salts of potassium were determined under identical internal and external conditions, except for the substitution of the halide. With isohydric, isotonic solutions of KI and KBr, the average survival times

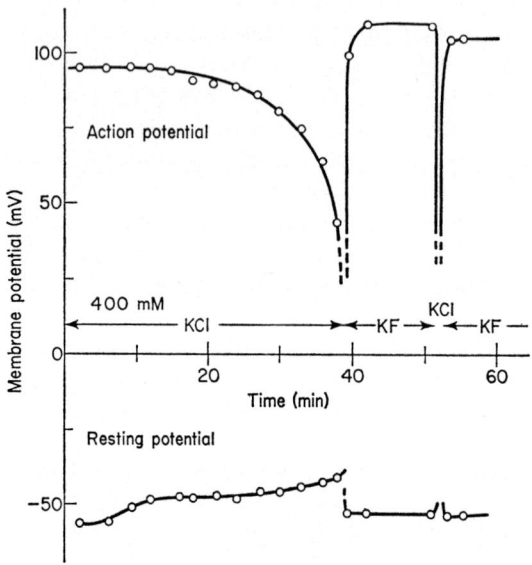

FIG. 5. Survival time and restoring ability with anions. The time-courses of the action potential "overshoot" and resting potential are indicated by the upper and lower curves, respectively. The breaks in each curve indicate those times when internal perfusion solution were changed. With an internal perfusion solution of 400 mM KCl, conduction was blocked in 38 min (the "survival time" for this axon). Conduction was restored within one minute by changing to 400 mM KF at this time, indicating the "restoring ability" of the anion. The rapid decline in action potential amplitude following reintroduction of the KCl solution (at 51 min) is evidence for the existence of changes in the membrane which are "masked" during KF perfusion (see text). The restoring ability of fluoride ion is demonstrated again at 52 min; action potentials could be elicted for hours with this internal medium. (Reprinted from Tasaki et al., 1965a).

were 4·9 min (10 axons) and 16·9 min (26 axons), respectively. With similar solutions of KF, the survival time was greater than several hours and could not be determined directly.

In the graph shown in Fig. 6, a linear relationship was obtained when the average survival times were plotted against the classic lyotropic numbers for the halide ions. The assignment of lyotropic numbers was an attempt to place Hofmeister's original qualitative description of the lyotropic series (Hofmeister, 1888) on a quantitative basis. These numbers were determined according to the relative abilities of particular ions to precipitate various sols (Büchner and Postma, 1931; Bruins, 1932; Voët, 1937) and have been extended to include all of the anions commonly used for internal perfusion of squid giant axons (Tasaki *et al.*, 1965a, Appendix). Many physico-chemical properties of macromolecules are affected by added salts in the order represented by the lyotropic numbers (e.g. swelling of gels, viscosity of sols and

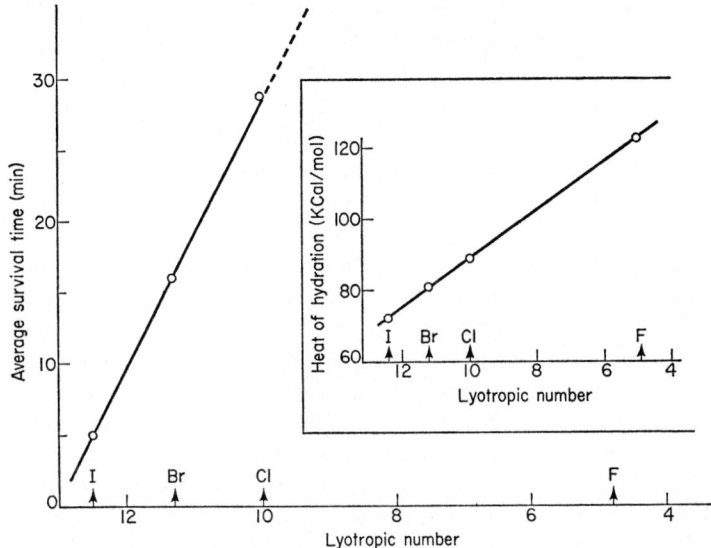

Fig. 6. Relationship between survival time and lyotropic number. The lyotropic numbers for the halides are indicated by their symbols on the abscissa. The average survival times for axons perfused with 400 mM potassium salts of these halides are indicated on the ordinate. The average survival times were 4·9 min for iodide (10 axons), 16·9 min for bromide (26 axons) and 27·4 min for chloride (18 axons); the average survival time for fluoride was greater than several hours and was not included on this graph. Inset: Relationship between heat of hydration and lyotropic number (Vöet, 1937). The exceptionally long survival time observed with fluoride salts might have been predicted from this diagram. (Reprinted from Tasaki *et al.*, 1965a).

zeta potentials of colloids). One example is shown in the inset in Fig. 6 where the linear relationship between the heats of hydration and the classic lyotropic numbers is demonstrated for the halides. The exceptionally long survival time observed with fluoride salts might have been predicted from the relative anion positions indicated by the lyotropic numbers.

Similar observations have been made for the remainder of the anions studied, as well as for the cations. It should be noted that the absolute lyotropic numbers are based on an arbitrarily selected constant and that only the relative values are of significance. Since the anion series and the cation series have different arbitrary constants, they have entirely different scales and are not directly comparable. But if the lyotropic numbers for anions and cations are assigned with appropriate constants, simple addition of the lyotropic numbers of the anionic and cationic members of a salt will predict the relative survival time (and restoring ability) observed with that salt quite accurately (Singer and Tasaki, 1965). Details of the method of determining relative lyotropic numbers and assigning appropriate constants are given elsewhere (Voët, 1937; McBain, 1950, pp. 132–140; Tasaki et al., 1965a, Appendix).

As implied in the foregoing section, identical lyotropic sequences can be obtained from examination of the relative restoring abilities of the ions concerned. An example of such a restoration produced by a change in anions is shown in Fig. 5 for two potassium salts (Tasaki et al., 1965a). When the internal perfusion solution (400 mM KCl) was promptly changed to an isohydric, isotonic solution containing 400 mM KF, the action potential was immediately restored (at 39 min). The simultaneous changes in resting potential are indicated in the figure; in other cases of anion restoration, no changes in resting potential were observed during recovery. The magnitude of the membrane conductance accompanying the described changes in excitability varied inversely with amplitude of the action potential and followed the same time course (see also, Adelman et al., 1966).

It should be noted that the amplitude of the action potential remained at a high level following substitution of KF for KCl but when the original KCl solution was re-introduced into the axon, the action potential immediately declined and conduction was blocked within 1 min. Since the average survival time with 400 mM KCl is 27·4 min, the axon had not returned to its original state following KF perfusion. A second restoration could be produced with KF substitution after the second KCl perfusion (at 52 min) and the action potential could then be maintained for hours. Thus, some alteration has taken place in the membrane during KCl perfusion which persists through the period of KF perfusion but is "masked" by the presence of the fluoride salt; i.e. the changes in the membrane produced by perfusion with a "less favorable" salt (KCl) were not sufficient to reduce the amplitude of the action potential when a "more favorable" perfusion solution (KF) was used. Since all other conditions are common to the two salts studied, the observed difference in excitability is attributed to the difference in the anions and fluoride is designated more "favorable" than chloride for maintaining excitability. These observations imply that neither the amplitude of the action potential nor the magnitude of the resting potential completely describe the physico-chemical state of the physiological membrane.

The relative "favorabilities" of different internally perfused anions and cations for maintaining excitability were determined by both the "survival time" and "restoring ability" methods and gave entirely consistent results. With a common cation, the anions studied formed the sequence:

F > HPO$_4$ > glutamate, aspartate > citrate > tartrate > propionate > SO$_4$ > acetate > Cl > NO$_3$ > Br > I > SCN,

where fluoride is the most favorable and thiocyanate is the least favorable internal anion for maintaining excitability. The analogous series obtained for cations with a common anion was:

Cs > Rb > K > NH$_4$ > Na > Li

where cesium is the most favorable and lithium is the least favorable internal cation for maintaining excitability. In general, divalent cations (Ba, Sr, Ca, Mg) in the internal perfusion fluid are much less favorable than univalent cations. By similar methods, the relative favorabilities of several organic cations for internal perfusion were determined to be:

tetramethylammonium > tetracaine, D-tubocurarine > strychnine,

where tetramethylammonium was the most favorable cation of this group. However, most members of this group were even less favorable than the divalent cations; i.e. very low concentrations of these organic cations rapidly produced irreversible conduction block.

The existence of these lyotropic sequences cannot be accounted for by the classic equivalent circuit model (Hodgkin and Huxley, 1952; Hodgkin, 1964). For example, the fact that the amplitude of the action potential depends on the nature of the internal anion cannot be explained. In the equivalent circuit model, the axon membrane is thought to behave essentially as a sodium-sensitive membrane electrode in the excited state where the amplitude of the action potential "overshoot" (E_a) is determined primarily by the Nernst equation for sodium ion:

$$E_a = 58 \log \frac{[Na]_e}{[Na]_i} \qquad (2)$$

where $[Na]_e$ and $[Na]_i$ are the external and internal concentrations of sodium ion, respectively. In this formulation, anion and divalent cation effects are assumed to be negligible (Hodgkin, 1964, p. 38) but it has already been demonstrated that anion effects are very significant. Moreover, the presence of external divalent cations is far more important than the presence of external univalent cations for the maintenance of excitability (Tasaki et al., 1967).

Furthermore, the magnitude of the action potential does not vary simply with the internal and external concentrations of sodium ion according to Eq. (2). When the external sodium ion is replaced with an equivalent amount of choline ion, the amplitude of the action potential decreases (Hodgkin and Huxley, 1952b). In these unperfused axons, the magnitude of the action potential "overshoot" (i.e. the magnitude of the action potential less the magnitude of the resting potential) approached the values predicted by the Nernst equation for sodium ion. However, when the internal concentration of sodium ion is varied, there are significant deviations from the slope predicted by the Nernst equation (Tasaki and Takenaka, 1964). In these internally perfused axons, excitability can be maintained by including some more favorable internal cation, such as potassium or rubidium, with the less favorable sodium ion in the perfusion solution. There is a significant "overshoot" when the internal and external concentrations of sodium ion are equal and Eq. (2) would predict an "overshoot" of zero.

The establishment of the lyotropic series and dilution effects has led to experiments in which excitation is possible when the gradients for sodium or potassium ions are reversed (Tasaki *et al.*, 1966a). Since sodium is a relatively unfavorable internal cation, these experiments were not possible before the relative favorability of fluoride and the effects of dilution were determined. In the example shown in Fig. 7, the internal perfusion medium contained 90 mM NaF and was potassium-free. The external medium contained hydrazine and calcium chlorides and was sodium-free. Propagated, all-or-none action potentials were obtained for more than 20 min when Eq. (2) would predict that the action potential should be inverted or that no excitation at all should take place.

It is important to note that the magnitude of the action potential overshoot depends on the absolute value of the resting potential. There is no ambiguity involved in determining the magnitude of the potential variation during excitation but there are many difficulties involved in interpreting the magnitude of the resting potential (see Section III-E). Even if the physiological membrane did behave as a potassium-electrode at rest and as a sodium-electrode during excitation, it would not be possible to test these possibilities unambiguously with the present experimental procedures for determining the potential difference across the resting membrane. (The ambiguity associated with determination of the resting potential is equivalent to the ambiguity encountered in definition of the single ion activity coefficient (Helfferich, 1962, p. 371).

The macromolecular approach to nerve excitation provides a reasonable qualitative interpretation of the existence of the lyotropic sequences. The physiological differences observed between neutral salts of the same concentration and valence are attributed to differences in the "affinity" of the ions concerned to charged groups on the membrane macromolecules. This interpretation is based on the work of Bungenberg de Jong (1949, pp. 283–285) and others (Eisenman, 1961) who showed that the (lyotropic) cation sequences are based on relative cation affinities to negative sites ("ion fixation effect").

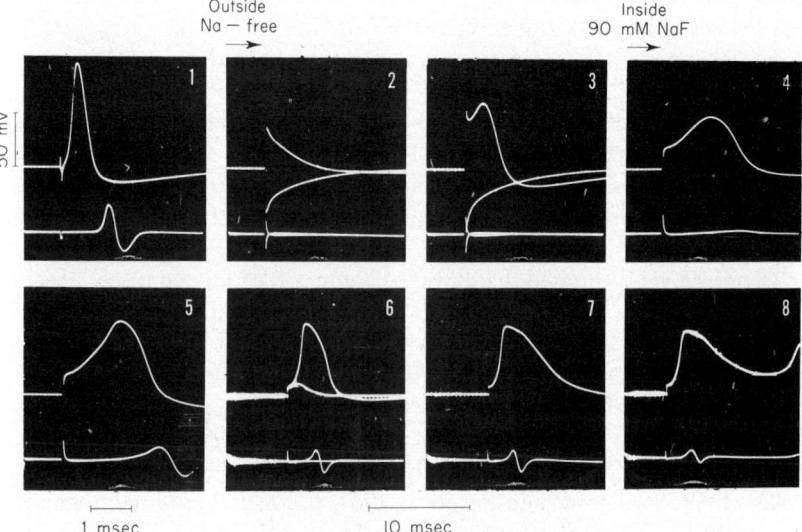

FIG. 7. Excitation with internal NaF solutions. The upper tracing in each record represents the potential variation obtained with an intracellular microelectrode. The lower tracing represents the potential variation conducted from the perfusion zone to the lateral pool and recorded with extracellular electrodes (see Fig. 1). Record 1 was obtained in natural sea water, prior to perfusion. All subsequent records were taken in sodium-free external media. Record 2 was obtained after the natural sea water was changed to a mixture of 0·3 M choline and 0·2 M calcium chlorides. No differences were observed between cathodal and anodal stimulation in the superposed upper tracing, i.e. there was no sign of excitation in this case. Record 3 was obtained after the external medium was changed to a mixture of 0·3 M hydrazine (HzCl) and 0·15 M calcium chlorides. Although there was no sign of impulse conduction (lower tracing), there was a sign of excitation in the perfusion zone (superposed upper tracing), i.e. the cathodal stimulus evoked a small action potential but there was no comparable response to anodal stimulation. Records 4 through 8 were obtained after internal perfusion was begun with a solution containing 0·09 M NaF and 0·01 M cesium phosphate buffer (pH 7·3). Record 4 shows a graded response which was not conducted. Record 5 shows a delayed but conducted impulse. Record 6 shows the all-or-none nature of the response; subthreshold and suprathreshold stimuli are superposed on the upper tracing. (Note the difference in time scale in this and subsequent records.) Records 7 and 8 were obtained about 30 min after the onset of internal perfusion with NaF; there was a gradual increase in the duration of the potential variation and a second (spontaneous) potential variation is just visible at the right of the record. (Reprinted from Tasaki *et al.*, 1966a).

The particular ion sequences observed are thought to be characteristic of the particular macromolecular charged group concerned. The cation order established for the abilities of several chloride salts to suppress "dicomplex formation" by mixtures of proteins and phosphate colloids follows the order: Li > Na > K; the anion order for several potassium salts follows the order: I > Br > Cl. In each sequence, the ion to the left of the inequality sign suppresses dicomplex formation more than the ion to the right (Bungenberg de Jong, 1949, p. 377; Bank and Hoskam, 1940). The observation of distinct ion sequences with the uni-univalent salts used for intracellular perfusion of squid giant axons is consistent with this interpretation and suggests that the charged group concerned is phosphate. (The external ion sequence suggests that the charged group on the external physiological surface of the membrane is likely to be carboxylate.)

Therefore, it is proposed that both the anion and cation sequences observed in squid axons are directly related to the effects of the ions concerned on the charged proteins and phospholipids within the membrane. The axon membrane is visualized as a macromolecular structure, primarily bound together by inter- and intramolecular salt linkages. In sufficient concentration, neutral salts can disrupt such linkages and alter the membrane structure so as to increase the number of charged groups available on the membrane (see Peller, 1959; Steinhardt and Beychok, 1964, p. 294). This would alter membrane permselectivity, increase membrane conductance and eventually abolish excitability. For example, when the affinity of a cation for a particular anionic site is great, the tendency to disrupt linkages inolving that site will be great, the survival time will be short and the restoring ability of that cation will be small. Since divalent cations and organic cations, such as procaine and strychnine, have very strong affinities to negatively charged colloids (Bungenberg de Jong, 1949, p. 301), the extremely unfavorable nature of these ions can also be explained.

There are three general mechanisms by which neutral salts can suppress dicomplex formation or disrupt salt linkages between macromolecules (Overbeek and Bungenberg de Jong, 1949, p. 226; see also von Hippel and Wong, 1964, 1965): (a) screening effects; (b) ion fixation effects and (c) salting-out effects. The simple screening effect (a) arises from the formation of a Debye-Hückel atmosphere of ions derived from the neutral salts around the charged groups of the macromolecules. This effect is strongly dependent on the concentration and valence of the ion considered but is insensitive to differences in the specific chemical properties of the ions. Screening can account for the effects of dilution (Section III-C) and for the marked unfavorability of internal divalent cations but cannot account for the lyotropic sequences (where differences are observed among ions of the same concentration and valence).

The ion fixation effect (b) refers to the bonding of ions derived from the neutral salts to various oppositely charged groups on the membrane macromolecules and depends on the chemical nature of both the ions and the macromolecules. Ion fixation can account for the lyotropic sequences and for the prediction of the nature of the charged group involved. The salting-out effect (c) is generally attributed to alterations of water structure (Frank and Wen, 1957); such alterations not only can affect salt linkages between macromolecules, but can also disturb hydrogen and hydrophobic bonding (Hamaguchi and Geiduschek, 1962). The salting-out effect is independent of the existence of charges on the macromolecules and does depend on the chemical nature of the ions concerned but tends to be less important in charged systems. Although the lyotropic sequences predicted by the salting-out effect are identical to those predicted by the ion-fixation effect, ion-fixation is expected to be more important in the charged, macromolecular membrane. Since salting-out and screening effects are relatively less important in a charged macromolecular system, the mechanism determining the lyotropic sequences in squid giant axons is most likely to be the ion fixation effect. Therefore, not only should the rate of disruption of inter- and intramolecular salt linkages increase with the concentration of the added ions but, at a given concentration, the rate of disruption should increase with the affinity of the added ions to the oppositely charged groups in the membrane.

In this regard, it is of interest that perfusion with relatively less favorable salts does not immediately reduce the amplitude of the action potential or increase the membrane conductance; there is almost always a plateau period during which the effects of the unfavorable salt are not apparent. If a very favorable salt is substituted during this plateau period, the disruptive effects of the unfavorable salt are "masked" until an unfavorable salt is introduced again. Empirically, the very gradual change in membrane conductance during the plateau period and the subsequent rapid increase in conductance (when excitability is suppressed) may be expressed by a simple continuous function of time:

$$g - g_0 = kt^n \qquad (3)$$

where g is the membrane conductance at time t; g_0 is the membrane conductance at the beginning of the perfusion period ($t = 0$); k and n are constants. The difference in the length of the plateau period with various ions is represented mainly by differences in the value of n in this empirical formula. An analogous equation can be used to describe the time course of the changes in action potential amplitude. For example, $n = 2$ with a perfusion solution containing 400 mM KBr and $n = 3$ for a similar solution of KCl; this is reflected by the relatively longer plateau period for both the membrane conductance and action potential with KCl (compared with KBr) as an internal perfusion medium.

This empirical equation can be given some physical meaning within the macromolecular model for the physiological membrane, i.e. an interconnected, macromolecular complex of proteins and phospholipids. Since ion fixation effects are most likely to be responsible for any disruption of inter- and intramolecular salt linkages, the process of breaking salt junctions may be visualized as a reaction of the type:

$$A^- + X^+ \ldots Y^- = A^-X^+ + Y^- \tag{4}$$

In this equation, A^- represents the added anion derived from the neutral salt in the perfusion fluid and $X^+ \ldots Y^-$ represents a junction between a positive side group (X^+) of one macromolecule and a negative side group (Y^-) of another.

For a given salt solution, the rate of breaking junctions is expected to remain constant as long as the number of broken junctions is far smaller than the number of intact junctions. Then the probability of finding a broken junction in a given portion of a macromolecule should increase directly with the time t and can be expressed by ht where h is a proportionality constant. It is assumed that in order to produce a physiologically significant (measurable) "damaged segment" in the membrane, n successive junctions must be broken. The probability of producing such damaged segments should increase as $(ht)^n$. If each damaged segment is assumed to increase the membrane conductance by Δg, and N is the total number of junctions involved, then the relationship between the conductance increase and the time would be given by Eq. (3) where $k = Nh^n\Delta g$. Eventually, irreversible loss of excitability is brought about by cooperative, progressive damage of intact portions of the membrane by the local currents flowing between adjacent intact and damaged segments of the macromolecules.

Experimentally, the empirical factor n is large for more favorable salt solutions and small for less favorable perfusion media. The apparent restoring ability of a favorable salt solution is attributed to the relatively large number of broken junctions needed to produce a detectable damaged segment in the membrane. For example, at the moment when conduction is blocked during perfusion with 400 mM KBr ($n = 2$) most of the damaged segments of the membrane consist of two successive broken junctions. If perfusion had been begun with a similar solution of KCl ($n = 3$), it would require damaged segments to consist of three successive broken junctions to be detected. Therefore, at the moment conduction is blocked during KBr perfusion, the damage would not be sufficiently extensive to be detected if the same axon were immediately perfused with the KCl solution; i.e. the changes produced by the KBr solution would now be masked in the presence of the KCl solution.

Processes similar to those described above are known to occur in a variety of macromolecules (see also Section IV). For example, when a protein with a helical configuration is converted to a random coil configuration, the conformational change begins at several randomly separated regions in the macromolecule. Initially there is no overt conformational change but careful physico-chemical analysis can reveal localized changes in configuration during this period. As the conformational change continues, these locally altered segments extend rapidly and new regions become affected as well. In general, there is a greater tendency for the conformational change to progress cooperatively from a previously changed segment than to begin again in a previously unaltered segment; however, both progressive and new disruptions occur and eventually the conformational change is complete (see Steiner, 1965, p. 156).

Thus, the breaking of junctions between or within macromolecules can proceed without any overt change until a critical number of segments has been altered and then the process may progress very rapidly. Unless carefully investigated, there is apparently a period of time (plateau period) during which no grossly detectable change takes place. This mechanism can account for the shape of the survival time curve; there is a plateau period during which there is no apparent change in the action potential or conductance and this period is followed by a rapid decline in the action potential, accompanied by a rapid increase in membrane conductance. The presence of a hidden change during the plateau period can be demonstrated by perfusing with a relatively unfavorable salt and then changing to a more favorable internal solution during the plateau period. While perfused with the more favorable salt, there is little or no change in the amplitude of the action potential for a long period of time. However, if the original, relatively unfavorable solution is re-introduced, the time course of the change in action potential resumes from the points at which the first substitution was made and continues as if that substitution had never been made (Tasaki et al., 1965a). This observation implies that the number of junctions broken during the initial perfusion with the relatively unfavorable salt (e.g. KBr, $n = 2$) were insufficient to produce any detectable change in the action potential (plateau period); i.e. the number of successive broken junctions was less than two at this time. These changes continued to be masked during perfusion with a favorable salt (e.g. KF, $n \gg 2$) but were revealed by returning to the original solution because the number of broken junctions rapidly increased to the critical value (of n) for the original solution and then became detectable.

The argument stated above shows that the macromolecular approach can account for the effects of concentration and pH both internally and externally. With the empirical Eq. (3), this macromolecular model can adequately describe the shape of the survival time curves observed with various internally perfused salts. In addition, a macromolecular approach can account for the particular lyotropic sequences, survival times and restoring abilities observed with the internal cations and anions (univalent, divalent, organic and inorganic). The external cations and anions will be considered in the next subsection.

It is worthwhile to note that lyotropic sequences have been observed for many biological phenomena, including studies of the swelling of cerebral cortex (Haldi et al., 1927), the depolarization of skeletal muscle (Höber, 1945, p. 289) and the beat frequency of ependymal cilia (Singer and Goodman, 1966). On a molecular level, similar series have been observed for many physico-chemical phenomena, particularly those involving conformations of biological macromolecules (e.g. Gortner et al., 1928, p. 188; Höber, 1928, p. 621; McBain, 1950, p. 132; Hamaguchi and Geiduschek, 1962; von Hippel and Wong, 1964, 1965).

4. Nature of External Anions and Univalent Cations

Relatively few external anions can be studied because of limitations imposed by the solubility product in an alkaline medium containing divalent cations. When studied by the methods used for the internal anions, no significant differences were observed among the different external anions (e.g. Br, Cl, SO_4). The lack of any differences among the external anions may be taken as further evidence for the supposition of a relatively large excess of fixed negative charge in the external superficial layer of the physiological membrane. As discussed in Section III-B, there is a considerable amount of circumstantial evidence that macromolecules have such fixed charges and that these charges are located predominately in the external superficial layer of the membrane (see also Section IV-A). With this property, the external surface of the membrane would tend to exclude all anions uniformly. Hence, it would be extremely difficult to detect any differences among anions even if more anions could be studied.

In contrast to the lack of differences among anions, the different effects of cations are very well known (Tasaki et al., 1965a). In general, the sequence observed for the alkali metals is:

$$K < Rb < Cs < Na \leqslant Li,$$

where sodium and lithium are the most favorable external univalent cations for maintaining excitability under the usual experimental conditions. From another point of view, the members of the sequence to the left are more favorable for producing membrane depolarization. These observations are complicated by the necessity for adjustment of the relative concentrations of the external univalent and divalent cations, as well as by the necessity for selection of a favorable internal perfusion medium. In any case, this particular lyotropic sequence suggests that the critical macromolecular charged group on the external surface of the membrane is probably carboxylate in nature. (In the previous subsection, it was shown that the critical group on the internal physiological layer is probably phosphate in nature.)

The effects of many organic cations on excitation in squid giant axons have been examined in terms of their respective abilities to substitute for sodium ion in the external medium (Tasaki et al., 1965b, 1966; Tasaki and Singer, 1965, 1966). Representative sequences of homologous nitrogenous cations are presented below:

(1) $R_4N < R_3NH < R_2NH_2 < RNH_3 < NH_4$ (R = $C_4H_9, C_3H_7, C_2H_5,$ or CH_3)

(2) $(C_4H_9)_n NH_m < (C_3H_7)_n NH_m < (C_2H_5)_n NH_m < (CH_3)_n NH_m$
($n + m = 4$)

(3) $RNH_3 < HNH_3 < H_2NNH_3$

More extensive series may be found elsewhere (Tasaki et al., 1965b; Tasaki and Singer, 1965). In general, the more that alkyl groups are substituted for hydrogen in the ammonium ion structure, the less favorable is the cation for sodium substitution (line 1); similarly, as the length of the alkyl chain increases, the less favorable is the cation (line 2). Although the size of the molecule appears to play some role in determining favorability, the presence of non-alkyl functional groups is important (line 3). The most favorable nitrogenous cation substitutes for sodium ion were hydrazine, hydroxylamine, guanidine and aminoguanidine. If the most favorable internal and external conditions are selected, depolarizing cations such as potassium and rubidium can substitute for sodium (Tasaki et al., 1966a). Furthermore, under these conditions, external univalent cations may even be eliminated entirely without eliminating excitability (Tasaki et al., 1966c).

When the impedance variations, membrane resistance, voltage-clamp curves and isotope flux data are examined, the nature of excitation in various sodium-free media does not seem to be significantly different from excitation in the usual sodium-containing media (Singer et al., 1966; Tasaki et al., 1965, 1966a, b).

Typical examples of the voltage-clamp curves obtained in a sodium-free system are shown in Figs 8 and 9 (Tasaki and Singer, 1966, 1967). (Representative action potentials and impedance variations in sodium-free external media are shown in Figs 7 and 10.) In these examples, the external medium contained 0·3 M hydrazinium chloride (HzCl) and 0·2 M CaCl$_2$; the respective internal media contained 0·1 M KF (Fig. 8) and 0·1 M RbF (Fig. 9).

When the commanding clamping pulse is delivered, the potential inside the axon is suddenly raised above the resting level; this results in a transient outward current (initial upward deflection in the inset, Fig. 8), followed by a relatively slow inward current (downward deflection). (The outward "capacitative current" is considered to represent a transient phenomenon associated with a transition of the membrane from one stationary state to a new quasi-stationary state.) The current-time and current-voltage curves obtained from axons in sodium-free media are not significantly different from those obtained in the usual sodium-containing media. The current-voltage curves

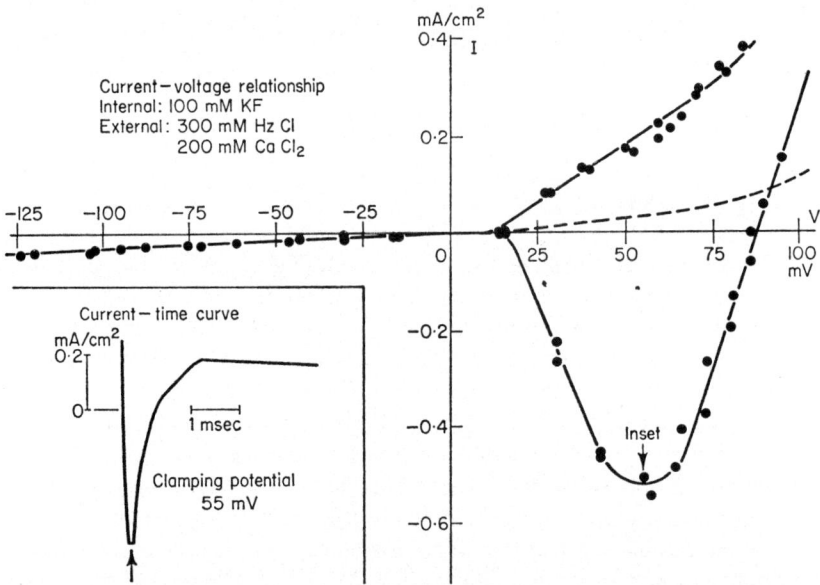

FIG. 8. Current-voltage relationship obtained in sodium-free media. The internal perfusion solution contained 0·1 M KF; the external medium contained 0·3 M HzCl and 0·2 M CaCl$_2$. The thick, v-shaped curve represents the relationship between the peak inward current (I) and the clamping potential (V). The thin, continuous curve shows the I–V relationship 4 msec after the onset of the clamping-potential pulse. The dashed line represents the I–V relationship determined after tetrodotoxin ($5 \cdot 10^{-7}$ gm/ml) was added to the external medium. The inward-directed current was completely abolished after addition of this poison. Inset: Typical current-time curve obtained in sodium-free media. The experimental conditions are given above. The clamping-potential was 55 mV in this case. The peak inward current (inset arrow) was plotted against the clamping-potential (55 mV) to give the I–V relationship (arrow in Fig. 8). (Reprinted from Tasaki and Singer, 1966a.)

in the figures (v-shaped) are determined by plotting the peak values of the inward currents obtained from the current-time curves at different clamping voltages. For example, the arrow in the I–V curve (Fig. 8) represents the peak inward current (0·5 mA/cm²) of the current-time curve (Fig. 8, inset, arrow) when the clamping voltage was 55 mV.

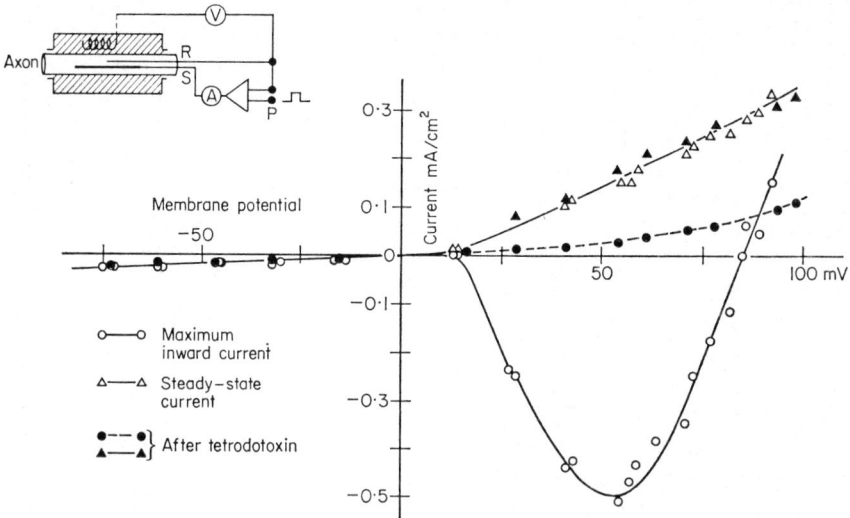

FIG. 9. Top: Schematic voltage-clamp arrangement for squid giant axon. Only the portion of the axon in the central pool of the perfusion chamber is indicated. The symbols R and S refer to the internal potential-recording and current-passing electrodes, respectively. The clamping-potential pulse is delivered from a square pulse-generator at P and the current is monitored at A; the potential variation is monitored at V. The positive current is directed from the operational amplifier (large triangle) through the current electrode, S, to the ground-electrode in the medium and finally back to the ground-terminal of the operational amplifier (not shown in the figure). Bottom: Current-voltage relationship in sodium-free and potassium-free media. The internal perfusion medium contained 0·1 M RbF; the external medium contained 0·3 M HzCl and 0·2 M CaCl₂. The other details are given in the legend of Fig. 8; the effects of tetrodotoxin are indicated by the broken line. (Reprinted from Tasaki et al., 1966.)

The initial linear segment of the I–V curve (from the left to just beyond the origin) indicates that the "resting" membrane obeys Ohm's law. In the region of large negative (internal) potential, there is no change in the slope of the I–V curve. Since external cations would be expected to be driven into the membrane throughout this region, it is assumed that in the physiological "resting state" the available fixed negative sites in the membrane are occupied predominantly by cations derived from the external medium. (For reasons to be discussed in Section IV-B, these sites confer cation-exchanger properties on the physiological membrane and are most likely to be occupied by external divalent cations in the resting state.)

The final linear segment of the I–V curve indicates that the membrane also obeys Ohm's law in the "active" or "excited" state. In this region of large positive (internal) potentials, the curve crosses the abscissa ($I = 0$) at approximately the peak value of the action potential ($V = 87$ mV in Fig. 8; $V = 84$ mV in Fig. 9). The intermediate segment of the I–V curve, with a large negative slope, will be discussed in Section IV-B which deals with the "two-stable-states" theory of excitation from a macromolecular point of view.

The other continuous curve serves to represent the current observed 4 msec after a (positive) potential pulse is applied; the curve described above (v-shaped) was obtained 1 msec after the onset of the clamping voltage. No differences were observed between the two current-voltage curves to the left of the origin but significant differences appear with large positive clamping voltages. These differences are attributed to the occupation of the critical (cation-exchanger) layer of the membrane by different cations at these two times. The implications of this observation for excitation theory will also be discussed in Section IV-B.

The broken lines in Figs 8 and 9 are of particular interest to physiologists. These curves represent the I–V relationships observed after addition of a very small amount of the puffer-fish poison, tetrodotoxin, to the external medium. This poison is known to suppress action potentials at very low concentrations (Furukawa et al., 1959) and has been used to analyze the currents observed during the excitation process. It may be seen that the inward current (usually obtained about 1 msec after the clamping pulse) is completely abolished by the application of tetrodotoxin. No differences are observed between the curves obtained before and after tetrodotoxin with negative clamping voltages or 4 msec after positive clamping voltages.

It has been postulated that there is a sodium-specific mechanism within the membrane responsible for the inward current and that this mechanism is specifically blocked by tetrodotoxin (Narahashi et al., 1960; Narahashi, et al., 1964; Nakamura, et al., 1964, 1965). However, since tetrodotoxin is capable of blocking the action potential and the inward current in these sodium-free media, either the mechanism responsible for action potential production is rather non-specific for all of the cations which can replace sodium or tetrodotoxin has a non-specific blocking effect on whatever mechanism is responsible for the production of an action potential and the inward current (see Watanabe, et al., 1967).

The fact that apparently normal excitation can take place in the absence of sodium has several implications for excitation theory. There is no doubt that under favorable conditions sodium ion is not essential for the process of excitation in squid giant axons. Since this is true for many other biological systems where sodium is the usual external univalent cation (see Tasaki *et al.*, 1965b; Tasaki and Singer, 1965), the property of sodium specificity does not appear to be necessary for excitable systems in general. There is little doubt that sodium is one of the most favorable external univalent cations for maintaining excitability under the usual experimental conditions and that, in the natural environment of the axon, sodium ion plays an important role in determining the action potential amplitude. But the fact that under favorable conditions external sodium salt can be entirely replaced with salts of other univalent cations or even with sucrose implies that the role played by the sodium ion is neither unique among cations nor indispensable for the maintenance of excitability.

For these reasons, it is not surprising that external application of tetrodotoxin is able to block the process responsible for excitation, regardless of which external cations are involved (Tasaki and Singer, 1966; Tasaki *et al.*, 1966a) and to abolish the associated inward current, regardless of which cations may be responsible for that current (Tasaki and Singer, 1965; Tasaki *et al.*, 1966b). In the macromolecular model, tetrodotoxin is assumed to bind strongly to the macromolecules in the external superficial layer of the membrane. Two mechanisms are responsible for the strong binding of large organic cations to macromolecules: (1) the strong ion fixation effect and (2) the strong short-range interaction conferred by the uncharged portion of the molecule. Therefore, only strongly depolarizing cations, which have very great affinity for membrane macromolecules, can compete with tetrodotoxin for a given anionic site. In fact, it generally does require a greater concentration of tetrodotoxin to block the excitation process when the more depolarizing cations are used in place of external sodium ion.

The effects of tetrodotoxin and sodium-substitution are thus consistent with the absence of any absolute specificity in the lyotropic sequence; all of the effects are quantitative, rather than qualitative. It would seem that a much more general mechanism than a "sodium-specific carrier" or a "sodium-specific pore" is required to understand the excitation process; the macromolecular model for excitation provides such a mechanism and will be discussed in detail in Section IV.

5. *External Univalent Cation/Divalent Cation Ratio*

The biological importance of external divalent cations and the univalent/divalent cation ratio has been stressed by physiologists for many years (Heilbrunn, 1952). Although external univalent cations do not appear to be essential for electrical excitation in some biological systems, such as crustacean muscle fibers (Fatt and Katz, 1953; Hagiwara *et al.*, 1964), it was thought that both univalent and divalent cations were required for excitation in squid giant axons (Tasaki, 1963; Hodgkin, 1964, p. 68; Tasaki and Singer, 1966). However, excitability has recently been demonstrated in squid axons in an external medium consisting solely of divalent cations; in this case, a very favorable univalent cation (Cs^+) was used internally (Tasaki et al., 1966c). When only univalent cations (e.g. Na^+) are used in the external medium, the axons become inexcitable within a short time. When divalent cations (Ca^{++}, Sr^{++}, Ba^{++}) are added to the medium, the excitability is restored immediately.

In general, a change in the univalent/divalent cation ratio affects the excitability. This has been observed under many conditions: (1) when sponteneous repetitive discharge occurs in alkaline media, addition of divalent cations to the external medium will restore normal excitability (Tasaki, et al., 1965a); (2) when depolarizing univalent cations are applied externally, addition of divalent cations to the external medium tend to restore normal polarization and excitability and (3) when establishing the optimum conditions for excitation in sodium-free external media the need for a balance between univalent and divalent cations is particularly apparent (Tasaki et al., 1965b). There are significant differences observed among the alkaline earth divalent cations under the usual experimental conditions. Magnesium is a very effective cation for the suppression of spontaneous firing but the other divalent cations can also serve.

In terms of the macromolecular hypothesis presented above, intramembrane fixation of univalent (depolarizing) cations are thought to produce drastic changes in the membrane structure (see Section III-C). Divalent cations actually tend to stabilize macromolecular systems with fixed negative charges by complex formation ("cross-linking between adjacent charges") (Michaeli, 1960; see Fig. 14, left). It should be pointed out that divalent and univalent cations compete for fixed negative sites in the external layer of membrane and divalent cations are generally favored in complex formation with organic anions (Meites, 1963, pp. 1–13 to 1–19 and pp. 1–39 to 1–46). A coordination compound of this kind can provide a rapidly reversible, highly selective mechanism for altering the cation distribution in the membrane (Martell and Calvin, 1952, p. 204); such alterations may be accompanied by conformational changes in the structure of the membrane which would markedly affect its physiological properties (see Sections IV-A and IV-B). The effects of heavy metal ions may be related to the ability of these metals to combine almost irreversibly with proteins (Steinhardt and Beychok, 1964, p. 265) so that ion-exchange processes are blocked or slowed markedly. These cation-exchanger properties are essential for the excitation theory based on macromolecular processes involving two stable states (e.g. two different conformations of the physiological membrane) and will be discussed further in Section IV.

D. PHYSICAL REQUIREMENTS FOR EXCITABILITY

1. *Hydrostatic and Osmotic Pressure*

The action potential appears to be almost insensitive to large increases of the osmotic pressure (Tasaki et al., 1965a). When the internal osmolarity is changed by varying the concentration of glycerol (from 25 to 6 volumes %) in the perfusion solution, there is only a very slight change in the duration and amplitude of the action potential. Addition of 1·0 M sucrose to the usual internal or external solutions produces no noticeable effects. (Use of sucrose or mannitol in place of glycerol has no noticeable effects.) Hypotonic solutions are known to suppress excitability rapidly and irreversibly; this effect may be attributed to disruption of the macromolecular complex by increasing the intramembrane swelling pressure (Section III-C).

The squid giant axon appears to be rather insensitive to changes in hydrostatic pressures as well. The hydrostatic pressure can be varied by raising and lowering the height of the perfusion fluid reservoirs to provide a pressure head from 60 cm to -10 cm of water; no significant differences in the resting or action potentials were observed (Tasaki et al., 1965a). As mentioned in Section II-B, axons become more sensitive to pressure changes after the axoplasm is removed; a change of a few centimeters of water may be sufficient to make the axon inexcitable. But at the moment conduction is blocked in axons perfused with unfavorable internal solutions, a few action potentials may be produced by a sudden rise of internal pressure of a few cm of water. In this regard, it is of interest that macromolecular systems in a metastable conformation can be converted into a stable conformation by an appropriate change in pressure. Since such pressure variations do not usually produce significant potential variations, this mechanism is not likely to be important under the usual experimental circumstances but may be very important in the consideration of the mechanism of mechano-receptors (Teorell, 1953).

2. *Temperature*

Most experiments on squid giant axons are carried out at temperatures of 10 to 25°C. Isolated axons remain excitable for long periods of time at low temperatures and can be stored for a day or two without serious damage. In general, raising the temperature raises the threshold for electric excitation in squid giant axons (Tasaki and Spyropoulos, 1957). Similar studies have been performed with invertebrate nerve from the clam (*Maia*) and with nerves from the frog and the toad (*Bufo Marinus*); these data offer several interesting applications for excitation theory.

The results obtained from *Maia* and frog medulated nerve (Abbott et al.,

1958; Hill and Howarth, 1958) demonstrated that the process of excitation in nerve is associated with a diphasic variation in the temperature. Specifically, it was found that conduction of an action potential is accompanied by the successive generation and absorption of heat. The results from the toad show that prolonged action potentials can be abolished during their plateau period by delivery of a short pulse of heat (Spyropoulos, 1961). In addition, the length of the plateau can be increased by cooling and under some conditions an action potential can be initiated by a sudden cooling. (The nature of "heart-like" action potentials with a prolonged plateau period will be discussed in Section IV-B.)

These observations suggest that initiation of an action potential is an exothermic process and that termination of excitation is an endothermic process; this is expected from the supposition that the mechanism involves a reversible physico-chemical change in the membrane from one state to another, followed by a return to the original state. In cation-exchangers, it it known that replacement of divalent cations with univalent cations is exothermic and the reverse process is endothermic (see Coleman, 1952).

In macromolecular terms, a similar process is encountered in the well-known temperature dependent helix-random coil transitions that occur with a variety of macromolecules (Katchalski *et al.*, 1964, pp. 470, 496; von Hippel and Wong, 1965). In fact, the temperature at which such systems undergo conformational change is often used as an index for the study of other physico-chemical determinants of conformation between two stable configurations of a macromolecule. The irreversible effects of high temperature in macromolecular systems are related to protein denaturation (Steinhardt and Beychok, 1964, p. 292). Irreversible temperature effects have been observed for many physiological processes, e.g. ciliary activity (Singer and Goodman, 1966), and are entirely consistent with the macromolecular interpretations of biological phenomena and particularly with the macromolecular approach to excitation theory.

3. *Light*

Although most giant axons do not contain any natural pigments, under special circumstances excitation can be induced by light radiation (Arvanitaki and Chalazonitis, 1958; Chalazonitis and Chagneux, 1961). In biological systems containing natural pigments, such as the snail heart (*Helix*) and the sea slug giant nerve soma (*Aplysia*), excitation can be produced by wavelengths of light corresponding to the peak absorption band of the pigment. The giant axons of the cuttlefish (*Sepia*) can be excited by light after intracellular injection of pigments (e.g. neutral red). Action potentials are produced by illuminating the stained axon with the wavelength of light corresponding to the peak absorption band of the pigment; such action potentials are often prolonged and exhibit oscillatory phenomena during their plateau period (see Section IV-B).

Since proteins macromolecules are known to have characteristic interactions with various radiations (Weber and Teale, 1965), it seems possible that light can induce a macromolecular conformation which leads to excitation. A study of the nature of the energy transfer from the pigment to the excitable membrane would be of fundamental importance, both for sensory physiology and for the excitation process in general (see also, Spyropoulos and Tasaki, 1960).

E. CHARACTERISTICS OF THE RESTING MEMBRANE

1. *Measurement of the "Resting Potential"*

The "resting potential" of a living cell is usually defined as the potential difference between an electrode within the cell and an electrode in the external medium. The internal electrode is usually inserted directly through the cell membrane into the interior. The most common electrodes used for this purpose are glass micropipettes, several microns or less in diameter, filled with a concentrated KCl solution (Ling and Gerard, 1949). Glass pipette electrodes containing a platinum wire are also used (Chandler and Hodgkin, 1965). A single microelectrode may be used for this measurement by recording the difference in potential levels observed with the electrode inside and outside of the cell, respectively.

However, there are several problems involved in assuming that this difference in potential levels is the "true" electrical potential difference across the cell membrane. Even if the problem of using two different electrodes is avoided by successive placement of a single electrode, there is no assurance that the junction potentials will be the same inside and outside of the cell. The media involved are very different and are not sufficiently dilute to assume that the junction potentials are negligible (see MacInnes, 1939, p. 220). Unfortunately, since there are also polyvalent ions in the system, it is not possible to properly estimate the junction potential.

In addition, there are special difficulties associated with the use of unplated platinum or silver wire electrodes for measurement of intracellular potentials. Even if "contamination" of the electrode by the intracellular protein is avoided by using the perfused giant axon, large variations in the recorded potential can be produced by small variations in the electric current, oxygen tension, and pH, particularly in the presence of multivalent ions (Ives and Janz, 1961, p. 120; Tasaki and Singer, 1968).

Furthermore, in most biological systems there are significant quantities of intracellular polyelectrolyte. These macromolecules can give rise to very large potential differences, in some cases, of the order of 100 mV, without the presence of a membrane (Tasaki and Singer, 1968). This "suspension effect" or "Pallmann effect" is well-known to physical chemists (Pallmann, 1930; Overbeek, 1952, p. 184; Sollner, 1953; Bates, 1964, p. 273) but is generally ignored by biologists. This difficulty does not exist in the internally perfused squid giant axon where the intracellular materials are removed by the perfusion but even this system is subject to the other limitations described above.

When the intracellular potential of the internally perfused squid giant axon (immersed in sea water) is measured with a KCl-filled glass microelectrode by the operational procedures described in Section II-D, the values obtained are generally in the range -40 to -60 mV with respect to the external medium (Tasaki et al., 1965a). Dilution of the internal salt solution produces a decrease in the resting potential. With a given salt solution, the variation in resting potential from axon to axon is not more than 10 mV or so. These results are in marked contrast to those reported by others (Chandler and Hodgkin, 1965) where resting potentials were measured with a low-impedance electrode including an unplated platinum wire. In the latter case, resting potentials varied by more than 30 mV from axon to axon under the same conditions and values as high as -100 mV were reported. The variability within the latter set of results is probably related to the special difficulties associated with the use of metal electrodes discussed above. However, in more general terms, the differences between the two sets of results are probably related to the ambiguities inherent in the accurate measurement of potentials in a biological system (Tasaki and Singer, 1968). Since the absolute value of the action potential overshoot depends on the absolute value of the resting potential, both Eqs (1) and (2) are very difficult to evaluate (see also Sections C2 and C3).

2. Membrane Resistance and Impedance

Measurement of the a.c. impedance of the squid axon membrane in the resting state was first performed with unperfused axons immersed in natural sea water (Cole and Curtis, 1939; see Sections II-E and II-F). The membrane was estimated to have a resistance of 1 to 2 K$\Omega \times$ cm^2 and a capacitance of about 1 μF/cm^2. These values are also obtained with the usual potassium-containing internal media in perfused axons. With internal perfusion media containing dilute RbF or CsF, membrane resistances tend to be somewhat higher in the resting state (1 to 5 KΩ); in addition, the resistance of the resting membrane in sodium-free, calcium-rich external media tends to be somewhat higher than in the usual sodium-containing media (Tasaki et al., 1966b). (The resistivities of the external medium and the axoplasm are of the order of 25 ohm $\times$ cm.)

3. Ion Fluxes

Transmembrane ion fluxes in internally perfused axons are measured by the use of radio-isotopes of the ions concerned (see Section II). To measure the influx of sodium ion, for example, the radio-isotope of sodium, either Na^{24} or Na^{22}, is added to the external medium containing non-radioactive sodium ions (Na^{23}). Under the condition that the internal perfusion fluid contains no sodium ion, the influx of the radioactive sodium traces the influx of the non-radioactive sodium faithfully. Similarly, the efflux of K^+ can be traced faithfully by the use of radioactive K^{42} ion, provided that the external fluid medium contains no potassium salt.

The following radio-isotopes are commonly used for measuring cation fluxes across the squid axon membrane: H^3-labelled polyatomic cations, C^{14}-labelled organic cations, Na^{22}, Na^{24}, K^{42}, Ca^{45}, Sr^{85}, Rb^{86}, Cs^{134}, etc. Radio-isotopes of various anions (such as Cl^{36}, $P^{32}O_4$, $S^{35}O_4$, Br^{82}, etc.) have been used to measure the fluxes of the corresponding anions.

In axons immersed in media containing a mixture of salts of a univalent cation and a divalent cation, the influx of the divalent cation is very small as compared with the influx of the univalent cation. It is also known that the fluxes of anions in internally perfused axons are far smaller than those of univalent cations. This behavior of the internally perfused squid giant axon is consistent with the supposition that the axon membrane has negative fixed charges (see Tasaki et al., 1966b).

In axons immersed in a solution containing NaCl and $CaCl_2$ and internally perfused with a KF (or K-glutamate) solution, the influx of Na^+ through the membrane at rest is of approximately the same magnitude as the efflux of K^+. (This approximate equality is expected from the condition of no membrane current.) When the membrane resistance is of the order of 1 to 2 $K\Omega \times cm^2$, the fluxes of the univalent cations are usually between 100 and 250 pmole $\times$ sec^{-1}. When the Na^+ and/or K^+ in this system is replaced with other univalent cations, there is a change in the magnitude of the interdiffusion fluxes. Substitution of Rb^+ or Cs^+ for internal K^+ tends to decrease the interdiffusion fluxes. Replacement of external Na^+ with hydrazinium ion also reduces the fluxes. The interdiffusion fluxes can also be reduced by diluting the salt solutions on either or both sides of the membrane or by increasing the divalent cation concentration in the external fluid medium.

F. CHARACTERISTICS OF THE PHYSIOLOGICAL MEMBRANE IN THE EXCITED STATE

1. Measurement of the "Action Potential"

The "action potential" of a nerve is usually defined as the magnitude of the potential variation which takes place during excitation. The action potential "overshoot" is defined as the magnitude of the action potential less the magnitude of the resting potential. As discussed earlier (Section III-C), the ambiguities involved in determining the resting potential limit the interpretation of the "overshoot" and Eq. (2) but there is no ambiguity in measuring the amplitude of the entire action potential.

In unperfused axons immersed in natural sea water, the magnitude of the action potential is 100 to 120 mV. In axons perfused with the usual potassium-containing media (e.g. 500 mM KCl), the action potential is of the same magnitude (Oikawa et al., 1961; Hodgkin, 1964, p. 38). However, when the internal medium is diluted and the more favorable anions are used (e.g. 50 mM KF), action potentials of about 150 mV can be obtained from axons immersed in natural sea water. In fact, when dilute RbF or CsF is used internally, action potentials close to 200 mV have been observed in axons immersed in artificial sea water (Tasaki et al., 1965a). The resting potentials of axons immersed in natural or artificial sea water are: -50 to -60 mV in unperfused axons, -40 to -60 mV with 500 mM KCl and -20 to -40 with 50 mM KF, internally.

Even if it could be assumed that the measured resting potential represents the "true" potential difference across the resting membrane unambiguously, the values for the action potential "overshoot" exceed the maximum value predicted by the Nernst equation for sodium ion (Eq. (2)) and the resting potential does not vary with the internal concentration of potassium as predicted by Eq. (1). Since action potentials of 120 mV can be obtained from axons internally perfused with 50 mM RbF and immersed in sodium-free artificial media containing 0·3 M HzCl and 0·2 M $CaCl_2$ and since such axons have "normal" resting potentials (Section III-C), it is apparent that Eqs. (1) and (2) do not adequately describe the behavior of the physiological membrane (Tasaki and Singer, 1966; Tasaki et al., 1965b, 1966a, b). Therefore, the excitable physiological membrane behaves neither as a potassium-electrode in the resting state nor as a sodium-electrode in the excited state.

It was recently found (Tasaki et al., 1966c) that when axons are internally perfused with dilute cesium solutions, action potentials of about 100 mV in amplitude can be obtained in external solutions of calcium or strontium salts (free of any univalent cation). This finding indicates that the process of action potential production is not merely a consequence of an increase in the "sodium-permeability" of the membrane.

2. *Membrane Resistance and Impedance*

The process of excitation in unperfused giant axons immersed in natural sea water is accompanied by a 100- to 200-fold increase in membrane conductance (Cole and Curtis, 1939). This marked increase in membrane conductance, or impedance-loss, takes place in perfused axons under a wide variety of experimental conditions including: sodium-containing and sodium-free external media, concentrated and dilute (isotonic) internal salt solutions and favorable and unfavorable ion compositions (Tasaki et al., 1965a; Tasaki et al., 1966a; Singer et al., 1966). The impedance-loss during excitation follows a roughly triangular time-course under all of these experimental conditions, although its magnitude is slightly diminished in sodium-free external media. Typical examples of action potentials and impedance variations in sodium-free external media are shown in Fig. 10 for axons perfused with 0·1 M RbF. (A typical voltage-clamp curve is shown in Fig. 9.)

3. Cation Fluxes

In an axon internally perfused with KF solution and immersed in a medium containing NaCl and $CaCl_2$, the simultaneous efflux of K^+ and influx of Na^+ are enhanced when the axon is subjected to repetitive stimulation. There is also a significant increase in the influx of Ca^{++} across the membrane during repetitive stimulation but the total quantity (expressed in mole $\times$ cm^{-2}) of Ca^{++} transferred is far smaller than that of Na^+ transferred during the same period (Tasaki et al., 1967). There is no detectable increase in the fluxes of anions during repetitive stimulation (Tasaki et al., 1961).

The "extra" efflux of K^+ associated with production of action potentials is of the order of 10 to 15 pmole $\times$ cm^{-2} per impulse. Since the condition of electroneutrality has to be satisfied by the internal perfusion fluid, the loss of K^+ in the perfusion fluid must be accompanied by a gain of an equivalent amount of Na^+. (Anion and divalent cation fluxes are negligible.) Although the time-resolution of these tracer measurements is limited, there seems no doubt that these enhanced cation fluxes take place during the approximately 1 msec-long period of reduced membrane resistance (Tasaki and Singer, 1966). It is of interest that the magnitude of the peak flux in the excited state is 100- to 300-fold the magnitude of the flux observed in the resting state under comparable conditions (see Figs 11 and 12). This change in isotope flux is of exactly the same order as the change in membrane resistance; the similarity is not surprising since both cation flux and membrane resistance are determined by the products of the ion mobilities and concentrations within the major diffusion barrier in the membrane. (This observation supports the assumption made in Section II-H that the time-course of the isotope fluxes is accurately reflected by the time-course of the impedance-loss during excitation.)

The existence of this univalent cation interdiffusion during excitation is important in the interpretation of the results obtained from voltage-clamp experiments (see also, Tasaki and Singer 1968). If the values obtained for the peak univalent cation interdiffusion flux are converted into current units by multiplication with the Faraday constant, they may be compared to the

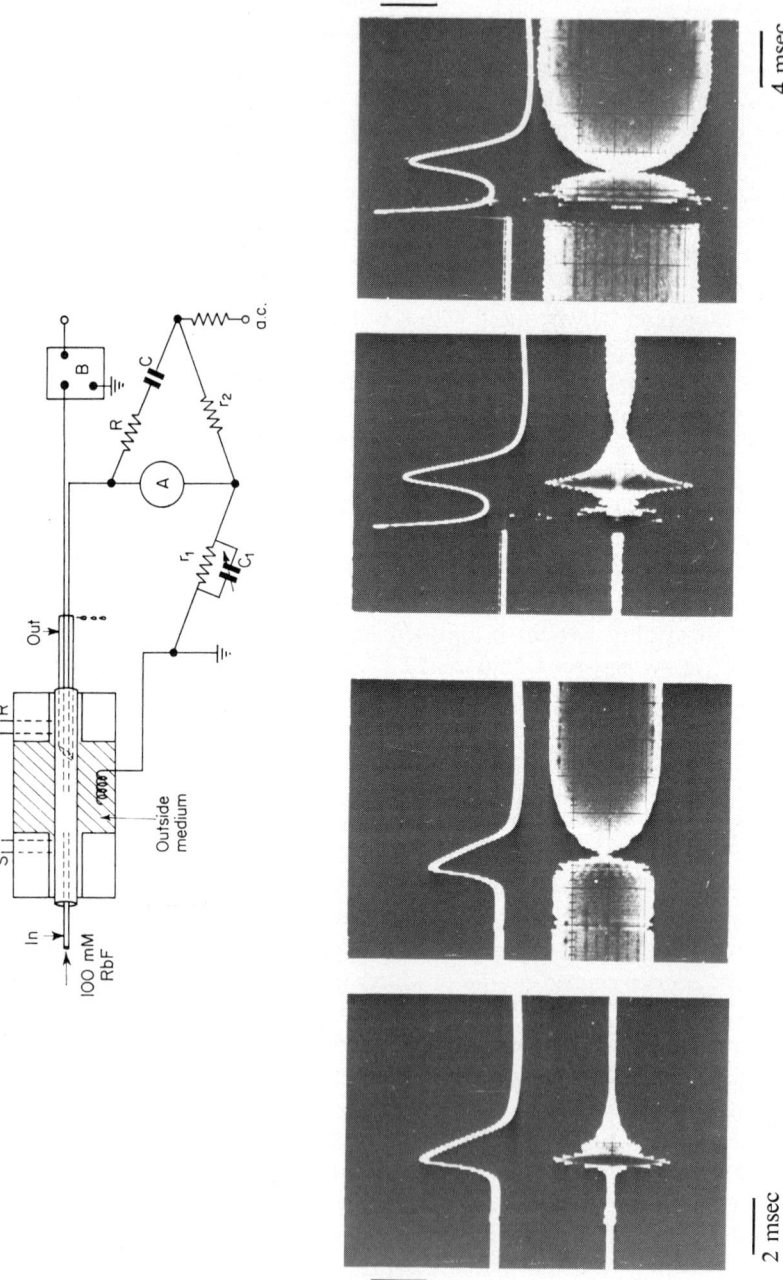

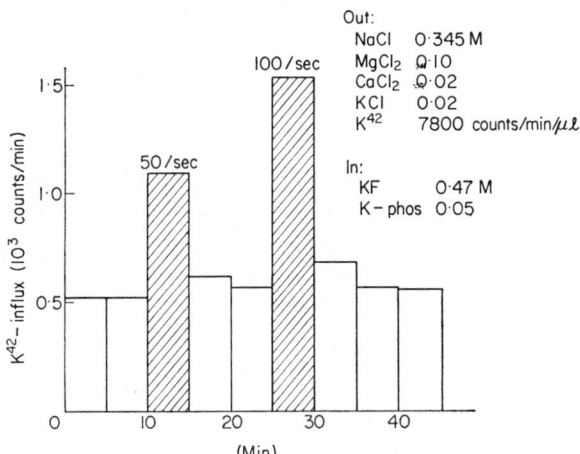

FIG. 11. Potassium (K^{42}) influx in squid giant axons. The compositions of the internal and external media are given in the figure. The periods of stimulation are indicated by the cross-hatching. Samples of the internal medium were collected at regular 5 min intervals and analyzed for K^{42} influx. The extra potassium influx was roughly proportional to the frequency of stimulation. (Reprinted from Tasaki et al., 1966b.)

FIG. 10. (Facing) A. Schematic impedance bridge arrangement for squid giant axon. The Lucite chamber is indicated at the left; the cross-hatched area represents the external medium. The positions of the paired external stimulating (S) and recording (R) electrodes are indicated. The ground electrode is represented by the coil in the external medium. (All of these extracellular electrodes were platinum wires). The positions of the inlet (In) and outlet (Out) cannulae after separation are indicated. The paired platinized impedance electrodes are positioned midway between the cannulae, concentrically within the outlet cannula and the axon. The smaller electrode (upper, in the figure) was led to the recording system (B); the larger electrode (lower) received the output of the a.c. oscillator (a.c.). (Further details are available in Section II-F.) B. Impedance and potential variations in sodium-free external media. In both sets of records the internal perfusion medium contained 0·1 M RbF. In each record, the potential variations are indicated by the upper trace and the simultaneous impedance variations are indicated by the lower trace. In each horizontal pair of records, the record at the left was obtained when the impedance bridge was balanced initially in the resting state. The record at the right was obtained when the bridge was at best balance in the excited state, but unbalanced in the resting state. Left: The external medium contained 0·3 M HzCl and 0·2 M CaCl₂. Right: The external medium contained 0·1 M guanidinium chloride, 0·2 M tetramethylammonium chloride and 0·2 M CaCl₂. (Note the difference in voltage and time scales between these two sets of records.) (Reprinted from Tasaki et al., 1966a.)

peak currents determined from the voltage-clamp. The following example may be cited to clarify this point.

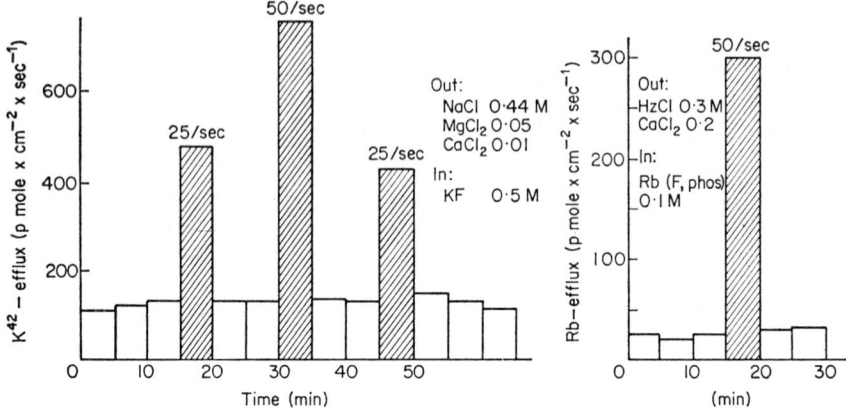

FIG. 12. Cation effluxes in squid giant axon. The compositions of both the external and internal solutions are given in the figures. The entire external medium was collected at regular 5 min intervals as a single sample. The periods of stimulation are indicated by the cross-hatching. Left: Potassium (K^{42}) efflux into sodium-containing media. The resting efflux remained almost constant throughout the period of investigation. Stimulation produced a distinct increase in the efflux which was roughly proportional to the frequency. Right: Rubidium (Rb ^{86}) efflux into sodium-free media. Note the change in the scale of the ordinate. Although both the resting efflux and the efflux during stimulation were diminished in sodium-free media, the relative constancy of the resting efflux and the distinct increase during stimulation were clearly demonstrated. (Reprinted from Tasaki et al., 1966b.)

With axons immersed in 0·3 M HzCl and 0·2 M CaCl₂ and internally perfused with 0·1 M RbF, the peak inward current observed by voltage-clamp was 0·34 mA/cm² (average). Under the same sodium-free conditions, the estimated peak value of the interdiffusion flux is 3·8 pmoles × cm² × msec⁻¹, which is equivalent to 0·37 mA/cm² (Singer et al., 1966; Tasaki et al., 1966b). Therefore, the peak interdiffusion flux and the peak inward current during excitation are of almost the same magnitude. Since the interdiffusion fluxes are significantly large, the currents observed by voltage-clamp during excitation must result from the difference between the efflux (of the internal cation) and the influx (of the external cation). The inward current can represent an increase in the influx of the external univalent cation, or a decrease in the efflux of the internal univalent cation, or a combination of the two possibilities.

This conclusion has several implications for excitation theory. In the equivalent circuit model it is assumed that the inward current observed during excitation is carried by the external univalent cation (sodium) and that the outward current is carried by the internal univalent cation (potassium) (Hodgkin, 1964, p. 57). Originally, the separation of the two currents was based on the voltage-clamp observation that substitution of choline for external cation (sodium) abolished the inward current but did not affect the outward current (Hodgkin and Huxley, 1952a). From this observation it was assumed that the Na^+-flux and the K^+-flux are separated in time. The argument developed above indicates that both the Na^+-flux and K^+-flux take place simultaneously and that a net membrane current appears when these two fluxes are not exactly the same in magnitude and opposite in direction.

In the equivalent circuit model it is proposed that there are separate mechanisms which determine the inward and outward currents during excitation (Hodgkin, 1964, p. 63). These mechanisms of increased and decreased ion movement are called "activation" and "inactivation", respectively, and occur successively in the following order: (1) sodium activation, (2) potassium activation, (3) sodium inactivation and (4) potassium inactivation. The successive increases in ion movements are attributed to successive increases in ion mobilities and are thought to account for the large (200-fold) impedance-loss observed during excitation. However, such a mechanism cannot account for the equally large increase in interdiffusion flux. In order for interdiffusion to increase, it is almost imperative for the mobilities of both species (e.g. K and Na) to increase at the same time; there can be no significant increase in the interdiffusion flux with successive increases in mobility because the flux is determined primarily by the mobility of the less mobile species (Robinson and Stokes, 1959, pp. 129–131; Helfferich, 1962, p. 302; Tasaki *et al.*, 1966b). Thus, the equivalent circuit theory is unable to account for non-specificity of the participating ions, for simultaneous increases in membrane conductance and interdiffusion flux. These observations can be accounted for by the macromolecular model for the excitation process and will be discussed in Section IV.

G. THE FLUX-RESISTANCE PRODUCT

Since the magnitudes of the interdiffusion flux and the membrane conductance are both determined by the products of the ion mobilities and concentrations in the membrane (Section III-F), it is not surprising that the membrane conductance is directly proportional to the rate of interdiffusion flux. If the interdiffusion flux and membrane resistance are measured under a variety of comparable conditions, it is found that the approximate relationship

$$rFJ \approx 20 \text{ to } 30 \text{ mV} \tag{5}$$

is applicable to the membrane of the squid giant axon in both the resting state and the excited state (Singer et al., 1966; Tasaki et al., 1966b). In this empirical relationship, F represents the Faraday constant (96,500 coulombs/equiv), J represents the univalent cation interdiffusion flux (equiv $\times$cm^{-2} $\times$ sec^{-1}) and r represents the membrane resistance (Ω $\times$cm^2) determined by voltage-clamp measurements.

This empirical relationship can be expressed more generally in the form:

$$rJ \approx \frac{RT}{F^2} \qquad (6)$$

where R represents the gas constant and T represents the absolute temperature. The rule of constancy for the flux-resistance product originates in the Nernst-Einstein equation relating the diffusion coefficient to mobility (Spiegler and Coryell, 1953). This relationship can describe the behavior of both living and artificial membranes; a more quantitative and theoretical discussion of Eq. (6) and its application can be found elsewhere (Tasaki and Kobatake, 1968).

IV. Excitation as a Physico-chemical Process

A. THE MEMBRANE AS A MACROMOLECULAR COMPLEX WITH CATION-EXCHANGER PROPERTIES

1. *Macromolecular Complex with Fixed Anionic Sites*

As described in Sections III-A,B, the physiological membrane of the squid giant axon can be visualized as a macromolecular complex of proteins and phospholipids. A highly schematic diagram of such a macromolecular complex is shown in Fig. 14, left, constructed solely on the basis of physiological observations. The major diffusion barrier of the excitable membrane is represented by a framework of macromolecules held together primarily by intra- and intermolecular salt linkages between adjacent oppositely-charged groups. Some cross-linking by divalent cations derived from the external medium, as well as hydrogen and hydrophobic bonds, may also play roles in maintaining the membrane structure.

The major diffusion barrier of the membrane may be subdivided into two physiologically distinct layers (see Fig. 3). The outer layer possesses relatively large numbers of fixed, negatively-charged side-groups on its component macromolecules; these negatively-charged groups are likely to be carboxylate in nature and confer cation-exchanger properties on this layer. The inner layer of the barrier contains a relatively small number of fixed anionic sites, which are likely to be phosphate in nature. The following observations may be offered to support this physiological subdivision:

(1) Most biological macromolecules are negatively-charged at physiological pH. The optimal pH of the external medium is more alkaline than the optimal pH of the internal medium. This suggests that the physiological membrane as a whole is likely to have a relative excess of fixed negative charges and that more of these charges are likely to be located at the external physiological surface of the membrane.

(2) No significant differences are observed among the external anions whereas there are distinct lyotropic sequences for the internal anions and for both internal and external cations (Section III-C). These observations suggest that there is a significant density of negative charge at the external layer of the membrane which tends to exclude any anion and that many different cations can act as counter-ions. This interpretation is supported by the fact that anion flux across the axon membrane is rather small (Tasaki et al., 1961).

(3) Dilution of the external electrolyte solution with isotonic non-electrolyte solution increases the magnitude of the resting potential considerably; similar dilution of the internal electrolyte solution alters the resting potential only slightly (Section III-C). For a ten-fold dilution externally, the increase is roughly 60 mV for salts of univalent depolarizing cations, about 30 to 40 mV for salts of non-depolarizing cations and about 30 mV for divalent cations (T. Teorell and C. S. Spyropoulos, 1960, unpublished data). This behavior is consistent with the presence of negatively-charged groups at the external layer which have almost ideal cation-exchanger properties (Tasaki et al. 1967).

2. Cation-Exchanger Properties

In the "resting state" it is assumed that the negatively-charged sites that comprise the external physiological layer (cation-exchanger) are occupied by counter-ions (predominately divalent cations) derived from the external medium (Fig. 13A); the relatively few fixed anionic sites present in the internal layer may be occupied by counter-ions (univalent cations) derived from the axon interior. There are several experimental observations supporting this assumption:

(1) In most macromolecular systems involving fixed negative charges, there is competition among all of the cations present for the anionic sites. Although there is a distinct order of preference for univalent cations, which is characteristic of the charged group concerned, the divalent cations are very strongly preferred over any of the univalent cations (see Section III-C).

(2) Addition of divalent cations to the external medium tends to stabilize the membrane in the resting state, whereas removal of divalent cations from the external medium leads to marked fluctuation of the membrane potential and to spontaneous excitation (Section III-C).

(3) Inward-directed applied currents, which tend to drive external cations into the membrane, produce almost no change in the transmembrane resistance or e.m.f. (Fig. 13B); outward-directed currents, which tend to drive internal cations into the membrane, produce large changes in the mem-

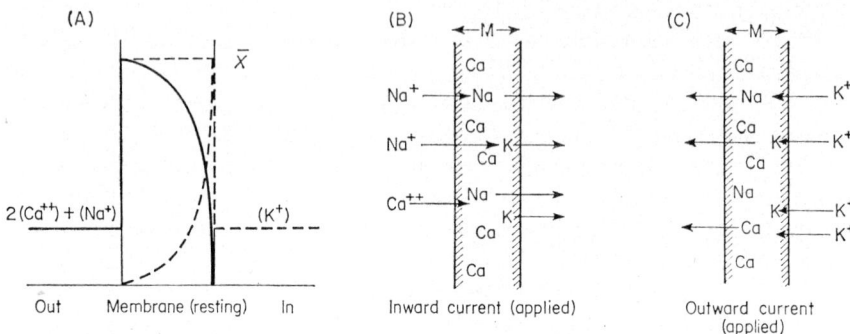

FIG. 13. Effects of applied currents on the resting membrane. The membrane is indicated by M in each case and the arrows indicate the direction of the applied currents. Figure 13A shows the ion concentration profiles in the limiting case for the resting membrane where the concentration of fixed negative charge (X) is assumed to be constant through the membrane. The external physiological layer of the membrane (cation-exchanger layer) is occupied primarily by (divalent) cations derived from the external medium (Out). The internal surface of the membrane is occupied primarily by (univalent) cations derived from the axon interior (In). Figure 13B shows the effect of an applied inward current which tends to drive external cations into the membrane. This (anodal) current produces very little change in the ion profiles and does not lead to excitation. Figure 13C shows the effect of an applied outward (stimulating) current. The movement of internal univalent cations into the membrane by this (cathodal) current can profoundly alter the ion profiles and leads to excitation (see text for details).

brane resistance and in the e.m.f. and lead to excitation (Fig. 13C). (The fact that there is univalent-divalent cation competition at the fixed negative sites of the external physiological layer, and the fact that (different) univalent cations can enter this layer from the external medium and the axon interior, provides the basis for the "two-stable-state" model for the excitation process described in Section IV-B.)

3. *Origin of the Membrane Potential*

In this macromolecular model, the membrane potential can be regarded as consisting of at least three components: (1) the potential difference between the external medium and the outer physiological layer of the membrane (cation-exchanger), (2) that between the outer and inner layers of the membrane and (3) that between the inner layer and the internal medium.

This approach is analogous to that used for the pH-dependent glass electrode. In the latter case, the potential variation (in response to pH changes in the medium) is known to arise at the boundary between the outer hydrated layer of glass and the external medium (Dole, 1941). In a similar manner, the potential variation of the macromolecular membrane (caused by changes in the external medium) is thought to arise primarily at the boundary between the outer physiological layer (cation-exchanger) and the external medium.

B. THE "TWO-STABLE-STATES" THEORY OF EXCITATION

1. *Initiation of Excitation*

The process of excitation in the macromolecular model of the physiological membrane is based on rapid, reversible transitions of the membrane macromolecules between two stable conformations which have different physiological properties (Fig. 14). Such transitions can be affected by physical factors (temperature, hydrostatic and osmotic pressure, light) and by chemical factors (enzymes and other non-electrolytes, concentration, mole fraction and nature of electrolytes, pH and total ionic strength). Under most experimental conditions, and presumably in the natural environment, the macromolecular transitions in the physiological membrane of the squid giant axon are thought to be produced by a change in the ratio of univalent cations to divalent cations in the outer (cation-exchanger) layer of the membrane.

The simplest way to alter the univalent/divalent cation ratio within the membrane is to change the ratio in the external medium; the effects of such a change have been described in several biological systems. For example, if the concentration of potassium ion in the external medium surrounding a squid giant axon is increased, a nerve impulse may be initiated. In the frog nerve fiber, a sudden increase in the external calcium ion concentration during the plateau of a prolonged action potential will terminate the action potential prematurely (Spyropoulos, 1961).

It is well known that a change in the univalent/divalent cation ratio can produce "phase transitions" or "conformational changes" in colloidal and macromolecular systems. For example, the sudden phase transition from an "oil-in-water" state to a "water-in-oil" state can be produced by a relative increase in the divalent cation concentration (Clowes, 1916). Similarly, intramolecular phase transitions of various macromolecules take place at critical ratios of the univalent and divalent cation concentrations (Michaeli, 1960; Katchalsky and Curran, 1965).

A more indirect method of producing a change in the univalent/divalent cation ratio in the excitable membrane is with an electric current (see Fig. 13). An inward-directed electric current can be used to drive external cations

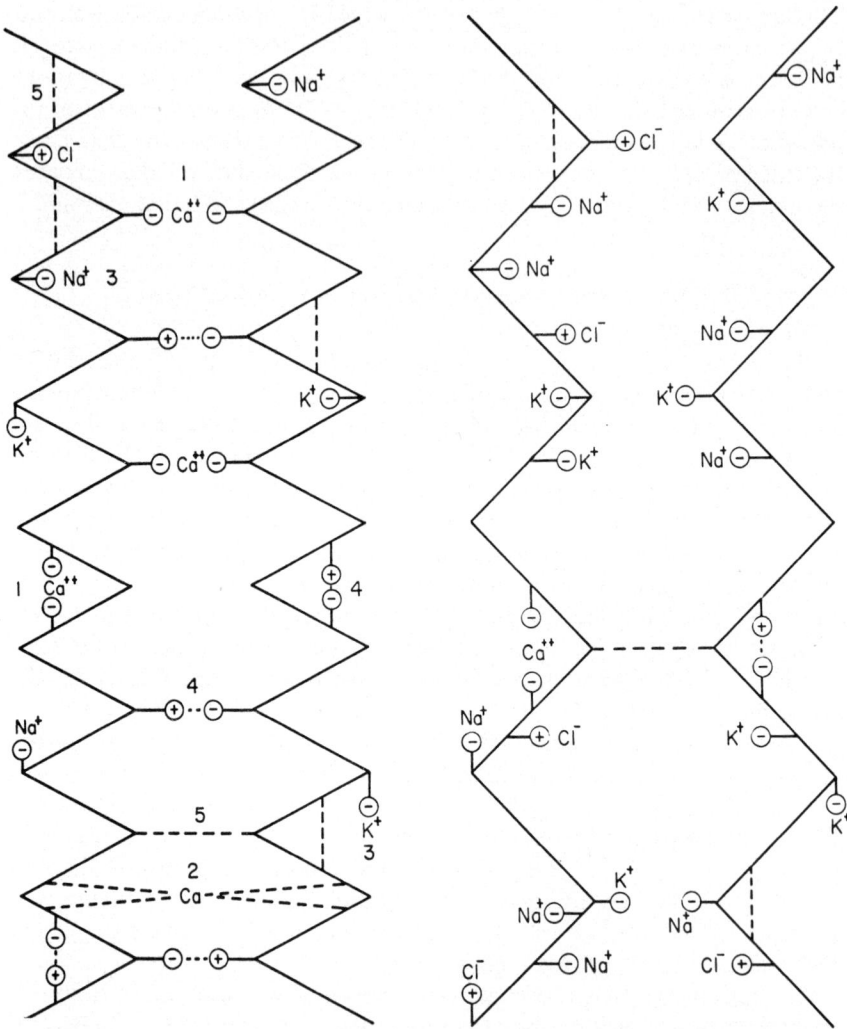

FIG. 14. Left: Schematic diagram of the macromolecular complex: "resting (stable) conformation." Two macromolecules in their resting conformation are indicated by the jagged lines. The various charged side-groups are indicated with representative counter-ions. Several of the different kinds of bonds which maintain the complex in the resting (high resistance, low cation interdiffusion) stable state are indicated: (1) intramolecular and intermolecular cross-linking involving divalent cations and adjacent fixed anionic sites (Michaeli, 1960); (2) coordination complex formation between divalent cations and negative macromolecular ligands; (3) excess fixed negative charges occupied by cations derived primarily from the external medium; (4) intra- and intermolecular salt linkages between adjacent

(e.g., Na$^+$, Ca^{++}) into the membrane; an outward-directed current can be used to drive internal cations into the membrane (e.g., K$^+$). These two currents produce entirely different effects on the excitable membrane; the former produces almost no change in the resting membrane resistance or in the e.m.f. across the membrane, whereas the latter produces depolarization followed by an action potential (see Sections III-C and IV-A). This difference provides an insight into the distribution of ions in the resting state (Fig. 13A) and into the mechanism for producing excitation.

In the "resting state" divalent cations derived from the external medium (usually Ca^{++} or Mg^{++}) form relatively stable complexes with the macromolecules comprising the cation-exchanger portion (outer layer) of the physiological membrane (Fig. 14, left). Hence, in the resting state these anionic sites are not available for transmembrane ion transport and membrane resistance is high. The usual, small external univalent cations (Li$^+$, Na$^+$) have little tendency to disrupt the macromolecular-divalent cation complexes (Section III-C and IV-A; see also, Martell and Calvin, 1952). However, the larger internal cations (K$^+$, Rb$^+$, Cs$^+$) have much greater "affinity" for anionic sites, particularly for the carboxyl groups which are likely to be the dominant anionic group at the external physiological surface of the membrane (Section III-C; see also, Bungenberg de Jong, 1949, p. 284). Therefore, it would be expected that, at comparable concentrations, the usual internal univalent cations would be more effective competitors for these anionic sites than the usual external univalent cations. When an inward-directed current is applied, there is no change in the resting membrane resistance because the anionic sites are already occupied predominately by divalent

---

oppositely-charged groups on the macromolecules and (5) intra- and intermolecular hydrogen and hydrophobic bonding. Right: Macromolecular complex "active (stable) conformation." An outward (stimulating) current displaces the divalent cations in the membrane with univalent cations derived from the axon interior; the alteration in the univalent/divalent cation ratio is responsible for the conformational changes that result. The conformational change is represented by the stretching and separation of the two macromolecules which comprise this representative complex. Disruption of some of the bonds described above result in the exposure of more (and possibly different) sites of fixed negative charge and markedly alters the physiological properties of the membrane. For example there are large increases in the functional cation-exchanger capacity, conductance and univalent cation interdiffusion. Cations derived from both the external medium and the axon interior compete for the sites available for transmembrane ion transfer but in the absence of a maintained outward current, the divalent cations are preferred at the sites of fixed negative charge. The subsequent alteration in the univalent/divalent cation ratio returns the macromolecular complex to its resting conformation. (Further details of the "two stable state" theory of excitation may be found in the text.)

cations derived from the external medium (Fig. 13B) and the usual external univalent cations (e.g. Na^+, Li^+) are not capable of displacing the usual divalent cations (e.g. Ca^{++}, Mg^{++}) from the anionic sites of the macromolecular complex. However, when an outward-directed current is applied (Fig. 13C), univalent cations derived from the axon interior (K^+, Rb^+) are driven into the membrane and these cations are much more capable of displacing the divalent cations from the macromolecular complex. Thus, the ratio of univalent/divalent cations is increased and a conformational change can take place (Michaeli, 1960) which alters the physiological properties of the membrane (Fig. 14).

Not only can this mechanism account for the use of "cathodal" currents to stimulate peripheral nerves, but it can also account for the simultaneous changes in membrane impedance and isotope flux during excitation and for the existence of significant univalent cation interdiffusion. When the divalent cations in the membrane are displaced by univalent cations, there is less cross-linking and many anionic sites are available for the transport of cations across the membrane (Fig. 14, right; see Peller, 1959); hence, ion mobilities will be high, isotope flux will be large and membrane resistance will be low.

2. *Spread of Excitation*

Some cation-exchanger systems (zeolite crystals, for instance) show strong mutual interaction between adjacent sites of fixed negative charge (Barrer and Falconer, 1956). The presence of a particular cation at one fixed negative site increases the probability of similar cations being brought to adjacent anionic sites. In a similar manner, displacement of a single divalent cation from one negative site by univalent cations tends to form a very small region or "patch" in the cation-exchanger portion of the membrane which is low in divalent cations. Thus, when a brief stimulating (outward-directed) current results in the displacement of a few divalent cations from the anionic sites by a few internal univalent cations, an "active patch" is formed in the physiological membrane. Cooperative and competitive interaction of the type discussed above tends to fill the negative sites around an active patch with univalent cations derived from both the internal and external medium. Since this portion of the membrane is now in the "active" or "excited" state, ion transport through this region increases and the local conductance is high. The local current which results is carried primarily by univalent cations. Since the potential level of the axon interior is negative with respect to the external medium, the local current is directed inward through the "active patch" and outward through the resting area. The outward-directed component of the local circuit tends to drive more univalent cations derived from the axon interior into resting areas of the membrane and results in the formation of more "active patches." In this manner, the local currents can serve as stimuli for the remainder of the resting membrane to be transformed into the "active state."

Not only can the "two-stable-state" model for excitation in a macromolecular membrane account for the cation-exchanger properties of the giant squid axon, but it can also account for the apparent "negative conductance" portion of the current-voltage relationship obtained from voltage-clamp experiments (Section III-C). The voltage-clamp serves as a powerful method for reducing electrical interactions between the "active" and "resting" regions of the physiological membrane. In the presence of voltage-clamping the local currents flowing inward through the active regions do not easily stimulate the resting regions. The intermediate segment (negative slope) of the current-voltage relationship (Figs 8 and 9) can then be attributed to the gradual increase in the total area of active (inward-current carrying) regions in the membrane produced by raising the clamping potential level (Tasaki and Spyropoulos, 1958; Tasaki and Singer, 1966). In summary, the initial linear segment (small positive slope) of the voltage-clamp curve represents the I-V relationship for the almost uniformly resting membrane; the final linear segment (large positive slope) represents the I-V relationship for the almost uniformly active membrane. In each of these "stable states" the current-voltage relationship of the physiological membrane obeys Ohm's law; the intermediate segment (with a negative slope) represents a mixture of membrane areas in the active and resting states with a gradual change in the total active area.

3. *Termination of Excitation*

As mentioned above, in the absence of a maintained outward current, the external divalent cations are strongly favored in the competition for the anionic sites. During excitation, the ionic environment on both sides of the membrane changes gradually as the result of enhanced cation interdiffusion. Eventually the divalent cations once again form stable complexes with the negative sites. This process progresses rapidly by cooperative interaction between adjacent charged sites; the subsequent decrease in the univalent/divalent cation ratio reverses the conformational process which initiated excitation and restores the physiological membrane to the resting state. Since the anionic sites occupied by divalent cations are not available for transmembrane ion transport, univalent cation interdiffusion is very low in the resting state and membrane impedance is high.

This mechanism is consistent with the observations that initiation of an action potential is exothermic and termination is endothermic (Section III-D). In known cation-exchangers, the process of exchange of divalent cations with univalent cations is exothermic, accompanied by a large change in the entropy and enthalpy of the system (Coleman, 1952); the reverse process is endothermic, as in the case for the termination of excitation. The assumption of an underlying change in macromolecular conformations is also supported by these findings.

Under special circumstances, termination of the excitation process is delayed or prevented. Action potentials with a prolonged plateau period often of several seconds or more in duration, have been observed in squid giant axons after internal perfusion with proteases, cesium and rubidium salts (Tasaki et al., 1965a; Tasaki et al., 1966a) and low concentrations of potassium (Adelman, 1965; Adelman et al., 1965). Prolonged action potentials have also been observed in the toad nerve fiber with external cooling, addition of transition cations such as Ni^{++} and increasing the external univalent/divalent cation ratio (Spyropoulos, 1961). Substitution of some depolarizing cations, divalent cations or large organic cations for external sodium ion in the squid giant axons will often produce similar action potentials (Tasaki et al., 1965b, 1966a, b, c). In the case of external depolarizing cations, termination of the action potential may not take place until more divalent cations are added to the external medium (Tasaki, 1959).

From the standpoint of the "two-stable-states" theory of excitation, the mechanism of prolongation of the action potential is as follows; when the membrane is in the excited state, there is an enhancement of the interdiffusion fluxes of cations across the membrane and, consequently, the composition of the electrolytes in the medium in the immediate vicinity of the membrane changes gradually. This gradual change in the electrolyte composition brings about a gradual fall in the membrane potential. At the moment when the electrolyte composition reaches a certain critical level, there is a transition of the membrane from the excited state to the resting state (see Subsection 3). Therefore, any agent which tends to reduce the interdiffusion fluxes during excitation (e.g., dilution of the internal electrolyte solution, introduction of tetraethylammonium ion into the membrane, etc.) is expected to prolong the action potential. (Since dilution of the external medium tends to increase the divalent cation concentration in the membrane, this procedure does not prolong the action potential.) When the interdiffusion fluxes across the resting membrane are enhanced, the alteration of the ionic environment of the membrane during excitation would be relatively slow and would tend to prolong the action potential.

According to the "two-stable-states" hypothesis, the theoretical form of the action potential is rectangular. The reduction of the plateau period under "normal" conditions (with the production of a triangular form for the action potential) is attributed to the rapidity with which the interdiffusion alters the ionic environment of the membrane.

C. SUMMARY

The equivalent circuit model for the excitation process has provided the stimulus for many fruitful investigations of the process by which nerve impulses are generated and propagated (Hodgkin and Huxley, 1952; Hodgkin, 1964). At present, however, there is no simple model of this type which is capable of explaining the recent data obtained from intracellularly perfused squid giant axons (see Cole, 1965). The inherent limitations and ambiguities of the equivalent circuit model are particularly apparent in dealing with various physico-chemical data. Therefore, a macromolecular approach to the excitation process was adopted to provide a new departure for interpreting both electrophysiological and physicochemical results.

The excitable membrane of the squid giant axon is visualized as a macromolecular complex of proteins and phospholipids; a relative excess of fixed negative charge at the external layer of the physiological membrane confers cation-exchanger properties on the system. Interactions between this macromolecular system and its physico-chemical environment may be thought of in terms of direct effects within the cation exchange process or in terms of indirect effects on the underlying macromolecular conformations.

The process of excitation involves a rapid, reversible cation-exchange process, based on transitions between "two stable (conformational) states" of the membrane macromolecules. In the "resting" stable state the anionic sites in the membrane are occupied primarily by divalent cations derived from the external medium; in the active (or "excited") stable state these sites are occupied predominantly by univalent cations. The conformation of the membrane macromolecules and the properties of the cation exchanger are determined primarily by the univalent/divalent cation ratio.

The excitation process is triggered by the transport of internal univalent cations into the membrane. These univalent cations are capable of competitive displacement of divalent cations from the anionic sites. The continued displacement of the divalent cations by local cooperative effects results in an increase in the local univalent/divalent cation ratio and produces an exothermic conformational change of the membrane macromolecules. This conformational change determines the density and nature of the anionic sites available (see Steinhardt and Beychok, 1964, p. 294; Peller, 1959) and hence determines the cation-exchanger properties of this particular conformation (the physiological "active" state).

The difference in the properties of the two macromolecular conformations ("active" and "resting") is responsible for the spread of the excitation process. When a local region of the membrane is transformed into its "active" conformation a local circuit is set up: Current is carried (inward) by external univalent cations through the "active" patch to the axon interior and current is carried (outward) by internal univalent cations through the "resting" regions into the external medium. The internal cations which are driven into "resting" regions of the membrane by the local circuits convert these regions to new "active" patches; eventually, if the local current is strong enough, the entire membrane is converted from the "resting" stable conformation to the "active" stable conformation, with a large increase in the number of anionic sites available for transmembrane ion transfer. Consequently, there is a large increase in conductance and univalent cation interdiffusion in the physiological "active" state.

Excitation is terminated when divalent cations derived from the external medium once again form stable complexes with the fixed anionic sites. This is associated with a reduction in the univalent/divalent cation ratio in the membrane macromolecules. This transition from the "active" (or "excited") state to the resting is endothermic. The reduction in the number of anionic sites available for transmembrane ion transport is associated with this transition and is responsible for the relatively high membrane resistance and low cation interdiffusion in the physiological "resting" state.

The "two-stable-states" theory for the excitation process and the macromolecular model for the physiological membrane of the squid giant axon provide one qualitative mechanism for the generation and propagation of the nerve impulse. It is hoped that future investigations will provide a more quantitative interpretation, based on this physico-chemical approach to the excitation process.

References

Abbott, B. C., Hill, A. V. and Howarth, J. V. (1958). *Proc. R. Soc.* **B148**, 149.
Adelman, W. J. (1965). *Proc. XXIII Int. Congress Physiol. (Tokyo)*, p. 59.
Adelman, W. J., Dyro, F. M. and Senft, J. (1965). *J. gen. Physiol.* **48**, 1.
Adelman, W. J., Dyro, F. M. and Senft, J. (1966). *Science, N.Y.* **151**, 1392.
Arvanitaki, A. and Chalazonitis, N. (1958). *Archs Sci. physiol.* **7**, 73.
Baker, P. F., Hodgkin, A. L. and Shaw, T. I. (1961). *Nature, Lond.* **190**, 885.
Baker, P. F., Hodgkin, A. L. and Shaw, T. I. (1962a) *J. Physiol.* **164**, 330.
Baker, P. F., Hodgkin, A. L. and Shaw, T. I. (1962b). *J. Physiol.* **164**, 355.
Baker, P. F., Hodgkin, A. L. and Meves, H. (1964). *J. Physiol.* **170**, 541–560.
Bank, P. F. and Hoskam, E. G. (1940). *Protoplasma* **34**, 188.
Barrer, R. M. and Falconer, J. D. (1956). *Proc. R. Soc.* **A236**, 227.
Bates, R. G. (1964). "Determination of pH." 435 pp., John Wiley and Sons, Inc., New York.
Buchner, E. H. and Postma, G. (1931). *Proc. K. ned. Akad. Wet.* **34**, 699.
Bungenberg de Jong, H. G. (1949a). *In* "Colloid Science" (H. R. Kruyt, ed.) Vol. 2, p. 232. Elsevier Publishing Co., New York.
Bungenberg de Jong, H. G. (1949b). *In* "Colloid Science" (H. R. Kruyt, ed.) Vol. 2, p. 335. Elsevier Publishing Co., New York.

Bungenberg de Jong, H. G. (1949c). *In* "Colloid Science", (H. R. Kruyt, ed.) Vol. 2, p. 259. Elsevier Publishing Co., New York.
Caldwell, P. C. (1956). *J. Physiol.* **132**, 35.
Caldwell, P. C. (1960). *J. Physiol.* **152**, 545.
Cavanaugh, G. M. (ed.) (1956). "Formulae and Methods IV of the Marine Biological Laboratory," 61 pp., Woods Hole, Mass.
Chalazonitis, N. and Chagneux, R. (1961). *Bull. Inst. océanogr. Monaco* No. 1223, 1.
Chandler, W. K. and Hodgkin, A. L. (1965). *J. Physiol.* **181**, 594.
Clowes, G. H. A. (1916). *J. phys. Chem.* **20**, 407.
Cole, K. S. (1965). *Physiol. Rev.* **45**, 340.
Cole, K. S. and Curtis, H. J. (1939). *J. gen. Physiol.* **22**, 649.
Cole, K. S. and Moore, J. W. (1960). *J. gen. Physiol.* **44**, 123.
Coleman, N. T. (1952). *Soil Sci.* **74**, 115.
Davison, P. F. and Taylor, E. W. (1960). *J. gen. Physiol.* **43**, 801.
Davson, H. and Danielli, J. F. (1943). "The Permeability of Natural Membranes." 361 pp., Cambridge University Press, Cambridge.
Deffner, G. G. (1961). *Biochem. biophys. Acta* **47**, 378.
Dole, M. (1941). "The Glass Electrode: Methods, Applications and Theory." 332 pp., John Wiley and Sons, Inc., New York.
Du Bois-Reymond, E. (1849). "Untersuchungen über Thierische Elektricität," Vol. II part (1), 608 pp., Verlag von G. Reimer, Berlin.
Eisenman, G. (1961). *In* "Symposium on Membrane Transport and Metabolism," p. 163. Publishing House of the Czechoslovak Academy of Science, Prague.
Elbers, P. F., Ververgaert, P. H. J. T. and Demel, R. A. (1965). *J. biophys. biochem. Cytol.* **24**, 32.
Erlanger, J. and Gasser, H. S. (1937). "Electrical Signs of Nervous Activity." University of Pennsylvania Press, Philadelphia.
Fatt, P. and Katz, B. (1953). *J. Physiol.* **120**, 171.
Frank, H. S. and Wen, W. -Y. (1957). *Discuss. Faraday Soc.* **24**, 133.
Furukawa, T., Sasaoka, T. and Hosoya, Y. (1959). *Japan J. Physiol.* **9**, 143.
Geren, B. B. (1954). *Expl. Cell Res.* **7**, 558.
Geren, B. B. and Schmitt, F. O. (1954) *Proc. natn. Acad. Sci. U.S.A.* **40**, 863.
Gortner, R. A., Hoffman, W. F. and Sinclair, W. B. (1928). *In* "Colloid Symposium Monographs." (H. B. Weiser, ed.) Vol. 5, p. 179. Chemical Catalog Book Co., New York.
Hagiwara, S., Chichibu, S. and Naka, K. (1964). *J. gen. Physiol.* **48**, 163.
Haldi, J. A., Rauth, J. W., Larkin, J. and Wright, P. (1927). *Am. J. Physiol.* **80**, 631.
Hama, K. (1959). *J. biophys. biochem. Cytol.* **6**, 61.
Hamaguchi, K. and Geiduschek, E. P. (1962). *J. Am. Chem. Soc.* **84**, 1329.
Heilbrunn, L. V. (1952). "An Outline of General Physiology." (3rd Ed.) 818 pp., W. B. Saunders Co., Philadelphia, Pennsylvania.
Helfferich, F. (1962). "Ion Exchange." 624 pp., McGraw-Hill Book Co., Inc., New York.
Hill, A. V. (1932). "Chemical Wave Transmission in Nerve." 74 pp., University Press, Cambridge.
Hill, A. V. and Howarth, J. V. (1958). *Proc. R. Soc.* **B149**, 167.
Höber, R. (1928). *In* "Colloid Chemistry." (J. Alexander, Ed.) Vol. II, p. 619, Chem. Cat. Book Co., New York.
Höber, R. (1945). "Physical Chemistry of Cells and Tissues." 676 pp., The Blakiston Co., Philadelphia, Pennsylvania.

Hodgkin, A. L. (1964). "The Conduction of the Nervous Impulse." 108 pp., Liverpool University Press, Liverpool.
Hodgkin, A. L. and Huxley, A. F. (1952a). *J. Physiol.* **116**, 449.
Hodgkin, A. L. and Huxley, A. F. (1952b). *J. Physiol.* **116**, 473.
Hodgkin, A. L. and Huxley, A. F. (1952c). *J. Physiol.* **116**, 497.
Hodgkin, A. L. and Huxley, A. F. (1952d). *J. Physiol.* **117**, 500.
Hofmeister, F. (1888). *Arch. exp. Path. Pharmak.* **24**, 247.
Ives, D. J. G. and Janz, G. J. (eds) (1961). "Reference Electrodes: Theory and Practice." 651 pp., Academic Press, Inc., New York.
Jirgensons, B. and Straumanis, M. E. (1962). "A Short Textbook of Colloid Chemistry." 489 pp., The MacMillan Co., New York.
Katchalski, E., Sela, M., Silman, H. I. and Berger, A. (1964). *In* "The proteins." (H. Neurath, ed.) Vol 2, p. 405. Academic Press, New York.
Katchalsky, A. and Curran, P. F. (1965). "Nonequilibrium Thermodynamics in Biophysics." 248 pp., Harvard University Press, Cambridge, Massachusetts.
Kauzmann, W. (1959). *Adv. Protein Chem.* **14**, 1.
Keynes, R. D. and Lewis, P. R. (1951). *J. Physiol.* **114**, 151.
Keynes, R. D. and Lewis, P. R. (1956). *J. Physiol.* **134**, 399.
Koechlin, B. A. (1955). *J. biophys. biochem. Cytol.* **1**, 511.
Kruyt, H. R. (ed.) (1949). "Colloid Science," Vol. II. "Reversible Systems," 753 pp. Elsevier Publishing Co., New York.
Kruyt, H. R. (ed.) (1952). "Colloid Science," Vol. I. "Irreversible Systems, Hydrophobic Colloids," 389 pp., Elsevier Publishing Co., New York.
Ling, G. N. (1962). "A Physical Theory of the Living State: The Association-Induction Hypothesis." 680 pp., Blaisdell Publishing Co., New York.
Ling, G. N. and Gerard, R. W. (1949). *J. cell. comp Physiol* **34**, 382
MacInnes, D A (1939) "The Principles of Electrochemistry " Reinhold Publishing Corp , New York.
Martell, A. E. and Calvin, M. (1952). "Chemistry of the Metal Chelate Compounds." 613 pp., Prentice-Hall, Inc., Englewood Cliffs, N.J.
McBain, J. W. (1950). "Colloid Science." 450 pp., Heath Co., Boston.
Meites, L. (1963). "Handbook of Analytical Chemistry." pp. 1–13 to 1–19 and 1–39 to 1–46, (Tables 1–17 to 1–20). McGraw-Hill Book Co., New York.
Michaeli, I. (1960). *J. polymer. Sci.* **48**, 291.
Moore, J. W., Narahashi, T. and Ulbricht, W. (1964). *J. Physiol.* **172**, 163.
Nakamura, Y., Nakajima, S. and Grundfest, H. (1964a). *Science, N.Y.* **146**, 266.
Nakamura, Y., Nakajima, S. and Grundfest, H. (1964b). *Biol. Bull.* **127**, 382.
Nakamura, Y., Nakajima, S. and Grundfest, H. (1965). *J. gen. Physiol.* **49**, 321.
Narahashi, T. (1963). *J. Physiol.* **169**, 91.
Narahashi, T., Deguchi, T., Urukawa, N. and Ohkubo, Y. (1960). *Am. J. Physiol.* **198**, 934.
Narahashi, T., Moore, J. W. and Scott, W. R. (1964). *J. gen Physiol.* **47**, 965.
Oikawa, T., Spyropoulos, C. S., Tasaki, I. and Teorell, T. (1961). *Acta physiol, scand.* **52**, 195.
Overbeek, J. Th. G. (1952). *In* "Colloid Science." (H. R. Kruyt, ed.), Vol. 1, p. 115. Elsevier Publishing Co., New York.
Overbeek, J. Th. G. and Bungenberg De Jong, H. G. (1949). *In* "Colloid Science," (H. R. Kruyt, ed.), Vol. 2, p. 184. Elsevier Publishing Co., New York.
Pallmann, H. (1930). *Kolloidchem. Beih.* 30, 334.
Peller, L. (1959). *J. phys. Chem.* **63**, 1199.

Robertson, J. D. (1960). *Prog. Biophys. biophys. Chem.* **10**, 343.
Robinson, R. A. and Stokes, R. H. (1959). "Electrolyte Solutions." 559 pp., Butterworths Publishing Ltd., London.
Rojas, E. and Luxoro, M. (1963). *Nature, Lond.* **199**, 78.
Schellman, J. A. and Schellman, C. (1964). *In* "The Proteins." (H. Neurath, ed.) Vol. 2, p. 1. Academic Press, New York.
Segal, J. (1958). "Die Erregbarkeit der Lebenden Materie." 252 pp., Veb Gustave Fischer Verlag, Jena.
Singer, I and Goodman, S. J. (1966). *Expl. Cell Res.* **43**, 367.
Singer, I. and Tasaki, I. (1965). *Fedn. Proc. Fedn. Am. Socs exp. Biol.* **24**, 585.
Singer, I., Tasaki, I., Watanabe, A. and Kobatake, Y. (1966). *Fedn. Proc. Fedn. Am. Socs. exp. Biol.* **25**, 569.
Sollner, K. (1953). *J. Colloid Sci.* **8**, 179.
Speigler, K. S. and Coryell, C. S. (1953). *J. phys. Chem.* **57**, 687.
Spyropoulos, C. S. (1961). *Am. J. Physiol.* **200**, 203.
Spyropoulos, C. S. and Tasaki, I. (1960). *Ann. Rev. Physiol.* **22**, 407.
Steinbach, H. B. (1941). *J. cell. comp. physiol.* **17**, 57.
Steinbach, H. B. and Spiegelman, S. (1943). *J. cell. comp. Physiol.* **22**, 187.
Steiner, R. F. (1965). "The Chemical Foundations of Molecular Biology." 468 pp., D. Van Nostrand Co., Inc., New York.
Steinhardt, J. and Beychok, S. (1964). *In* "The Proteins." (H. Neurath, ed.), Vol. 2, p. 139. Academic Press, New York.
Takenaka, T. and Yamagishi, S. (1966). *Proc. Japan. Acad.* **42**, 521.
Tasaki, I. (1959). *J. Physiol.* **148**, 306.
Tasaki, I. (1963). *J. gen. Physiol.* **46**, 755.
Tasaki, I. and Kobatake, Y. (1968). Appendix I and II in Tasaki, I. "Nerve Excitation: A Macromolecular Approach," p. 20, Charles C. Thomas, Springfield, Illinois.
Tasaki, I. and Shimamura, M. (1962). *Proc. natn. Acad. Sci. U.S.A.* **48**, 1571.
Tasaki, I. and Singer, I. (1965) *J cell comp. Physiol,* **66**, Part II, Suppl. 2, p. 137.
Tasaki, I. and Singer, I. (1966). *Ann. N.Y. Acad. Sci.* **137**, 792.
Tasaki, I. and Singer, I. (1968). *Ann, N.Y. Acad. Sci.* **148**, 36.
Tasaki, I. and Spyropoulos, C. S. (1957). *In* "The Influence of Temperature on Biological Systems." (F. S. Johnson, ed.) p. 201. Waverly Press, Baltimore, Maryland.
Tasaki, I. and Spyropoulos, C. S. (1958). *Am. J. Physiol.* **193**, 318.
Tasaki, I. and Takenaka, T. (1964). *In* "The Cellular Functions of Membrane Transport." (J. F. Hoffman, ed.) p. 95. Prentice-Hall Inc., Englewood Cliffs, N.J.
Tasaki, I., Teorell, T. and Spyropoulos, C. S. (1961). *Am. J. Physiol.* **200**, 11.
Tasaki, I., Watanabe, A. and Takenaka, T. (1962). *Proc. natn. Acad. Sci. U.S.A.* **48**, 1177.
Tasaki, I., Singer, I. and Takenaka, T. (1965a). *J. gen. Physiol.* **48**, 1095.
Tasaki, I., Singer, I. and Watanabe, A. (1965b). *Proc. natn. Acad. Sci. U.S.A.* **54**, 763.
Tasaki, I., Luxoro, M. and Ruarte, A. (1965c). *Science, N.Y.* **150**, 899.
Tasaki, I., Singer, I. and Watanabe, A. (1966a). *Am. J. Physiol.* **211**, 746.
Tasaki, I., Singer, I. and Watanabe, A. (1966b). *J. gen. Physiol.* **50**, 989.
Tasaki, I., Watanabe, A. and Singer, I. (1966c). *Proc. natn. Acad. Sci. U.S.A.* **56**, 1116.
Tasaki, I., Watanabe, A. and Lerman, L. (1967). *Am. J. Physiol.* **213**, 1465.
Teorell, T. (1953). *Prog. Biophys. biophys. Chem.* **3**, 305.

Teorell, T. (1962). *Biophys. J.* **2**, 27.
Tobias, J. M. (1955). *J. cell. comp. Physiol.* **46**, 183.
Vöet, A. (1937). *Chem. Rev.* **20**, 169.
von Bruns, E. M. (1932). *Proc. K. med. Akad. Wet.* **34**, 699.
von Bruins, E. M. (1932). *Proc. K. ned. Akad. Wet.* **35**, 107.
von Hippel, P. H. and Wong, K. -Y. (1964). *Science, N.Y.* **145**, 577.
von Hippel, P. H. and Wong, K. -Y. (1965). *J. biol. Chem.* **240**, 3909.
Watanabe, A., Tasaki, I., Singer, I. and Lerman, L. (1967). *Science, N.Y.* **155**, 95.
Weber, G. and Teale, F. J. W. (1965). *In* "The Proteins," (H. Neurath, ed.) Vol. 3, p. 445. Academic Press, New York.
White, A., Handler, P. and Smith, E. L. (1964). "Principles of Biochemistry," 1106 pp., (3rd Ed.) McGraw-Hill Book Co., New York.
Winter, S. S. (1956). *J. Chem. Ed.* **33**, 246.
Young, J. Z. (1936a). *Q. Jl. microsc. Sci.* **78**, 367.
Young, J. Z. (1936b). *Cold Spr. Harp. Symp. quant. Biol.* **4**, 1.

Addendum

Quite recently, excitation of squid giant axons immersed in media entirely free of univalent cations has been extensively studied by Watanabe, A., Tasaki, I. and Lerman, L. (*Proc. natn. Acad. Sci., U.S.A.* **58**, 2246, 1967). Axons internally perfused with dilute (5–50 mM) sodium salt solutions were found capable of producing all-or-none action potentials when immersed in external media containing 30 to 100 mM $CaCl_2$ (or $SrCl_2$) as the sole electrolyte species. Under these conditions, with the usual concentration gradient for sodium across the axon membrane reversed, prolonged action potentials with overshoots of approximately 20 mV were observed. Similar results were obtained from axons perfused with dilute solutions of the salts of cesium, lithium, guanidine or methylguanidine. The presence of the salt of a divalent cation externally, and that of a univalent cation internally, was found to be sufficient for excitation. These experimental findings cannot be explained in terms of a theory which maintains that a specific increase in sodium permeability and influx of Na-ion into the cell is the primary cause of excitation or that the peak of the action potential is determined solely by the Na-ion concentration across the membrane. On the other hand, the findings are consistent with the "two-stable-states" theory which proposes that a conformational change of membrane macromolecules, associated with a cooperative ion exchange process at negatively charged sites of the macromolecules, is the primary physico-chemical event in excitation.

Author Index

Numbers in italics refer to the text page on which the complete reference may be found.

A

Abbott, B. C., 385, *406*
Abrahamsson, S., 128, *199*
Abramson, M. B., 254, *285*
Adelman, W. J., 370, 404, *406*
Agin, D., 271, *288*
Ahrens, E. H., Jr., 27, 29, 31, 32, 34, *65*
Akimov, I., 335, *346*
Allen, B. T., 198, *199*
Allison, A. C., 23, 29, *66*
Aloof, S., 24, 26, *64*
Alvord, E. C., Jr., 49, *66*
Amesz, J., 307, *342*
Andersen, D. E., 101, *121*
Anderson, H. M., 27, *68*
Anderson, P. J., 272, *285*
Anderson-Cedergren, E., 166, *201*
Andrews, E. L., 28, 33, 34, *67*
Apostolakis, M., 28, 34, *65, 68*
Arnold, W., 303, 312, 313, 317, 334, 335, 338, *342, 343*
Arnon, D. I., 307, 311, 312, *342*
Arvanitaki, A., 386, *406*
Ashhurst, D. E., 166, *199*
Ashworth, L. A. E., 10, *64*
Attwood, D., 226, *231*
Auguenstine, L. G., 290, 312, *342*
Autilio, L. A., 49, *67*

B

Babini, B., 29, *64, 66*
Bahr, G. F., 162, *199*
Baird, S. L., 323, *344*
Baker, P. F., 348, 349, 351, 364, 366, *406*
Balasubramanian, D., 267, *285*
Balint, J. A., 31, 60, *64*
Ballard, H. S., 51, 53, *66*
Bangham, A. D., 166, *199*, 214, 218, 224, 226, *231*, 247, 250, 251, 265, 266, 277, 278, 279, 281, 282, 283, *285, 287*, 340, *342*
Bank, P. F., 374, *406*
Bar, R. S., 235, *285*
Barlow, J. S., 197, *200*
Barnola, F. V., 254, *288*
Barr, R., 307, *343*
Barratt, M. D., 153, *199*
Barrer, R. M., 402, *406*
Basford, J. M., 10, *64*
Bassham, J. A., 291, *343*
Bates, R. G., 388, *406*
Bauman, A. J., 18, 32, 56, 60, *67*
Baxter, C. F., 16, *68*
Bay, Z., 313, 334, *342*
Beament, J. W. L., 263, 264, *285*
Bear, R. S., 103, 115, 117, *121, 122*
Beare, J. L., 226, *231*
Bearn, A. G., 230, *232*
Bejnarowiecz, E., 182, 183, 185, 186, 187, 188, *202*
Bellamy, W. D., 338, *342, 344, 346*
Benedetti, E. L., 161, *199*
Benson, A. A., 256, *285*, 297, 301, 303, 304, *342, 346*
Bentley, H. P., Jr., 26, *64*
Benton, D. P., 85, *121*
Berger, A., 363, 386, 399, *408*
Bernal, J. D., 247, *285*
Berry, J. F., 16, 26, *65, 66*
Beveridge, J. M. R., 24, 27, 31, 32, 34, 35, *65, 66*
Beychok, S., 177, 184, *199*, 362, 363, 374, 384, 386, 405, *409*
Bieri, J. G., 28, 33, 34, *67*
Bieth, R., 49, *67*
Biggins, J., 170, *201*, 295, 297, 305, 306, *345*
Billimoria, J. D., 27, 53, *64*

411

Bisalputra, T., 251, *288*
Bishop, N. I., 307, *342*
Blaisie, T. K., 294, 305, *342*
Blank, M. L., 18, 32, *68*
Blasie, J. K., 158, 161, *199*, 251, 253, *285*
Blaurock, A. E., 158, 161, *199*, 251, 253, *285*, 294, 305, *342*
Blomstrand, R., 23, 51, 53, *64*
Blout, E. R., 180, *199*
Bock, R. M., 165, *199*, 306, *343*
Boeyrers, J. C. A., 196, *200*
Bonomolo, A., 30, *64*
Booth, D. A., 56, *64*
Boscoe, G., 30, *67*
Bose, M., 330, *344*
Bourgès, M. C., 115, *122*, 147, 148, *201*
Boyle, E., 230, *231*
Bradley, D. F., 313, *342*
Bradlow, B. A., 23, 24, *64*
Brady, G. W., 101, *121*
Bragdon, J. I., 230, *231*
Branton, D., 172, 173, *199*, 251, 262, *285*
Brenner, S., 166, *199*
Brierley, G., 144, *200*, 296, 306, *343*, *344*
Brodie, A. F., 307, *343*
Brotz, M., 49, *66*
Brown, P. K., 314, 325, 327, *343*
Brunner, J. R., 56, *64*
Buchner, E. H., 369, *406*
Bungenberg de Jong, H. G., 363, 367, 373, 374, 401, *406*, *407*, *408*
Burge, R. E., 157, *200*
Butler, W. H., 335, *343*
Butler, W. L., 303, *345*
Byrne, P., 103, 104, *121*, 130, 131, 132, 133, 199

C

Cahill, G. F., Jr., 59, *67*
Caldwell, P. C., 358, *407*
Calvin, M., 291, 297, 301, 302, 303, 305, 310, 313, 335, 336, 338, *342*, *343*, *345*, *346*, 384, 401, *408*
Camiscoli, J. F., 335, *346*
Campbell, D. J., 10, *64*
Cardew, M. H., 328, 330, *343*
Cardi, E., 30, *64*
Carpenter, G. B., 101, *121*
Carroll, K. K., 34, *64*
Carter, H. E., 16, *64*

Carver, J. P., 180, *199*
Casley-Smith, J. R., 54, *64*
Cassim, J. Y., 185, 187, *199*
Casu, A., 11, 37, 39, 41, 55, *65*
Cavanaugh, G. M., 353, *407*
Century, B., 28, 29, 33, 34, *66*, *69*
Chagneux, R., 386, *407*
Chalazonitis, N., 386, *406*, *407*
Chance, B., 42, *67*, 329, *343*
Chandler, W. K., 387, 388, *407*
Chapman, A., 247, *286*
Chapman, D., 72, 98, 103, 104, *121*, 129, 130, 131, 132, 133, 134, 135, 136, 137, 138, 139, 140, 141, 142, 143, 145, 148, 151, 152, 153, 164, 174, 189, 198, *199*, *200*, 202, 204, *231*, 239, *285*, 326, 327, 339, *343*, *345*
Chappell, J. B., 256, *285*
Chargaff, E., 213, *232*
Cherry, R. J., 327, *343*
Chichibu, S., 383, *407*
Ciliento, G., 331, *343*
Clayton, R. K., 311, 313, 314, 333, 334, *343*
Clejan, L., 24, *68*
Clements, J. A., 267, *286*
Clowes, G. H. A., 399, *407*
Clunie, J. S., 91, 96, *121*, 247, 252, *286*
Cockbain, E. G., 229, *231*
Coggiola, E. L., 230, *232*
Colacicco, G., 272, *286*
Cole, K. S., 236, *286*, 348, 354, 355, 388, 390, 405, *407*
Coleman, N. T., 386, 404, *407*
Coleman, R., 157, 158, *200*
Collin, D. T., 133, *199*
Collins, F. D., 226, *231*
Colmano, G., 338, *346*
Commoner, B., 329, *343*
Condrea, E., 24, 26, *64*
Cope, F. W., 326, *343*
Corkill, J. M., 91, 96, *121*, 247, 252, *286*
Cornwell, D. G., 235, *285*
Corsini, F., 29, *64*, *66*
Coryell, C. S., 396, *409*
Craigee, R., 163, *199*
Crane, F. L., 166, *199*, 212, 213, *231*, 307, *343*
Criddle, R. S., 165, *199*, 306, *343*
Crowley, 2, J.4, 27, 30, 31, 32, *64*

Cumings, J. N., 27, 54, *64*
Curran, P. F., *408*
Curtis, H. J., 348, 354, 388, 390, *407*
Cuzner, M. L., 49, *64*
Cunningham, W. P., 166, *199*

D

Danielli, J. F., 233, 234, 235, 236, 237, 241, 253, 258, 268, 269, *286, 288*, 360, *407*
Danon, D., 26, 28, 34, 35, *65*
Das, M. L., 212, 213, *231*
Dauvillier, P., 198, *201*
Davidson, B., 178, 179, 180, 181, 182, 184, *200*
Davies, J. T., 244, *286*
Davis, K. M. C., 327, *343*
Davison, A. N., 49, *64*
Davison, P. F., 358, *407*
Davson, H., 233, 234, 235, 236, 269, *286*, 360, *407*
Dawson, R. M. C., 206, 212, 214, 218, 220, 222, 224, 226, *231, 232*, 250, *285*
Day, A. J., 54, *64*
Deamer, D. W., 235, *285*
Deffner, G. G., 358, *407*
de Gier, J., 26, 28, 29, 30, 31, 33, 34, *64, 66, 67, 68*, 138, 140, 141, 143, 151, 189, *199*, 204, *231*
Deguchi, T., 382, *408*
de Haas, G. H., 26, *68*, 145, *202*, 226, *231*, 269, *288*
del Castillo, J., 277, *285*
Demel, R. A., 144, 192, *199*, 266, 267, 277, *286, 288*, 357, *407*
de Pury, G. G., 226, *231*
Dervichian, D. G., 72, 115, *121, 122*, 147, *199*, 204, *231*, 276, *286*, 340, *343*
de Vries, A., 23, 24, 26, *64, 66*
Dewey, M. M., 158, 161, *199*, 251, 253, *285*, 294, 305, *342*
Dickerson, R. E., 242, *286*
DiMarco, N., 30, *66*
Dimopoullos, G. T., 26, *68*
Dingle, J. T., 265, 266, 267, *285, 286, 287*, 326, *343*
Dinning, J. S., 28, 34, *65*
Dobbins, J., 27, *65*
Dodge, J. T., 23, *64*
Dole, M., 399, *407*

Doneda, G., 30, *66*
Doscher, T., 91, *121*
Doty, P., 187, *201*, 226, *231*
Dougherty, R. M., 263, 266, *286*
Dourmashkin, R. R., 263, 266, *286*
Dowben, R. M., 56, *64*
Du Bois-Reymond, E., 347, *407*
Duggar, B. M., 339, *343*
Dutton, H. J., 339, *343*
Duysens, L. N. M., 310, 311, 313, 317, 333, *343*
Dyro, F. M., 370, 404, *406*

E

Eckles, N. E., 10, *64*
Eden, E. G., 22, 23, *67*
Eder, H. A., 230, *231, 232*
Eisenman, G., 373, *407*
Ekholm, J., 35, *65*
Elbers, P. F., 160, *200*, 357, *407*
Eley, D. D., 212, 226, *231*, 269, *286*, 322, 324, 325, 327, 328, 330, 333, 335, 337, *343*
Elfvin, L. G., 251, *288*
Elkes, J. J., 209, *231*
Elsbach, P., 54, *65*
Emerson, R., 312, *343*
Emmelot, P., 161, *199*
Erbland, J., 39, *66*
Erlanger, J., 348, *407*
Evans, W. H., 54, *65*
Eyring, H., 332, *346*

F

Falconer, J. D., 402, *406*
Farnard, R., 85, *121*
Farnsworth, P., 26, 28, 34, 35, *65*
Farquhar, J. W., 23, 27, 29, 31, 32, 34, 35, *65*
Fasman, G., 178, 179, 180, 181, 182, 184, *200*
Fatt, P., 383, *407*
Feldman, G. L., 17, 18, 32, 50, 56, *65, 68*
Feldman, L. S., 50, 56, *65*
Fernandez-Moran, H., 162, 166, *200*, 255, *286*, 292, *343*
Finean, J. B., 10, *65*, 103, *121*, 127, 156, 157, 158, 162, 163, *200*, 237, 238, 249, 269, *286*
Finke, S. R., 54, *65*

Firkin, B. G., 53, 54, *65*
Fitch, C. D., 28, 34, *65*
Fleischer, B., 10, 11, 37, 39, 41, 55, *65*, 165, *200*, 247, *286*
Fleischer, S., 6, 10, 11, 15, 17, 18, 23, 25, 36, 37, 39, 41, 53, 55, *65*, 67, *68*, 144, 153, 165, *200*, 204, 213, 225, 226, *231*, 247, *286*, 290, 296, 305, 306, 307, 330, *343*, *344*, *345*
Fluck, D. J., 142, 164, *199*
Folch, J., 20, *65*
Folch, Pi, J., 110, *121*
Foley, K. A., 331, *344*
Förster, R., 323, *344*
Förster, Th., 317, *344*
Frank, H. S., 375, *407*
Fraser, M. T., 337, *344*
Frazer, A. C., 209, *231*
Frazer, M. J., 210, *231*
Freeman, N. K., 19, *68*
French, C. S., 310, *345*
Frey, S., 249, *287*
Frey-Wyssling, A., 161, 167, *200*, *201*, 297, *344*
Friedel, G., 75, *121*
Furukawa, T., 382, *407*

G

Gaffron, H., 312, *344*
Gail, M. H., 187, *202*
Gaines, G. L., 338, *342*, *344*, *346*
Galli, C., 18, 32, 48, 60, 61, *68*
Gallot, B., 85, 86, 87, 91, 93, 97, 98, 119, *121*, *122*
Gammack, D. B., 226, *231*, 247, 254, *286*
Gandal, C. P., 27, *68*
Gasser, H. S., 348, *407*
Geiduschek, E. P., 375, 378, *407*
Gellhorn, A., 26, 28, 34, 35, 54, *65*, *66*
Gent, W. L. G., 247, *286*
Gerard, R. W., 387, *408*
Geren, B. B., 159, *200*, 357, *407*
Gibbons, I. R., 314, 325, 327, *343*
Gibbs, J. H., 187, *200*
Gjone, E., 26, *65*
Glaeser, R. M., 160, *200*
Glaser, A., 28, 34, *65*
Glassman, H. N., 28, *66*
Glauert, A. M., 165, *200*, 245, 247, 248, 249, 250, 252, 253, 266, 281, *286*, *287*
Glomset, J. A., 27, *65*
Glover, J., 10, *64*
Goedheer, J. C., 303, 337, *344*
Gofman, J. W., 163, *200*
Goldstein, D. A., 254, *286*
Gompel, C., 166, *200*
Goodman, J. F., 91, 96, *121*, 247, 252, *286*
Goodman, S. J., 378, 386, *409*
Goodwin, T. W., 301, 305, *344*
Gorchein, A., 184, *202*
Gordon, A., 179, 181, 182, 183, 184, 186, 187, 188, 189, *202*
Gordon, S., 49, *66*
Gorter, E., 234, 235, *286*
Gorter, H. F., 147, *200*
Gortner, R. A., 378, *407*
Gould, R. G., 10, *64*, *65*
Green, C., 10, *64*
Green, D. E., 10, *65*, 153, 165, *199*, *200*, 204, 213, 225, 226, *231*, 256, *286*, *287*, 290, 296, 304, 306, 307, 329, 330, 337, *343*, *344*
Green, D. K., 153, *199*
Green, W. G., 157, 158, *200*
Greenberg, L. D., 28, 34, *65*
Greenfield, N., 178, 179, 180, 181, 182, 184, *200*
Greenwood, F. L., 16, *64*
Gregor, H. P., 254, *285*
Gregson, N. A., 49, *64*, 247, *286*
Grendel, F., 147, *200*, 234, 235, *286*
Grey, C. T., 307, *343*
Grimmer, G., 28, 34, *65*, *68*
Grourke, M. J., 187, *200*
Grundfest, H., 382, *408*
Guisti, P., 331, *343*
Gulik-Krzywicki, T., 74, 84, 88, 92, 94, 103, 106, 107, 110, 111, 112, 113, 117, 119, 120, 121, *122*
Gulyaev, B. A., 338, *345*
Gurd, F. R. N., 230, *231*
Guzzo, A. V., 333, *346*

H

Haah, E. D., 213, *231*
Habeeb, A. F. S. A., 229, *231*
Hagerman, J. S., 10, *65*
Hagins, W. A., 314, 325, *344*

Hagiwara, S., 383, *407*
Haldi, J. A., 378, *407*
Hama, K., 357, *407*
Hamaguchi, K., 375, 378, *407*
Hanahan, D. H., 23, *64*
Hanahan, D. J., 24, 26, 29, 30, 31, 32, 35, *65*, *68*, *69*, 229, *232*
Hanai, T., 237, 247, 269, 274, 275, *286*
Handa, N., 56, *65*, *69*
Handa, S., 56, *65*, *69*
Handler, P., 360, *410*
Harbury, H. A., 331, *344*
Harkins, W. D., 101, *122*
Harris, R. J. C., 263, 266, *286*
Harrison, A., 251, *288*
Hartley, G. S., 115, *122*, 252, *286*
Hartshorne, N. H., 292, *344*
Harvey, C. C., 28, 29, 33, 34, *66*, *69*
Harvey, E. N., 235, *286*
Haslewood, G. A. D., 258, *286*
Hauser, H., 212, 214, 218, 220, 224, *231*
Havel, R. J., 230, *232*
Hawkins, J. K., 284, *286*
Haydon, D. A., 205, 228, *231*, 237, 247, 250, 258, 259, 262, 267, 269, 274, 275, 283, *286*, *287*, *288*
Hayes, T., 160, *200*
Hayes, T. L., 163, *200*
Hays, R. M., 255, *287*
Healy, T. W., 268, *287*
Hecht, S., 314, *344*
Hechter, O., 239, 241, 242, 244, 255, *287*
Hedge, D. G., 212, 226, *231*, 269, *286*, 337, *343*
Heemskerk, C. H. T., 26, *65*
Heick, H. M. C., 55, *65*
Heilbrunn, L. V., 383, *407*
Helfferich, F., 363, 372, 395, *407*
Heller, D., 8, 16, 18, 32, 50, 56, 60, *67*, *68*
Hemmington, N., 222, 224, 226, *231*
Henninger, M. D., 307, *343*
Hill, A. V., 348, 385, 386, *406*, *407*
Hill, J. G., 24, 27, 31, 32, 34, 35, *65*, *66*
Hille, B., 283, *287*
Hillier, J., 160, *200*
Hills, G. J., 249, *287*
Höber, R., 348, 378, 384, *407*
Hochstrasser, R. M., 321, *344*
Hodge, A. J., 292, 297, 301, *344*
Hodgkin, A. L., 348, 349, 351, 357, 358,

360, 364, 365, 366, 371, 372, 387, 388, 390, 395, 405, *406*, *407*, *408*
Hoffman, D. A., 101, *122*
Hoffman, J. F., 160, *200*
Hoffman, W. F., 378, *407*
Hofmeister, F., 368, *408*
Hokin, L. E., 26, *66*
Hokin, M. R., 26, *66*
Hollocher, J. C., 329, *343*
Holman, R. T., 33, *66*
Holt, A. S., 301, 338, *344*
Horne, R. W., 166, *199*, 247, 249, 250, 251, 266, 281, *285*, *287*
Horwitt, M. K., 28, 29, 33, 34, *65*, *66*, *69*
Hoskam, E. G., 374, *406*
Hosoya, Y., 337, *344*, 382, *407*
House, H. D., 197, *200*
Houtsmuller, U. M. T., 28, 34, *68*, 145, 202, 269, *288*
Howarth, J. V., 385, 386, *406*, *407*
Howe, P. G., 85, *121*
Hrachovec, J. P., 54, *66*
Huang, C., 235, 273, 274, 277, *287*
Huang, C. H., 274, *288*
Hubbard, R., 310, *344*
Hughes, E. W., 101, *122*
Hulcher, F. H., 49, *66*
Hultin, H., 247, *286*
Husson, F., 75, 91, 92, 93, 95, 96, 97, 106, 108, 109, 118, 119, *122*, 157, *200*, 251, 252, 255, 261, 262, 263, 266, *287*
Huxley, A. F., 348, 365, 371, 372, 395, 405, *408*
Hyono, A., 337, *344*

I

Ingraham, J. L., 197, *200*
International Tables for X-ray Crystallography, 76, 77, *122*
Introzzi, P., 30, *66*
Irani, V. J., 27, 53, *64*
Irie, R., 56, *66*, *69*
Isenberg, I., 323, 331, *344*
Ives, D. J. G., 387, *408*
Iwanaga, M., 56, *66*, *69*

J

Jackson, C. M., 35, *65*
Jacobs, E. E., 301, 335, 338, *344*

Jacobs, M. H., 28, 66
Jahn, T. L., 326, *344*
James, A. T., 23, 26, 29, 34, 66
Janssen, E. T., 27, *65*
Janz, G. J., 387, *408*
Jennings, W. H., 303, 314, 325, *344, 345*
Jensen, L. H., 128, 129, *200, 202*
Jensen, W. N., 26, *69*
Jirgensons, B., 360, *408*
Johnston, P. V., 197, *200*
Jones, J. W., 24, 27, 30, 31, 32, *64*
Jones, K. W., 325, *343*
Jurtshuk, P. Jr., 226, *231*

K

Kamat, V. B., 138, 140, 141, 143, 151, 187, 189, *199, 202,* 204, *231*
Kamimura, M., 56, *69*
Kaplan, J. G., 210, *231*
Kaplan, N. O., 229, *232*
Kaplan, T. G., 337, *344*
Karaca, M., 24, 53, *66*
Karlson, U., 166, *201*
Karnovsky, M. L., 54, *66, 68*
Karreman, G., 329, *344*
Kasha, M., 321, *344, 345*
Katchalski, E., 363, 386, 399, *408*
Katchalsky, A., *408*
Kates, M., 23, 26, 29, 34, *66,* 226, *231*
Katz, B., 383, *407*
Katz, E., 313, 324
Katzman, R., 254, *285*
Kavanau, J. L., 206, *231,* 262, *287*
Kauzmann, W., 229, *231,* 367, *408*
Kayden, H. J., 54, *65*
Kearns, D. R., 322, 324, 330, 335, 336, *344*
Kennedy, R., 27, *65*
Kerridge, D., 249, *287*
Keynes, R. D., 358, *408*
Kidson, C., 54, *66*
Kies, M. W., 49, *66*
Kinsky, S. C., 266, 277, *286, 287, 288*
Kiso, N., 56, *69*
Klenk, E., 56, *66*
Klibansky, C., 23, *66*
Klouwen, H., 296, 306, *343, 344*
Knudson, A., 48, 60, 61, *68*
Kobatake, Y., 379, 391, 394, 396, *409*

Koechlin, B. A., 358, *408*
Kögl, F., 29, *66*
Korey, S. R., 49, *66*
Korn, E. D., 10, *66,* 152, 162, 163, 164, 165, *200*
Kowalksy, A., 192, 194, *200*
Kraut, J., 129, *200*
Kritchevsky, G., 8, 11, 16, 17, 18, 19, 20, 32, 37, 39, 41, 48, 50, 55, 58, 60, 61, *65, 67, 68*
Kromhour, R., 301, 338, *344*
Kropf, A., 310, *344*
Kruyt, H. R., 208, *232,* 348, *408*
Kuckmak, M., 24, *68*
Kuksis, A., 24, 27, 31, 32, 34, 35, *65, 66*
Kummerow, F. A., 28, 33, 34, *69*
Kuriyama, S., 337, *344*
Kunkel, H. G., 230, *232*
Kyriakides, E. C., 60, *64*

L

Laatsch, R. H., 49, *66*
Labes, M. M., 330, *344*
Ladbrooke, B. D., 138, 139, 140, 141, 143, 148, 149, 150, *199, 200*
Lamborg, M. F., 229, *232*
La Mer, V. K., 268, *287*
Lands, W. E. M., 198, *200*
Lapous, D., 198, *201*
Larkin, J., 378, *407*
Larsson, K., 109, 114, *122,* 129, *200*
Lauenstein, K., 56, *66*
Leaf, A., 255, *287*
Leathes, J. B., 144, *200*
Lee, J., 23, 24, *64*
Lee, W. W., 81, 82, *122*
Lees, M., 20, *65,* 110, *121*
Lehninger, A. L., 290, 296, 314, *345*
Lenard, J., 181, 182, 183, 185, 186, 188, 189, *200*
Lerman, L., 382, 391, 397, *409, 410*
Leslie, R. B., 328, 333, 339, *343, 345*
Lester, R. L., 296, 307, 329, *344, 345*
Leveille, G. A., 26, *66*
Levene, R. J., *199*
Levis, G. M., 53, 54, *66*
Lewis, P. R., 358, *408*
Lichtenthaler, H. K., 297, 305, *345*
Lieber, E., 8, 16, 18, 32, 50, *67, 68*
Liebman, P. A., *345*

AUTHOR INDEX

Lietman, P., 31, 60, *64*
Limbrick, A. R., 157, 158, *200*
Lindemann, B., 254, *287*
Lindgren, F. T., 163, *200*
Ling, G. N., 348, 387, *408*
Litvin, F. F., 338, *345*
Livingston, R., 317, *346*
Locke, M., 263, *287*
Loewenstein, W. R., 261, *287*
Lovern, J. A., 203, *232*
Lovtrup, S., 39, *66*
Lucy, J. A., 165, *200*, 234, 245, 246, 247, 248, 249, 250, 252, 253, 255, 256, 259, 260, 261, 262, 263, 265, 266, 267, 276, 281, *285, 286, 287*, 326, *343*
Lumry, R., 301, 310, 333, *345*
Lundegårdh, H., 310, *345*
Luxoro, M., 353, 360, *409*
Luzzati, V., 72, 74, 75, 83, 84, 85, 86, 88, 90, 91, 92, 93, 95, 96, 97, 101, 102, 103, 106, 107, 109, 110, 111, 112, 113, 115, 117, 118, 119, 120, 121, *122*, 157, *200*, 251, 252, 253, 255, 261, 262, 263, 266, *287*
Lyman, R. L., 34, *66*
Lynch, V. H., 310, *345*
Lyons, L. E., 322, *345*

M

Mabis, A. J., 129, *200*
McBain, J. W., 81, 82, 91, 101, *122*, 370, 378, *408*
McConnell, D. G., 256, *287*, 305, 307, *343*
McConnell, H. M., 196, *200*
McCree, K. J., 335, 338, *345*
McElroy, F. A., 55, *66*
McGlynn, S. P., 323, *345*
Macheboeuf, M., 147, *199*
MacInnes, D. A., 387, *408*
Maclagan, N. F., 27, 53, *64*
MacLennan, D. H., 225, *232*, 256, *287*
McRae, E. G., 321, *345*
McSwain, B. D., *342*
Maddy, A. H., 174, *200*, 228, *232*, 277, *287*
Madeley, J. R., 226, *231*
Makita, A., 56, *69*
Malcolm, B. R., 174, *200*, 228, *232*, 270, *287*

Malhotra, S. K., 251, *287*
Mammon, Z., 24, 26, *64*
Mandel, P., 49, *67*
Manfredi, G., 29, *64*, *66*
Manning, W. M., 339, *343*
Mantzos, J. D., 53, 54, *66*
Marcus, A. J., 51, 53, *66*
Maricic, S., 328, *345*
Marinetti, G. V., 16, 19, 20, 22, 23, 26 39, *66, 67, 68*
Markham, R., 249, *287*
Marks, P. A., 54, *66*
Marr, A. G., 197, *200*
Marrian, G. F., 258, *286*
Marsden, S. S., 91, *122*
Martell, A. E., 384, 401, *408*
Martin, D. B., 26, *67*
Matalon, R., 153, *200*, 209, 218, 225, *232*
Matsumoto, M., 26, 56, *66, 69*
Mattoon, R. W., 101, *122*
Maupin, B., 26, *67*
Mazanowska, A. M., 225, *232*
Meduri, D., 30, *66*
Mednieks, M., 182, 183, 185, 186, 187, 188, *202*
Meek, E. S., 303, 317, *342*
Meites, L., 384, *408*
Mel, H., 160, *200*
Menke, W., 297, 310, 340, *345*
Meves, H., 364, 366, *406*
Michaeli, I., 384, 399, 400, 402, *408*
Middleton, E., Jr., 54, *65*
Miller, L., 230, *232*
Miller, N., 277, 278, 279, 282, 283, *285*
Minari, O., 230, *232*
Miras, C. J., 53, 54, *66*
Mitchell, C., 23, *64*
Mitchell, P., 255, *287*
Mohrhauer, H., 33, *66*
Monsen, E. R., 34, *66*
Montfoort, A., 28, 34, *68*
Moody, M. F., 293, 294, 305, 326, *345*
Moon, H. D., 28, 34, *65*
Moor, H., 166, 167, 168, 170, 172, 173, *199, 200, 201*, 291, 297, *345*
Moore, J. W., 355, 364, 382, *407, 408*
Morgan, H. R., 26, *68*
Morrison, A., 151, *199*
Mueller, G. C., 257, *287*
Mueller, P., 153, *201*, 273, 276, *287*

Mueller, P. S., 54, *65*
Mühlethaler, K., 167, 168, 170, 171, *200*, *201*, 291, 297, *345*
Mulder, E., 26, 27, 28, 34, *66*, *67*, *68*, 145, *202*, 269, *288*
Mulder, I., 26, 29, *64*, *66*
Müller, A., 74, 99, *122*
Mulliken, R. S., 323, *345*
Munck, A., 258, *287*
Murphy, J. R., 10, 11, *67*
Mustacchi, H., 75, 91, 92, 97, 118, 119, *122*
Myers, D. E., 212, *231*

N

Naka, K., 383, *407*
Nakajima, S., 382, *408*
Nakamura, Y., 382, *408*
Nakayama, F., 23, 51, 53, *64*
Narahashi, T., 364, 382, *408*
Nash, H. A., 271, *287*
Nazir, D., 21, *67*
Nelson, G. J., 19, 20, 22, 25, 27, 28, 35, *67*, *68*
Nelson, R. C., 335, *345*
Neuberger, A., 184, *202*, 225, *232*
Nichols, A. V., 230, *232*
Nicolaides, N., 18, 56, 60, *67*
Nilsson, I. M., 23, 51, *64*
Nilsson, S. E. G., 251, *287*, 294, *345*
Nishimura, S., 56, *69*
Norton, W. T., 49, *67*
Notario, A., 30, *66*, *67*
Nothman, M. M., 27, *67*
Nussbaum, J. L., 49, *67*
Nyberg, S. C., 130, *201*
Nyhan, W. L., 31, 60, *64*

O

O'Brien, J. S., 18, 32, *67*
Ohkubo, Y., 382, *408*
Oikawa, T., 348, 390, *408*
Okaya, Y., 129, *201*
Okey, R., 34, *66*
Oliveira, M. M., 26, *67*
Olson, R. A., 303, *345*
O'Neal, R. M., 54, *68*
Orchen, M., 49, *66*
Orii, T., 230, *232*
Orlando, J. M., 327, 328, *346*
Orlando, R. A., 327, 328, *346*
Osimi, Z., 23, *66*
Oster, G., 292, *345*
Ostwald, R., 33, *67*
Overbeek, J. Th. G., 363, 367, 374, 388, *408*
Owens, N. F., 131, 132, 145, *199*, 239, *285*, 326, *343*

P

Packer, L., 290, *345*
Padberg, G., 56, *66*
Paganelli, C. V., 254, *288*
Palade, G. E., 159, *201*
Pallmann, H., 388, *408*
Palmer, K. J., 103, 115, 117, *121*, *122*, 153, *201*, 208, 213, *232*
Paolucci, G., 29, *64*, *66*
Papahadjopoulos, D., 229, *232*
Pappajohn, D., 26, 32, 35, *65*
Park, R. B., 170, *201*, 295, 297, 305, 306, 312, *345*
Parker, F., 27, *65*
Parpart, A. K., 28, *66*
Parsons, D. F., 42, *67*, 130, *201*
Parsons, E. A., 27, *68*
Pawlowski, R., 271, *288*
Paysant, M., 26, *67*
Pascher, I., 128, *199*
Pearlstein, R. M., 313, 334, *342*
Pearse, A. G., 305, 326, *345*
Peller, L., 374, 402, 405, *408*
Penkett, S. A., 138, 140, 141, 143, 151, 152, 189, *199*, 204, *231*, 239, *285*
Pennell, R. B., 21, *67*
Perutz, M. F., 189, *201*, 259, *288*
Peters, R. A., 260, *288*
Pethica, B. A., 205, *232*, 239, 247, 262, 268, 272, 275, 279, *285*, *288*
Phillips, G. B., 26, 27, 54, *65*, *67*
Philpott, D. E., 56, *64*
Pierce, L. E., 26, *69*
Pifat, G., 328, *345*
Pirenne, N. H., 314, *344*
Pittman, J. G., 26, *67*
Plack, P. A., 28, 33, 34, *67*
Polonovski, J., 26, *67*
Pon, N. G., 297, 305, 312, *345*
Ponder, E., 22, *67*, 154, *201*
Porte, D., 230, *232*

AUTHOR INDEX

Porter, K. R., 159, *201*
Postma, G., 369, *406*
Pravdic, G., 328, *345*
Privett, O. S., 18, 32, *68*
Proger, S., 27, *67*
Puddington, I. E., 85, *121*
Pullman, A., 323, *345*
Pullman, B., 323, *345*
Putseiko, E., 335, *346*
Pysh, E. S., 179, *201*

R

Rabinowitch, E., 301, 303, 335, 338, *344, 345*
Raderecht, H. J., 26, *67*
Raper, J. H., 247, *286*
Rapoport, S. M., 26, *67*
Raulin, J., 198, *201*
Rauth, J. W., 378, *407*
Ray, A., 333, *346*
Redfearn, E. R., 307, *346*
Redwood, W. R., 274, 275, *286*
Reed, C. F., 22, 23, 29, *67*, 69
Rees, K. R., 265, *285*
Reich, M., 212, *232*
Reid, M. E., 28, 33, 34, *67*
Reiss-Husson, F., 72, 92, 93, 101, 102, 104, 105, 106, 107, 114, 117, *122*, 253, *287*
Renold, A. E., 59, *67*
Rerat, A., 198, *201*
Rezza, E., 30, *64*
Richardson, S., 247, *286*
Ricotti, V., 30, *67*
Rideal, E. K., 266, *288*
Riemersma, J. C., 163, 164, *201*
Riordan, D. F., 197, *200*
Rivas, E., 74, 92, 103, 106, 110, 111, 112, 113, 117, 119, 120, 121, *122*
Robertson, J. D., 159, 160, 161, 162, 163, 168, 188, *201*, 251, *288*, 357, *409*
Robinson, C., 185, 187, *201*
Robinson, G. W., 316, *346*
Robinson, R. A., 395, *409*
Rodrigues, A., 277, *285*
Roelofsen, B., 26, 31, *67*
Roheim, P. S., 230, *232*
Rojas, E., 360, *409*
Romero, C. A., 277, *285*
Rosenberg, B., 327, 328, 335, *346*

Roots, B. I., 197, *200*
Rosa, C. G., 251, *288*
Rosenzweig, A., 26, 54, *67*
Rouser, G., 6, 8, 11, 15, 16, 17, 18, 19, 20, 21, 22, 23, 24, 25, 32, 36, 37, 39, 41, 48, 50, 53, 55, 56, 58, 60, 61, *65*, *66, 67, 68*
Rowe, C. E., 27, *68*
Ruarte, A., 353, *409*
Rubenstein, R., 23, 24, *64*
Rudin, D. O., 153, *201*, 273, 276, *287*
Rumsby, M. G., 249, *286*
Ruska, C., 166, 173, *201*
Ruska, H., 166, 173, *201*
Rutiloni, C., 30, *64*

S

Safier, L. B., 51, 53, *66*
Sakagami, T., 230, *232*
Salem, L., 129, *201*, 204, 226, 228, *232*
Salsbury, N. J., 134, 135, 198, *199*
Salvioli, G. P., 29, *64, 66*
Sanchez, V., 277, *285*
Sarkar, P. K., 187, *201*
Sasaoka, T., 382, *407*
Sauberlich, H. E., 26, *66*
Saunders, L., 226, *231*
Sauer, K., 303, 335, *346*
Sbarra, A. J., 54, *68*
Scanu, A., 182, *201*
Schellman, C., 360, 362, *409*
Schellman, J. A., 360, 362, *409*
Schmidt, W. J., 293, *346*
Schmitt, F. O., 103, 115, 117, *121, 122*, 153, *201*, 208, 213, *232*, 357, *407*
Schölzel, E., 26, *67*
Schrader, G. T., 26, *68*
Schulman, J. H., 145, 153, *200, 201*, 209, 210, 218, 225, 226, *231, 232*, 266, *288*, 337, *344*
Schultze, M., 293, *346*
Scott, W. R., 382, *408*
Segal, J., 348, *409*
Segerman, E., 262, *288*
Sehr, R., 330, *344*
Sekuzu, I., 226, *231*
Sela, M., 363, 386, 399, *408*
Senft, J., 370, 404, *406*
Sessa, G., 266, 283, *288*

Seufert, W. D., 276, *288*
Shah, D. O., 145, *201*
Shanes, A. M., 268, *288*
Shannon, A., 33, *67*
Shaw, T. I., 348, 349, 351, 364, *406*
Shechter, E., 180, *199*
Shellman, C., 177, *201*
Shellman, J. A., 177, *201*
Sherwood, H. K., 335, 338, *342*
Shimamura, M., 364, *409*
Shipley, G. G., 103, 104, *121*, 130, 131, 132, 133, *199*
Shlaer, S., 314, *344*
Shockley, J. W., 26, *66*
Shortman, R. C., 27, 54, *64*
Shotlander, V., 265, *285*
Siakotos, A. N., 17, 18, 21, 22, 24, 25, 32, 54, *68*
Silman, H. I., 363, 386, 399, *408*
Simon, G., 8, 15, 16, 19, 20, 50, *68*
Sinclair, W. B., 378, *407*
Sineshchekov, A., 338, *345*
Singer, I., 348, 349, 350, 351, 352, 353, 355, 356, 360, 361, 362, 363, 364, 365, 367, 368, 369, 370, 371, 372, 373, 377, 378, 379, 380, 381, 382, 383, 384, 385, 386, 387, 388, 389, 390, 391, 393, 394, 395, 396, 397, 403, 404, *409*, *410*
Singer, S. J., 181, 182, 183, 185, 186, 188, 189, *200*
Sjöstrand, F. S., 159, 160, 161, 166, *201*, 245, 251, 254, 255, *288*, 291, 292, 294, 295, *346*
Skou, J. C., 265, 267, *288*
Skoulios, A. E., 75, 83, 84, 85, 86, 87, 89, 90, 91, 92, 93, 97, 98, 99, 100, 112, 115, 118, 119, *121*, *122*, *123*
Skrbic, T., 27, *64*
Slauterback, D. B., 296, *343*
Sloane-Stanley, G. H., 20, *65*
Sloviter, H. A., 26, *68*
Small, D. M., 115, *122*, 147, 148, *201*
Smith, E. L., 360, *410*
Smith, L. M., 19, *68*
Snart, R. S., 327, 332, 335, *343*, *346*
Sollner, K., 388, *409*
Solomon, A. K., 254, *286*, *287*, *288*
Soule, D. W., 26, *68*
Spegt, P. A., 85, 87, 88, 89, 90, 91, 93, 94, 115, *122*

Speigler, K. S., 396, *409*
Spiegelman, S., 358, *409*
Spikes, J. D., 301, 310, 333, *345*
Spitzer, H. L., 60, *64*
Spyropoulos, C. S., 348, 363, 385, 386, 387, 390, 397, 399, 403, 404, *408*, *409*
Standish, M. M., 277, 278, 279, 282, 283, *285*, 340, *342*
Stanley, G. H. S., 110, *121*
Stauff, J., 91, *122*
Stearns, R. S., 101, *122*
Steel, R. H., 329, *344*
Stefanini, M., 24, 53, *66*
Steim, J., 181, 183, 186, 187, *201*
Stein, O., 27, *68*
Stein, W. D., 236, 237, 241, 253, *288*
Stein, Y., 27, *68*
Steinbach, H. B., 358, *409*
Steiner, R. F., 377, *409*
Steinhardt, J., 362, 363, 374, 384, 386, 405, *409*
Steinmann, E., 161, *200*, 297, *344*
Stewart, H. B., 55, *65*, *66*
Stewart, H. C., 209, *231*
Stoeckenius, W., 10, *65*, 104, 108, *122*, 153, 160, 163, 164, 165, *200*, *201*, 234, 251, *288*
Stokes, R. H., 395, *409*
Stotz, E., 16, 39, *66*
Straumanis, M. E., 360, *408*
Stuart, A., 292, *344*
Styrer, L., 329, *346*
Sundaralingam, M., 128, 129, *202*
Suzuki, M., 54, *68*
Suzuki, S., 56, *69*
Svennerholm, L., 39, *66*
Swartz, M. N., 229, *232*
Swisher, S. N., 22, 23, *67*
Szarkowski, J. W., 297, *345*
Szeinberg, A., 24, *68*
Szent-Györgyi, A., 290, 323, 324, 329, 331, *344*, *346*

T

Tait, G. H., 225, *232*
Takenaka, T., 349, 351, 352, 353, 357, 360, 361, 362, 364, 365, 366, 367, 368, 369, 370, 372, 377, 378, 379, 384, 385, 388, 390, 391, 404, *409*
Tanaka, S., 26, *68*

Tardieu, A., 84, 88, 94, 107, *122*
Tasaki, I., 348, 349, 350, 351, 352, 353, 355, 356, 360, 361, 362, 363, 364, 365, 366, 367, 368, 369, 370, 371, 372, 373, 377, 378, 379, 380, 381, 382, 383, 384, 385, 387, 388, 389, 390, 391, 393, 394, 395, 396, 397, 403, 404, *408, 409, 410*
Taylor, C. B., 10, *64*
Taylor, E. W., 358, *401*
Taylor, F. H., 205, 228, *231*
Taylor, J., 237, 247, 250, 259, 262, 267, 269, 274, 275, 283, *286, 287*
Taylor, J. L., 258, *288*
Teale, F. J. W., 329, 333, *346*, 387, *410*
Teorell, T., 348, 362, 363, 385, 390, 397, *408, 409, 410*
Terenin, A. N., 335, *346*
Thompson, G. A., Jr., 35, *68*
Thompson, H. P., 159, *201*
Thompson, T. E., 235, 273, 274, 277, *287, 288*
Thompson, W., 42, *67*, 220, *231, 232*
Thomson, T. J., 323, *344*
Tien, H. T., 153, *201*, 273, 274, 276, *287, 288*
Tisdale, H., 165, *199*, 306, *343*
Tjaden, M., 27, *65*
Tobias, C., 160, *200*
Tobias, J. M., 271, 272, *287, 288*, 360, *410*
Tollin, G., 322, 333, 336, *346*
Treloar, L. R. G., 118, *123*
Trurnit, H. J., 338, *346*
Tsou, K-C., 251, *288*
Tsujimoto, H. Y., *342*
Tsujo, K., 337, *344*
Turner, D. A., 26, 31, 60, *64, 65*
Turner, J. C., 23, 27, *68*
Tweet, A. G., 338, *342, 344, 346*
Tzagoloff, A., 225, *232, 256, 286, 287*, 296, 304, 337, *344*

U

Ulbricht, W., 364, *408*
Ullman, H. L., 51, 53, *66*
Ullmer, D. D., 184, *202*
Urry, D. A., 332, *346*
Urry, D. W., 182, 183, 185, 186, 187, 188, *202*, 332, *346*
Urukawa, N., 382, *408*

V

Vallee, B. L., 184, *202*
van Deenen, L. L. M., 26, 27, 28, 29, 30, 31, 33, 34, *64, 65, 66, 67, 68*, 127, 145, 191, *202*, 226, *231*, 266, 267, 269, 277, *286, 288*
van den Berg, D., 26, 30, *68*
van den Berg, H. J., 273, *288*
van den Berg, J. W. O., 27, *67*
Vandenheuvel, F. A., 10, 12, *68*, 127, 144, 196, *202*, 238, 240, 269, *288*
van Gastel, C., 26, 30, *64, 68*
van Iterson, W., 165, *202*
van Niel, C. B., 290, 313, *346*
van Senden, K. G., 26, 28, 33, 34, *64*
van Zutphen, H., 277, *288*
Vatter, A. E., 338, *344*
Vaughan, M., 26, *67*
Verloop, M. C., 30, *64*
Ververgaert, P. H. J. T., 357, *407*
Vikrot, O., 27, *68*
Villegas, R., 254, *288*
Vincent, J. M., 98, 99, 100, 112, 119, *123*
Vöet, A., 369, 370, *410*
Voigt, K. D., 28, 34, *65, 68*
Vold, M. J., 85, *123*
Vold, R. D., 85, 91, 121, *123*
von Bruins, E. M., 369, *410*
von Hippel, P. H., 374, 378, 386, *410*
von Sydow, E., 74, *123*
Vorbeck, M. K., 19, 20, *68*

W

Wainio, W. W., 212, *232*
Wald, G., 310, 314, 325, 327, *343, 346*
Waldner, H., 167, *201*
Walker, B. L., 28, 33, *69*
Walker, D. A., 131, 132, 145, *199*, 239, 285, 326, *343*
Wallach, D. F. H., 54, *66*, 174, 176, 179, 181, 182, 183, 184, 186, 187, 188, 189, *202*
Warner, D. T., 239, *288*
Watanabe, A., 349, 350, 355, 356, 364, 371, 372, 373, 379, 381, 382, 383, 384, 388, 389, 390, 391, 393, 394, 395, 396, 397, 404, *409, 410*
Watkins, J. C., 243, 244, 277, 278, 279, 282, *285, 288*, 340, *342*
Watson, W. C., 26, 28, 34, 35, *69*
Watson, W. F., 317, *346*

Watts, R. M., 26, 32, 35, *65*
Ways, P., 24, 26, 27, 29, 30, 31, 32, 54, *64, 67, 69*
Weber, G., 329, *346*, 387, *410*
Weber, M. M., 307, *343*
Weier, T. E., 251, *288*, 297, 303, *346*
Weisman, R. A., 162, 164, *200*
Weissman, G., 277, 278, 279, 282, 283, *285, 288*
Weissman, S., 283, *288*
Wen, W-Y., 375, *407*
Westcott, W. C., 153, *201*, 273, 276, *287*
Westerman, M. P., 26, *69*
Wetlaufer, D. B., 267, *285*
Whatley, M., 16, *68*
Wheeldon, L., 235, 273, 274, *287*
White, A., 360, *410*
Whittaker, V. P., 166, *202*, 250, *287*
Whittam, R., 236, 255, 282, *288*
Whittington, S. G., 137, *202*
Wigglesworth, V. B., 162, *202*
Williams, G., 42, *67*
Williams, G. R., 329, *343*
Williams, J. H., 24, *69*
Williams, R. H., 27, *65*
Williams, R. M., 138, 139, 140, 141, 143, 148, *199, 200*
Williams, W. J., 53, 54, *65*
Willis, M. R., 325, *343*
Willmer, E. N., 256, 258, 282, *288*

Wilson, D., 42, *67*
Wilson, K. M., 267, *288*
Winter, S. S., 366, *410*
Witter, R. F., 24, *69*
Witting, L. A., 33, 34, *68*
Witting, L. F., 28, 34, *65*
Wohl, K., 312, *344*
Wolken, J. J., 301, 305, 310, 339, *346*
Wong, K-Y., 374, 378, 386, *410*
Wood, P. M., 307, *343*
Worthington, C. R., 158, 161, *199*, 251, 253, *285*, 294, 305, 342
Wosilait, W. D., 307, *346*
Wright, P., 378, *407*

Y

Yamagishi, S., 357, *409*
Yamakawa, T., 56, *65, 66, 68*
Yamomoto, A., 17, 18, 20, 32, 48, 60, 61, *68*
Yang, T. J., 185, 187, *199*
Yin, E. T., 229, *232*
Yokoyama, S., 56, *69*
Young, J. Z., 348, 356, *410*

Z

Zahler, P. H., 174, 176, 181, 184, 186, 187, 188, 189, *202*
Zaidman, J., 24, *68*
Zanetti, A., 30, *67*

Subject Index

A

Acceptors
  in charge transfer complexes, 323
  in exciton energy transfer, 317
Acetylcholine, action on membranes, 243
Action potential (*see also* Excitability)
  definition of, 389
  dependence on internal KCl, 364-5
  effect of enzymes on, 361
  effect of pH on, 362-3
  effect of pressure on, 385
  effect of sodium ion on, 372-3
  effect of tetrodotoxin on, 382-3
  heat changes during conduction of, 386, 403
  inducement by light, 386
  magnitude of, 390
  overshoot, 371-2, 390
  prolongation of, 404
  recording of, 352-3
  restoration by anions, 367-71
  survival time of, 367-8
Active patches during excitation, 402
Adsorption
  of counter ions in double layers, 207-8
  of phospholipase D at lipid-water interfaces, 223
  of proteins, at lipid monolayers, 209, 212, 226
    at oil-water interfaces, 209-211
Alcohols, effect of on phospholipid dispersions, 283
γ-Aminobutyric acid, action of on membranes, 243
Amnion cells, phospholipids of, 54
Anaesthetics, in monolayers, 265, 267
  in phospholipid dispersions, 283-4
Analysis of lipids, 16-20
Animal cells, lipids of, 7, 55
Anionic sites on axon membrane, 396-400

Anions,
  activation of phospholipase D by, 222
  diffusion of in phospholipid dispersions, 278-82
  effect of on excitability, 367-71, 378
  inhibition of phospholipase C by, 219
  initiation of phospholipase B by, 214-18
  inhibition of enzymatic triphosphoinositide hydrolysis by, 221
  lyotropic series of, 368-70, 374
Anisotropy in light spectroscopy, 292
Antibiotics
  in bilayers, 277
  in monolayers, 266
  in phospholipid dispersions, 283
Antigen-antibody reactions on bilayers, 277
Antioxidant, use of in lipid storage, 42
Aorta, phospholipids of, 50
Area S, definition of, 77-8
  dependence upon polar group concentration, 93, 97, 107, 110, 114, 119
  of anhydrous sodium soaps, 85-6, 89
  of lipid-water systems, 104-7, 110-112
  of soap micelles, 102-3
  of soap-water phases, 91, 95, 97-99
Autolysis of lipids in postmortem degradation, 13
Axon (*see also* Excitability)
  anatomy of, 356-7
  biochemistry of, 358-62
  cation flux in, 391-5
  current and ion flux in, (*see also* Voltage clamp), 394-5
  dissection of, 349
  effect of salt concentration on, 363-7
  effect of temperature on, 385
  electrical stimulation of, 352
  electron microscopy of, 357-9

Axon contd.
  ion flux in, 389, 394, 395
  isotope flux in, 351, 352, 355, 356
  membrane
    capacitance of, 388
      charged groups of, 360, 361
      current, voltage clamp technique, 354
      impedance of, 354
      macromolecular model of, 361, 362, 366, 367, 396–400
      resistance of, 353, 388, 390–3
    perfusion of, 349–52
    resting potential, 387, 388
    sensitivity to internal pressure, 351, 385
  sources of, 349
Axoplasm, 356–9
  removal of, 350, 351, 359

B

Bacterial cells, lipids of, 7
Barium soaps, phases of, 89
Bilayers,
  as model membranes, 273–7, 338, 339
  dimension and electrical properties of, 273, 274
  effects of detergents on, 276, 277
  formation of, 273
  incorporation of cholesterol in, 275, 276
  of purified lecithin, 274
  permeability of, 274, 275
  pores in, 275, 276
Bimolecular leaflet,
  Danielli-Davson model, 234, 235
  depolarization of, 239–42
  effect of odorants on, 244
  -micelle equilibrium, 245, 246
  permeability of, 235, 236, 239–42
  pharmacological action on, 243
  phospholipid-cholesterol interactions in, 237–40
  pores in, 236, 237
  stability of, 242, 243
  steroids in, 257–8
Bioenergetics, historical development, 312–4
Birefringence, 292, 293
  of chloroplasts, 303
  of phospholipids, 133
  of rod outer segments, 293, 294
Bound water in phospholipids, 140
Bacillis subtilis membranes, ORD and CD parameters of, 183
Brain, gangliosides in, 57
  lipids, effect of acclimatization temperature on, 197
  lipid-water phases 105, 106, 109, 110
  nuclei, phospholipids of, 44
  phospholipids of, 18, 48, 49
  phospholipids, autolysis of, 14
  phospholipids, effect of age on, 48
  phospholipids, metabolic abnormalities of, 60

C

Caesium soaps, phases of, 97, 98
Cadium soaps, phases of, 89
Calcium,
  activation of phospholipase A by, 224
activation of phospholipase D by, 222–4
  effect of on phospholipid lamellae, 208
  soaps, phases of, 89, 94
Cardiolipid, formula of, 80
Cardiolipin (*see also* Diphosphatidyl glycerol)
  complexes with cytochrome c, 212, 213
β-Carotene, function and structure of, 309, 310
  in monolayers, 338
Carotenoids, conduction of, 327
  in micellar dispersions, 339
Cations,
  diffusion of in phospholipid dispersions, 227–82
  effect of on axon excitability, 369–72, 378–83
  exchange of in axon membrane, 382, 384, 397, 398
  flux of in excited axons, 391–5
  flux, measurement of, 355, 356, 389
  inhibition of phospholipase B by, 216–8
  inhibition of phospholipase D by, 223

SUBJECT INDEX

Cations *contd.*
  initiation of phospholipase C by, 219
  lyotropic series of, 369, 370, 373, 374
  maintenance of enzymic hydrolysis of triphosphoinositide by, 220
  permeability of membranes to, 254
  ratio of in axon excitation, 383, 384, 399–403
  stabilization of macromolecular membrane by, 384, 396
Cell membranes, electrical properties of, 236
Cells, lipid classes of, 7
Cephalin, (*see also* Phosphatidylethanolamine, phospholipids)
  reaction with histones, 213
Ceramide, 7
  aminoethylphosphonate, 8, 46, 49, 50
  polyhexosides, 8, 11, 53, 56, 60
Cerebrosides, 8, 80 (*see also* Glycolipids)
  abnormal metabolism of, 60
  occurrence of, 41, 48, 49, 53
Cerebroside-water system, lamella phase of, 105
Ceroid, 58
Characterization of lipids, 15
  of organelles and membranes, 15
Charge carrier generation, 321, 322
  migration and energy transfer, 321, 322
    in mitochondria, 329–33
  separation in photosynthesis, 313
  transfer complexes, formation of, 322–3
    in lamellar systems, 323–5
    in mitochondrial semiconduction, 330–2
    in photosynthetic conduction, 336
    of cytochrome *c*, 332
    of pyridine nucleotides, 331
    of quinones, 333
Chloranil, in protein semiconductivity, 327, 328
Chlorophyll, 309, 311
  in chloroplast lamellae, 301–3, 311
  in monolayers, 337, 338
Chloroplast lamellae, composition of, 305
  model of, 302, 303
Chloroplasts, birefringence of, 303
  conduction in, 334–6
  dichroism of, 303
  electron microscopy of, 161, 170–2, 297–301
  energy transfer in, 333, 334
  e.s.r. signals in, 334
  fluorescence polarization of, 303
  luminescence of, 334
  pigment arrangement in, 301–3
  structure of, 297–301
chloroplastin, micellar dispersion of, 339–40
Cholesterol-cephalin model membrane, 271
Cholesterol,
  effect of any lipid transitions, 151
  in bilayers, 275, 276
  in erythrocytes, N.M.R. spectra of, 192, 193
  in membranes, 10, 127, 151
  in monolayers, 144–7, 258, 259, 265, 269
  in myelin, 156, 158
  in phospholipid dispersions, 282, 283
  -lecithin myelin figures, 248–50
  occurrence of, 22, 37, 39, 62
  -phospholipid complexes, 10, 237–40
    interactions, 144–51
    interactions in water, 147–51
    -water system, D.S.C. studies of, 148–50
      N.M.R. studies of, 150–2
      phase diagram of, 147, 148
      X-ray spacings of, 148, 150
Choline, structure in phospholipid derivatives, 128–9
Circular Dichroism (C.D.), 176–89
  of membranes, 181–4
    effect of environment on, 187, 188
    interpretation of, 184–9
    phosphatide activity in, 186, 187
  of membrane models, 188, 189
  of peptides, 177
  of polypeptides, 178–80
  of proteins, 179, 181
Coagel phase-in brain lipids, 109, 110
  in soap water systems, 81, 97–101
Cockroach, lipid secretion and water uptake by, 293–5

Column chromatography,
  of lipids, 15–20
  of erythrocyte lipids, 23–5
  use of cellulose, 18, 19
  use of Sephadex, 17, 21
  use of silicic acid, 19
Conductance of axon membrane, 390, 391
  relationship with ion flux, 395, 396
Conduction band, 319, 321
β-Conformation in myelin, 175, 176
  ORD and CD parameters of, 178, 180
Conformation changes in axon membranes, 399–402
  of lipid paraffin chains, 73, 74, 118
    in anhydrous soaps, 82, 85, 89, 90
    in biological lipids, 103, 110–3, 115–7
    in soap-water systems, 97
    in the gel phase, 81, 99
  of protein, effect of acetyl choline on, 243
Counter ions, 207, 208
Crystalline lipid structures, nomenclature of, 75
Cubic phase of anhydrous soaps, 85
  of phosphatidylcholines, 139
  of soap-water systems, 91, 96
  of surfactants, 252
Cyclic photophosphorylation, 312
Cytochromes in mitochondrial electron transfer, 329
  structure of, 309, 311
Cytochrome c, complexes with phospholipids, 153, 212, 213
  charge transfer complexes of, 332
Cytochrome oxidase, phospholipid binding of, 256

D

Danielli-Davson membrane model, (see also Bimolecular leaflet), 234–7
Dehydration of lipids during negative staining, 250
Depolarization in membrane models, 239–42
Desoxycholate, action of on membranes, 194
Detergents (see also Soaps)
  action of on bilayers, 276
  formulae of, 79

Differential Scanning Calorimetry (D.S.C.)
  of lecithin-cholesterol water systems, 148–50
  of phospholipid-water systems, 139–42
Differential Thermal Analysis (D.T.A.)
  of phospholipids, 133, 136
Diffusion from phospholipid dispersions, 277–83
  in erythrocyte membranes, 282
Digalactosyl lipid, structure of, 308
1,2-Dilauroylphosphatidylethanolamine, structural features of, 129
Diphosphatidyl glycerol, 7, 8
  in membrane repeating units, 11
  occurrence of, 39, 43, 51, 62
Dispersion of phospholipids,
  action of anaesthetics and narcotics on, 283
  action of antibiotics on, 283
  action of steroids on, 282, 283
  diffusion of ions from, 277–82
  effect of dicetylphosphoric acid on, 278, 279, 281
  pore structure in, 279, 281
  structure of, 278–80
  swelling of, 277
Donors in charge transfer complexes, 323
  in exciton energy transfer, 317
Double cannulation, 349–51
Drugs, action of on monolayers, 267

E

Ehrlich ascites carcinoma membranes,
  ORD and CD parameters of, 181, 183
Electric birefringence, 292, 293
Electrical double layer,
  in lipid-protein interactions, 227, 228
  of lipid-water systems, 206–8
Electrical properties of cell membranes, 236
Electroendosmosis of lipids, 207
Electrodes, for measurement of action potentials, 352, 353
  for measurement of axon resistance, 353, 354
  for measurement of resting potential, 387
  for stimulation of axons, 352

Electron density of erythrocytes, 157, 158
Electron microscopy,
  fixation reactions, 161–5
  freeze etching, 167–74
  membrane changes during staining, 297, 300
  negative staining, 166–7
  of chloroplasts, 161, 170–2, 297–301
  of erythrocytes, 160, 161
  of lecithin cholesterol mixtures, 248–51
  of lipid-water systems, 104, 108
  of membranes, 159–74, 251, 252
  of microsomes, 38, 40
  of mitochondria, 38, 40
  of nuclei, 38, 40
  of phospholipids in water, 142
  of plasma membranes, 38, 40, 160
  of rod outer segment membranes, 294, 295
  of root tip cells, 172, 173
  of squid axon, 357, 358
  of yeast cells, 168–70
Electron Spin Resonance in chloroplasts, 334
Electron transfer in mitochondria, 314, 329, 330
Electronic transitions, 315, 316
Electrophoresis of lipids, 207
Electrophoretic mobility and phospholipase activity, 215–7, 219
  of oil droplets, effect of protein on, 210, 211
  relationship with ζ potential, 214
Electrostatic fields,
  at lipid-protein interfaces, 227
  at lipid-water interfaces, 206
Electrostatic interactions, between lipids and proteins, 204, 227–8
  in lipid lamellae, 205, 206
Endoplasmic reticulum, phospholipids of, 42, 50, 62
Endothelial cells, phospholipids of, 49, 50
Energy conversion in organelles, 291
  levels, 314, 315
  transfer and charge migration, 321, 322
    by excitons, 317–320

Förster mechanism, 316–318, 320
  in chlorophyll monolayers, 338
  in mitochondria, 328
  in photosynthesis, 312, 313
  in photosynthetic unit, 333, 334
  in rod outer segments, 314, 325
  rates, 320
  role of membranes in, 320, 321
Enzymatic esterification of phospholipids, 198
  postmortem degradation of lipids, 13–5
Enzyme activity,
  at oil–water interfaces, 210, 211
  dependence on ζ potential, 214, 215, 219, 224
Enzyme-substrate, complexes at lipid-water interfaces, 213, 214
  reactions on bilayers, 277
Enzymes in micellar membranes, 255, 256
Equivalent circuit model, 348, 363, 364, 371, 394, 395, 405
Erythrocyte composition, 21, 22
  fatty acids, 28–35
    effect of diet on, 34
    human, 29–32
    species variation of, 31–4
  glycolipids, 22, 55, 56
  lipid disorders, 60
  membranes, diffusion in, 282
    effect of desoxycholate on, 194
Erythrocyte membrane
  effect of lysolecithin on, 194
  effect of phospholipase C on, 191
  effect of temperature on, 192
  effect of urea on, 194
  electron microscopy of, 160
  hydration and structure of, 157, 158
  infrared spectra of, 174–6
  n.m.r. spectra of, 189–96
  O.R.D. and C.D. parameters of, 181, 183
  X-ray diffraction of, 157, 158
Erythrocyte permeability, 28
  phospholipids, 23–28
    exchange with plasma, 26–28
    human, 23–5
    species variations, 25
  plasmalogens, 35

Excitability
  effect of anions on, 367–71, 378
  effect of cations on, 371, 378–84
  effect of ionic strength on, 363–4
  effect of light on, 386
  effect of pressure on, 385
  effect of temperature on, 385, 386
  effect of tetrodotoxin on, 380–3
  in sodium-free media, 379–81
  interpretation of ionic effects, 374–8
  Ohm's Law characteristics of, 381, 382
  pH optimum of, 362, 363
  time dependence of, 375–8
Excitation, cation flux during, 391–5
  equivalent circuit model of, 348, 363, 364, 371, 394, 395, 405
  heat changes during, 403
  initiation of, 399–402
  impedance loss during, 390, 391
  macromolecular model of, 396–9, 405
  prolongation of, 404
  spread of, 402, 403
  termination of, 403, 404
  Two Stable States theory of, 399–404
Exciton bands, 318, 319
  coupling, 320
  migration, 319, 320

F

Fatty acids
  determination of, 17
  distribution in phospholipids, 127
  of erythrocytes, human, 28–31
    species variations, 29, 33, 34
  of erythrocyte phospholipids, 31, 32, 35
  of leukocyte glycolipids, 53
  of platelets, 51, 53
  unsaturation, and fluidity in membranes, 196
    and growth temperatures, 197
    and incorporation into phospholipids, 197, 198
Ferredoxin, in photosynthesis, 311
Flow birefringence, 292
Fluorescence, 315, 316
  in energy transfer systems, 317, 320
  of chlorophyll monolayers, 338
  of chloroplasts, 303
  polarization, 293
Flux-resistance product, 395, 396
Form anisotropy, 292
  birefringence, 292, 293
  dichroism, 292
Förster energy transfer, 316–8, 320
  in photosynthetic units, 333, 334
Freeze-etching, 167–74

G

Gangliosides, 8
  arrangement in membranes, 11
  in Tay Sachs disease, 60
  occurrence of, 22, 56, 57
Gaucher's disease, 58, 60
Gel phase,
  effect of cholesterol on, 149, 151
  in brain lipid, 109, 110
  in soap-water systems, 81, 97–101
  lamellar structure of, 98, 100
  order in, 99
  orthorhombic structure of, 99
Glomeruli, phospholipids of, 49
Glucose permeability of phospholipid dispersions, 283
Glutamic acid and membrane permeability, 243
Glycerol phospholipids,
  metabolic disorders of, 58
  postmortem degradation of, 13
  types of, 7
Glycerol, structure in phospholipid derivatives, 129
Glycolipids,
  arrangement in membranes, 11
  occurrence of, 7, 22, 53, 55–7, 61
  types of, 8, 56
Grana, structure of, 297–300

H

Halobium membranes, ORD parameters of, 183
Heart, endoplasmic reticulum, phospholipids of, 43
  lipids, autolysis of, 14
  mitochondria, electron micrographs of, 38, 40
    lipid content of, 39
    phospholipids of, 41

Heart, nuclei phospholipids of, 44
  phospholipids, 45, 52
Heat changes during excitation, 386, 403
α-Helix, in membranes, 182–9
  in myelin, 175
  in polypeptides, 178–80, 270
  in proteins, 179, 181, 270
Hexagonal phase,
  chain dependence of, 114
  equations of, 77
  in brain lipids, 109
  in mitochondrial lipids, 110, 111
  in lysolecithin-water systems, 104, 107
  of anhydrous soaps, 84, 88, 94
  of lipid-water systems, 104, 106–111, 114, 255
  of soap-water systems, 90, 91, 96–8, 101, 102
  type I, 91, 104
  type II, 91, 104, 114
Histone-cephalin complexes, 213
Homiothermic organisms, lipids of, 197
Hydration and semiconductivity of proteins, 327
Hydrocarbon chain, (see also Paraffin Chain)
  interactions in membranes, 12
  mobility, 135, 136
    effect of cholesterol on, 149–51
    in membranes, 196, 197, 261, 262
    in monolayers, 146, 147
  packing in phospholipids, 129, 130, 137, 138
Hydrophobic interactions between lipids and pigments, 311, 340, 341
β-Hydroxybutyrate dehydrogenase requirement for sonicated lecithin, 226

I

Imidazole pump and electron transport, 332
Impedance, changes during axon excitation, 390–3
  measurement of in axon membrane, 354, 388
Infrared spectroscopy
  of membranes, 174–6
  of phospholipids, 132, 133
Insects, lipid secretion and water uptake by, 263–5

Interdiffusion flux, 391, 394, 395
  relationship with axon conductance, 395, 396
Intrinsic anisotropy, 292
  birefringence, 292
  dichroism, 292
Ion fixation effects, 374, 375
Ion flux in axons, 389, 391–6
Ionic strength, effect of on excitability, 363–367
Isotope flux in axons, 351, 352, 355, 356, 389, 391–6

K

Kidney
  glycolipids of, 56
  phospholipids of, 45, 52
  endoplasmic reticulum, phospholipids of, 42
  mitochondria, lipid content of, 37, 39
  phospholipids of, 41

L

Lamellae, in chloroplasts, 295–301
  pigment arrangement in, 301–3
Lamellar organization, 75
Lamellar phase, dimensions in lipid-water systems, 105, 107, 109
  equation of, 77
  Lα high temperature form, 110, 119–21
  Lα low temperature form, 111, 119–21
  Lβ form, 112, 120, 121
  Lγ form, 113, 120, 121
  of anhydrous soaps, 82, 83, 85–9
  of brain lipids, 109
  of mitochondrial lipids, 110–3
  of phospholipid-cholesterol-water systems, 147, 148
  of soap-water systems, 90, 92, 95–98
  thermodynamic properties of, 118
Lamellar structure, in myelin, 156–8, 174
  stabilization of, 205, 206
Lamellar systems,
  bioenergetics of, 304, 340, 341
    charge transfer complexes in, 323–325
  semiconduction of, 324, 325
Lattice dimensions, determination of, 76–8

Lecithin, (see also Phosphatidylcholine, phospholipids),
  bilayers of, 274, 275
  -cholesterol dispersions, electron microscopy of, 248–50
  -cholesterol interactions, in monolayers, 144–7
    in water, 147–51
  -cytochrome c interactions, 213
  dispersions of, 277–84
  formula of, 79
  -protein model complex, 243
  -water system, phases of, 105, 107
  -water interface, phospholipase A activity at, 224
    phospholipase B activity at, 214–8
    phospholipase C activity at, 218–20
    phospholipase D activity at, 222–4
Leucodystrophy, 61
Leukocytes, lipids of, 53
Light, effect on axons, 386, 387
Limiting transition temperature, 139
Lipid,
  analysis and characterization, 15, 16
  arrangement in membranes, 10–2, 174, 175
  autolysis, 13
  bilayers, 273–7
  chromatography, 16–20
  classes, 7–9
  composition of leukocytes, 53
    of platelets, 51
  content of mitochondria, 37, 39
    of subcellular organelles, 36
  distribution in erythrocytes, 21
  exchange reactions, 229, 230
  extraction, 20, 21
  formulae, 79, 80
  metabolic disorders, 57–61
  monolayers, adsorption of proteins on, 209, 212, 226
    effect of chain unsaturation on, 143
  occurrence, 7
  phase diagrams, 103–15
  -protein interactions, 152–3, 225–8
    in membranes, 128, 229
    in monolayers, 152
  secretion by cockroaches, 263–5
  synthesis and membrane formation, 230
  -water dispersions, formation of myelin figures, 141–3
    interfaces, electrostatic conditions at, 206–8
Lipid-water systems, (see also Soap-water phases),
  dependence on chemical structure, 97, 116
  dimensions of, 95, 106, 116
  factors determining phase structure, 114
  hexagonal phase of, 104, 106–10
  lamellar phase of, 105, 107, 109–13
  of natural lipids, 103–15
  paraffin chain conformation in, 115
  relationship to membranes, 117
  water uptake in, 109
Lipoproteins, bonding in, 204
  formation of membranes from, 225–30
  sub-units in membranes, 247
    in mitochondria and chloroplasts, 256
Liquid-paraffin phases, 73, 74
Lithium soaps, phases of, 85, 87
Liver, endoplasmic reticulum, phospholipids of, 42
  mitochondria, lipids of, 37, 39, 41
  nuclei, light micrograph of, 38
    phospholipids of, 43
  plasma membranes, electron micrographs of, 38, 40
    phospholipids of, 45, 46, 52
Local currents during excitation, 402
Lung, glycolipids of, 56
  phospholipids of, 45, 52
Lyotropic mesomorphism of phospholipids, (see also Lipid-water systems and soap-water systems), 143, 144
  of surfactants, 252
Lyotropic numbers, assignment of, 368, 369
  sequences,
    and axon excitability, 368–79
    and equivalent circuit model, 371, 372
    and macromolecular model, 372–8
    of anions, 368–71, 374
    of cations, 369–71, 374, 378, 379
Lysolecithin, (see also Lysophosphatidylcholine),

Lysolecithin *contd.*
  effect of on erythrocytes, 194
  effect of on myelin, 247
  formula of 79
  micelles of 206
  release from monolayers, 225
  -water system, 104, 106, 107
Lysohosphatides, occurrence of, 44, 45
  formation during autolysis, 13, 14
Lysophosphatidylcholine, 7
  exchange with phosphatidylcholine, 26, 34
  in erythrocytes, fatty acids of, 35
  occurrence of, 25, 26, 54
Lysophosphatidylethanolamine, formula of, 79

M

Macromolecular membrane model, 360–2
  action of salts on, 374–8
  anionic sites of, 396, 397, 400
  bonds in, 400
  cation exchange in, 397, 398
  charge groups of, 374, 379
  disruption of salt links in, 374, 376, 377
  effect of temperature on, 386
  effect of tetrodotoxin on, 383
  ionic effects in, 300, 367, 384
  lyotropic sequences in, 372–4
  Two-Stable-States Theory of, 399–404
Magnesium soaps, phases of, 89
Melting points of phospholipids and soaps, 131
Membrane conductance, time dependence of, 375–378
  current in axons, 354, 355
  impedance of axons, 354, 388, 390, 391
  potential, 387, 388, 398, 399
  resistance of axons, 353, 354, 388, 390, 391
Membranes,
  bioenergetic processes in, 290–2
  electrical characteristics of, 236
  electron microscopy of, 159–74
  formation of, 225–30
  hydrocarbon chain fluidity in, 196–8, 261, 262
  infra red studies of, 174–6
  isolation of, 9, 15
  lipid disorders of, 57–61
  nuclear magnetic resonance spectra of, 189–96
  ORD and CD studies of, 176–89
  release from monolayers, 225
  role in energy transfer, 320, 321
  role of cholesterol in, 151
  X-ray diffraction studies of, 154–9
Mesomorphic phases, nomenclature of, 75
Metabolic disorders in lipid metabolism, 57–61
Micellar model membranes, 244, 245
  enzymes and proteins in, 255, 256
  hormone action on, 256–61
  permeability of, 253–5
Micellar model bioenergetic membranes, 339, 340
Micellar solutions of soaps, 101–3, 116
Micellar structure, in membranes, 245, 251–5
  in mitochondria, 256
Micelle-bimolecular leaflet equilibrium, 245, 246
Micelles, dimensions in membrane models, 246, 247
  in insect lipid secretion, 263–5
  in lecithin cholesterol dispersions, 248–51
  in membranes during cytoplasmic streaming, 262
  in surfactants, 252, 253
  interaction with membrane proteins, 247
Micrococcus laidlawi, ORD of membranes, 183
Microsomes, electron micrographs of, 38, 40
  lipid content of, 36, 37
  phospholipids of, 42
Millipore filters, as model membranes, 271
Mitochondria, electron microscopy of, 38, 40, 166, 256
  lipid composition of, 39, 41, 62
  lipid content of, 37, 39
Mitochondrial cristae (M.C.),
  composition of, 306
  lipids, phase diagram of, 110, 111

Mitochondrial lipid-water,
  membranes, charge migration in, 329–33
    electron transfer in, 312, 314
    electronic processes in, 295, 296
    energy transfer in, 328, 329
      structure of, 11, 251, 255, 256, 295, 296
        ORD and CD parameters of, 183
    protein, composition of, 306
      reaction with lecithin, 226
    system, 106, 110–3, 119–21
Model membranes,
  bilayers, 144, 273–7, 338, 339
  bimolecular leaflets, 234–44
  for bioenergetic processes, 336–40
  from ORD studies, 188, 189
  micellar, 244–61, 339, 340
  millipore filters, 271
  oil water interfaces, 271, 272
  phospholipid dispersions, 277–84, 339, 340
  polarization and permeability of, 239–42
Monoglycerides, formulae of, 79
Monoglyceride-water system, lamellar phase of, 105
Monolayers,
  as bioenergetic model membranes, 337–8
  as model membranes, 267, 268
  adsorption of protein on, 152, 153, 209, 212, 270
  effect of anaesthetics on, 265, 267
  effect of antibiotics on, 266
  effect of drugs on, 267
  effect of lytic agents on, 266
  effect of phospholipase on, 218, 225
  effect of retinol and vitamin A on, 265, 266
  effect of vitamins on, 265, 267
  of chlorophyll, 338
    and vitamin K_1, 338
  of phospholipids, 143
  of rhodopsin, 337
  of steroids, 258, 259
  permeability of, 268
  phospholipid-cholesterol interaction in, 144. 147, 269
  phospholipid polar group arrangement in, 268, 269
Muscle, phospholipids of, 46
Myelin,
  absence of gangliosides in, 57
  action of lysolecithin on, 247
  cerebrosides of, 48
  figure formation 141–3, 248, 249
  glycolipids of, 56
  infra red spectra of, 174
  phospholipids of, 48
  phospholipids, effect of age on, 47
  structure of, 156, 157
  X-ray diffraction of, 155–7

N

Neat phase in anhydrous soaps, 85, 86
Negative staining, 166, 167
  effect on lipid structure, 250
  of lecithin cholesterol mixtures, 248–251
Nernst equation, 371
Nerve, *see* Axon
Niemann-Pick disease, 60
Nuclear Magnetic Resonance, (N.M.R.) spectroscopy,
  of erythrocytes, 189–96
    assignment of peaks, 189–91
    lipid structure, 191, 192
    erythrocyte-lipid extracts, 192, 193
  of lecithin-cholesterol-water systems, 150–2
  of phospholipid transitions, 134, 135, 139, 140
Nuclei, lipid content of, 37
  micrographs of, 38, 40, 169
  phospholipids of, 43, 63

O

Oblique phase of anhydrous soaps, 83, 86
Odorants, action on membrane, 244
Oestrogen in membranes, 257
Oil-water interfaces, adsorption of protein on, 209–11
Optical Rotary Dispersion (ORD), 176–89
  of membranes, 181–4
    effect of environment on, 187, 188
    intepretation of, 184–9
    model system for, 187

ORD of membranes *contd.*
  phosphatide activity in, 186–7
  models of, 188, 189
  of peptides, 177
  of polypeptides, 178–80
  of proteins, 179, 181
Ordered structures, nomenclature of, 75
Organelles, characterization of, 15
Organization,
  short range, 73–5
  long range, 75, 76
Orthorhombic phase of anhydrous soaps, 84, 86
Osmium tetroxide,
  fixation and membrane structure, 160, 165
  fixation of phosphatidylcholines, 142
  location in fixed tissues, 161–3
  reaction with lipids, 163
  reaction with olefins, 163, 164
  reaction with saturated phospholipids, 164, 165
Osmotic pressure, effect of on excitability, 385
Oxidative phosphorylation, comparison with photosynthesis, 312

P

Pancreas nuclei, phospholipids of, 44
Paraffin chains (*see also* Hydrocarbon chains), conformation of, 73–5
  conformation in membranes, 117
  disorder in, 115
Paraffin-in-water structures, 76
Partial specific volumes of lipid-water systems, 81
Pentadecanoic acid, A′ form of, 74
Peptidases, effect of on action potential, 361
Perfusion of axons, 350, 351, 359
Permeability,
  of bilayers, 274, 275
    effect of detergents on, 276
  of bimolecular leaflets, 235, 236, 239–43
  of erythrocytes, 28
  of micellar model membranes, 253–5
  of monolayers, 268
  of phospholipid dispersions, 277–82
    effect of narcotics on, 283
    effect of steroids on, 282, 283

pH at lipid-water surfaces, 208
pH effect on axon excitability, 362, 363
Phase changes in membranes, 261, 265
  induced by hormones, 262
Phase diagram, construction of, 72
  of brain lipid-water system, 109
  of cholesterol-phospholipid-water system, 147, 148
  of mitochondrial lipids, 110, 111
  of mixed lipids, 114
  of soaps, 81, 82
  of soap-water systems, 96, 97
  temperature dependence of, 114
Phase dimensions, concentration dependence of, 114–6
Phase, H, 88, 89, 94, 107
Phase, Q, 88, 89, 94, 107
Phase, R, 88, 89, 94, 107
  sequences, 94, 95
  structure, crystallographic verification of 119
    paraffin chain dependence of, 114, 116
Phase, T, 88, 89, 94
  transitions, in phospholipids, 132–8
    induced by cation changes, 399
  in phospholipid-cholesterol-water systems, 149–51
  in phospholipid water systems, 139–44
Phosphatidic acid, 7
Phosphatidylcholines, (*see also* Phospholipids)
  D.S.C. of, 140–2
  D.T.A. of, 136
  in model membranes, 144
  in monolayers, 143, 145, 146
  structure of, 7, 79, 308
  -water systems, 139–43
Phosphatidylethanolamine, (*see also* Phospholipids), birefringence of, 133
  charge on, 205
  complexes with cytochrome c, 212, 213
  D.T.A. of, 133
  hydrolysis by phospholipase A, 225
  infra red spectra of, 132
  monolayers of, 143
  N.M.R. spectra of, 134, 135
  structure of, 7, 80, 308

SUBJECT INDEX

Phosphatidylethanolamine contd.
  thermotropic mesomorphism of, 132-6
  -water phases, 104, 106, 107
  X-ray diffraction of, 129-31, 133, 134
Phosphatidyl glycerol, (see also Phospholipids), structure of, 308
Phosphatidylinositol, (see also Phospholipids), formula of, 80
  complexes with cytochrome c, 212, 213
Phosphatidylserine, (see also Phospholipids), and cation exchange in membranes, 271, 272
Phosphodiesterase, 220, 221
  activity and ζ potential, 221
Phospholipase activity, effect of ζ potential on, 214
  use of ^{32}P labelled phospholipids, 214
  with sonicated lipids, 226
Phospholipase, effect of on action potential, 361
  -substrate complex at interfaces, 213
Phosphalipase A, action on monolayers, 225
  action on phosphatidylethanolamine, 225
  effect of ether on activity, 224
Phospholipase B, 214-8
  activity and ζ potential, 215-7
  action on monolayers, 218
Phospholipase C, 218-20
  activity and isoelectric point of, 219
  activity and ζ potential, 219, 220
  effect of on erythrocytes, 191
Phospholipase D, 222-4
  effect of pH on activity, 223
  effect of ζ potential of lecithin on, 223
Phospholipid-cholesterol interactions, 144-51
Phospholipid-cytochrome c complexes, 212, 213
Phospholipid dispersions as model membranes, 277-84
  exchange reactions, 27
  lamellae, hydration of, 208
  micelles in water, 206
  -water system (see also Lipid-water systems), 139-44
    interface, electrostatic conditions at, 205-9
Phospholipids, characterization of, 15
  chromatography of, 17-20
  fatty acid composition of, 31-5
  hydrocarbon chain fluidity in membranes, 196-8
  in membrane repeat units, 11
  metabolic disorders of, 57-61
  nomenclature of, 7
  occurrence in membranes, 127
  of amnion cells, 54
  of bovine organs, 44, 45
  of brain and myelin, 47-9
  of endoplasmic reticulum, 42
  or erythrocytes, 22-6
  of leukocytes, 53
  of liver, 46
  of lung, 45
  of mitochondria, 39-42
  of muscle, 46
  of nuclei, 43
  of plasma, 26
  of platelets, 51-3
  of the sea anemonae, 50
  of vascular structures, 49
  phase changes of (see also Thermotropic mesomorphism and Soaps, anhydrous phases of), 132-8
  postmortem autolysis of, 13
  single crystal studies of, 128-31
Phosphorescence, 315
Photoconductivity, 321, 322
  in carotenoids, 327
  in chloroplasts, 335, 336
  in visual function, 326-8
Photosynthesis, comparison with oxidative phosphorylation, 312
  semiconduction in, 335, 336
  energy migration and charge separation in, 212, 313
Photosynthetic lamellae, (P.S.) composition of, 305, 306
  pigments in, 301-3
  structure of, 297-301
Photosynthetic processes, electronic effects in, 334
Photosynethic unit, energy transfer in, 333, 334
Phthalocyanine, structure of, 328

Physiological axon membrane, layers of, 361, 362, 396, 397
Plant cells, lipid classes of, 7
Plasma, exchange reactions in, 26
  membrane, glycolipids in, 55
    lipid content of, 36, 37, 56
      ORD of, 184
      structure of, 160, 161, 235
    phospholipids of, 26
Plasmalogens, 7, 35
Plastoquinones, structure of, 307, 309, 310
  orientation in lamellae, 302, 311
Platelets, phospholipids and fatty acids of, 51–3
Poikilothermic organisms, lipid composition of, 197
Polar group area (see Area S)
Polar groups,
  of steroids in monolayers, 257, 258
  of steroids in phospholipid dispersions, 260
  organization in phospholipids, 130, 131
  orientation, in bilayers, 275
    in lipid monolayers, 268, 269
    in pores, 279
  selective sodium binding by, 254
Pores
  in bilayers, 275
  in bimolecular leaflets, 236, 237, 239–42
  in micellar model membranes, 254
  in phospholipid dispersions, 279–81
Polypeptide conformation, 178–80
Polypeptides in monolayers, 270
Portmortem degradation of lipids, 13
Potassium,
  concentration and membrane potential, 364–67
  flux during axon excitation, 391–95
  permanganate fixation, 159–62
  soaps, phases of, 85–7, 96–100
Pressure, effect of on excitability, 385
Progesterone in lipid monlayers, 259
Proteases,
  effect of on axons, 360, 404
  in axon perfusion, 357
Protein adsorption,
  at lipid monlayers, 209, 212, 225, 226

  at oil–water interfaces, 209–11
Protein conformation, in membranes, 184–9
  infra red spectra of, 175, 176
  negative staining of, 166, 167
  ORD and CD of, 179–81
Protein,
  in axons, 360, 361
  in bilayers, 277
  in model membranes, 235–7, 239, 241–3, 255, 256
  -lipid interaction in membranes, 225, 227–9
  mitochondrial, reaction with lecithin, 226
  semiconduction of, 327
  unfolding at lipid-water interfaces, 211, 210
Proton Magnetic Resonance (P.M.R.) (see N.M.R.)

Q

Quantasomes, composition of, 305, 306
  energy transfer in, 333, 334
  in chloroplasts, 297
Quinones, long chain structures of, 307–9, 310
  orientation in lamellae, 302, 311

R

Radiationless transitions, 315
Random conformation, ORD and CD spectra of, 178, 180
Rectangular phase in soap-water systems, 93, 94, 96
  of anhydrous soaps, 83, 86
Resistance,
  of axon membrane, measurement of, 353, 354, 388, 390–3
  of excited axon and isotope flux, 391
Resonance transfer (see Energy transfer)
Resting potential
  effect of salt concentration on, 363–7
  in axons, measurement of, 353, 387, 388
Restoring ability in nerve conduction, 367

Retina outer segments, structure of, 158, 159
  particles in, 253
Retinal, photoisomerization of, 326, 327
Retinol,
  action on membranes, 266, 267
  action on monolayers, 265
Retinene, structure of, 309, 310
Rhodopsin
  in monolayers, 337
  in visual excitation, 325, 326
  micellar dispersions of, 339, 340
Rhombohedral phase of anhydrous soaps, 84, 88
Ribbon structure in anhydrous soaps, 83, 85, 89, 90
Rods, semiconductivity of, 328
Rod outer segments, (R.O.S.),
  birefringence of, 293, 294
  composition of, 305
  dichroism of, 294
  electron microscopy of, 294, 295
  emergy transfer in, 314, 325, 326
  sub-units in, 294, 295
  structure of, 293–5
Rod-like,
  micelles of soaps, 101–3
  phases of anhydrous soaps, 88, 89
Root tip cells, electron microscopy of, 172, 173
Rubidium soaps, phases of, 85, 97–9

S

Saponin, interactions with lecithin-cholesterol mixtures, 250
Salt links, disruption of, 374–7
Sea anemone, phospholipids of, 50
Schwan cells, 356–9
Semiconduction, 321, 322
  of charge transfer complexes, 330–2
  in chloroplast, 334–6
  in electron transfer chain, 330
  in lamellar membrane, 324, 325
  in visual functions, 326–8
  of carotenoids, 327
  of proteins, 327, 331
  of sheep's rods, 328
Serum lipoproteins, lipid exchange reactions of, 229, 230
Singlet state, 315

Soaps, anhydrous phases of, 82–90
  body-centred cubic, 85, 88
  body-centred tetragonal, 84, 88
  hexagonal, 84, 88
  lamellar, 82, 83, 85, 87–9
  oblique and rectangular, 83
  orthorhombic, 84
  rhombohedral, 84, 88
Soaps,
  anhydrous rod-like structures of, 88, 89
  formulae of, 79
  micellar solutions of, 101–3
  of divalent cations, anhydrous phases of, 87–90
  of monovalent cation, anhydrous phases of, 85–7
  phase diagram of, 81, 82
Soap-water phases, 90–97
  sequences of, 94, 95
Soap-water systems
  dimensions of, 95
  gel phases of, 97–101
  cubic phase of, 91
  hexagonal phases of, 90, 91
  lamellar phase of, 90
  phase diagrams of, 96
  polar group area of, 92, 93, 97
  relationship to chemical structure, 97
Sodium flux during axon excitation, 391–5
Sodium ions, in membrane excitability, 382, 383
  effect of, on action potential, 371, 372
Sodium soaps, phases of, 85–7, 96–8, 101, 102
Spacings
  equations of, 77
  of phosphatidylethanolamines, 130, 131
Sphingolipids
  interaction in membranes, 12
  metabolic disorders of, 58–61
  postmortem degradation of, 13
  types of, 7
Sphingomyelin, 8
  fatty acids of, 29, 32, 35
  formula of, 80
  metabolic disorders of, 60
  occurrence of, 14, 18, 23–25, 27, 41, 43–54

Sphingomyelin-water system, lamellar phase of, 105
Sphingosine, 7
Spleen, phospholipids of, 45
Steroids (see also Cholesterol)
  at water-heptane interfaces, 258
  in bimolecular leaflets, 257, 258
  in membranes and physiological action, 256, 259
  in micellar membranes, 259, 260
  in monolayers, 258, 259
  in phospholipid dispersions, 282
  occurrence of, 9
  specificity in micellar membranes, 260, 261
Sterol esters, occurrence of, 8, 22
Stroma, 297–301, 303
Strontium soaps, phases of, 89, 94
Structure
  analysis, 76–8
  elements, 76
  parameters, determination of, 76–8
  Types I and II, 76
Structural protein
  in membranes, 188, 304
  in mitochondria, 295, 306
Subneat phase in anhydrous soaps, 85, 86
Subunits
  in bioenergetic lamellae, 304
  in chloroplasts, 297
  in membranes, 158, 161, 168, 170, 173, 247, 251–3
  in mitochondria, 296
  in rod outer segments, 294
Sulphatides, 8
Superwaxy phase in anhydrous soaps, 85, 86
Surface potential
  of cephalin-cholesterol monolayers, 269
  of lipid-water systems, 205–8
Survival time in nerve conduction, 367–371
Symmetry systems, equations of, 77

T

Tay-Sachs disease, 58, 60
Temperature, effect of on excitability, 385, 386
Tetrodotoxin, effect of on axon excitability, 380–3
Thermotropic mesomorphism (see also Soaps, anhydrous phases of)
  hydrocarbon chain structure in, 135, 136
  of egg yolk lecithin, 136
  of phosphatidylethanolamine, 132–6
  of phosphatidylcholine, 136
  theoretical model of, 137, 138
Thin layer chromatography
  lipid identification, 15, 23–5
  methods, 15–18
  quantitative lipid analysis, 17
Three-dimensional conformation of paraffin chains, 73, 74
Thylakoids, electron microscopy of, 170, 171
Thymus, phospholipids of, 44
Trifluoroacetic acid, action on membranes, 194
Triglycerides
  occurrence of, 8
  X-ray structure of, 129
Triphosphoinositide,
  phosphomonoesterase, 220–1
  enzymatic hydrolysis of, 220, 221
  micelles in water, 220
Triplet state, 315
Trypsin, effect on action potential, 361
Two-Stable-States Theory of excitation, 399–404

U

Ubiquinone
  function of, 326
  structure of, 307, 309
Ultrasonication of lipids, effect of on phosolipase activity, 226
Unit membrane, electron microscope evidence for, 10, 159–65, 297, 300

V

Vascular structures, phospholipids of, 49
Vertebrates, phospholipids of, 51, 52
Vision (see also Rod outer segments)
  excitation processes in, 326
Visual function, conductivity in, 326–8

Vitamin A, in membranes and monolayers, 265-7
  in monolayers of chlorophyll, 338
  structure of, 307, 309, 310
Voltage-clamp in sodium-free media, 379-82, 403
  ohmic behaviour of, 381, 382
  technique, 354, 355, 380, 381

W

Water, binding by phosphatidylcholines, 140, 141
  -in-paraffin structures, 76
  uptake by insects, 263-5
    by natural lipids, 109
  structure in membrane models, 239-42, 255
Waxy phase in anhydrous soaps, 85, 86

X

X-ray
  crystal studies of phospholipid derivatives, 128-30
  diffraction, of erythrocytes, 157, 158
    of histone-cephalin complexes, 213
  of membranes, 154-9
  of myelin, 155-7
  of phospholipids, 130, 131
  of retina outer segments, 158, 159, 253
  of surfactants, 252, 253
  spacings, of lecithin-cholesterol water systems, 148, 150
    of phospholipids, 133-5
    of phospholipid-water systems, 138, 139
  scattering technique, 81

Y

Yeast
  electron microscopy of, 167-70
  lipids of, 55

Z

Zeta (ζ) potential, 207
  and phospholipase activity, 215-7, 291-21, 223-25
  relationship to electrophoretic mobility, 214
Zinc soaps, phases of, 89